Key Cultivation Techniques for Main Fruits and Vegetables in the Lancang-Mekong Region (Yunnan)

澜湄地区（云南）主要果蔬栽培关键技术

陈霞 汪骞 丁仁展 桂敏 主编

中国农业科学技术出版社

图书在版编目（CIP）数据

澜湄地区（云南）主要果蔬栽培关键技术/陈霞等主编. -- 北京：中国农业科学技术出版社，2024.12

ISBN 978-7-5116-6845-5

Ⅰ.①澜… Ⅱ.①陈… Ⅲ.①果树园艺②蔬菜园艺 Ⅳ.①S6

中国国家版本馆 CIP 数据核字（2024）第 108823 号

责任编辑　施睿佳　姚　欢
责任校对　王　彦
责任印制　姜义伟　王思文

出 版 者	中国农业科学技术出版社
	北京市中关村南大街 12 号　邮编：100081
电　　话	（010）82106631（编辑室）（010）82106624（发行部）
	（010）82109709（读者服务部）
网　　址	https://castp.caas.cn
经 销 者	各地新华书店
印 刷 者	北京建宏印刷有限公司
开　　本	185 mm × 260 mm　1/16
印　　张	38.5
字　　数	798 千字
版　　次	2024 年 12 月第 1 版　2024 年 12 月第 1 次印刷
定　　价	128.00 元

━━━▅▂ 版权所有·侵权必究 ▂▅━━━

《澜湄地区（云南）主要果蔬栽培关键技术》
编委会

主　　编：陈　霞　汪　骞　丁仁展　桂　敏

编　　委：（按姓氏笔画排序）

于　菲　于诗如　万　红　王连润　方成刚

龙荣华　尼章光　朱彩华　刘学敏　刘家迅

苏　俊　杜　磊　李石开　李坤明　杨光柱

吴丽艳　何英云　张　晏　张芮豪　陈　瑶

胡华冉　钟秋月　袁　艺　高　婷　陶　婧

黄文静　龚亚菊　梁艳萍　董　莉　鲍　锐

黎志彬　薛　娜

植保校正：尹可锁　郭志祥　毛　佳

翻　　译：姚　黎　卞露白　韩思琪　王江艳

前 言

澜湄地区横跨亚洲东南部，包括中国（云南）、缅甸、泰国、老挝、柬埔寨和越南6个国家。该地区具有独特的地理区位优势、丰富的动植物资源和良好的生态环境。气候以热带、亚热带季风气候为主，有明显的干湿季节变化，年均温差异较小，日均温差异较大，为该地区的农业发展提供了得天独厚的自然条件。澜湄地区由澜沧江-湄公河一水相连，地势总体为北高南低，山河相间呈南北纵列分布，北部多高原山地，南部沿海地区多为平原地形，因此农业生产上存在着天然的产业互补优势。

作为传统农业国家，农业在澜湄地区经济发展中占有重要地位。然而，长期以来，该地区经济相对落后，农业生产技术和现代化水平有待提高，农业发展面临诸多挑战。土地利用率低、优良品种缺乏、现代化生产技术落后等问题制约了区域农业的发展，因此澜湄地区国家都面临着发展经济和改善民生的共同要务。

澜湄地区农业合作历史悠久，在多边合作机制下建立了稳定、友好、互助、互惠、互信的新型农业合作关系，各方农业交流互访密切、合作项目较多，合作成绩显著。澜湄地区已然成为"一带一路"沿线最为活跃的地区之一。"同饮一江水，命运紧相连"，农业是澜湄地区重要的经济支柱，农业的发展与社会发展、人民生活水平、区域经济发展息息相关。为促进澜湄地区农业发展，各国需要不断加强合作，充分利用"一带一路"和"澜沧江-湄公河"合作机制，进一步加强地区间农业生产技术的合作与交流，促进产业协作，优化双方生产要素布局，提升资源配置效率，增强贸易互补，通过人才培训培养、技术交流示范、共建示范基地等深度合作，促进区域农业生产技术进步和农业生产效率的整体提升。

云南地处低纬高原，地形地貌复杂，气候多样，具有从寒温带到热带的多种气候类型，为农产品的多样化生产提供了便利的自然条件。由于政府的重视及现代农业科技的推广应用，自2000年以来，云南农业综合生产效率实现了较大提升。其中水果、蔬菜在云南低纬高原特色产业中占有重要地位，果蔬产业发展成为云南低纬高原地区巩固拓展脱贫攻坚成果同乡村振兴有效衔接的重要支柱产业。果蔬也是澜湄地区的重要农产品，果蔬产业的发展和澜湄地区农业农村可持续发展及农业贸易与投资合作密切相关。

本书以澜湄地区云南果蔬产业为背景，介绍了主要栽培水果、蔬菜的关键生产技术（包括苹果、梨、桃、草莓、蓝莓、李、猕猴桃、人参果、香蕉、杧果、柑橘、葡萄、菠萝蜜、西番莲、西瓜共15种温带、亚热带水果，番茄、辣椒、茄子、菜豆、速生叶菜、南瓜、黄瓜、苦瓜、洋葱共9种主要蔬菜），围绕适宜栽培的生态气候条件、土壤条件、主要栽培品种及特性、栽培管理关键技术、主要病虫害管理技术以及周年管理措施等进行了详细介绍。

本书由云南省内一批长期从事果蔬研究的专业人士共同编写，他们都在各自的研究领域内有着深入的研究和广泛的实践经验。本书基于各位作者多年的经验积累和生产实际，力求将专业知识和经验智慧分享给广大的读者群体，通过分享云南果蔬产品的栽培种植技术经验，为澜湄地区其他近似气候区域果蔬产业的发展提供技术参考，为各地相关果蔬栽培技术培训、交流与合作提供参考依据。

本书由云南省科技厅重点研发计划国际科技合作专项"智汇云南"计划－培训班"越南北部四省（老街、奠边、莱州、河江）山地果蔬栽培技术培训"项目（202103AL140010）提供经费支持；此外得到了"地理标志农产品关键技术研究与应用"项目（202102AE090051）、"临沧市冬春蔬菜产业发展科技创新综合示范"项目（202204AC100001-A06）、云南种子种业联合实验室项目（202205AR070001）及云南省农作物品种推广后补助项目的支持。

云南省农业科学院园艺作物研究所（以下简称"园艺作物研究所"）为本书著作权单位，是专业从事低纬高原水果、蔬菜资源研究、品种选育、高效栽培、质量控制等领域研发的省级公益性科研单位，为中国园艺学会理事单位、云南省园艺学会法人单位。园艺作物研究所设有果树创新团队、大宗蔬菜创新团队、特色蔬菜创新团队、设施农业创新团队等研发团队。在果蔬资源收集鉴定评价与开发利用、果蔬新品种选育、果蔬栽培技术研究与集成、国内外交流与合作等方面都取得了显著的成绩。园艺作物研究所将持续以"特色果蔬种质资源创新、基因发掘与新种质创制""果蔬产业转型升级及科技链关键技术集成与支撑"等领域为重点研究方向，聚焦云南省果蔬优势产业发展，以"错季、优质、周年生产"为目标，构建绿色化、标准化、设施化、数字化的配套生产技术体系，为果蔬全产业链提供科技支撑和技术服务，推动云南省低纬高原特色产业持续发展。

在本书写作中，由云南省农业科学院园艺作物研究所陈霞、汪骞、桂敏、方成刚及西南林业大学丁仁展负责前言和第一章的编写；果树栽培方面，由黄文静负责低纬高原苹果栽培关键技术的编写，陈霞、何英云、苏俊负责梨栽培关键技术的编写，于菲负责桃栽培关键技术的编写，万红负责草莓栽培关键技术的编写，王连润负责蓝莓栽培关键技术的编写，李坤明、梁艳萍负责李栽培关键技术的编写，陈瑶、陈霞负责猕猴桃栽培关键技术的编写，刘家迅负责人参果栽培关键技术的编写，张晏负责西番莲栽培关键技术的编写，杨光柱负责香蕉栽培关键技术的编写，朱彩华、汪骞负责西瓜栽培关键技术的编写；蔬菜栽培方面，由桂敏、张芮豪、杜磊、钟秋月负责辣椒栽培关键技术的编写，龚亚菊、吴丽艳负责茄子栽培关键技术的编写，龚亚菊、鲍锐负责番茄栽培关键技术的编写，袁艺、李石开负责菜豆栽培关键技术的编写，龙荣华、朱彩华负责黄瓜和苦瓜栽培关键技术的编写，高婷负责南瓜栽培关键技术的编写，黎志彬、薛娜负责洋葱栽培关键技术的编写，陶婧、胡华冉负责叶菜类蔬菜栽培关键技术的编写。此外，云南省农业科学院热带亚热带经济作物研究所尼章光负责杧果栽培关键技术的编写，云南省绿色食品发展中心董莉负责葡萄栽培关键技术的编写，云南省红河热带农业科学研究所刘学敏负责菠萝蜜栽培关键技术的编写，会泽县农业技术推广中心于诗如负责柑橘栽培关键技术的编写。陈霞、汪骞分别对果树、蔬菜栽培技术的稿件进行初审和统稿，云南省农业科学院农业环境资源研究所尹可锁、郭志祥、毛佳负责植保部分的审定，云南省农业科学院国际农业研究所姚黎以及云南民族大学2022级英语笔译硕士研究生卞露白、韩思琪、王江艳负责全书稿件的汉译英工作。在此对所有作者的辛勤付出表示感谢。

本书的写作参考了大量专家学者的研究成果，重要的参考书目及文献均在书后列出，谨此表示谢忱。感谢为本书翻译和校正的各位专家。由于编者水平有限，书中存在不足之处，敬请各位读者批评指正。

编　者

2024 年 8 月 31 日

❖ Preface ❖

The Lancang-Mekong region spans across Southeast Asia, including six countries: China (mainly in Yunnan), Myanmar, Thailand, Laos, Cambodia, and Vietnam. This region has unique geographic advantages, abundant animal and plant resources, and a favorable ecological environment. The climate is mainly tropical and subtropical monsoonal, with distinct wet and dry seasons and small annual temperature differences, providing favorable natural conditions for agricultural development. The six countries along the Lancang-Mekong River are connected by the same water system, with the terrain generally being higher in the north and lower in the south. The region is characterized by a north-south distribution of mountains and rivers, with most highlands and mountains in the north, and coastal plains in the south predominantly. As a result, there are natural complementary advantages in agricultural production.

As a traditional agricultural region, agriculture plays an important role in the economic growth of the Lancang-Mekong region. However, the region has been relatively economically underdeveloped for a long time, and there is a need to improve agricultural production techniques and modernization. Challenges such as low land use efficiency, lack of superior varieties, and outdated production technologies have constrained regional agricultural development, making it a shared priority for the countries in the Lancang-Mekong region to develop their economies and improve people's livelihoods.

The Lancang-Mekong region, with a long history of agricultural cooperation among the six countries, has established a new-model relationship of agricultural cooperation under a multilateral cooperation mechanism, featuring stability, amity, mutual assistance, mutual benefit, and mutual trust. Upon this, the agricultural exchanges and visits among the countries are frequent, with many cooperative projects yielding significant achievements. The Lancang-Mekong region has become one of the most active areas along the "Belt and Road" initiative.

"Shared River, Shared Future". Agriculture is an important economic pillar for the countries in the Lancang-Mekong region, and its agricultural development is closely related to social development, people's livelihoods, and regional economic growth. In order to promote agricultural development in this region, countries need to continuously enhance cooperation, fully utilize the "Belt and Road" and "Lancang-Mekong" cooperation mechanisms, further strengthen inter-regional cooperation and exchanges in agricultural production technology, promote industrial cooperation, optimize production factors, improve the efficiency of resource allocation, and enhance trade complementarity. Through in-depth cooperation such as talent training and cultivation, technology exchange and demonstration, and jointly building demonstration bases, the overall improvement of regional agricultural production technology and agricultural production efficiency can be promoted.

Yunnan is located on a low-latitude plateau with complex topography and diverse climates, ranging from cold-temperate to tropical, which provides favorable natural conditions for the diversified production of agricultural products. Due to government attention and the widespread application of modern agricultural technology, Yunnan's overall agricultural production efficiency has been greatly improved since 2000. Fruits and vegetables play a dominant role among Yunnan's characteristic agricultural industry in low-latitude plateau regions, and the development of the fruit and vegetable industry has become an important pillar industry to consolidate and expand the achievements in poverty alleviation with efforts to promote rural revitalization in the low-latitude plateau region of Yunnan. Fruits and vegetables are also important agricultural products in the Lancang-Mekong region, and their development is closely related to the sustainable development of agriculture and rural areas, as well as agricultural trade and investment cooperation in the Lancang-Mekong region.

This book focuses on the fruit and vegetable industry in Yunnan of China in the Lancang-Mekong region, and it introduces the key production technologies for major cultivated fruits and vegetables (including 15 temperate and subtropical fruits such as apple, pear, peach, strawberry, blueberry, plum, kiwifruit, ginseng fruit, banana, mango, citrus, grape, jackfruit, passion fruit, and watermelon, and 9 main vegetables such as tomato, pepper, eggplant, green bean, leafy vegetables, pumpkin, cucumber, bitter gourd, and onion). It provides detailed information on suitable ecological and climatic conditions for cultivation, soil conditions, main cultivars and characteristics, key techniques for cultivation and management, management techniques for major diseases and pests, as well as year-round management measures.

The book is collectively written by a group of professionals in Yunnan who have been engaged in fruit and vegetable research for a long time. They have deep research and extensive practical experience in their respective fields. Based on the authors' years of experience and production practices, the book aims to share professional knowledge and experience with a wide range of readers. By sharing the cultivation and planting technology experience of fruit and vegetable products in Yunnan, it aims to provide technical references for the development of the fruit and vegetable industry in similar climate areas with the Lancang-Mekong region, and to serve as a reference for fruit and vegetable cultivation technology training, exchange, and cooperation in different regions.

This book is funded by the project "Yunnan Intelligence Union Program"-Training Course on "Training of Mountainous Fruit and Vegetable Cultivation Techniques in Four Northern Provinces of Vietnam (Lao Cai, Dien Bien, Lai Chau, Hoa Giang)" (202103AL140010), which is a special project of International Science and Technology Cooperation of Key Research and Development Program of Yunnan Provincial Department of Science and Technology and also supported by "Research and Application of the Key Technologies of Geographical Indication Agricultural Products" project (202102AE090051) and "The Comprehensive Science and Technology Innovation Demonstration of Winter and Spring Vegetables Industry Development for Lincang City" project (202204AC100001-A06), "the Yunnan Seed and Breeding Industry Joint Laboratory Project (202205AR070001) " and "the Post-Subsidy Program for Crop Variety Promotion in Yunnan Province.".

The Horticulture Research Institute of Yunnan Academy of Agricultural Sciences (hereinafter referred to as "Horticulture Research Institute") holds the copyright of this book. It is a provincial public welfare scientific research institution specializing in the research and development of fruit and vegetable resources, variety breeding, efficient cultivation, quality control in the low-latitude plateau, and is a member of the Chinese Society of Horticultural Science and the legal entity of the Yunnan Society of Horticultural Science. The institute innovative teams for research and development, specializing in fruits, common vegetables, special vegetables and protected agriculture. The institute has achieved significant progress in the collection, identification, evaluation, development and utilization of fruit and vegetable resources, new variety breeding, research and integration of cultivation technology, domestic and international exchanges and cooperation. It will continuously concentrate on the research areas of "germplasm resources innovation, gene discovery and new germplasm creation of characteristic fruit and vegetable" and "transformation and upgrading of the fruit and vegetable industry and key technology integration and support for the science and technology chain", focusing on the development of Yunnan's advantageous fruit and vegetable industries. With the goal of "staggered, high-quality, year-round production", the institute will construct a supporting production technology system that is green, standardized, facility-based, and digital, which provides scientific and technological support and technical services for the entire fruit and vegetable industry chain, and promotes the sustainable development of

Yunnan's characteristic industries on the low-latitude plateau.

In writing this book, the members of Horticulture Research Institute have done a lot of contribution. The preface and the first chapter were completed by Chen Xia, Wang Qian, Gui Min and Fang Chenggang as well as Ding Renzhan from Southwest Forestry University. In terms of fruit tree cultivation, Huang Wenjing completed the writing of key techniques for apple cultivation in low-latitude plateau, Chen Xia, He Yingyun and Su Jun for pear, Yu Fei for peach, Wan Hong for strawberry, Wang Lianrun for blueberry, Li Kunming and Liang Yanping for plum, Chen Yao and Chen Xia for kiwifruit, Liu Jiaxun for ginseng fruit, Zhang Yan for passion fruit, Yang Guangzhu for banana, Zhu Caihua and Wang Qian for watermelon; as for vegetable cultivation, Gui Min and Zhang Ruihao, Du Lei and Zhong Qiuyue were responsible for the writing of key techniques for pepper cultivation, Gong Yaju and Wu Liyan for eggplant, Gong Yaju and Bao Rui for tomato, Yuan Yi and Li Shikai for common bean, Long Ronghua and Zhu Caihua for cucumber and bitter gourd, Gao Ting for pumpkin, Li Zhibin and Xue Na for onion, Tao Jing and Hu Huaran for fast-growing leafy vegetables. In addition, the writing of key mango cultivation techniques was completed by Ni Zhangguang of the Tropical and Subtropical Cash Crops Research Institute of Yunnan Academy of Agricultural Sciences; the grape was completed by Dong Li of the Development Center of Yunnan Green Food; the jackfruit was completed by Liu Xuemin of Honghe Tropical Agriculture Research Institute of Yunnan province; and the citrus was completed by Yu Shiru of the Huize Agricultural Technology Promotion Center. The manuscript review and compilation of fruit and vegetable cultivation techniques were respectively completed by Chen Xia and Wang Qian, and the approval of the plant protection section was completed by Yin Kesuo, Guo Zhixiang and Mao Jia of Agricultural Environment and Resources Institute of Yunnan Academy of Agricultural Sciences. The translation of the manuscript from Chinese to English was completed by Yao Li of the International Agricultural Research Institute, Yunnan Academy of Agricultural Sciences, together with Class of 2022 postgraduates: Bian Lubai, Han Siqi and Wang Jiangyan in the Master of Translation and Interpreting (MTI) program from Yunnan Minzu University. Here we would like to thank all authors for their hard work.

In addition, this book draws on extensive research findings from numerous experts and scholars. The most cited references are listed at the end of the book, for which we express our heartfelt gratitude. We would like to thank the experts who translated and proofread this book. Due to the editors' limited expertise, there may be deficiencies in this book, and we sincerely welcome criticism and correction from our readers.

Editor
August 31st, 2024

本书术语表和缩略
Terms and Acronyms in this Book

1. 单位换算 Units

1 mu ≈ 0.067 hectare (ha)

2. 农药剂型缩略 Abbreviations of Pesticide Formulations

AS　Aqueous Solutions 水剂

EC　Emulsifiable Concentrates 乳油

EW　Emulsion in Water 水乳剂

GR　Granules 颗粒剂

ME　Microemulsion 微乳剂

SC　Suspension Concentrates 悬浮剂

SE　Suspoemulsion 悬乳剂

SG/WSG Water-Soluble Granule 水溶性粒剂

SL　Soluble Concentrate 可溶液剂

WDG　Water Dispersable Granules 水分散性颗粒

WG　Wettable Granules 水分散粒剂

WP　Wettable Powders 可湿性粉剂

目 录

第一章 云南省果蔬产业发展概况 　　001

第二章 水果 　　009

　　低纬高原苹果栽培关键技术 …………………………… 011
　　梨栽培关键技术 ………………………………………… 021
　　桃栽培关键技术 ………………………………………… 030
　　草莓栽培关键技术 ……………………………………… 038
　　蓝莓栽培关键技术 ……………………………………… 047
　　李栽培关键技术 ………………………………………… 058
　　猕猴桃栽培关键技术 …………………………………… 073
　　人参果栽培关键技术 …………………………………… 084
　　香蕉栽培关键技术 ……………………………………… 095
　　杧果栽培关键技术 ……………………………………… 106
　　柑橘栽培关键技术 ……………………………………… 118
　　葡萄栽培关键技术 ……………………………………… 128
　　菠萝蜜栽培关键技术 …………………………………… 136
　　西番莲（百香果）栽培关键技术 ……………………… 145
　　西瓜栽培关键技术 ……………………………………… 154

第三章 蔬菜   167

番茄栽培关键技术 …………………………………… 169

辣椒栽培关键技术 …………………………………… 179

茄子栽培关键技术 …………………………………… 191

菜豆栽培关键技术 …………………………………… 201

速生叶菜栽培关键技术 ……………………………… 211

南瓜栽培关键技术 …………………………………… 220

黄瓜栽培关键技术 …………………………………… 230

苦瓜栽培关键技术 …………………………………… 240

洋葱栽培关键技术 …………………………………… 250

❖ Contents ❖

Chapter One Overview of the Development of Fruit and Vegetable Industry in Yunnan 261

Chapter Two Fruits 271

Key Techniques for Apple Cultivation in Low-latitude Plateau ············ 273
Key Techniques for Pear Cultivation ············ 287
Key Techniques for Peach Cultivation ············ 298
Key Techniques for Strawberry Cultivation ············ 309
Key Techniques for Blueberry Cultivation ············ 322
Key Techniques for Plum Cultivation ············ 336
Key Techniques for Kiwifruit Cultivation ············ 353
Key Techniques for Ginseng Fruit Cultivation ············ 368
Key Techniques for Banana Cultivation ············ 383
Key Techniques for Mango Cultivation ············ 397
Key Techniques for Citrus Cultivation ············ 411
Key Techniques for Grape Cultivation ············ 426
Key Techniques for Jackfruit Cultivation ············ 437
Key Techniques for Passion Fruit Cultivation ············ 447
Key Techniques for Watermelon Cultivation ············ 459

Chapter Three Vegetables 475

Key Techniques for Tomato Cultivation ······················· 477
Key Techniques for Pepper Cultivation ······················· 491
Key Techniques for Eggplant Cultivation ······················· 507
Key Techniques for Common Bean Cultivation ······················· 521
Key Techniques for Fast-growing Leafy Vegetables Cultivation ············· 534
Key Techniques for Pumpkin Cultivation ······················· 547
Key Techniques for Cucumber Cultivation ······················· 559
Key Techniques for Bitter Gourd Cultivation ······················· 572
Key Techniques for Onion Cultivation ······················· 585

第一章

云南省果蔬产业发展概况

云南位于中国西南边陲，北纬21°8′~29°15′，东经97°31′04″~106°11′48″，南北长约877.5 km，东西宽约721.4 km，总面积约39.43万 km²。云南省内主体部分属于云贵高原西部，称为滇东高原，西北部则为青藏高原南延部分，称为滇西纵谷，地势上总体呈北高南低倾斜的阶梯形高原地貌，有海拔高差数百至千米以上的高山峡谷、山地、高原、丘陵、盆地、河谷冲积平原等地形地貌，其中山地和高原面积约占云南省总面积的94%，土壤以山地红壤为主。

云南整体属于亚热带高原季风气候，全省年平均气温为14.5 ℃，最热月（7月）均温为19~22 ℃，最冷月（1月）均温为6~8 ℃，整体表现为年温差较小而日温差较大，冬季较温暖。境内六大水系以西北—东南向或南北向为主，多数地区年均降水量约1 100 mm，年降水量中等但季节分配不均，有干湿季节之分，干季为11月至翌年4月，雨季为5—10月，冬干夏湿，冬暖夏凉。地理区位、大气环流和复杂的地理环境共同造就了云南丰富的气候带特征，从北到南纵跨温带到热带6个气候类型带，丰富的气候类型赋予了云南丰富的植物资源和小区域作物种植的多样化特点。

云南作为典型的低纬高原地区，其特殊的地理区位和气候条件为云南农产品的生产提供了充足的阳光、优质的水源、清洁的空气、肥沃多样的土壤，这些自然优势赋予了云南农产品"安心云品"的天然原生态属性。云南主要的农产品有粮食、蔬菜、水果、茶叶、中草药、坚果、鲜切花及各类畜禽产品，农产品出口多年来保持相对稳定发展，其中水果、蔬菜在农产品贸易中一直占有极其重要的地位，是云南农产品出口的晴雨表。

1 云南水果产业发展概况

1.1 云南果树种植面积与产量

云南水果种类丰富，主要以温带、亚热带及热带水果为主，现已查明的果树资源有133种，约占全国果树种类的60%。国家统计局和云南省统计局数据显示（图1-1），

2010年以来，云南省果园整体面积呈逐年增长趋势，2011年果园面积超过500万亩[①]，经过10年的发展，2021年总面积超过1 000万亩（约1 062.62万亩），产量也由2010年的407.14万t增长至2021年的1 142.60万t。2019—2021年，云南省果园面积连续3年稳居全国第7位，产量居第12～13位。

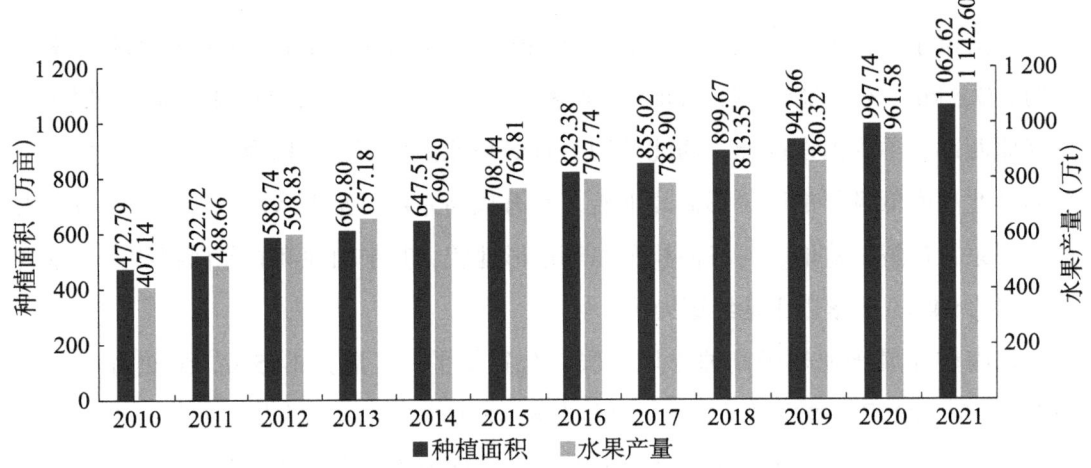

图1-1　2010—2021年云南果树种植面积与产量变化趋势
（数据来源：国家统计局、云南省统计局）

1.2　区域布局

云南果树种植分布具有明显的气候区域特征，且云南独特的"立体气候"使得绝大多数水果与其他产区水果存在上市时间差，鲜果可实现"四季生产、周年供应"。部分水果在云南有"早熟更早、晚熟更晚"的特点。种植面积与产量规模相对较大的品类为香蕉、柑橘、杧果、葡萄、梨、苹果、桃、西瓜、石榴、菠萝等，一些特色新兴水果如甜柿、枇杷、人参果、猕猴桃、杨梅、荔枝等近年来种植规模逐渐扩大。其中以苹果、梨等为代表的温带水果面积约占38.5%，以柑橘、葡萄等为代表的亚热带水果面积约占28.2%，以杧果、香蕉等为代表的热带水果面积约占33.3%，初步形成三足鼎立之势。

1.3　贸易情况

自2015年以来，云南省果品稳居全省农产品出口第一大品类，出口额均保持在全省农产品出口总额的40%左右。云南省是全国果品出口第一大省，占全国果品出口额

---

① 1亩≈667 m^2，15亩=1 hm^2。全书同。

的 25% 以上。2021 年，出口额达到 18.7 亿美元，占全省农产品出口总额的 43%，占全国果品出口总额（75.1 亿美元）的 25%。柑橘、葡萄等果种出口量常年居全国第 1 位。

1.4 云南水果产业特点

近年来，云南省发展特色优势水果、推广精细化生产技术、建立标准示范基地、培育品牌拓展市场等工作，云南水果产业在助推区域经济发展、巩固拓展脱贫攻坚成果同乡村振兴有效衔接中的地位越发凸显。全省水果生产区域化格局基本形成，如滇东北、滇西北的苹果产区，滇东的蓝莓、梨产区，红河流域的葡萄、香蕉、菠萝、石榴产区，滇中的桃、草莓产区等。果业品牌打造已基本形成，截至 2020 年年底，全省水果累计有效认证绿色食品 531 个、有机产品 225 个。同时，云南水果生产也面临着基础配套设施薄弱、规模化程度低、技术推广滞后、采后设施落后等问题，导致云南水果生产环境优势还未完全转化为品质优势。

1.5 发展对策

加大水、电、路等基础设施建设，增加水肥一体化等生产设施的投入，配套建设保障性水源工程和基本冷链贮运设施；加强良种繁育、生产、销售体系的建设，强化标准化关键生产管理技术的集成研究与应用；强化企业、农民专业合作社、产业联盟等新型经营主体的带头和示范作用，支持和巩固龙头企业引领作用，推动地理标志农产品等品牌效益的充分发挥；强化全产业链的科技覆盖度，促进水果产业提质增效，保障产业的可持续健康发展。

2 云南蔬菜产业发展概况

2.1 云南蔬菜种植面积与产量

蔬菜是云南省种植面积第一大的经济作物，是云南优势特色明显和产业基础较好的生物产业，长期以来在全国南菜北运、西菜东运、蔬菜出口中发挥了不可替代的重要作用。

国家统计局和云南省统计局数据显示（图 1-2），2010 年以来，云南省蔬菜种植面积与产量均呈现逐年增长趋势：2015 年种植面积超过 1 500 万亩，2021 年已近 2 000 万亩（约 1 937.37 万亩）；2013 年产量超过 1 500 万 t，2017 年开始超过 2 000 万 t，

2021年已近3 000万t（约2 748.36万t）。云南省蔬菜面积常年稳居全国第10位，产量常年稳居全国第11位。

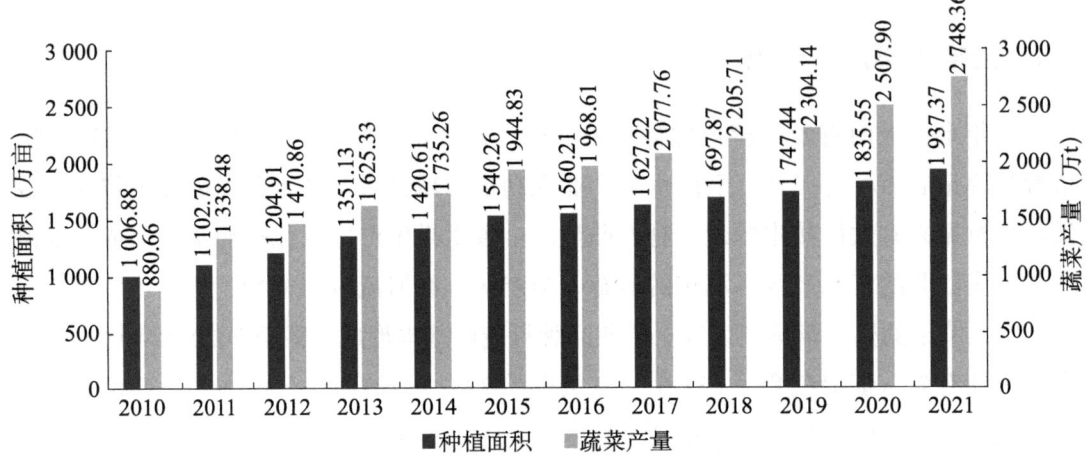

图1-2 2010—2021年云南省蔬菜种植面积和产量变化趋势
（数据来源：国家统计局、云南省统计局）

2.2　区域布局与产量结构

云南省依托低纬度立体气候，充分发挥"天然温室""天然凉棚"优势，形成了以滇西及滇西南的保山市、普洱市、西双版纳州、德宏州、临沧市及中北部低热河谷区为重点的冬春蔬菜优势产区，主要生产茄果类、洋葱、苦瓜、豇豆、菜豆；以滇东北的昭通市，滇西北的大理州、丽江市、怒江州和迪庆州，滇东南的文山州为重点的夏秋蔬菜优势产区，主要生产结球甘蓝、大白菜、萝卜等喜凉蔬菜和辣椒、番茄；以昆明市、曲靖市、玉溪市、楚雄州和红河州北部为重点的常年蔬菜优势产区，主要生产普通白菜、生菜类和葱蒜类蔬菜。2019年，云南省冬春、夏秋和常年蔬菜优势产区播种面积分别占全省的25.11%、23.77%和51.12%，实现全省蔬菜产品错峰上市，保障了国内蔬菜市场周年均衡供应。

近几年《云南统计年鉴》数据显示，云南省蔬菜呈现"白菜类＞根茎类＞叶菜类＞茄果类＞葱蒜类＞豆类＞瓜菜类＞甘蓝类＞水生菜类"的产量结构。2021年，云南省蔬菜总产量为2 748.86万t，其中，白菜类以586.1万t占全省蔬菜产量的21.32%，根茎类以437.90万t居其次。

2.3　产业特点

得益于近年来云南省"绿色食品牌"重点产业打造和"千亿级产业"等政策的支

持，云南省蔬菜种植规模发展迅速，设施化水平不断提高，2019年的设施栽培面积约199.6万亩，占全省蔬菜播种面积的11.39%；优势产区逐步形成，基本实现不同类型蔬菜产品周年均衡供应；蔬菜的种类繁多，播种面积在20万亩以上的蔬菜品类有23个；龙头企业的引领作用及新型农民专业合作社及社会化服务组织的作用不断加强，广大农户共同参与形成蔬菜产业发展共同体；蔬菜销售渠道不断拓宽，云南省已经逐步成为全国重要的"南菜北运""西菜东运"的知名"菜园子"。

2.4 存在问题和发展对策

一是设施装备水平不高，仍然需要进一步提高，建议依托现有设施设备，针对不同地区和蔬菜品类加强生产机械的改造升级、新设备研发和推广应用，以进一步适应现代产业发展的需要，增强抵御自然灾害的能力。

二是蔬菜加工多停留在分拣、包装等初级加工上，精深加工不足，建议加强加工设施设备建设，提高采后处理能力和水平，并根据蔬菜品类和价值开展加工技术改造和生产线升级，不断加大新型蔬菜加工制品技术研究和产品开发力度。

三是蔬菜创新品种的培育及制种能力不足，建议加强蔬菜创新品种的育种研究，扩大蔬菜品类的自主创新能力，加大新品种试验示范基地和工厂化育苗设施建设，大力推进蔬菜健康种苗生产和工厂化育苗设施建设。

四是打造蔬菜区域品牌、企业品种和产品品牌的力度不够，"云菜"在社会上的知名度和影响力不足，建议大力引进国内外大型蔬菜加工、流通企业，培育一批大型蔬菜企业集团，加大蔬菜品牌和产品的宣传推介力度，提高"云菜"的整体社会影响力。

第二章

水 果

第二集

目次

低纬高原苹果栽培关键技术

1 产业发展概况

苹果在中国有两千多年的栽培历史,目前我国主要种植在辽宁、河北、山西、山东、陕西、甘肃、四川、云南、西藏等地区。

1.1 自然分布情况

云南及其周边地区是我国苹果属植物资源最为丰富的地区之一,其资源种类占全国苹果属植物资源种类的70%以上,在已知的资源中,有9种为该地区特有种且类型多样,目前已发现的云南苹果属种质资源有21种,其中16种为野生型、5种为栽培型。

2021年云南省苹果种植面积140.8万亩,总产量约134.7万t,居全国第12位,以果实品质优良,早中熟品种价格优势稳定,果品外观色泽艳丽,成熟早,甜脆爽口,内在品质好而著称,云南已成为我国优质苹果南方生产基地。

1.2 种植情况

云南苹果种植主要分布:滇东北产区(昭通市、曲靖市)、滇西北产区(丽江市、大理州、迪庆州)、滇中产区(昆明市)、滇东南产区(红河州),海拔2 000~2 700 m贫困山区半山区,地形地貌复杂的冷凉高地区域,年均气温5~18 ℃,年均降水量500~800 mm,土壤微酸性或中性土壤,排灌良好。主栽品种有富士优系、金冠优系、嘎啦优系等。

2 主要栽培品种及特性

2.1 晚熟红富士

【品种来源】日本。

【适宜区域】苹果种植区域均可种植,适宜在海拔2 000~2 400 m的区域种植。

【特征特性】树势强健，树姿半张开。幼树或健壮枝条有明显的腋花芽结果习性。初结果期的树，长果枝和腋花芽占有一定的比例，但很快会转向以短果枝结果为主，盛果期短果枝结果约占70%。晚熟。花紫红色。平均单果重220 g，最大果重650 g。果形扁圆至近圆形。果皮中厚而韧，果面鲜红色，阳面有红霞和条纹，有片红和条红，果点小，果面光滑、无锈、果粉多。果肉乳黄色，肉质松脆，汁液多，酸甜适中，可溶性固形物含量14.5%~15.5%，品质极佳，丰产性、耐贮藏性好。

【生产表现】结果早，丰产。乔砧树4~5年开始结果，矮砧树3年开始结果，5年后进入盛果期。在加强综合管理、采用促花措施后，定植第二年见花。容易形成大小年。坐果率较高，在正常授粉情况下，花序坐果率达70%左右，花朵坐果率16.2%~40%。果台枝较细，连续结果能力较差。抗病性强。在昭通10月上旬成熟，在丽江10月中下旬成熟。自花结实率较低，定植时常需要配置一定比例的授粉树，异品种授粉结实率高。生理落果和采前落果轻。成熟前无裂果现象。丰产期一般亩产3 000~4 000 kg。

2.2 烟富3号

【品种来源】山东省烟台市果树科学研究所选育。

【适宜区域】苹果种植区域均可种植，适宜在海拔2 000~2 400 m的区域种植。

【特征特性】晚熟品种。果形近圆形，有光泽，果点小，果面光滑、无锈、果粉多，树冠上下、内外着色均好，偏红，全红果比例可达80%左右，果皮中厚而韧，果肉乳黄色，肉质松脆，汁液多，风味酸甜适中，品质极佳。可溶性固形物含量14.8%~15.4%。

【生产表现】果实生育期190 d。平均单果重250~300 g，丰产性、抗病性能好，耐贮藏。摘袋后5~7 d即达全红，尤其在秋季高温、昼夜温差小时，比其他富士品种有明显的着色优势。在昭通10月中旬成熟，在蒙自9月成熟。

2.3 秦冠

【品种来源】西北农林科技大学选育。

【适宜区域】苹果种植区域均可种植，适宜在海拔1 800~2 200 m的区域种植。

【特征特性】以丰产优质的金冠作母本，以结果早、极丰产、抗性强的鸡冠作父本进行杂交选育而成。晚熟品种。树皮光滑，多年生枝暗红褐色，一年生枝褐色，节间长，皮孔大而密、椭圆形，茸毛少。花红色。果实圆锥形，平均单果重200~250 g，丰产性好。底色黄绿，阳面有暗红晕及断续红条纹，常带有白色锈，成熟时，可以达到全面暗红色，果面光滑、蜡质较多，果点明显，果皮较厚韧，果肉乳黄色，肉质脆稍致密，可溶性固形物含量 16.5%，酸含量 0.19%，维生素 C 含量 2.31 mg/100 g。较耐贮存。

【生产表现】树势较强，生长较健壮，干性极强。萌芽率高，易成花，花芽饱满，不容易形成大小年，坐果率高，连续挂果力强。抗旱、抗寒、抗病虫能力强。在昭通 10 月中下旬至 11 月上旬成熟。叶片于 11 月下旬开始大量落叶。

2.4　金帅

【品种来源】云南最早引进品种，原产美国。

【适宜区域】苹果种植区域均可种植，适宜在海拔 1 800～2 200 m 的区域种植。

【特征特性】中熟种，平均单果重 150 g 以上，圆锥形，顶部稍有棱突；果梗细长，果皮薄，稍粗糙，色绿黄，稍贮藏后变为金黄，采收晚时阳面偶有淡红色晕；果肉黄白色，肉质细密；刚采收时脆而多汁，味浓甜芳香，品质极佳。晚采果实果肉淡黄色，鲜食风味极佳，耐贮性稍差。丰产稳产性能好。

【生产表现】树势较中庸，枝条生长一般，干性强，萌芽率高，成枝率弱，树枝较开张，容易成花，不容易形成大小年，花白色，坐果率较高，连续挂果力强，抗病性强，易生果锈，果实在昭通 8 月下旬成熟，叶片于 11 月中旬开始大量落叶，落叶整齐一致，丰产期果树亩产 4 500 kg 左右。

2.5　嘎啦

【品种来源】新西兰。

【适宜区域】苹果种植区域均可种植，滇中地区较好，适宜在海拔 1 600～1 800 m 的区域种植。

【特征特性】该品种属中、早熟品种。树势强健、萌芽率高、易成花、成枝力强。节间平均长度为 2.7～3.3 cm，易形成短枝。长、中、短枝及腋花芽均可结果，幼龄树腋花芽结果所占比率高，盛果期树以短果枝结果为主。果实圆锥形，稍带五棱，平均单果重 170 g，着色指数 65%～85%，具浓红条纹。果肉黄白色，肉质细脆，汁液多，酸甜适度，香味浓，含可溶性固形物 14.58%、总糖 11.16%、总酸 0.22%，果肉硬度 7.0～8.3 kg/cm²，较耐贮藏，常温下可贮藏 1 个月，冷藏条件下可贮藏 3 个月。

【生产表现】连续结果力强、丰产性强、分批成熟、易感白粉病。品种自花结实率高，着色较好，在曲靖、丽江无须套袋即能生产出色泽艳丽的优质果。

2.6 红元帅（红蛇）

【品种来源】原产于美国加利福尼亚州。

【适宜区域】苹果种植区域均可种植，适宜在海拔 1 600～2 200 m 的区域种植。

【特征特性】果形端正，高桩，萼部五棱明显，平均单果重 200 g，大者可达 500 g 以上，而且果肉质脆、汁多、味甜、有芳香。生长至初上色时，出现明显的断续红条纹，随后出现红色霞，充分着色后全果浓红色，并有明显的紫红色粗条纹，果面富有光泽，十分鲜艳夺目；果点浅褐色或灰白色，果肩起伏不平；果肉黄白色，肉质脆，质中粗，较脆，汁多，味甜，可溶性固形物含量 14%，有浓郁芳香，品质上等。

【生产表现】连续结果力强、丰产性强，果实生育期 145 d 左右，一般在 9 月中旬成熟，较耐贮藏，采后贮藏 1～2 个月后为最佳食用期。

2.7 乔纳金

【品种来源】美国纽约州农业试验站用金冠 × 红玉杂交育成。

【适宜区域】苹果种植区域均可种植，适宜在海拔 1 600～2 200 m 的区域种植。

【特征特性】果实圆锥形，平均单果重 220～250 g；底色绿黄或淡黄，阳面大部有鲜红色霞和不明显的断续条纹；果面光滑有光泽，蜡质多，果点小，不明显；果肉乳黄色，肉质松脆，质中粗，汁多，风味酸甜，稍有香气，含可溶性固形物 14% 左右，品质优，果实较耐贮藏。

【生产表现】苗木栽后 3～4 年结果，7～8 年进入大量结果期，丰产。不仅是鲜食品种，也是制汁及加工的优良品种。

3 栽培管理技术

3.1 园地选择

选择生态条件良好、远离污染源并具有可持续生产能力的农业生产区域，地势比较平坦的丘陵地带，灌溉条件良好。在山地建园，坡度最好在 25° 以下。针对防御台风、保持水土、防风固沙等问题，营造防风林系。防风林树种选择适应性强、生长迅速、寿命长，与果树无相同病虫害或中间寄主，经济价值较高的树种。新建苹果园周边 5.0 km 内禁止有柏类植物。园地土壤肥沃的标准是土质疏松透气、保墒条件好、活土层深 0.6 m、有机质含量达到 0.8% 以上。

3.2 高效栽培技术模式

3.2.1 苹果乔化栽培技术模式

苹果乔化栽培技术模式在云南已有近80年的历史，目前是昭通苹果产区主要的栽培模式。根据主栽品种情况配置相应的授粉树。可根据行间距合理套种蔬菜、花生等作物，以豆科作物为主，不仅能增加果农收入，还能做到种养地相结合。

注意事项：苹果乔化栽培应注意栽培密度，树形应根据密度来选择。果园生草或种植绿肥时，应播种在行间离树1 m处，11月前必须全部压青或翻耕，冬季苹果园不能有绿肥生长，更不能在苹果园留种。

3.2.2 苹果矮化密植集约栽培技术模式

矮化密植集约栽培是未来苹果产业发展的方向，因树体小、操作方便、连续结果能力强、果品质量高、省工省力而逐渐受到世界各国重视。目前，该技术采用优质矮化自根砧M9-T337二年生大苗定植，安装格架系统设施、水肥一体化设施，定植株行距$1.0 m \times 3.5 m$，亩植192株，干高1 m，冠幅$0.8 m \times 1.2 m$，株高$2.5 \sim 3.0 m$，采用细长纺锤形整形。宽行生草，窄行微滴灌，苹果园采用立架栽培，整形以每年拉枝为主，修剪上多用疏除、长放两种手法，很少用短截。树形以主干直接着生结果枝组的圆柱形树形为主。种植品种有皇家嘎啦、米奇拉、红元帅、华硕、蜜脆、粉红女士、烟富3号等。

通过宽行密植、格架栽培，实现"四省、两高、两早"，即省水60%、省肥70%、省土地70%、省人工80%，高品质、高产出，早挂果、早回报。使用该栽培模式，可有效推动云南苹果生产标准化、规模化、产业化、机械化。

3.3 修剪技术

冷凉高地苹果园可以选择无立柱矮化密植修剪快速成形技术。萌芽后及时抹去60 cm以下的萌芽。顶芽选留1个健壮的芽作中央领导枝，其下$2 \sim 3$个相邻的萌芽及时抹除。侧枝萌芽后长至30 cm时用牙签撑平。

第一年冬剪：秋季将主干上的新梢拉至$95° \sim 120°$，将冬季修剪延迟到2月下旬至3月上旬。修剪时，侧枝枝条全部留马牙形重短截，对中干延长头轻短截或不短截，80 cm以下位置的侧枝全部去除。

第二年冬剪：秋季将主干上的新梢拉至$95° \sim 120°$，冬季继续对侧枝枝条全部留马牙形重短截，对中干延长头轻短截，树高基本达到3 m左右。

第三年冬剪：选留生长势中庸、角度大的一年生枝条作主枝，不打头。每年需要对所有长度超过 1 m、基部粗度超过主干粗度 1/4 的侧枝视情况去除，25 cm 以上的侧枝拉到水平以下，保持树体生长势缓和，树形成形并开始结果。

3.4 水肥管理

萌芽期至开花期：一般 3—4 月是苹果根系生长的第一高峰期，地上部开始生长前后，到新梢进入生长高峰时变缓。追肥遵循"适、浅、巧、匀"的技术要求，一般每年 2 次。第一次追肥在萌芽前，以氮肥为主、磷肥为辅。第二次追肥在萌芽后至开花前，一般以氮、磷肥为主，可选用磷酸二铵，采用穴施，结合浇芽萌动水进行追肥，每亩追施尿素 10 kg、磷酸二铵 20 kg、硫酸钾 5 kg、含氨基酸有机冲施滴灌肥 10 kg。

开花期至坐果期：每亩施磷酸二铵 60～70 kg，也可追施土壤调理剂、腐殖酸有机肥或生态有机钾、沼液等。此期如遇干旱，要及时浇水，保证土壤湿度。对花量大、树势弱、发芽前未施氮肥的树，应追施尿素。12～13 年生树株施尿素 0.5～1 kg，20 年生的大树株施尿素 1～1.5 kg，施肥后浇水。对于树势旺、花量适中或偏少的树，花前只浇水不施氮肥。

果实膨大期：6 月上旬至 7 月上中旬，此次追肥是为了满足果实膨大、枝叶生长和花芽分化的需要。此次追肥以钾为主，选用穴施或"井"字沟浅施，每亩施氯化钾 40 kg、磷酸二铵 5 kg。一般产量 2 000～2 500 kg 的苹果园，全年应追施纯氮 18～23 kg、纯磷 20～25 kg、纯钾 25～30 kg。8—9 月，果实迅速膨大，此时以氮、磷、钾肥为主。亩施尿素 30 kg、磷酸二氢铵 12 kg、硫酸钾 30 kg、含氨基酸有机冲施滴灌肥 20 kg。

果实成熟期：果实成熟前 20～30 d，进行叶面喷施，时间选在阴天或晴天的 9:00 以前和 16:00 以后进行，肥料在叶片上保持湿润时间长，吸收时间长。果实着色初期至着色期，每间隔 10～15 d，进行 1 次叶面肥的喷施；中晚熟品种适当加喷次数，一般在采果前 15～20 d 停止施用。

采后：采后追肥可促进根系生长发育、增加树体营养积累、提高花芽质量、提高养分的利用率、增强树势和树体的抗冻性。施肥时应以有机肥为主，配施适量的氮磷钾复合肥、中微量元素肥及根系专用肥。其中，氮肥施用量应占全年施肥量的 30% 左右，磷、钾施用量占全年施肥量的 60%，即氮肥（N）10～15 kg、磷肥（P_2O_5）10～12 kg、钾肥（K_2O）18～25 kg。最佳施肥时间为 9 月初至 10 月中下旬。采果后，叶面再喷施 1 次 0.5% 的氮肥，促进根系生长及树体储存营养，有利于来年果树的丰产。

3.5 主要病害防控

3.5.1 腐烂病

【为害情况】主要为害结果树枝干及果实。枝干症状表现有溃疡型、枯枝型两种。溃疡型表现为酒糟味、小黑点、冒黄丝。枯枝型多发生在2～4年生小枝上，常呈现黄褐色与褐色交错的轮纹状斑。果实发病初期病斑呈褐红色，后期病斑中部形成黑色小颗粒。

【发生规律】发病高峰期在早春，5月发病盛期结束。

【防治方法】对已经感染腐烂病的主干、主枝应及时治疗。常用治疗方法有两种：常规刮治涂药法、不刮皮划道涂药法。无论采用哪种方法，病疤部位均需涂抹药剂，大病疤必须桥接。最新治疗方法：用生物菌肥与净土按照质量比1∶1混合，加水稀释搅拌成糊状，涂抹于刮治之后的病疤上，厚1 cm，外用塑料薄膜包裹严实，3个月后解开，结合地下施生物菌肥，强土壮树，提高肥效，增强抗性，增产提质效果更好。

3.5.2 白粉病

【为害情况】发病迅速，容易反复发生。该病主要为害叶片及嫩梢，尤其是新梢顶部叶片容易感染。白粉病萌发的一个重要条件就是湿度，连天阴雨导致果园湿度过大，为白粉病发生提供了条件，所以如果花期出现几次降雨，后期白粉病一定会加重。

【发生规律】由子囊菌引起的真菌病害，每年可集中发病2次，春季萌芽期（4—6月）和秋季秋梢生长期（8月底）。

【防治方法】花前要做好白粉病预防工作，花期遇雨一定要加强防治。发芽前树上喷布3°Bé石硫合剂，开花前喷40%腈菌唑悬浮剂3 000倍液或10%醚菌酯水乳剂1 000倍液；落花后树上喷10%己唑醇悬浮剂1 500倍液或25%吡唑醚菌酯悬浮剂2 000倍液1～2次。

3.5.3 锈病

【为害情况】主要为害幼叶、叶柄、新梢及幼果等绿色幼嫩组织。叶片初染后，叶片表面出现橘黄色斑点，随着病程发展，患病叶片背面出现葱根状菌孢。叶柄受害后病部橙黄色，纺锤形，膨大隆起，其上也出现黄色小点和菌孢。幼果被害多在萼洼附近形成圆形、橙黄色斑点，后期病斑变褐色。

【发生规律】锈病是由担子菌引起的真菌病害。早春气温高、降雨多时易于发生。

【防治方法】从苹果展叶期开始每隔10～15 d喷施1次代森锰锌、百菌清、多菌灵等保护性杀菌剂，连续喷施2~3次，保护叶片不受锈病菌侵染。4月下旬至5月上旬，在树上喷30%苯醚甲环唑悬浮剂2 000倍液或43%戊唑醇悬浮剂3 000倍液。果园周围5 km范围内禁止种植柏树。

3.5.4 黑星病

【为害情况】由担子菌引起的真菌病害。主要为害叶片、果实、花、芽和嫩梢。

【发生规律】6—7月是发病盛期，叶片初发病表现为直径12 mm的近圆形白色病斑，随着病程发展，病斑扩大，呈近圆形或放射状，边缘明显，正反面出现黑褐色至黑绿色霉层，数个接近的小病斑也可连成大病斑，其边缘呈不规则状，后期病斑增厚，叶片不平整，甚至扭曲，病叶无病斑部位失绿，导致早期落叶。果实上的病斑呈疮痂状，初期呈近圆形斑，淡黄绿色，逐渐扩展，病果表面也有黑色或黑绿色霉层，边缘颜色略深，随着果实生长而凹陷、龟裂，致果实畸形。

【防治方法】冬季彻底清扫落叶、病果，带出果园，集中深埋。发现病叶立即摘除并装入塑料袋带出果园深埋，防止二次传播；幼果期使用苯醚甲环唑、嘧菌酯和吡唑醚菌酯等安全的药剂，每次喷药时加上芸苔素内酯和氨基酸；果实膨大期，推荐使用苯醚甲环唑＋氟唑菌酰羟胺、克菌丹＋戊唑醇等；套袋前10 d喷1次药。要想取得满意的防治效果，必须做到以下几点：防早治小；防治要有连续性，注意间隔5~7 d喷1次药，连喷2次；树上、地面同时喷，不留死角；保证喷水量，只有喷雾细致才能保证效果。

3.6 主要虫害防控

3.6.1 橘小实蝇

【为害情况】主要为害果实。成虫产卵时在果皮形成伤口，把虫卵产在果实内，等幼虫长大直接啃咬果肉，果实内部开始腐烂。

【发生规律】橘小实蝇在14~34 ℃都能够生长，最适温度为25~30 ℃。春季高温干旱、夏季温润少雨的地方有利于虫害的发生。

【防治方法】做好果园的清理，及时清除园区烂果、病果，集中深埋或者用药物浸泡后销毁。有条件的地方可进行果园灌溉2~3次，消灭土里的幼虫、蛹，还有刚出蛹的成虫。营造适宜橘小实蝇天敌如黄金小蜂、隐翅虫、蚂蚁等生存的环境，利用性诱剂或糖醋液诱杀成虫。果实转色成熟时，可选用高效氯氟氰菊酯、氰戊菊酯、阿维菌素、乙基多杀菌素等药剂喷雾防治。

3.6.2 苹果蚜虫

【为害情况】 为害幼叶、嫩梢。

【发生规律】 苹果蚜虫不转移寄主,当年气候对其影响较大。冬季温度高于往年的平均温度、早春温度回升较快、新梢生长旺盛、少雨干旱等有利于蚜虫的发生。繁殖速度极快,一年发生10代以上。虫卵一般会在芽旁和芽腋处越冬,越冬卵在果树的萌芽期开始孵化,幼虫集中在芽处取食,在苹果展叶后爬到小叶上。5月下旬开始出现被害梢,6—7月是为害盛期。在蚜群内会产生有翅蚜进行迁飞传播。7月下旬有翅蚜迁飞到杂草上以孤雌胎生的方式进行繁殖,10月又产生有翅蚜飞往果园,交尾产卵。

【防治方法】 发生初期,可选用75%吡蚜·螺虫酯水分散粒剂5 000倍液、22%氟啶虫胺腈悬浮剂8 000倍液、70%吡虫啉水分散粒剂5 000倍液或15%吡蚜酮可湿性粉剂4 000倍液进行喷雾处理,可有效降低果园内蚜虫数量,且对蚜虫天敌及其后代无持续影响。

3.6.3 苹果绵蚜

【为害情况】 为害枝条、新梢、叶腋、果洼和外露根系,形成伤口感染其他病害。

【发生规律】 一年发生12~18代,在树干伤疤裂缝和近地表根部上越冬。

【防治方法】 4—5月,扒土露根,撒药后原土覆盖。生长期可喷施药剂或进行枝干轻刮皮、药剂涂抹。

休眠期果树:可结合叶螨类、介壳虫的防治,在果树发芽以前喷布含油量5%的矿物油乳剂,可以杀死越冬的蚜卵。

药剂涂干:对水源较远、取水困难的未结果果树尤其适用。涂药后用塑料布或废报纸包扎好。

3.6.4 叶螨类

【为害情况】 为害植物的叶子和果实,使叶片严重变薄、变白,甚至脱落,影响植株的正常生长。

【发生规律】 一年发生7~9代。冬卵在短果枝、果台和二年生以上的小枝条分杈、叶痕、芽轮及粗皮等处越冬。冬卵孵化相对集中,正常为10~20 d。因此,冬卵孵化期盛期是药剂防治的第一个关键时期。

【防治方法】

石灰水:此法主要针对木本植物,深秋待植物落叶进入休眠之后,及时清理枯叶杂草、树干死皮，然后给树干刷上石灰水，从而杀死寄生在树干表面的虫卵。

硫磺粉：在深秋，清理土面上的枯枝烂叶和杂草，然后均匀地撒少量硫磺粉，并翻动土壤使硫磺粉均匀散布在表层土中。

石硫合剂：深秋植物休眠后进行，用石硫合剂粉剂300~500倍液，然后均匀喷洒树干，并均匀喷洒土面至0.5~1 cm表层土湿润即可。

螺螨酯：在早春进行，取20%螺螨酯乳剂4 000~6 000倍液，均匀喷洒植株叶片正反面、枝干，并进行灌根。

生物防治：一般在春末夏初还未发生螨害时释放捕食螨，选择在阴天或者傍晚（要避免雨天）释放，直接将纸袋钉在树干上。每袋防治有效期一般是一个季度。

化学防治：可选用73%炔螨特乳油1 500倍液、34%螺螨酯悬浮剂4 000倍液+43%联苯肼酯悬浮剂1 800~2 500倍液、20%阿维·联苯肼悬浮剂2 000~2 500倍液、2.5%高效氯氟氰菊酯乳油4 000倍液、5%噻螨酮乳油3 000倍液喷雾。每隔7~10 d喷1次，连喷2~3次。

4 采收

适宜时间为晴天、无露水。果面着色60%以上即可采收。轻拿轻放，避免损伤，放置于阴凉干燥处。

参考文献

姜中武，于青，宋来庆，等，2010.成龄苹果园优质高效标准土肥水综合管理技术 [J].烟台果树（3）：32-35.

李丙智，2010.矮砧苹果建园与幼树整形修剪技术 [J].西北园艺（10）：13-16.

刘汉涛，于海军，刘萍，等，2022.现代栽培模式苹果园建立与管理的关键技术 [J].落叶果树，54（5）：84-85.

王玉玲，2010.苹果优质高效栽培技术 [J].河北果树（2）：8-11.

张建，2022.苹果园病虫害综合治理技术 [J].世界热带农业信息（1）：49-50.

梨栽培关键技术

1 产业发展概况

梨，蔷薇科梨属，多年生落叶乔木或灌木果树。梨果实营养丰富，果肉脆嫩多汁，酸甜可口，风味极佳。鲜梨富含蛋白质、脂肪、碳水化合物、钙、磷、铁、胡萝卜素、维生素 B_1、维生素 B_2、维生素 B_3、维生素 C 等多种营养物质，具有"百果之宗"的美誉。传统医学认为梨果具有生津润燥、清热化痰、清心润肺，以及治热咳、消渴、热病津伤、便秘等功效，经常食用可预防泌尿、消化系统疾病等。梨果富含膳食纤维，热量低，果皮中含有丰富的植物营养素，其中黄酮醇类物质的含量位居水果前列，具有较好的抗氧化、消炎作用，可能对心脏病、2 型糖尿病、癌症等有一定的预防作用，故梨果也被称为健康果品。

云南是中国梨的原产地之一，梨树栽培在云南至少有 1 200 年的历史，梨树在云南省 127 个县市，海拔 450～3 400 m 的地区都有分布，其中以海拔 1 800～2 200 m 范围为最适宜种植区域。截至 2022 年，云南省梨栽培面积约 102 万亩，以滇中地区分布最多，面积最大；其次为滇西北及滇东北，滇西南最少。梨果因皮色不同分为红皮梨、褐皮梨、绿皮梨，在云南这 3 种皮色的梨都有分布和栽培，其中以红皮梨尤为出名，种植面积约 29 万亩。云南为中国特色梨四大优势区域之一的红皮梨优势区域，独特的低纬高原地理条件造就了独特的气候条件，使云南生产的红皮梨具有颜色艳丽、风味浓郁、甜酸适中、脆爽多汁的特点，深受国内外消费者青睐，适宜进行鲜食高端果品的打造。云南的梨品种和类型有 400 余个，其中 320 个是长期在当地自然条件影响及栽培下不断选择出来的地方品种，有些具有较高的加工价值，目前云南的梨加工产品有泡梨、梨罐头、梨膏、梨醋、梨酒、梨饮料等。

2 主要栽培品种及特性

2.1 美人酥

平均单果重 275 g，纵径 9.8 cm，横径 9.5 cm，倒卵圆形；果实阳面鲜红艳丽；

果心中，5心室；果肉淡黄白色，肉质中，松脆，汁液多，味酸甜，微涩，有微香；含可溶性固形物13.2%；品质中上等，常温下可贮藏10 d。树势强，树姿半开张，萌芽力强，成枝力强，丰产。一年生枝红褐色；枝端被毛，叶片披针形，长9.5 cm，宽7 cm，叶尖渐尖，叶基宽楔；花蕾浅粉红色，每花序7～9朵花，平均8朵；雄蕊31～34枚，平均33枚；花冠直径3.8 cm。在安宁地区，果实8月中下旬成熟。

2.2　满天红

平均单果重280 g，纵径8.5 cm，横径8.7 cm，近圆形；果皮阳面鲜红色；果心中，5心室；果肉乳白色，肉质中，松脆，汁液多，味酸甜，微涩，有微香；含可溶性固形物13.8%；品质中上等，常温下可贮藏10 d。树势强，树姿半开张，萌芽力强，成枝力中，丰产。一年生枝红褐色，顶端被毛；叶片披针形，长9.3 cm，宽8.1 cm，叶尖渐尖，叶基宽楔；小叶紫红色；花蕾浅粉红色，每花序7～9朵花，平均8朵；雄蕊30～31枚，平均30枚；花冠直径3.7 cm。在安宁地区，果实8月中下旬成熟。

2.3　早白蜜

平均单果重160 g，纵径8.5 cm，横径8 cm，卵圆形或近圆形；果皮黄绿色，阳面有淡红晕；果心中，5心室；果肉乳白色，肉质细，松脆，汁液多，味甜，有微香；含可溶性固形物11.2%；品质上等，常温下可贮藏10 d。树势中，树姿半开张，萌芽力强，成枝力中，丰产。一年生枝红褐色；叶片卵圆形，长9.5 cm，宽6.5 cm，叶尖渐尖，叶基宽楔；花蕾浅粉红色，每花序6～8朵花，平均7朵；雄蕊30～38枚，平均35枚；花冠直径3.8 cm。在安宁地区，果实6月下旬成熟。

2.4　云红梨1号

平均单果重320 g，纵径6.2 cm，横径5.8 cm，长圆形；果皮85%以上为红色；果心中，5心室；果肉白色，肉质致密，汁液中，味甜，有微香；含可溶性固形物13.6%；品质中等，果实硬度较高，贮藏性好，常温下可贮藏30 d以上。树势强，树姿直立，萌芽力强，成枝力强，丰产。一年生枝红褐色；叶片披针形，长11 cm，宽8 cm，叶尖渐尖，叶基心形；花蕾浅粉红色，每花序6～8朵花，平均7朵；雄蕊18～21枚，平均20枚；花冠直径4.1 cm。在安宁地区，果实10月上旬成熟。

2.5　彩云红

树势中庸，树姿开张，萌芽力强，成枝力弱，丰产。一年生枝红褐色，幼叶微红，成叶卵圆形，长 11.5 cm，宽 7.3 cm，叶尖渐尖，叶基宽楔；花梗无毛，花蕾浅粉红色，每花序 5～11 朵花，多 8 朵；花冠直径 3.45 cm，雄蕊 29～31 枚，平均 30 枚，花药紫红色。平均单果重 240 g，纵径 8.0 cm，横径 8.3 cm，近圆形；萼洼较深，脱萼；果皮底色黄色，阳面有橘红晕，着色面积大，自然着色率 80% 以上，上色时间快，着色后无返绿现象；果心小，5～7 心室；果柄粗短，平均 2.1 cm，自然情况下果实基本不易脱落。果肉脆爽多汁、味道纯甜无涩味，性状稳定遗传。果心小，可食率 95% 以上，含可溶性固形物 13.8%。果实耐贮运，常温下可贮藏 21 d，贮藏期间果皮底色变亮，红色更加鲜艳；机械损伤果皮和果肉短时间内无明显褐变。在安宁地区，果实 8 月中下旬成熟。

2.6　宝珠梨

云南地方品种，在呈贡、晋宁等地有栽培。平均单果重 258 g，纵径 10.5 cm，横径 10.0 cm，近圆形；果皮绿色，较厚，果心大，5 心室；果肉白色，质中粗，松脆，汁液多，味甜，无香味；含可溶性固形物 12.3%；品质中等。树势强，树姿开张，萌芽力、成枝力都强，定植 6～7 年结果，有隔年结果的现象。一年生枝灰褐色；叶片椭圆形，长 12.2 cm，宽 7.2 cm，叶尖渐尖，叶基宽楔；花蕾边缘浅粉红色，每花序 5～7 朵花，平均 6 朵；雄蕊 24～33 枚，平均 27 枚；花冠直径 4.1 cm。在呈贡地区，果实 9 月下旬成熟。

2.7　红雪梨

产于巍山县海拔 1 800～2 300 m 地区，果实大，近圆形，平均单果重 280 g，最大单果重 600 g。果皮表面粗糙绿黄色、黄色，阳面 1/2 以上覆红色晕。果梗长，长 4.6 cm，梗洼浅小，萼洼深广，萼片脱落，底部褐色。果肉黄白色，质中粗，较脆，有石细胞，味甜酸，汁中多，含可溶性固形物 12.3% 左右，品质中等。初采时稍带涩味，经贮后郁香，11 月上旬成熟，可贮至翌年 2—3 月。树生长势强，大树圆形或尖圆形，树高可达 8 m 以上。萌芽力、成枝力强，一年生枝红褐色；叶片椭圆形，长 10.9 cm，宽 7.3 cm，叶尖渐尖，叶基楔形。以中、短果枝结果为主，腋花芽亦能结果。丰产，适应性强，抗病性稍弱。

3 栽培管理技术

3.1 种植规划与园区建设

3.1.1 建园

适宜的栽培区域主要为海拔 1 800～2 200 m 的平地或缓坡地，园区要求背风向阳，排灌方便，适宜的土壤 pH 值范围为 5.5～6.5，呈微酸性；土质以砂壤土或壤土为佳，轻质黏土可以用秸秆等混合有机质适当改良后进行栽培。建园时根据土壤质地和栽培模式选择开挖宽度及深度 60～80 cm 的定植塘或定植沟，消毒晾晒后，每亩一次性施入有机肥 4～6 t，与底层土充分混匀后覆盖上层熟土高至离地面 20 cm 左右，完成幼苗定植和覆盖后灌大水沉实。定植初期 5 d 左右浇 1 次水，确保成活。

3.1.2 授粉品种配置

梨为配子体型自交不亲和果树，大多数品种表现为自花授粉不能结实，需要选择授粉品种，可采用配置授粉树或者人工辅助授粉。授粉树配置有 3 种：等行配置、不等行配置、棋盘式配置。

3.2 整形修剪

3.2.1 梨圆柱形树形

圆柱形树形为单主干树形，树高 3.0～3.5 m，中干强壮直立，干高 50～60 cm，无二级侧枝和明显分层，直接在中干上均匀着生 20～25 个小枝，长度控制在 1.2 m 以内，小枝直接着生花芽，进入结果盛期后，进行中心干落头。

整形关键技术是刻芽、撑枝及枝条配置。根据南方气候及品种特点，株行距以（1.2～1.5）m×（3.0～3.5）m 为宜，主干长至 1.6 m 左右开始进行刻芽。刻芽选择萌芽前后 7～10 d 进行，刻芽区间为顶芽以下 30 cm 至根颈以上 60 cm 的干中部分，在芽上方 0.5～1.0 cm 处，宽度为 1/2 树干，深至木质部。撑枝目的在于开张枝条的角度，缓和生长势，促进花芽分化，提早开花结果。撑枝时间选择在 5—6 月至木质化之前，新梢长度 20 cm 左右最佳，利用牙签或小竹片进行开角支撑，一般要求小枝基角分支角度为 45°，腰角为 60° 左右。盛果期树体以配备 20～25 个小枝为宜。小枝要求充分利用空间，在主干上呈近似螺旋状分布，同侧小枝需间距 20 cm 以上，要求小枝不重叠、无叉枝，粗度以小于着生处主干粗度 1/3 为宜。

修剪在果树落叶后和萌动前都可进行，单主干形修剪有"三看"：一看主干上的结果枝组是否分布合理，无相互遮挡，螺旋状均匀分布在主干上；二看是否只有 1 个顶梢，否则只保留 1 个长势较弱的顶梢；三看结果枝，确保单轴延伸。

3.2.2 梨开心形树形

主干高为 60～80 cm，无中心干，从 60 cm 处往上均匀地排列着伸向四周的 3～4 个主枝，主枝与地面夹角成 45°；主枝的两侧间隔 25 cm 左右交叉分布着一级侧枝，在一级侧枝上面均匀分布中小型结果枝组或者结果枝。整个叶幕厚度 200～250 cm，株行距 3 m×4 m，每亩栽植 50 株左右。

该树形形成快、结果早、易更新换种，但整形修剪成本较高。修剪要点：该树形修剪以放为主，只对生长过壮或细弱枝组基部进行不同程度的回缩，以调节长放枝组间的生长势。对于生长旺盛的长条，可用竹竿绑缚拉枝开角，主枝开角 60°，辅养枝开角 70°。基轴长度 30 cm 左右，每个基轴一般分生两个长放枝组，再加上直接着生在中心干上的无基轴枝组，全树一共形成 10～12 个长放枝组。最上部的两个枝组，要求垂直伸向行间，下部枝组开张角度 70°，最上部两个枝组反弓弯拉倒成 90° 角。

3.3 水肥管理

采用测土配方施肥，根据土壤营养元素的亏缺情况进行肥料种类及用量的配置。化肥一般作为追肥，具有含量高、肥效快、使用方便等优点，但存在肥效短、成本高、污染环境等不足；有机肥通常用作基肥，具有肥效慢、肥效长、种类多、益于土壤改良等特点。因此，要根据树体养分情况配合施用化肥和有机肥。基肥一般在秋季采果后尽早施用，利于根系及时吸收和养分贮存，用量尽量做到"斤①果斤肥"，可适当拌入少量氮、磷肥。追肥通常根据树体养分诊断情况结合物候期进行，分为花前追肥、果实膨大期追肥、花芽分化期追肥，肥料可采用地面撒施、叶面追肥或使用水肥一体化技术。追肥通常结合灌溉进行。土壤含水量为田间持水量的 60%～80% 时，最适宜梨树的生长。把握好梨树需水的几个关键时期，如树体萌动期、幼果形成期、果实膨大期、花芽分化期、果实采摘后、土壤封冻之前，对实现优质丰产具有重要作用。

3.4 主要病虫害防控

原则是预防为主，防治结合，加强肥水管理，重视有机肥并注意平衡施肥，保证

---

① 1 斤 =0.5 kg，全书同。

树体营养供给，提高树体抵抗力；合理负载，控制产量；保持园区通风透光，增加园区生物多样性和天敌昆虫；对机械伤口通过涂抹乳胶进行隔离，减少病菌的入侵途径等。在休眠期用 3～5°Bé 石硫合剂或晶体石硫合剂 20～30 倍液喷雾，可有效防止梨黑星病、赤星病、褐斑病等病害的发生。采用物理防治、化学防治和生物防治相结合，综合控制病虫害的发生和蔓延。经调查，在云南梨主产区常见病虫害有 8 种。

（1）梨黑星病：梨黑星病主要为害梨树的各种绿色幼嫩组织，从落花后到果实成熟前，均可造成为害。形成长椭圆形、梭形或长条形病斑，产生黑色霉层；病组织枯死，逐渐凹陷，呈疮痂状，叶片易脱落，后期形成裂果。主要发生在 5—7 月，降水量多，病害更为严重。在梨树发芽后开花前，树上喷洒 25% 戊唑醇乳油 5 000 倍液、25% 苯醚甲环唑水乳剂 6 000 倍液、12.5% 烯唑醇可湿性粉剂 2 000～3 000 倍液，以杀灭病部越冬后产生的分生孢子。

（2）梨黑斑病：梨黑斑病主要为害叶片和新梢，病斑呈近圆形或多角形，中心灰白色至灰褐色，外围有黄色晕圈，造成大量病果和提前落果。春季梨树发芽前，枝干上喷洒石硫合剂，杀灭树上的越冬病菌。在 6 月中下旬及 7—8 月降水量多时，再喷 3% 多抗霉素可湿性粉剂 300 倍液、35% 氟菌·戊唑醇悬浮剂 3 000 倍液，连续喷施 2～3 次。

（3）梨炭疽病：梨炭疽病主要为害果实，造成梨果腐烂。病斑呈水渍状，软腐凹陷，病斑不断扩大，从果肉到果心，果肉呈圆锥形腐烂。病害的发生和流行与降雨有密切关系，4—5 月多阴雨发病早，6—7 月阴雨连绵，则发病重。可选用 50% 多菌灵可湿性粉剂 600～800 倍液喷雾、75% 百菌清可湿性粉剂 500～800 倍液喷雾、450 g/L 咪鲜胺水乳剂 800～1 500 倍液喷雾、250 g/L 嘧菌酯悬浮剂 800～1 500 倍液喷雾。隔 7～10 d 喷施 1 次，连续喷施 2～3 次。

（4）梨锈病：通常在 4 月开始发生，与雨季不遇则自然消亡。主要为害叶片，严重时也为害果实。发病部位橙黄色，典型症状是病斑上长出性孢子器和锈子器，呈羊胡子状。可选用 25% 粉唑醇悬浮剂 3 000 倍液、25% 三唑酮可湿性粉剂 2 000 倍液、430 g/L 戊唑醇悬浮剂 3 000 倍液，连续喷施 2～3 次。

（5）梨小食心虫：主要为害果实，膨大后的果实受害更严重。先在果肉浅层为害，逐渐向果心蛀入，并排便其中，形成"豆沙馅"。采用性诱剂诱杀雄成虫，或释放赤眼蜂，每 3～5 d 释放 1 次，连续释放 3～4 次，每亩释放 3 万～5 万只；化学防治可选用 10% 高效氯氟氰菊酯乳油 1 000～1 500 倍液喷雾、1.8% 阿维菌素乳油 1 000～1 500 倍液喷雾、8 000 IU/μL 苏云金杆菌菌剂 800～1 000 倍液喷雾。隔 7～10 d 用药 1 次，连续用药 2～3 次。

（6）梨木虱：主要以若虫刺吸叶片汁液，也可为害果实。叶片受害后叶脉扭曲，叶面皱缩，产生黑斑，严重时叶片变黑，提早脱落。5月上旬出现第一代成虫，药剂防治的关键时期是越冬成虫出蛰期。可用药剂24%阿维·毒死蜱悬浮剂4 000倍液、20%噻虫胺悬浮剂2 000倍液、20%氰戊菊酯乳油3 000倍液、5%阿维菌素微乳剂5 000倍液喷雾，连续用药2～3次。

（7）果实蝇：在果实内产卵，可造成烂果、落果，使果实失去经济价值，损失严重。虫情监测，选定橘小实蝇发生动态监测区，挂诱捕器，每天观察诱捕器内虫口数，当虫口数<5头/瓶，1亩梨园悬挂1个诱捕器；当虫口数5～10头/瓶，1亩梨园悬挂3个诱捕器；当虫口数>10头/瓶，1亩梨园悬挂5个诱捕器并配合其他的防治措施。其他物理防治措施：性诱剂、糖醋毒液、实蝇信息素粘虫板防治。化学药剂防治：可选用2.5%高效氟氯氰菊酯乳油1 000倍液、2%阿维菌素乳油4 000倍液、75%灭蝇胺可湿性粉剂4 500倍液、5%氯氰菊酯乳油1 000倍液、2.5%溴氰菊酯乳油1 000倍液等。

（8）金龟子类：金龟子为害期为3—6月。金龟子具有群集性，即1株梨树可有10～20只。啃食叶片和嫩芽，为害严重。通过安装杀虫灯对金龟子进行诱捕，对诱捕的金龟子进行了鉴定分类，发现为害梨树和梨果的金龟子主要来自鳃金龟科（Melolonthidae）鳃金龟属（*Melolontha*），共有4个种：华阿鳃金龟、棕色鳃金龟、毛黄脊鳃金龟、华胸突鳃金龟。可采取冬季翻耕、灯光诱杀的物理防治，化学防治可选用5%氯虫苯甲酰胺悬浮剂2 000～2 500倍液喷雾、15%茚虫威乳油1 000～1 500倍液喷雾、5%甲氨基阿维菌素苯甲酸盐微乳剂2 000～3 000倍液喷雾。隔7～10 d用药1次，连续用药2～3次。

3.5 采收与采后处理

3.5.1 采收期的确定

采收期与果实的产量、品质及耐贮性有着密切的关系。采收过早，不仅产量受影响，而且果实含糖量低，风味淡，色泽差，品质劣，贮藏过程中易失水皱皮，果心易褐变；采收过晚，梨果衰老过快，对二氧化碳敏感程度增强，黑心、黑皮等生理病害发生的概率增加。根据距销售目标市场的距离、贮藏预期及果实的生长发育情况来综合确定果品的采收时间，生产中采用的指标如下。

（1）果皮底色：果实在接近成熟时，果皮叶绿素逐渐分解或转化，底色逐渐呈现出来，果皮底色由绿色开始变为浅绿色或绿黄色，果面略带蜡质，出现光泽时，表明果实即将成熟，可以采收。

（2）硬度和可溶性固形物：果实硬度和可溶性固形物（通常指糖分）是影响口感的两个主要因素。随着果实成熟度的增加，果实硬度逐渐降低，可溶性固形物含量增加。可采硬度和可溶性固形物因品种不同而存在较大差异，早熟品种可溶性固形物至少9%，中熟品种11%以上，晚熟品种12%；通常认为短期通风贮藏的鸭梨采收时硬度在3.81 kg/cm² 以上，用于冷藏的果品硬度在4.67 kg/cm² 以上。

（3）种子颜色：随着果实的成熟，种子从尖端开始逐渐由白色变为褐色、花籽或全黑。

（4）果柄脱落难易：生长后期，果柄基部形成离层，容易采摘。

3.5.2 采收方法

分期采收，根据果实成熟度和人工情况分2～3次采收。树体外围、见光向阳处的果实先采收，果实内膛、遮光处的果实后采收。梨果水分多，果皮薄，肉质脆，采收时戴上手套，轻拿轻放，剪去过长过粗的果柄，减少采摘和运输过程中的机械损伤。

3.5.3 采后处理

对果实表面及萼洼与梗洼处的污物进行简单清除，有条件的可用清洁池进行清洗并干燥，进行机械分级或人为分级后再根据销售目的市场进行包装，并在尽量短的时间内进行果品预冷，排除田间热，并入库冷藏，等待销售。冷藏温度根据销售市场的远近及预计销售时间进行设定，一般设为1～4 ℃，湿度保持在90%左右，注意定期进行通风换气。运输尽量采用冷链运输。

参考文献

曹玉芬，刘凤之，胡红菊，等，2006. 梨种质资源描述规范和数据标准[M]. 北京：中国农业出版社：20-35.

何英云，陈衫艳，李永平，等，2020. 云南梨产业发展现状及对策分析[J]. 中国果菜，40（12）：63-66.

李秀根，张绍铃，2020. 中国梨树志[M]. 北京：中国农业出版社：205-307.

苏俊，陈霞，李林，等，2016. 红色砂梨新品种'彩云红'[J]. 园艺学报，43（S2）：2687-2688.

舒群，仇明华，2005. 云南梨果业现状与梨产业化开发[J]. 农业科技（1）：25-27.

张绍铃, 2013. 梨学 [M]. 北京: 中国农业出版社: 417-419.

张绍铃, 陶书田, 周应恒, 2010. 梨生产、加工及贸易现状与产业发展基本趋势 [M]. 北京: 中国农业出版社: 1-7.

张绍铃, 谢智华, 2019. 我国梨产业发展现状、趋势、存在问题与对策建议 [J]. 果树学报, 36(8): 1067-1072.

张玉星, 2011. 果树栽培学总论 [M]. 4版. 北京: 中国农业出版社: 58-99.

桃栽培关键技术

1 产业发展概况

桃（Prunus persica L.）是蔷薇科李属桃亚属的多年生中型落叶乔木。桃在我国栽培历史悠久，自古是象征吉祥长寿的果品。桃果皮色泽鲜艳，外形多样，口感或多汁可口，或清脆爽口，桃营养丰富，果实富含糖、蛋白质、脂肪、有机酸、钙、磷、铁、维生素 C、维生素 B 等。除了能作为鲜食水果外，桃的果实还能加工成果酱、果脯、干果、果汁、果酒、罐头等。此外，据《本草纲目》等中医著作记载，桃树的树根、树叶、花、果、仁都可以入药。目前，桃树干上分泌的桃胶也已逐渐成为消费者喜爱的营养保健食品。

桃原产我国西部，云南省也是桃的起源地之一。桃喜光、适应性强，云南全省均可种植桃。目前，云南省桃生产主要分布在山区、半山区。截至 2021 年年底，云南省桃种植面积 83.46 万亩，总产量 84.20 万 t，种植面积居全国第七位。桃种植已成为云南省山区、半山区农业产业结构调整以及农民增收的重要的途径之一。特别是云南 4—5 月上市的早熟桃、11—12 月成熟的晚熟桃，成熟期正值外省及国外桃产区露地桃的生产淡季，市场竞争优势明显。云南省桃产业发展存在空间、时间上的优势，具有较大的市场前景。

2 主要栽培品种及特性

云南省桃种植品种较多，从种类来看，以毛桃和油桃为主，种植蟠桃的较少。从成熟期来看，4 月下旬至 7 月上旬成熟的早熟、中早熟桃和 9 月下旬至 12 月上旬成熟的晚熟桃占比较多。目前主要栽培的桃品种有以下几种。

春雪：在云南省，商品名有紫桃、红桃、红雪桃等。此前为美国选育的早熟桃新品种，1998 年山东省果树研究所从美国引入。此品种树姿开张，生长势强，果实圆形，果顶凸，茸毛短而稀，两半较对称。大型果，平均单果重 180 g，最大单果重 280 g。果皮不套袋为紫色，套袋为浅红色。果肉白色，汁液多，肉质硬脆，可溶性固形物含量在12%以上，风味甜、香气浓。黏核，耐贮运，昆明地区5月中旬至6月上旬成熟，泸西、石林及滇南地区5月初成熟。适合在冬春霜冻少，光照资源好的地区栽培。

霞脆：中熟桃品种。江苏省农业科学院园艺研究所选育。树体生长健壮，树姿半开张。2月下旬始花，3月中旬为盛花期，花粉可育，花期不整齐，坐果率高、丰产性较好。蚜虫、缩叶病为害较轻。有一部分畸形花，但大多数能自我疏除，结果后畸形果率<10%。个体差异明显。果实椭圆形，果面90%着色，果肉白色，风味浓，品质较好。果肉脆，可溶性固形物含量在12%以上。平均单果重可达150 g以上。在云南中部地区6月上中旬开始成熟，成熟后挂树期长，采摘期30～45 d。耐贮运。适应性好，无特殊病虫为害。丰产稳产，盛果期树亩产1 000 kg以上。适合在云南中部，雨季开始较晚的地区栽培。

极早518：早熟桃品种。昆明石林地区4月底至5月初采摘上市。果实近圆形，平均单果重90～120 g，最大单果重可达180 g。果面着鲜红色、全红，色彩艳丽。果肉白色，肉细，汁多，味甜清香，肉质脆硬，耐贮运。成熟后再挂树20 d左右。坐果率高，丰产性好。

中油5号：早熟桃品种。中国农业科学院郑州果树研究所选育。果实大，极丰产，适应性强，果肉白色，硬溶质，肉质细密，味甜，香气中等。果实发育期72 d左右。平均单果重166 g，最大单果重270 g。果皮底色绿白，大部分果面或全部着玫瑰红色，十分美观，品质优。黏核。昆明石林地区5月中旬至5月下旬成熟。

鹰嘴蜜桃：又名鹰嘴桃、开远鹰嘴桃、开远蜜桃。中熟桃品种。国家农产品地理标志产品。树势强，树姿开张。果顶凸出，如鹰嘴，果实圆形，果实两半较对称，果实成熟时果皮青绿色，阳面呈紫红色，果肉为绿色、白绿色，果汁多，可溶性固形物含量在16%以上，耐贮运。适合在冬春光热资源丰富、气候干燥、雨季来临较晚的地区栽培。

中华寿桃：晚熟桃品种。中华寿桃引入丽江表现不好，改用当地光核桃为砧木嫁接后，品质产量得以大幅度提升，经过种植企业带动大面积推广，形成品牌。树势强，丰产，平均单果重320 g，最大单果重1 000 g以上。果实近圆形，果顶凸，缝合线深，果肉乳白色，黏核，可溶性固性物含量16%。果实适合在冬季气温低，需冷量能满足桃树需求的地区栽培。

中华冬桃2号：特晚熟品种。由开远市东洪绿色产业有限责任公司引进，云南省范围内都有栽培。生长势强，树姿直立，花芽多，自花结果率高，需要大量疏果。果实近圆形，果顶微尖，缝合线明显，果实两半都对称，平均单果重220 g，最大单果重

435 g。果肉青白色，近核处红色素较多，果肉脆，离核，可溶性固形物含量在16%以上。果实耐贮运，早产丰产性好。适合在11月雨水少、光照条件好，无霜或霜冻较轻的地区栽培。

佛都冬桃1号：特晚熟品种。中华冬桃2号芽变，由云南省农业科学院园艺作物研究所和宾川县佛都冬桃专业合作社联合选育。树势开张，生长势强；萌芽力、成枝力高。花芽多，以中、短果枝结果为主、自花结实能力强，坐果率高，丰产性能好。目前栽培地区以大理州宾川县及周边地区为主，2月下旬至3月上旬萌芽，3月中旬开花，花期持续10~16 d，12月中旬果实成熟，果实发育期为260 d，1月上旬落叶，年生育期为321 d。果实大，平均单果重220 g，最大单果重589 g。果实圆形，果顶平，套袋果着色后果皮底部白色，果实阳面鲜红色，着色面积50%以上。果肉白色，不溶质，近核处鲜红色。离核。商品性好。果实可溶性固形物含量14.39%，肉质清脆，品质好。果实耐贮运。适合在11—12月雨水少、温暖、光照条件好，无霜或霜冻较轻的区域栽培。无霜期短、秋冬温度低的地区不适合栽培。

3 栽培管理技术

3.1 种植规划与园区建设

桃是喜光植物，而云南省处于低纬高原，光照丰富，桃园选在不遮光的地区，都可以满足生产条件。按照云南省地理和气候条件分析，桃园应建立在海拔1 600~2 400 m、交通便利、可以满足冬春季节浇水需求的坝区、半山区、山区。拟建立桃园的地区需要土壤疏松。山谷、低洼地、沼泽和其他容易积水、冬春霜冻严重、倒春寒发生频繁、果实成熟期间阴雨天多以及处于冰雹带的地区都不适合优质桃生产，不建议选择建桃园。

桃园应配备能通过农事作业及采收运输车辆的道路。与道路规划结合设置排灌系统。平地建园以南北行向为宜，山区丘陵地栽植应根据地形地势决定桃园面积的大小，行向与等高线一致。桃园坡度>15°的地块，要先将其修成梯面宽度为2.5~10 m的梯田，梯田应略向内倾，并在内侧设置排水沟。

根据目标市场、栽培地环境条件等选择所栽植品种的类型。处于休眠期的桃树需要一段低于或等于7.2 ℃的小时数累积，才能正常开花结果。云南省及以南的地区，大多数桃栽培区冬季温度较高，应选择低需冷量品种进行栽植。建议苗木从正规育苗机构获得，以保证质量和纯度。

3.2 整形修剪

大型生产性桃园（50亩及以上）推荐采用宽行窄株的栽培模式，以便于农机操作，环境通风透光，减少病虫害，提高果品质量。农户家庭小型桃园的栽培模式可以操作方便、通风透光为主。云南省桃种植适宜用开心形和"Y"形两种树形。根据种植栽培要求、立地条件、水肥条件、管理水平选择树形和栽植密度。采用开心形树形，种植株距为 3~4 m，行距为 4~5 m；采用"Y"形树形，株距为 2 m，行距为 4~5 m。桃树定干高度 60 cm 左右，树高不超过 3 m，结果部位控制在 2.5 m 以下。

桃现代化栽培以长枝修剪技术为主，即不进行短截，不培养大型结果枝组，树体由主干、结果枝与小型结果枝组成。修剪强度、修剪部位、修剪方式等根据桃园地理气候特点、经营模式等来选择。根据桃园所在地区生长季降水频率，进行 2~4 次夏季修剪，以疏除直立枝、过密枝，保持树体通风透光。冬季修剪在桃落叶休眠后进行，以疏除直立枝、过密枝、病虫枝，适当调整树形。

3.3 水肥管理

桃相对于其他一些果树，耐旱性极强，但要达到令人满意的经济产量，栽培中是不能够亏缺水分的，特别是花期、幼果期、果实膨大期等关键时期，需要充足的土壤水分供给。萌芽至开花前：灌一次足水，水量以能渗透地面深度 60 cm 左右为宜。花期还需要适当的空气湿度，空气湿度太低，柱头干燥，授粉不良，影响受精。硬核期：灌水量以湿润土层 50 cm 为宜；此期间如有降水的地区，可根据实际情况确定。果实膨大期：应注意均匀灌水，特别是油桃园，避免剧烈的水分变化，水分变化剧烈容易引起果实裂果。在果实成熟期降雨较多、空气湿度大的地区，还应特别注意病虫害的发生。灌溉需要根据当地的经济条件、水源情况、水利设施条件以及地形等综合考虑，可采用树盘浇水、沟灌、喷灌、滴灌等多种方式，如果条件允许，尽量使用滴灌，提高水分利用效率。

桃需要合理调控氮、磷、钾肥施用水平，合理增加有机肥施用量，依据土壤肥力和早中晚熟品种及产量水平，早熟品种的需肥量比晚熟品种少 20%~30%；注意钙、镁、硼和锌的配合施用；云南省桃园理想产量为每亩 1 500~2 000 kg，需要每年每亩施用有机肥 3~4 m³、氮肥（N）12~16 kg、磷肥（P_2O_5）7~9 kg、钾肥（K_2O）17~20 kg。基肥以秋季为宜，以有机肥为主，桃果膨大期前后是追肥的关键时期，以高钾复合肥为主；全部有机肥、30%~40% 的氮肥、100% 的磷肥及 50% 的钾肥作基肥于桃果采摘后的秋季采用开沟方法施用；其余 60%~70% 的氮肥和 50% 的钾肥分别在春季桃树萌芽期、硬核期和果实膨大期分次追施（早熟品种1～2次、晚熟品种2～3次）。追肥可使用水肥一体化系统叶面喷施或进行滴灌。干旱地区可使用地膜将单个树盘或整行树盘条状覆盖，采用穴贮肥水技术进行水肥管理。无论采用哪种方式，施肥后都需要配合灌水，促进肥料的运输和吸收。

3.4 主要病虫害防控

现代桃园病虫害防控管理，按照"以防为主、综合防治"的原则，开展主要病虫害预测预报，以物理防治、农业措施、生物防控为主，选用高效、低毒、低残留农药，将害虫为害控制在经济允许水平之下。防治时所选药剂应符合相关规定，应轮换、交替使用农药，避免抗药性产生。

3.4.1 疮痂病

以农业防治为基础，抓好药剂防治；主要控制果实受害，尽可能套袋，加强套袋前药剂防治；因病害潜育期在20 d以上，药剂防治要早。

在品种选择上，尽量避免种植在雨季成熟的品种，选择果实成熟期降雨较多的品种栽培时需要进行套袋，防御病菌侵入。桃园保持通风透光；避免积水，及时清除病果、病枝，将病残体深埋或烧毁，减少田间菌源。

果实膨大期至成熟前20 d进行防治，可选用咪鲜胺、苯醚甲环唑、代森锰锌或甲基硫菌灵等药剂交替使用2～3次，每次间隔10 d左右。需要进行套袋的果实，必须提前20 d施药2次。

3.4.2 褐腐病

与疮痂病相似，通过农业措施与药剂防治相结合的方式进行防控；主要控制果实受害，加强套袋前药剂防治；药剂防治时选用高效、低毒、低残留农药，确保果实安全。

果园保持通风透光；避免积水，及时清除病果、病枝，将病残体深埋或烧毁，减少田间菌源。

防治时间为落花后，可使用的药剂有咪鲜胺锰盐、唑醚·氟酰胺、唑醚·代森联等，交替使用2～3次，每次间隔10 d左右。果实成熟中后期，根据降雨情况，继续轮换使用上述药剂。

3.4.3 缩叶病

在清除病叶基础上,抓好药剂防治,关键是早春及时喷药。

及时摘除病叶,并将其深埋或烧毁,减少田间菌源。

增施叶面肥,促进叶片生长,有利于恢复树势。

在叶芽吐绿和花芽露红但未展开前,喷药1~2次,间隔7~10 d。可使用的药剂有石硫合剂、咪鲜胺、苯醚甲环唑、多菌灵或甲基硫菌灵等。

3.4.4 细菌性穿孔病

在农业防治的基础上,抓好药剂防治。

清除田间病叶、病枝、病果,将其深埋或烧毁,减少田间菌源。

选择适当密度和树形,增强树体通风透光;及时排出积水,降低田间湿度;合理施肥,多施有机肥、复合肥,避免偏施氮肥。

展叶后喷药3~4次。可用的药剂有春雷·喹啉铜、中生菌素、喹啉铜和噻菌铜等,用药间隔在10 d左右。

3.4.5 桃蚜

叶芽萌发时,持续干旱会导致桃蚜发生较为严重。加强桃园土、肥、水管理。实行园内生草或种草,丰富果园生物多样性,以保护食蚜蝇、蚜茧蜂、瓢虫等蚜虫天敌。桃树益害比达1∶30以上时避免喷洒杀虫剂。桃叶芽萌动、越冬卵孵化盛期用药效果较好。在桃叶芽萌动期,结合多种病虫害防治,喷施3~5°Bé石硫合剂;越冬卵孵化始盛期,喷洒苦参碱、印楝素等生物药剂。发现蚜虫为害,结合春季修剪,剪除有蚜枝叶,集中销毁。建议使用吡虫啉、啶虫脒、甲氨基阿维菌素苯甲酸盐等低毒农药,如对吡虫啉等药剂产生抗性时,可选择噻虫嗪、吡蚜酮、螺虫乙酯和高效氯氟氰菊酯等药剂。

3.4.6 桑白蚧

加强土、肥、水管理,增强树势;结合冬季修剪,剪除介壳虫寄生严重的枝条,集中烧毁;用硬刷将枝条上的介壳虫刷掉,刷除部位主要是2~3年生枝条。

实行园内生草或种草,丰富果园生物多样性,保护利用桑白蚧的自然天敌(软蚧蚜小蜂、红点唇瓢虫、李斑唇瓢虫和日本方头甲等),在介壳虫成虫期不要盲目施药,避免杀伤天敌。

花芽萌动期进行化学防治，可采用的药剂有：使用石硫合剂防治休眠期桑白蚧；出蛰为害期、若虫孵化盛期使用高效氯氰菊酯等药剂进行防治。

3.4.7 桃一点叶蝉

桃一点叶蝉在桃园附近的常绿植物和园内越冬，冬季清园后应彻底清理落叶和杂物，集中烧毁；成虫出蛰前及时刮除翘皮，结合冬春季病虫防治，给周边常绿植物寄主上喷施石硫合剂或其他杀虫剂。

加强土、水、肥管理，增强树势；合理冬剪、夏剪，防止果园行间及树体密闭。

果园生草，引诱桃一点叶蝉的主要天敌，如草间小黑蛛、异色瓢虫、七星瓢虫、龟纹瓢虫、大草蛉和蜘蛛等。天敌数量少时也可以适时进行释放天敌。

在越冬成虫迁入期、1代若虫孵化盛期、若虫盛发期3个时期进行化学防治。对桃树及田间生草进行药剂喷雾，利用阴天或晴天下午喷药效果较理想。可选药剂：吡虫啉、阿维菌素、联苯菊酯、异丙威、甲氰菊酯等。高温季节也可选用啶虫脒、噻虫嗪内吸性强的药剂。

3.4.8 康氏粉蚧

防治要从冬季清园开始。在采收完成以后，要及时清除果园落叶杂草，特别是病果、病叶、病枝，必须集中深埋或烧毁。人工将树体上残留虫体刮除。冬季修剪时，疏除过密枝，保证果园通风透光。树体落叶后，进行树体涂白时，在涂白液中加入百菌清和氰戊菊酯。冬季清园，全园喷施石硫合剂1~2次。

果实套袋前施用高效氯氰菊酯（加少量洗衣粉），用药后及时套袋。如时间间隔长或遇降雨需重新用药。

3.5 采收与采后处理

桃果实需要适时采收，即根据其品种特性、用途、消费市场距离、贮运方式来选择合适的采摘时间。采收选择在一天内比较凉爽的时间段进行。

采收时最好戴手套，轻拿轻放，顺序堆码在采摘容器中，防止碰伤果实外皮。采摘力度适中，避免折断果柄，避免破坏树体主干及枝条。采摘后的果实需要放置于通风透气卫生的场所，不能日晒雨淋或被鸟兽伤害污染。

采收后的果实要根据桃的不同类型特点进行分级，并按要求检验之后再进行包装。要尽量选择避免桃果实互相磕碰造成机械损伤的包装材料。包装外箱上要标明桃果实的等级、重量、规格、数量等产品特性，并贴上产地标签。

采摘下来的桃果实根据售卖需要可进行短期贮藏，但不建议长期贮藏。雨季成熟桃以就近快消费为主，不建议以任何方式贮藏。特早熟桃、早熟桃、特晚熟桃适时采摘，经过适当预冷，再进行长途运输。

参考文献

曹尚银，谢深喜，卢晓鹏，等，2018.中国桃地方品种图志[M].北京：中国林业出版社.

陈玉星，唐明文，刘新文，等，2020.优质丽江雪桃标准化生产管理技术[J].南方农业，14（27）：24-25，29.

国家桃产业技术体系，2016.中国现代农业产业可持续发展战略研究·桃分册[M].北京：中国农业出版社.

马之胜，贾云云，王越辉，等，2014.冀中南地区桃产业技术[M].北京：中国农业出版社.

王力荣，2012.中国桃遗传资源与品种图谱[M].北京：中国农业出版社.

王力荣，朱更瑞，2005.桃种质资源描述规范和数据标准[M].北京：中国农业出版社.

于菲，张晏，陆琳，等，2017.特晚熟桃新品种'佛都冬桃1号'[J].园艺学报，44（S2）：2627-2628.

袁波，熊明国，殷曼，2017.桃规模生产与经营管理[M].北京：中国农业科学技术出版社.

❖ 草莓栽培关键技术

1 产业发展概况

草莓（*Fragaria* × *ananassa* Duch.）是一种世界各地广泛栽培的小浆果，色泽鲜艳，柔软多汁，酸甜适口，营养丰富，是当今世界十大水果之一，被视为果中珍品，享有"早春第一果"和"水果皇后"的美誉，具有较高的经济价值、营养价值和文化休闲价值。在世界卫生组织公布的十佳食品榜上位居第二。经过育种学家的不断选育，草莓栽培品种已有2 000多个，因其适应性较强，目前世界上绝大多数国家都有栽培。中国是世界草莓第一生产大国和消费大国，据统计，2018年中国草莓栽培面积为256万亩，产量500万t。中国地域广阔，草莓栽培方式为露地栽培与保护地栽培并存，保护地栽培在华北地区以日光温室促成栽培和塑料大棚半促成栽培为主，南方以塑料大棚促成栽培或半促成栽培为主，云南省海拔1 700 m以下的地区以露地栽培为主。

2 对环境条件的要求

影响草莓生长发育的环境条件主要有土壤、温度、水分、光照等。

（1）土壤。草莓对土壤的适应性较强，大部分土壤都能种植草莓。但在种植的时候最好选择肥沃、疏松的微酸性土壤，才能生产出更加优质的草莓。沼泽地、盐碱地、重度黏性土壤上草莓易生长不良。草莓根系对土壤中的肥料浓度耐受力较差，土壤中电导率越低，草莓的根系生长得越好。

（2）温度。草莓能够较快适应温度的变化，有一定的耐寒性，休眠期可耐受 $-15 \sim -10\ ℃$ 的低温，也有一定的耐热性，夏天可耐受 $40 \sim 45\ ℃$ 高温。草莓匍匐茎的发生一般在 $20 \sim 25\ ℃$ 较多，低于 $15\ ℃$ 或超过 $28\ ℃$ 匍匐茎抽生慢且数量少。草莓花芽分化一般以 $10 \sim 17\ ℃$ 为宜，温度过高或过低均不利于花芽分化。温度过高，果实成熟期较短，果实偏小且发酸；温度过低，成熟期较长，果实较大且甜，但成熟期延迟，影响产量。

（3）水分。草莓的根系大多分布在20 cm以内的土层中，既不耐涝也不抗旱，在生产过程中要注意及时浇水和排水。不同生育期对水分的需求不同，一般花芽分化期田间含水量在60%左右为宜，开花期70%左右，果实膨大期80%左右。灌水不足，会抑制生长并导致产量下降；灌水过多，不仅会导致果实硬度和风味等品质下降，影响根系的生长，还易引发病虫害。不仅土壤湿度对草莓生产影响较大，空气湿度也同样重要，一般空气湿度以60%左右为宜，开花期不要超过80%；湿度过大会影响散粉受精，容易产生畸形果，还易助长灰霉病的发生；湿度过小，影响植株生长和果实外观。

（4）光照。草莓喜光耐阴，光照强度、光照时长、光谱成分都影响着草莓的生长发育。光照充足，植株健壮，叶片大且浓绿，果实品质好，产量高；光照不足时，植株生长较弱，叶柄细长，花朵小而少，果实成熟期延迟，果实小，品质差，产量低。草莓叶片光补偿点一般为0.5万～1.0万 lx，光饱和点一般为2万～3万 lx。在种植中要保证充足的光照，但长时间过强的光照也可能灼伤叶片。光照时长影响着植株生长、花芽分化、休眠等，一般低温、短日照条件有利于花芽分化，较高温度、长日照条件有利于匍匐茎的发生。

3 主要栽培品种及特性

根据光周期反应和结果期的不同，一般将草莓栽培品种分为3个不同类型，即短日照型、长日照型和日中型。目前我国栽培的草莓主要为一季型的冬季草莓，又叫短日照草莓，采收时间主要集中在12月至翌年5月。夏季草莓也叫连续结果型即四季草莓，主栽的为日中型品种，其对光周期不敏感，在4～29℃温度条件下可持续开花结果，在温凉区域可实现夏秋鲜果生产，果实采收期为5—11月。

按照品种来源和特点，又可分为欧美品种和亚洲品种。欧美品种的草莓个头明显比亚洲品种大得多。欧美品种的草莓因为味道偏酸，更多是用于深加工。亚洲品种味道较甜，适合鲜食。

目前，我国栽培较多的品种主要有以下几种。

3.1 冬季草莓（短日照）品种

章姬： 日本品种，果实长圆锥形，鲜红美观，果形端正整齐。平均单果重18 g，可溶性固形物含量为9%～14%。果实充分成熟后品质极佳。由于该品种果实柔软多汁，耐贮运性较差，因此长距离运输时，需在7成熟时采摘。该品种对黄萎病、灰霉病抗性强，对白粉病、炭疽病抗性弱，栽培时需注意防治。

红颜： 日本品种，植株生长势旺，株型直立。果实圆锥形，果实表面和内部均呈鲜红色，有光泽，果心红色，果形端正整齐，着色一致，畸形果少；平均单果重21 g，酸甜适口，香味浓，品质极佳。果实硬度适中，耐贮运性较好。该品种各花序发生的时间间隔短，连续结果性强，丰产性好。耐热、耐湿和耐旱能力较弱，育苗困难，耐低温能力强，在冬季低温条件下连续结果性好。

香野：植株高大，较直立，长势旺盛，花序较长。休眠浅，成花容易，花量大，连续结果能力强，早熟丰产。果实圆锥形，平均单果重25 g，最大果重超过100 g，大果有空心现象。果皮橙红色，果肉米黄色，肉质脆嫩，香味浓郁，带蜂蜜味，含糖量为10%～12%，口感极佳，但温度较高时由于生长期较短，品质明显下降。果实硬度较大，耐贮运性好。植株抗性强，对炭疽病、白粉病的抗性较强。但匍匐茎数量偏少，育苗系数较低。

圣诞红：韩国品种，株型直立。果实表面平整，光泽强，果面呈鲜艳红色。一二级花序果平均单果重35.8 g，最大单果重64.5 g。花萼下果面着色中等，宿萼反卷，绿色，种子微凸果面，颜色黄绿兼有，密度中等。果肉橙红，髓心白色，无空洞。果肉细、质地绵，风味甜，可溶性固形物含量10%～13.1%，耐贮运性中等。

甜查理：美国品种，叶片近圆形较厚，叶缘锯齿较大钝圆，植株健壮，根系发达。高抗灰霉病和白粉病，对其他病害抗性也很强，很少有病害发生，适应性广，休眠期较短（45 h左右）。果实圆锥形，大小整齐，畸形果少，表面深红色有光泽，种子黄色，果肉粉红色，香味浓，可溶性固形物含量11.9%，硬度大、耐贮运。

桃薰：日本品种，是世界上第一个草莓白色果栽培种，果实心形，果面白里透粉，果肉白色，果实柔软多汁极具风味，有浓郁的水蜜桃香味；植株直立，叶片背面密生茸毛，匍匐茎发生性强。该品种亦是现今草莓栽培种中唯一的十倍体材料，生产上的缺点是花芽分化周期长、果实成熟期晚，另外，第一序第一二级果实畸形率高。

妙香7号：山东农业大学选育的品种，果实圆锥形，平均单果重35.5 g，比对照品种红颜高25.9%；果面鲜红色，富光泽；果肉鲜红细腻，香味浓郁，硬度高。促成栽培条件下，平均亩产3 427 kg。

白雪公主：北京市农林科学院选育的品种，株型小，生长势中等偏弱。果实中等大小，最大单果重48 g，果实圆锥形或楔形，果面白色，果实光泽强，种子红色，平于果面，果肉果心白色，果实空洞小，可溶性固形物含量9%～11%，风味独特，白粉病抗性强。

越心：浙江省农业科学院选育的品种，植株生长势强，直立，耐低温弱光，匍匐茎抽生能力强，浅休眠，早熟，适合促成栽培。果形中等大小，呈短圆锥形或心形。果面平整，浅红色，着色均匀，可溶性固形物含量12%～14.5%。风味极佳，酸甜适口，香味诱人，果皮较厚，耐贮运。抗炭疽病、灰霉病、白粉病能力强。

宁玉：江苏省农业科学院选育的品种，植株半直立，生长势强，匍匐茎抽生能力强。花序长短适中，坐果率高。果实圆锥形红色，光泽强，风味香甜，可溶性固形物含量10.7%。一二级序平均单果重24.5 g，连续开花坐果性强，早期产量占有率为40.5%～57.9%。耐热耐寒性强，冬季休眠浅不易矮化，春季高温不易疯长。白粉病、炭疽病抗性中等。

粉玉：杭州市农业科学研究院选育的品种，果实圆锥形，果面粉红色，果肉白色紧实，髓心空洞无或小；第一花序一级果平均单果重28.0 g，二级序果平均单果重17.1 g；可溶性固形物含量13.1%～18.0%；果实硬度适中，耐贮运性好；早熟品种，连续成花能力强，不断果。

3.2 夏季草莓（日中型）品种

蒙特瑞：美国加州大学品种，高产。促成栽培条件下果实上市早，植株生长势旺盛，平均单果重33 g，最大单果重60 g。果实品质极佳。抗病性强。

圣安德瑞斯：美国加州大学品种，果实圆锥形，平均单果重32.3 g，最大单果重125.6 g。果面红色，有光泽，果实产量高，硬度大，耐贮运。该品种耐高温或喜高温，较抗白粉病、黄萎病。

波特拉：美国加州大学品种，2008年育成，强日中型品种，连续开花结果能力极强，高产。果形均匀度高，平均单果重集中在10～25 g。

4 栽培管理技术

4.1 定植

一般选用穴盘苗，双行"品"字形定植，株距18～25 cm，定植时确保幼苗弓背朝外，这样植株开花后花序会朝外生长，便于昆虫授粉和采摘。如果选用裸根苗，建议选择阴天或气温较低的时段定植，定植后1周内适时向植株叶片喷水，避免植株脱水干枯。不管选用穴盘苗还是裸根苗，定植后要浇透水1次，确保土壤水分充足。

4.2 整形修剪

（1）打除老叶。植株定植成活，萌发2片以上新叶后，要及时开展第一次打叶。

打除老叶的操作贯穿整个种植期，但不同的阶段打叶有不同的要求。第一花序萌发前，此阶段气温尚高植株生长快，为促进第一花序的萌发和第二花序的分化，单株保留4片叶子；第一花序萌发后，为保障植株有足够的营养支撑后续开花结果的大量能量消耗，单株需保留6～7片叶子；春节后气温明显升高，植株生长加快，单株保留5～6片叶子。

（2）侧芽管理。植株定植成活后，整个冬春生产期间侧芽会不断地萌发，特别是桃熏、玉兔等白果品种侧芽数量较多。较多的侧芽会导致植株营养分散，花多果小，商品价值降低。栽培上建议不留侧芽仅保留1个头（生长点），如要增加前期产量，最多留1个侧芽保留2个头（生长点）。

（3）匍匐茎和花序管理。匍匐茎是营养生长的器官，挂果生产过程中匍匐茎的萌发是无谓的营养消耗，要及时打除。花序管理也是生产过程中重要的一环，生产过程中多数品种都需要疏花，以便集中植株养分，提高平均单果重。单花序疏花后保留果实最多7个（1个一级果，2个二级果和4个三级果），如果植株长势偏弱，需保障后续花序的产量，疏花后单花序保留3～5个果实。此外，果实采收结束后，要及时打除老化的花序，以便植株生长。

4.3 水肥管理

水肥管理措施应根据植株发育阶段、气候条件和生产目的等因素来确定。幼苗定植成活后一般要适当控制水肥，不让植株生长过旺导致花芽分化延迟。确定花芽分化完成后，要及时增加水肥促进营养生长以便支撑后续的植株开花结果。从季节上说，冬季气温低，植株蒸腾量少，应减少灌溉水量；夏天则相反，需增加灌溉水量。从生产实践上看，植株是否需水不完全取决于土壤是否湿润，判断是否该浇水的重要标志是要看植株叶缘在早晨是否"吐水"。如果清晨叶片边缘有水滴，即出现泌溢或"吐水"现象，可以认为水分充足，根系吸收功能较强；相反，则表示缺水或吸收水分能力较差，需及时浇水。从植物生长发育的角度看，草莓生长包括了营养生长、花芽分化、开花结果等过程。草莓种植过程中对肥料的要求比较高，首先氮、磷和钾肥的比例要比较合理，按1:（0.25～0.4）:（1.3～1.8）（元素质量比）比较合适；此外，钙、镁等中量元素供应要充足，钙肥量占氮肥的50%～80%，镁肥占氮肥的15%～20%；硼、铁等微量元素不能缺乏。

4.4 病虫害防治

4.4.1 病害防治

4.4.1.1 白粉病

白粉病是草莓种植过程中的第一大病害，属于真菌病害，病菌孢子主要靠空气传播，病菌随病株、病叶等残存于土壤中越冬，在 15～25 ℃的条件下传播最快。高温、干燥环境不易发病。白粉病主要为害叶片，也可侵害叶柄、花、花梗及果实。叶片被侵害初期发生大小不等的暗斑，不久叶背面产生薄霜似的白粉状物质，后期呈红褐色病斑，叶片卷缩、枯黄。幼果受害，果面上覆盖白色粉状物，果实停止发育，失去商品价值。严重时，整个植株死亡。防治措施主要包括农业防治、物理防治和药剂防治。

（1）农业防治措施。高温闷棚，消灭棚内白粉病初传染源；选用抗（耐）病品种；或选用干净种苗，通过高架栽培、基质覆膜、大棚侧窗、顶窗、换气扇等联动管理，控制棚内温湿度小环境，抑制白粉病的发生和流行；采用膜下滴灌，控制水量，肥料配方适当控制氮肥；及时摘除老叶病叶，增加植株周边的通风性。

（2）物理防治措施。采用硫磺熏蒸，每亩安装 10～15 个 40 W 硫磺熏蒸器，每天晚上熏蒸 1～2 h。

（3）药剂防治措施。分时期选用不同种类药剂进行白粉病防治。①开花结果期以前，选用广谱保护性杀菌剂如代森锰锌、多菌灵、苯菌灵、甲基硫菌灵等，对草莓白粉病及其他病害有较好的保护预防作用。②开花结果期—果实膨大期（或发病期）选用内吸性、速效性好、持效期长的杀菌剂，如三唑类（三唑酮、腈菌唑、四氟醚唑、氟菌唑、戊唑醇等）、硫磺、多·硫、吡唑醚菌酯、吡唑萘菌胺+嘧菌酯等药剂防治。③果实采摘期选用速效性好、有治疗或铲除作用的杀菌剂，如氟吡菌酰胺+肟菌酯、乙嘧酚磺酸酯、啶酰菌胺+醚菌酯（翠泽）、氟唑菌酰胺+吡唑醚菌酯等，控制病害扩展蔓延。采果期发病较轻时选用生物杀菌剂蛇床子、葡聚烯糖、醚菌酯等以及生物干扰素类药剂（嘧啶核苷类抗菌剂），对草莓花、果实采摘安全无害，不会抑制生长。在有滴灌设施栽培的草莓，栽培过程中可在滴灌液中，每亩加入 25% 吡唑醚菌酯悬浮剂 200 mL，随滴灌液进行滴灌，利用吡唑醚菌酯很好的内吸传导性，药剂能传输到植株的各个部位防控病害，但在果实采摘期应减少使用以避免影响果实转色。

注意事项：药水要充足，植株叶片内部要打透，三唑类杀菌剂对植株有抑制作用，冬季不能连续使用，夏季繁苗可用来抑制植株徒长。

4.4.1.2 灰霉病

草莓灰霉病是真菌性病害。该病发育最适宜的温度为18～23 ℃，空气相对湿度80%以上也有利于发病。寒潮频繁、大棚内地膜积水、连续阴雨天气、种植过密、通风透光不良等，都会加重该病的发生和蔓延。该病主要为害叶片、花、果柄和果实，多从伤口或枯死部位进行侵染繁殖，先为害枯枝老叶，然后危及果实。花器和果实一旦染病，很快发生腐烂，并迅速传播，对产量影响很大，重者可减产40%以上。

防治措施如下。

（1）要及时小心地将病叶、病花、病果等摘除，放进塑料袋，带出棚、室外进行妥善处理，以免病菌侵染到其他植株上。

（2）控制大棚内的温湿度。草莓进入开花至果实膨大期，白天温度在25 ℃以上，夜间控制在12 ℃以上时，尽量延长放风时间，使大棚内相对湿度保持在60%～70%。

（3）少施氮肥。

（4）药剂防治。可以用啶酰菌胺、嘧霉胺、异菌脲、氟硅唑、多抗霉素、腐霉利等交替喷雾。每隔7 d喷1次，连续喷3次。如遇阴雨天，用腐霉利烟剂进行烟熏。

4.4.1.3 炭疽病

炭疽病是草莓土传（种传）的主要病害，属于真菌性病害，炭疽病一旦发病会造成植株死亡和绝收。病菌在病组织或落地病残物中越冬。在同一块地连续种草莓，病原菌越积越多，这是发病越来越重的根本原因。种苗带病传播也是发病的主要原因。

防治措施如下。

（1）选择干净种苗，每批苗采购前要进行抽样，横向和纵向切开缩短茎检查切面是否有褐色病斑。

（2）田块消毒，种植田块定植前应用棉隆、石灰氮等消毒剂消毒。发病地块未来3年不能再种植草莓。

（3）定植后，发现有病株要及时拔除，并用吡唑醚菌酯、咪鲜胺、唑醚·氟酰胺、苯醚甲环唑、嘧菌酯、二氰·吡唑酯、溴菌腈等喷雾防治。

4.4.1.4 细菌性角斑病

细菌性角斑病是近年云南省草莓种植过程中主要病害之一，主要为害植物叶片，侵染时在叶片下表面出现水浸状不规则形病斑，病斑扩大时受细小叶脉所限呈角形叶斑，故亦称角斑病或角状叶斑病，病斑照光呈透明状，严重时有菌脓现象。细菌性角斑病的另一种表现比较隐蔽，主要为害维管束，早期没什么症状，不容易发现，等表现出叶片发黄发蔫症状后基本已是晚期，拔出这些发蔫的苗子，有时轻轻一用力就会拔断，可以看到苗的根部表皮完整，髓心已空，维管束变成红褐色，还有脓液一样的液体流出，脓液没有明显的臭味。有些植株的叶片上会出现角状叶斑。

防治措施如下。

（1）在草莓移栽前清除田间及四周杂草，集中烧毁或沤肥；深翻地灭茬，促使病残体分解，减少病源和虫源。及时防治害虫，减少植株伤口，减少病菌传播途径。

（2）合理轮作，水旱轮作最好，选用排灌方便的田块，开好排水沟，降低地下水位，达到雨停无积水；大雨过后及时清理沟系，防止湿气滞留，降低田间湿度，这是防病的重要措施，因为很多病害都是高湿引起的，要注意。

（3）发病初期，可用络氨铜、松脂酸铜（细菌、真菌兼治）、枯草芽孢杆菌、氯溴异氰尿酸、中生菌素、噻菌铜、喹啉铜、噻唑锌等喷叶淋根，每 5 d 进行 1 次，连续 2～3 次。很多情况下叶斑病是真菌型和细菌型混发，可用苯醚甲环唑+中生菌素，甲基硫菌灵+中生菌素或春雷霉素等防治。

4.4.2 虫害防治

4.4.2.1 红蜘蛛

红蜘蛛体型较小，不容易被识别，其喜欢在未展开的幼叶或叶背面吸取汁液，需仔细辨认或借助放大镜才能发现。红蜘蛛为刺吸式口器，受害植株叶片正面首先表现出团状小黄点，为害严重后植株矮化早衰，叶片红褐色干枯，最严重时受害植株叶片上可见白色网状物。考虑到果品的食用安全性，开花前后采用不同的防治策略。

（1）开花挂果前。以化学防治为主。联苯肼酯+噻螨酮、丁氟螨酯（或乙唑螨腈）+噻螨酮、阿维菌素+联苯肼酯等组合防治。要均匀喷雾 90% 以上的叶背，建议加有机硅等助剂，每 6 000 株苗药水喷雾量不少于 60 kg。化学防治前还应打除植株上的老叶，放于密闭的口袋内带出棚外，在一定程度上可减少虫口密度，同时老叶打除可增加植株的通透性，提高喷药防治的效果。

（2）开花挂果后。以生物（天敌）防治为主。每月投放 1 次捕食螨（加州小绥螨），以虫防虫。投放数量按 5 瓶/亩（每瓶 2 万头）；每株选 1 张较大平展的叶子，将虫子含麦皮点施于叶面。投放天敌以后，棚内尽量减少杀虫剂的使用。天晴干燥的季节可在棚内喷雾增加湿度，在一定程度上可抑制红蜘蛛的繁殖和为害。

4.4.2.2 蚜虫和蓟马

蚜虫主要吸取草莓的幼叶、花、心叶和叶背面的汁液，受影响的叶子卷曲和变形，草莓的生长受到阻碍。蓟马主要为害草莓的雄蕊、花瓣和幼果，导致授粉不良、果实僵硬或畸形，影响产量和质量。

虫害高发期，可使用乙基多杀菌素+吡丙醚、噻虫嗪、啶虫脒、吡虫啉、吡蚜酮、呋虫胺、螺虫乙酯、氟啶虫酰胺等药剂进行喷雾防治；采摘期可选用苦参碱、乙基多杀菌素、多杀霉素等生物源药剂进行防治。

蓝莓栽培关键技术

1 产业发展概况

蓝莓（Blueberry）学名越橘，又称蓝浆果，为杜鹃花科（Ericaceae）越橘属（*Vaccinium*）植物。蓝莓营养丰富，富含多种维生素及微量元素，还富含果胶物质、超氧化物歧化酶、花青苷等其他果品中少有的特殊成分，具有解除眼睛疲劳、改善视力、延缓脑神经衰老等保健功能，是优秀的保健果品。蓝莓颜色鲜艳，风味独特，既可鲜食，又可加工成多种食品及保健品，深受消费者喜爱，已被联合国粮食及农业组织列为人类五大健康食品之一，在国内外市场上引起了广泛的重视，在行业内占据重要的产业地位。

近20年来，随着人们生活水平的提高，以及农业产业结构调整的深入，蓝莓产业得以迅猛发展。2020年世界蓝莓栽培面积达126 144 hm^2，总产量达850 886 t。近年来，全球蓝莓栽培地区及面积不断增加。

蓝莓以其独特的营养价值、较高的经济效益、丰富的产品形式和备受关注的健康理念成为发展较快的新型果树产业，发展潜力巨大。目前，蓝莓产业已成为继草莓产业之后的世界第二大浆果产业，全球蓝莓产业发展呈现产地多元化的发展趋势。

2 主要栽培品种及特性

2.1 高丛蓝莓

包括北高丛蓝莓、南高丛蓝莓和半高丛蓝莓3种类型。一般树高1.5～3.0 m，适宜在有机质含量丰富、含水量充足及pH值在5.5以下的砂质土壤中生长。该类群果实较大，品质佳，鲜食口感好，果实可用于鲜食或加工。

2.1.1 北高丛蓝莓

喜冷凉气候，抗寒力较强，有些品种可抵抗-30 ℃低温，北高丛蓝莓低温需求量较南高丛蓝莓高，适于沿海湿润地区及寒地栽培。主要优良品种及其特性如下。

2.1.1.1 蓝丰（Bluecrop）

1952年美国新泽西州杂交选育，中熟品种。果实大，天蓝色，果粉厚，肉质硬，果蒂痕小而干，带清淡芳香味，未完全成熟时略偏酸，风味佳，贮藏性好。采收期稍有裂果及落果现象。树体生长健壮，树冠开张，幼树时枝条较软，抗寒、抗旱能力强，对土壤适应能力强，果实丰产性好且稳产，为优良的鲜食品种。

2.1.1.2 云蓝（Yunlan）

2015年云南省农业科学院园艺作物研究所及大连大学现代农业技术研究中心选育，中熟品种。果实中大，淡蓝色，果粉较多，果实圆形较多，少数球形，果肉硬，果蒂痕极小且干，味甜，风味极佳，耐贮藏。树势较强，树冠开张。果实丰产性较好，抗寒性和耐热性较强，为优良的鲜食品种。

2.1.2 南高丛蓝莓

喜湿润、温暖气候条件，低温需求量较低，适于温暖地区栽培。主要优良品种及其特性如下。

2.1.2.1 佛罗里达蓝（Flordablue）

1976年美国佛罗里达大学选育，中晚熟品种。果实中大，带香味，果蒂痕小而干，果肉硬度及酸度中等。树势中庸，树冠开张。低温需求量为150～300 h，果实丰产性好。

2.1.2.2 夏普蓝（Sharpblue）

1976年美国佛罗里达大学选育，中熟品种。果实中大，中等蓝色，带香味，果蒂痕小而湿。多汁，适宜制作鲜果汁，不耐运输。生长势中强，树冠开张，低温需求量为150～300 h。土壤适应性强，果实丰产性好，果实及树体特性与佛罗里达蓝极相似。

2.1.3 半高丛蓝莓

半高丛蓝莓是由高丛蓝莓和矮丛蓝莓杂交获得的品种类型。一般树高0.5～1.0 m，果实比矮丛蓝莓大但比高丛蓝莓小，抗寒力强，可抗-35 ℃低温，适宜于寒冷地区栽培。主要优良品种及其特性如下。

2.1.3.1 北陆（Northland）

1968年美国密歇根大学农业试验站选育，中早熟品种。果实中大，蓝色，圆形，果粉厚，果肉紧实，多汁，酸度中等，风味佳，果蒂痕中等且干，成熟期较集中。树体生长健壮，树冠中度开张，成龄树高可达1.2 m左右。土壤适应性较广，果实极丰产，抗寒，为寒冷地区栽培的优良品种。

2.1.3.2 北村（Northcountry）

1986年美国明尼苏达大学选育，中早熟品种。果实中大，亮蓝色，甜酸，风味佳。树势中等，树高约1.0 m，冠幅1.0 m左右，耐寒性很强，能抗 -37 ℃低温。早果，丰产性好，一般单株产量可达1.0～2.5 kg。叶片小、暗绿色，秋季叶色变红，树姿优美，适宜作观赏品种，为高寒山区栽培的优良品种。

2.2 兔眼蓝莓

该类群树体高大，树高一般在3.0 m以上，寿命长，抗湿热，对土壤条件要求不严，抗旱性好但较不抗寒，-27 ℃低温时易受冻害。一般低于7.2 ℃ 的冷温需求量达18～35 d。适宜气候较温暖、湿度较大的地区栽培，寒冷地区栽培时需考虑花期霜害及冬季冻害。主要优良品种及其特性如下。

2.2.1 芭尔德温（Baldwin）

1985年美国佐治亚州选育，晚熟品种。果实中大，暗蓝色，果粉少，肉质硬，酸味中等，风味佳，果蒂痕小且干。树势强，直立，树冠大。果实丰产性好，冷温需求量为18～25 d，抗病性强。果实成熟期可延续6～7周，适宜庭园及观光园栽培。

2.2.2 乌达德（Woodard）

1960年美国佐治亚州选育，早熟品种。果实中大，亮蓝色，扁圆形，果粉厚，果蒂痕大而干。果实完熟后风味极佳，但完熟前果味偏酸，果实质软，不适宜鲜果远销。幼树期树势弱，树冠开张型，成年期后树体生长旺盛。低温需求量低，春季温度升高后很快开花，易受霜害。为保证结实，以弱修剪为主。

2.3 矮丛蓝莓

该品种类群由原产地的野生种或其繁衍种培育而来，主要特点为树体矮小，一般高0.3～0.5 m。抗旱能力较强，抗寒能力很强，能抗 -40 ℃低温，在瘠薄地、岩石裸露的丘陵地及平地等均能正常生长。栽培管理方面要求简单，极适宜于高寒山区大面积商业化栽培。因果实较小，主要作为加工原料，大面积商业化栽培时需考虑与果实的加工能力配套发展。主要优良品种及其特性如下。

2.3.1 美登（Blomidon）

加拿大农业部肯特准尔研究中心从野生矮丛蓝莓品种中杂交选育，中熟品种。果实圆形，淡蓝色，果粉厚，带清淡香味，风味佳。树体生长势强，果实丰产，5年生树平均株产 0.83 kg，最高可达 1.59 kg。抗寒性极强，为高寒山区发展的首选品种。

2.3.2 芝妮（Chignecto）

加拿大选育，中熟品种。果实近圆形，蓝色，果粉厚。叶片狭长，树体生长旺盛，易繁殖，果实较丰产，抗寒性强。

2.3.3 芬蒂（Fundy）

加拿大选育，中熟品种。果实略大于美登，淡蓝色，带果粉，果实丰产。

3 栽培管理技术

3.1 种植规划与园区建设

3.1.1 园地选择

基地应避开交通要道，距离公路 60 m 以上，周围 3 km 以内没有工矿企业的直接和间接污染源。选择排水良好的平地或向阳缓坡地，园地坡度 <15°，土壤中性偏酸，pH 值以 4.0～5.5 为宜。园地要求土层深厚、土壤肥沃、质地疏松、通气状况良好、有机质含量 ≥3%，有灌溉条件。

3.1.2 土壤改良

定植前一年进行深翻，深度以 20～25 cm 为宜。清除大石块和树根等杂物，整平土地。水湿地、草甸、沼泽地应先清林，设置排水沟。起台田栽植，以台面高 25～30 cm、宽 1 m 为宜，宜在定植前 1 年结合整地进行。蓝莓对土壤酸碱度要求较严，当土壤 pH 值 > 5.5 时，需采取施用硫磺粉或柠檬酸等措施降低 pH 值，pH 值每降低 1 个单位，需施硫磺粉量为 65 kg/亩。将硫磺粉均匀撒入土壤，深翻混匀。当土壤 pH 值 < 4.0 时，可通过全园施石灰粉调节 pH 值，pH 值每升高 1 个单位，需施石灰粉量为 500 kg/亩。将石灰粉均匀撒入土壤，深翻混匀。当土壤有机质含量 < 3% 时，可掺入草炭土、腐烂树皮、腐熟农用秸秆粉末等有机物料，改善土壤结构，增加有机

质含量，同时草炭、松针、锯末等酸性基质还可提高土壤酸度，按园土∶有机物料＝1∶1的比例混合填入定植穴。

3.1.3 品种选择与配置

露地栽培时宜选择北高丛、半高丛或矮丛蓝莓中果实大、质优、高产、适应性强的品种。北高丛、半高丛蓝莓需配置授粉树，矮丛蓝莓可单品种建园。授粉品种选择与主栽品种花期一致且花粉量大的优良品种，主栽品种与授粉品种配置比例为1∶1或2∶1，隔行或隔株栽植。

3.1.4 栽植时期与密度

3.1.4.1 春栽

土壤解冻后至苗木萌芽前进行，一般3月下旬至4月上旬进行栽植。

3.1.4.2 秋栽

落叶后至土壤封冻前进行。北高丛蓝莓株行距一般为1.5 m×3.0 m，半高丛蓝莓株行距一般为1 m×2.0 m，矮丛蓝莓株行距一般为0.6 m×1.5 m，可计划密植，郁闭后间伐。

3.1.5 苗木选择与栽植

选择2～3年生、根系发达完整、主茎直径＞0.6 cm、株高＞30 cm、有3～5个分枝、无病无伤的优质大苗建园。远距离运输的苗木，栽植前用清水浸根12 h，剪除折伤枝、枯死枝。挖长×宽×深＝0.3 m×0.3 m×0.4 m的定植穴。将园土、有机物料按1∶1比例混合均匀后回填，灌水沉实。在灌水沉实的定植穴上挖20 cm×20 cm小穴。定植时将苗木栽入小穴，埋穴深3/4的土，轻轻踏实，做出容水穴，立即灌水约0.5 kg，待水完全下渗后再覆土1次，使苗木原土印与地面齐平。按行作畦进行漫灌，一次浇透水，待水完全渗入后覆土，覆土厚度约3 cm。

3.1.6 埋土防寒

在寒冷地区，封冻前需进行埋土防寒，当气温平均在5 ℃时，土壤夜冻日化时开始埋土，当最低气温降到-3 ℃时应开始进行埋土防寒。将枝条轻弯压倒，埋土厚度10～15 cm，将株丛地上部完全埋入土中。

3.2 整形修剪

3.2.1 修剪时期

分为休眠期修剪和生长期修剪。休眠期修剪在秋季落叶后至早春萌芽前进行，生长期修剪在春、夏季进行，以休眠期修剪为主。

3.2.2 修剪方法

3.2.2.1 高丛、半高丛蓝莓修剪方法

（1）幼树期。苗木定植后为迅速扩冠，可保留所有枝条自然生长。第一年冬剪时选留 5～6 个健壮基生枝，截留 40～50 cm 培养主枝，疏除密生枝、细弱枝及花芽。

（2）初果期。第二年冬剪时，主枝延长枝留 50～60 cm 短截，疏除密生枝、细弱枝，结果枝进行疏花，健壮枝留 4～5 个，中庸枝留 2～3 个，细弱枝不留花芽。第三年夏剪时，疏除多余基生枝，对旺长枝、延长枝和徒长枝重摘心促萌，增加结果枝数量。冬剪时疏除过高的徒长枝、交叉枝、重叠枝、密生枝和衰弱枝，结果枝进行疏花。

（3）盛果期。更新复壮骨干枝和结果枝组。对结果枝组采取回缩、缓放和短截等方法，按标准留花芽。清理基生枝，疏除交叉枝、重叠枝、密生枝和衰弱枝，对旺长枝进行摘心。

（4）衰老期。重回缩，疏除衰老主枝，必要时进行平茬。选留基生枝培养新主枝，更新复壮结果枝组。

3.2.2.2 矮丛蓝莓修剪方法

矮丛蓝莓修剪采用平茬，从植株基部将地上部分锯掉。剪除部分可用于覆盖园内地面，一般每 2 年平茬 1 次。

3.3 水肥管理

3.3.1 水分管理

通常，幼年蓝莓果园应始终保持最适宜的水分条件，以土壤相对持水量控制在 60%～70% 为宜，土壤相对持水量低于 60% 时，需进行适当灌溉。成年蓝莓果园在果实发育阶段和果实成熟前应适量减少水分供应量，果实采收后恢复至适宜的水分供应量，使土壤相对湿度恢复至 60%～70%，中秋至晚秋时期内应减少水分供应，以利于植株在气温降低后及时进入休眠。入冬前可考虑灌 1 次封冻水。此外，还可考虑采用松针、锯末、碎秸秆等进行土壤覆盖，保持土壤湿度。

灌溉可采用沟灌、畦灌等方式，较大规模生产时一般采用滴灌或喷灌进行，果园应设置排水系统，做到能蓄能排。

3.3.2 施肥方法

以有机肥为主、化肥为辅。栽植前和栽植后每3~5年进行1次土壤理化性状分析，根据测土配方确定施肥量。不使用含氯、钙及硝酸盐的化肥。每年施肥2次，分别在开花前后和果实采收后进行。开花前后施肥以速效肥为主，通常选用硫酸铵及果树专用复合肥，复合肥氮（N）、磷（P_2O_5）、钾（K_2O）比例通常为1∶1∶1，每亩施复合肥20~30 kg，辅施少量腐熟有机肥；果实采收后以营养全面的腐熟农家肥或商品有机肥为主，每亩施有机肥2 000~2 500 kg，辅施少量果树专用复合肥和硫酸铵。

土质疏松的砂质土壤果园可全园撒施；壤土、黏土果园一般采用条状沟施及穴施；果树表现缺素症时可叶面喷施缺乏元素的肥料，有条件的果园可结合滴灌施肥。

3.4 主要病虫害防控

蓝莓病虫害主要为害叶片、茎秆、根系及花果，造成树体生长发育受阻、产量下降、果实商品价值降低，严重时甚至失去商品价值，给生产造成巨大损失。为害蓝莓的主要病害有枝枯病、溃疡病、根癌病、灰霉病等；为害蓝莓的常见虫害有鳞翅目夜蛾类、灯蛾类、螟蛾类、尺蛾类、鞘翅目金龟子类、叶甲类、半翅目蝽类、双翅目蝇类等。

蓝莓病虫害防治应遵循"预防为主，综合防治"的原则，从病虫害的综合防治角度来考虑，具体措施可分为五大类。

3.4.1 植物检疫

蓝莓植物检疫包括两个方面的内容：第一是对外检疫，防止将蓝莓危险性病、虫、草等有害生物随同蓝莓植株和产品（如种苗等）由国外传入或由国内传出；第二是对内检疫，当蓝莓危险性病、虫、草害等有害生物已由国外传入或在国内局部地区发生，则需根据法律采取措施，将其限制、封锁在一定范围内，防止传播蔓延至未发生的地区，并采取积极有效的措施，力争彻底肃清或消灭。

3.4.2 农业防治

通过调整栽培措施或人为改变蓝莓生长条件，如加强栽培管理、增施有机肥、科学修剪、间作等，使树体健壮，提高树体抗病能力等，从而改变蓝莓的生长状态或环境条件，以减少或直接消灭病虫害，调控病虫害的种群数量。在蓝莓上应用的农业防治措施主要有改进和利用耕作制度，加强田间管理、采用设施栽培、利用抗病品种等。蓝莓主要病虫害及防治方法可参见表2-1。

表2-1 蓝莓主要病虫害及防治方法

防治对象	防治时期	农业防治或物理防治	化学防治
真菌、细菌、越冬害虫及虫卵	1月下旬至2月上旬	清园。剪除病虫枝，清理越冬虫体、虫蛹，清除落叶和杂草并销毁或深埋	选用29%石硫合剂水剂50倍液、50%多菌灵可湿性粉剂600～800倍液喷雾，选用80%代森锰锌可湿性粉剂600倍液，喷施树干
地下害虫	3月上旬	①人工捕杀成虫 ②园内散养鸡捕杀	选用40%辛硫磷乳油1 000倍液对地面喷雾；选用15%毒死蜱颗粒剂3 000 g/亩撒施
僵果病	①入冬前 ②早春 ③开花前	①清园 ②早春浅耕和施用尿素	选用磷酸二氢钾+微量元素800倍液、450 g/L咪鲜胺水乳剂800～1 500倍液、25%吡唑醚菌酯悬浮剂1 500倍液喷雾
炭疽病	①冬剪时 ②嫩梢期 ③落花后	冬剪时剪除病虫枝并销毁	选用50%多菌灵可湿性粉剂600～800倍液、75%百菌清可湿性粉剂500～800倍液、450 g/L咪鲜胺水乳剂800～1 500倍液、250 g/L嘧菌酯悬浮剂800～1 500倍液喷雾，隔7～10 d喷施1次，连续喷施2～3次
白粉病	6—7月	①清园 ②降低果园湿度	选用25%嘧菌酯悬浮剂1 000～1 500倍液、25%苯醚甲环唑1 500倍液、25%吡唑醚菌酯1 500倍液喷雾，隔7～10 d喷施1次，连续喷施2～3次
果蝇	6—8月	使用糖醋液和黄板等诱杀	选用0.1%阿维菌素饵剂200倍液进行点喷诱杀；选用60 g/L乙基多杀菌素悬浮剂1 500～2 500倍液喷雾
蛴螬	4月上旬	①黑光灯诱杀 ②利用小卷叶蛾线虫防治	选用5%阿维·毒死蜱颗粒剂2 000～3 000 g/亩拌毒土施用；选用15%毒死蜱颗粒剂3 000 g/亩撒施
蚜虫	3—4月	①利用天敌如瓢虫、寄生蜂等防治 ②悬挂黄色粘虫板、杀虫灯	选用70%吡虫啉水分散粒剂5 000倍液、70%啶虫脒水分散粒剂4 000倍液、21%噻虫嗪悬浮剂4 000～5 000倍液喷雾
美国白蛾	5—10月	①人工摘除网幕 ②树干绑草把诱集树下幼虫 ③黑光灯诱杀	选用10%高效氯氟氰菊酯乳油1 000～1 500倍液、1.8%阿维菌素乳油1 000～1 500倍液、8 000 IU/μL苏云金杆菌菌剂800～1 000倍液喷雾，隔7～10 d用药1次，连续用药2～3次
鸟害	5—7月	使用防鸟网或驱鸟器	—

3.4.3 生物防治

蓝莓上使用的生物防治方法有以下几种。

3.4.3.1 利用天敌昆虫防治害虫

如在寄生蜂成虫发生期不喷施任何化学农药，通过保护和利用寄生蜂防治瘿蜂，人工释放桑尺蠖脊腹茧蜂防治尺蠖；通过提供庇护场所或人工助迁释放大红瓢虫、黑缘红瓢虫等防治吹绵蚧、草履蚧等。

3.4.3.2 利用微生物防治害虫

如采用K84（放射形土壤杆菌）菌悬液浸苗或在定植前或发病后烧根，预防根癌病发生和为害等。利用木霉菌防治灰霉病，利用苏云金杆菌防治为害叶片、花器和果实的夜蛾类害虫，利用球孢白僵菌防治地下害虫等。

3.4.3.3 利用昆虫激素防治害虫

如利用性激素诱杀害虫。

3.4.3.4 利用其他有益生物

如通过饲养鸡、鸭等防治害虫。

3.4.4 物理和机械防治

蓝莓栽培常用的物理和机械防治方法如下。

3.4.4.1 捕杀法

根据害虫的发生规律和习性，利用人工和机械捕杀蓝莓毒蛾、象鼻虫、花象甲等害虫。

3.4.4.2 诱杀法

根据害虫的趋向性和某种特性，采用适当的方法将其诱集、杀灭。如用糖醋液诱杀果蝇、黑光灯或频振灯配高压电网诱杀毒蛾、黄板控制蚜虫、蓝板控制蓟马等。

3.4.4.3 温湿度的利用

利用病虫害对温湿度的适应性，通过控制温湿度减少病虫害的种群数量。

3.4.4.4 新技术的应用

如利用辐射杀灭害虫。

3.4.5 农药防治

蓝莓生产中的食品安全需要做到在未受污染的环境中（土壤、水、大气）科学使用农药，严格遵守农药禁限用规定，在必要时用药及最适时期用药，恰当选择农药并使用正确的施药方法和技术，在掌握用量的同时，注意控制使用次数和安全间隔期，以防人、畜中毒等。做到安全使用农药，保证使用的外来化学物质在蓝莓产品中符合消费国的标准。

3.5 采收与采后处理

3.5.1 采收时期

当果皮颜色呈黑色或紫黑色时即为成熟。蓝莓的成熟期不一致，需分批采收，高丛、半高丛蓝莓盛果期2～3 d采收1次，初果期和末果期4～6 d采收1次，矮丛蓝莓成熟期较一致，先成熟的果实不脱落，可待果实全部成熟后采收。

一般供鲜食、运输距离较短且贮藏条件好的果实在达到充分成熟前采收；用于加工的果实在充分成熟后采收。一般晴天早晨最宜采收，夏季气温较高的地区采收时间宜选择温度较低的早晨至中午或傍晚。

3.5.2 采收方法

高丛、半高丛蓝莓采收前清洗、消毒、晒干采收用具，戴橡胶手套进行采摘，采摘时轻摘、轻放，对病果、畸形果应单独放置。按照先冠外、后冠内，先上层、后下层的顺序进行。矮丛蓝莓用梳齿状人工采收器进行采收。大规模园区机械采收时，高丛、半高丛蓝莓用手持式电动采收机采收，矮丛蓝莓用大型梳齿状采收器加配摇动装置采收。

3.5.3 预冷和贮藏

常温条件下蓝莓容易变质，为了保持果实的品质及具有较长的货架供应期，鲜果应进行低温贮藏。果实采收后于10～12 ℃预冷10～12 h，使果实温度降至10 ℃以下，预冷后放入1～3 ℃冷库贮藏或速冻后贮藏于-18 ℃以下环境中。

3.5.4 包装和运输

为避免蓝莓的品质在运输过程中受损，通常采用较浅的透气筐篮、纸箱、果盘等盛装果实，采用小包装，多层次存放，避免重挤压及颠簸，鲜销鲜食果实常选用带有透气孔的泡沫盒子、纸箱等盛放，每盒装果量在1.5 kg以内，加工用的果实采用较大的透气容器包装直接运输至加工厂。

参考文献

贾云霞，李立强，穆旭东，等，2021. 蓝莓栽培技术规程 [J]. 果树资源学报，2（5）：49-51.

苏佳明，沙玉芬，段小娜，等，2007. 蓝莓主要优良品种简介 [J]. 山西果树（1）：17-18.

万红，王连润，2016. 蓝莓常见病虫害及其防控技术 [M]. 昆明：云南科技出版社.

王连润，陶磅，孔令明，等，2013. 昆明地区蓝莓生产技术 [J]. 林业实用技术（5）：50-51.

吴林，2016. 中国蓝莓35年：科学研究与产业发展 [J]. 吉林农业大学学报，38（1）：1-11.

李栽培关键技术

1 产业发展概况

李为蔷薇科（Rosaceae）李属（*Prunus* L.）落叶小乔木，在中国栽培已有 3 000 多年之久，是中国栽培历史最久的果树之一。2021 年云南李栽培面积接近 45 万亩，产量约 30 万 t。李在云南各地均有栽培，相比其他大宗水果的生产来说，李较耐粗放管理，栽培技术简单，早结丰产，易获得大量的果实，对于经济欠发达地区、边远民族地区的农民收入增加、经济发展和农村生产生活条件改善，具有重要的意义。在栽培模式上，除绥江县、砚山县等几个种植面积大的主产县以李农民专业合作社露地种植为主外，全省主要是以各地农户分散露地种植为主。

2 适宜气候环境条件

李树对环境条件适应性强，云南省除南部低海拔、高温、高湿地区外，海拔 1 100 m 以上地区都有分布。平地、坡地、山地种植均能正常生长结果。

2.1 温度

李树对温度的要求因种类和品种而异。李树生长季节的适温为 20~30 ℃，开花期最适温度 12~16 ℃，李树花期易受低温冻害，不同发育时期的有害低温也不相同，如花蕾期 -5.5~-1.1 ℃就会受害；花期和幼果期为 -2.2~-0.5 ℃。

2.2 水分

李树是浅根性果树，抗旱性中等，喜潮湿。李树对水分的要求在一年中不同时期也不同。新梢旺盛生长和果实迅速膨大时，需水最多，对缺水最敏感。花期干旱或水分过多，常会引起落花落果。花芽分化期和休眠期则需适度干旱。李园土壤湿度如能保持田间持水量的 60%~80%，李树就能正常生长发育。当土壤绝对含水量为 10%~15% 时，地上部停止生长，低于 7% 时则根系停止活动。因此，在干旱地区栽培李应有灌溉条件，在低洼、黏重的土壤上种植李树要注意雨季排涝。李树宜栽种在地下水位低、无水涝危害的地方。空气湿度对李树生长也有很大影响，空气过于干燥，会加强蒸腾强度，如果丧失正常含水量50%以上的水分，枝条就会干枯；花期干旱、空气湿度小会影响授粉受精；冬季天气干旱会使枝条特别是组织不充实的秋梢严重失水枯死。

2.3 光照

李树比较喜光。通风透光好的果园和树体，果实着色好、糖分高，枝条粗壮，花芽饱满。

2.4 风

李树对风的抗性弱，强风会给李树带来不良影响。花期如遇干热风，会吹干柱头，妨碍授粉；风速 >10 m/s 时，会使枝干折断，果实脱落；冬季干燥的北风，也会带来冻害。因此，特别是在受台风影响大的地区栽种李树，应选择避风处，并提前在园子四周栽种速生防护林。

2.5 土壤

李树对土壤要求不是十分严格。中国李对土壤的适应性强，黑钙土、红壤土、黄土，均适宜李树生长，但以土层深厚、湿润肥沃、水利条件好、地下水位低的砂质壤土最好。一般李树对土壤pH值的适应性强，在pH值4.7～7.4的中性偏酸土壤上均能生长良好。

2.6 坡向

因太阳辐射强度和日照时数不同，不同坡向的水热状况和土壤理化性质也有较大差异。李树喜欢背风向阳的坡地，坡向一般以南坡较好，因南坡温度较高，光照也强，在冬季南坡受来自西北方向的冷凉干燥气流影响小，受光最佳的坡向为南偏东 5°，但高寒地区以南偏西 5° 为宜。

3 主要栽培品种及特性

云南大约有100多个李品种，其中大部分是长期栽培过程中形成的地方品种。以下主要介绍部分新推广应用的品种。

3.1 早金玉

早熟鲜食品种。果实椭圆形，平均单果重 45.5 g，可溶性固形物含量 12.1%，可滴定酸含量 0.94%，维生素 C 含量 2.8 mg/100 g，完全成熟时果实表面覆盖红晕，果肉橘黄色，采前不落果、不裂果，核小，果实可食率 94.4%。

3.2 红美

早熟鲜食品种。果实圆形，平均单果重 50 g，果实红色，果肉红色，汁液多，果肉致密，纤维中，味甜微酸，无涩味，可溶性固形物含量 12.6%，可滴定酸含量 0.95%，维生素 C 含量 5.0 mg/100 g，品质上等，核小，综合性状优良。

3.3 红峰

晚熟鲜食品种，平均单果重 100 g 左右，果实圆形、红色，果肉黄色，汁液多，果肉致密，纤维中，味甜微酸，无涩味，香味浓郁，可溶性固形物含量 13.3%，品质上等；核小，果实可食率 98.5%；综合性状优良；丰产。

3.4 金红

晚熟鲜食品种，平均单果重 40 g，果实椭圆形、黄色，果肉黄色，汁液中，果肉致密，纤维中，味甜微酸，无涩味，香味浓郁，可溶性固形物含量 12.8%，可滴定酸含量 1.07%，维生素 C 含量 2.86 mg/100 g，品质上等；离核；综合性状优良。

3.5 国峰 2 号

大果型晚熟鲜食品种。果实圆形，平均单果重 115 g，果肉黄色，近皮红色，肉质硬脆，可食率高，汁液多，风味浓郁，半离核，可溶性固形物含量 13.2%，可溶性糖含量 7.4%，可滴定酸含量 0.8%，维生素 C 含量 3.1 mg/100 g，果实发育期 117 d，8 月末成熟。该品种具有优质、耐贮、果实大、丰产、栽培适应性强等特点。

3.6 国峰 7 号

大果型晚熟鲜食品种。果实扁圆形，平均单果重 100 g；果实整齐度好，果肉黄色，近皮红色，肉质硬脆，风味浓郁，半离核，可溶性固形物含量 18.0%，在设施栽培条件下可溶性固形物含量可达 21%，可滴定酸含量 1.6%，维生素 C 含量 6.0 mg/100 g，果实发育期 115 d 左右，8 月末成熟。该品种具有风味浓郁、耐贮运、抗寒性强、早果性和丰产性好等特点。该品种适合设施栽培或露地栽培。

3.7 国峰 17 号

大果型晚熟鲜食品种。果实圆形，平均单果重 118.5 g，果肉黄色，近皮红色，肉质硬脆，可食率高，汁液多，风味浓郁，半离核，可溶性固形物含量 14.8%，可滴定酸含量 1.1%，维生素 C 含量 5.1 mg/100 g，果实发育期 120 d 左右，9 月初成熟。该品种具有优质、耐贮、果实大、外观美、抗寒性较强等特点。

3.8 绥江半边红李

该品种树势中庸偏强，树姿半开张，树冠杯状形，分枝力较强。花期 3 月上中旬，果实成熟期 7 月上中旬。果实扁圆形或近圆形，平均单果重 30.7 g，果肉脆，肉厚，汁液中等，充分成熟后肉软，味甜有清香味，品质优良，离核，可溶性固形物含量 11.43%，在昆明室内常温下可放置 14 d 左右，耐贮性强。

4 园地选择与规划

4.1 园地选择

李树以土层深厚、湿润肥沃、水利条件好、地下水位低的砂质壤土种植最好。一般平地、丘陵、山地、沙滩均可以栽植李树。在山地建园时，应优先选择背风向阳的南坡地，然后要进行整地；在山坡地低洼处建园时，选择开花较晚的品种较好，能够避免晚霜的危害。

4.2 果园规划设计

首先进行果园的测绘，然后根据土壤状况、地形以及气象、水文资料，确定建园范围、防风林、道路、灌溉系统和建筑物。

4.2.1 小区划分

根据地形划成若干小区。地势较为平坦的果园小区可大些，山地果园小区可小些。平地果园的小区最好是南北向，以利于果园获得较均匀的光照。山地果园的小区应水平设置，长边与等高线平行，以利于水土保持。

4.2.2 道路系统

主路、小路和支路组成。主路要求位置适中，宽 4~6 m，以便于运送肥料和果品。山地果园主路可环山而上，呈"之"字形。道路设置时应与防风林、水渠相结合，尽量少占果园，以道路占果园总面积的 3%~5% 为宜。

4.2.3 排灌系统

包括干渠、支渠和输水沟。干渠应设在果园高处，以便能控制全园。支渠多沿小区边界设置，再经输水沟将水引入果树盘内渗入土壤。有条件的地方，应大力推广喷灌、滴灌，节约用水，减少水分的损失。山地或丘陵果园应搭建蓄水设施。地下水位较高的果园，应挖明沟进行排水，以免雨季长期积水对果园造成危害。

4.3 整地和改土

平地建立李园，可按规划设计的株、行距，开挖定植穴，施入有机肥以备栽植。若是在沙地建园，首先要进行土壤改良，方法是给沙中掺土和有机肥，用黏土 1 份、沙土 2~3 份，再混入一定数量的有机肥，拌匀后填入栽植坑，每年进行扩穴、掺土、施肥，可有效地改变土壤的物理状况。山地建园时，结合整修梯田和鱼鳞坑进行土壤改良工作。

5 栽植

5.1 方式、密度

长方形栽植：好处是行距大于株距，通风透光好，便于管理。
正方形栽植：特点是株行距相等，光照好，管理方便。
等高栽植：适宜于山地果园，按照一定的株、行距将果树栽植在同一条等高线上。

在土壤条件好、田间管理水平一般的果园，株行距可采用 3 m×4 m 或 3 m×5 m，山地、沙滩等土壤瘠薄的地方可采用 2 m×4 m。

5.2 配置授粉树

部分品种自花结实率较低，因此在建园时除主栽品种外，还应配置一定数量的授粉树。授粉树品种应与主栽品种花期相近，花粉量多且与主栽品种亲和性好。授粉树配置的比例是2行主栽品种、1行授粉品种，或3行主栽品种、1行授粉品种。也可以按株8∶1比例配置。

5.3 栽植方法

5.3.1 开挖定植穴

一般穴深80 cm、直径100 cm，挖坑时表土放一边，心土放在另一边，种植前对定植穴适当晾晒和用石灰消杀。然后将有机肥与土壤拌匀，回填时先填心土，再填表土，灌水沉实。在原定植穴中央标记好定植点位置。

5.3.2 种植

春季栽植时，在定植点挖一小穴，将李树苗放在定植穴中央，使根系舒展，然后培土，土深以苗木原来在苗圃内生长时留下的土印为准。填土时要把苗木轻轻向上提动，把根系舒展开，边填土边踩实，使土壤与根系充分结合。并在树干周围修树盘，灌足定根水。待水完全下渗后，在树盘上撒上1层细土，并将苗木扶直。

5.4 栽后管理

5.4.1 定干

栽植后及时定干，一般干高50～60 cm，再留20 cm的整形带，共留70～80 cm。要求整形带内芽体饱满，其余的不充实枝芽及时剪除。

5.4.2 堆土防寒

在冬季严寒地区，为防止冬春发生冻害，可在入冬前离苗木50 cm高的西北面，堆一个月牙形土堆防寒，等开春苗木萌芽后再撤除土堆。

5.4.3 灌水

秋季栽植的李树，入冬前要灌封冻水，水分下渗后及时松土。开春树木萌芽前也需及时灌水，以利于芽萌发。

5.4.4 检查成活率及补苗

秋季栽植的李树，在开春树木萌芽时，及时检查成活情况，发现死苗时，要及时补植同龄苗。

5.4.5 防治病虫害

早春苗木发芽时，易受金龟子和蚜虫为害，要注意观察，及时进行人工捕捉或药剂防治。

6 修剪整形

6.1 树形

6.1.1 自然开心形

树体结构为主干上有 3 个主枝，层内距 10～15 cm，以 120° 平面夹角均匀分布，开张角度为 45° 左右。每个主枝上留 1～2 个侧枝，无中心干，干高 30～50 cm。

6.1.2 "Y"形

适于宽行密植，株行距一般为 1.5 m×4 m。主干高 40 cm 左右，无中央干，在主干上分生 2 个较大主枝，斜向行间，呈 45° 角，形似"Y"。小枝直线或小弯曲延伸，其基部外斜侧或背后，可留 1～2 个侧枝，中上部则配置各类枝组，丰满紧凑。成行后，树高约 3 m，冠厚一般不超过 2.5 m，树冠向行间伸展较长，宽度一般为 3 m。栽后 4～5 年成形，通风透光好，果实品质佳，也便于作业。

6.1.3 细长纺锤形

适用于栽培密度较高的李园，树体结构为干高 50～60 cm，树冠直径 3 m 左右，在中心干上培养 10～12 个主枝，主枝与中心干夹角 70°～90°，主枝近似水平，向四周伸展，主枝在中心干上没有明显层次，主枝保持 10～15 cm 间距，同侧主枝间的垂直距离不少于 50～60 cm 间距，下层主枝长 1～2 m，上层主枝逐渐缩短，外形呈纺锤形，在各主枝上直接配置中小型结果枝组。

生产上根据品种特性、立地条件，栽培管理技术来灵活选择适宜的树形。一般来说，干性较弱、分枝力强的品种可采用自然开心形或小冠疏层形；而对于树势直立、长势强的品种一般以杯状形或"Y"形为宜；对于干性强、成枝力较差的品种，适合采用细长纺锤形。

6.2 修剪

6.2.1 幼树修剪

以整形为主，应轻剪各级延长枝，充分利用二次、三次枝培养主枝和侧枝，使各级枝条尽快成形、扩大树冠。李树以短果枝和花束状果枝结果为主，宜用轻剪放长枝，缓和生长势，促其萌发短枝，再根据花芽量和结果的需要短截修剪，这样可达到早结果、早丰产的目的。对竞争枝、过密枝进行疏除。

6.2.2 盛果树修剪

修剪的目的为平衡树势、复壮枝组、延长结果年限。修剪要以疏除为主、短截为辅。对过密枝、直立枝、重叠枝、交叉枝进行适当回缩或短截；对没有更新价值的徒长枝，由基部剪除；对树冠外围和上层的强壮枝，疏密留稀，去旺留壮；对延长枝中度短截，继续扩大树冠和维持树势；对结果壮果枝刺激后促使其抽出壮枝；对多年生枝回缩等都能取得良好的效果。

7 肥水管理

7.1 追肥

7.1.1 发芽前或开花前追肥

此时追肥，对提高受精率、减少落花落果、促使新梢旺盛生长有一定作用。施肥的方法是在树冠外缘，挖长 60 cm、宽 20 cm、深 40 cm 的 3 条沟，施肥量每株树（初果期）为 0.4～0.7 kg，以氮、磷、钾肥为主，比例为 1∶2∶1。

7.1.2 幼果膨大期追肥

主要是促进幼果膨大，减少落果，促进叶片生长，增大光合作用的面积。以速效氮肥为主，适当增施一些磷酸二氢钾复合肥料，每株树 0.5 kg。也可进行根外追肥，喷施 0.4%～0.5% 尿素溶液，使叶片增绿，枝条迅速生长，果实加速发育。

7.1.3 采果后追肥

结合施有机肥，追施磷、钾肥，这样才能获得丰产。

7.2 基肥

7.2.1 时间与种类

秋季施基肥。以有机肥为主，人畜粪尿、豆饼、油饼、秸秆、杂草、落叶都可以用作基肥，但必须经过充分发酵腐熟才能施用。

7.2.2 施肥量

要根据树龄、生长势、结果量、土壤肥力状况以及历年施肥情况而定。定植第一年的小树，每年施入 50 kg 左右基肥，进入结果期后，要做到"斤果斤肥"，甚至"一斤果二斤肥"。

7.2.3 方法

环状沟施肥法：在树冠外缘挖一环状沟，宽 30～40 cm、深 40～60 cm，将肥料施入沟中后覆土。此种方法操作简单，肥力集中有效。

条状沟施肥法：在树冠两侧挖宽 30～40 cm、深 50～60 cm 的 2 条沟，将肥施入后覆土，翌年改变开沟方向，交替施肥。此法适宜于面积较大、已到盛果期的大树。

全园施肥法：是将肥料撒在全园，然后通过翻耕入土。此法适宜于密植园，翻耕程度要深，不然容易造成根系上移。

7.3 灌水

7.3.1 花前灌水

春季云南气候干燥，对李树萌芽、开花、坐果极为不利。花前灌水可使花芽充实饱满，保证花芽有一定的水分和养分，为授粉良好和提高坐果率打好基础。花前灌水可结合追肥同时进行。

7.3.2 幼果膨大期灌水

幼果膨大期正是根、茎、叶、花、果实同时生长发育阶段，又是各器官旺盛生长时期，因此营养生长与生殖生长之间往往会出现较大矛盾。在天气干旱的条件下，还会出现各器官对水分需求的矛盾，即地上部蒸腾过大与地下部根系对水分吸收输送不足而失去平衡的矛盾等。这个阶段水分不足，不仅抑制新梢生长，而且影响果实发育，甚至引起落果。

7.3.3 封冻前灌水

结合深耕施肥灌水，时间可以从秋末直至冬季封冻前都可以进行，但宜早不宜迟。深耕时若能加入有机肥，不仅能有效地改良土壤，还能促进根系及时吸收，从而延长了根系的活动时间。在深耕时应尽量少损伤根系，翻出的根系不要外露时间过长，深耕后要及时灌水。

8 病虫害防治

8.1 春季病虫害防治措施

8.1.1 早春清园

为了大幅度降低越冬病菌和越冬害虫数量，必须在早春对园内外进行一次清园，包括刮树皮、摘除病虫果、捡拾落地果，并集中烧毁，同时对园外越冬寄主，如玉米秸秆进行彻底清除，集中烧毁。

8.1.2 发芽前防治

一是喷施 5°Bé 石硫合剂 1 次，或 45% 石硫合剂结晶 300～500 倍液，或 1∶1∶100 波尔多液，可以清除树枝、干上越冬的李褐腐病、穿孔病、炭疽病、李袋果病等病菌及害虫。

二是喷施 95% 蚧螨灵乳油 100～150 倍液，可以杀死越冬虫卵，且对天敌安全。也可以根据越冬蚜虫、介壳虫初发期与李树的萌芽期基本吻合的特点，在发芽开花前，喷施 2.5% 溴氰菊酯乳油 2 500～3 000 倍液、30% 氰戊·马拉硫磷乳油 2 500 倍液或 10% 吡虫啉可湿性粉剂 1 200～1 500 倍液。

三是防治叶螨类害虫，可在发芽期喷施 20% 四螨嗪悬浮剂 2 000～2 500 倍液。

四是防治流胶病，可在发芽前刮除病部但不要伤及好皮，然后在伤口处涂 0.5% 小檗碱可溶液剂或者稀释 5 倍液涂抹病部，3 d 涂 1 次，连涂 2 次。对于病情严重的园区及不便于涂抹的病株，使用 0.5% 小檗碱可溶液剂 30～60 倍液喷于严重病株的主干、侧干、枝条部位。或发芽前将流胶部位病组织，用竹片（竹片不易伤及健康组织）细致刮除，露出嫩皮，直至木质部，伤口涂抹 45% 石硫合剂结晶 30 倍液，然后涂 21%

过氧乙酸水剂 3～5 倍液保护。药剂防治还可用 30% 戊唑·多菌灵悬浮剂 1 100 倍液、50% 多菌灵可湿性粉剂 800 倍液、50% 异菌脲可湿性粉剂 1 000 倍液或 50% 腐霉利可湿性粉剂 1 200 倍液，防治较好。

8.1.3 开花前的防治

开花前主要防治对象是流胶病、穿孔病、缩叶病和囊果病。对开花前出现的流胶病可在刮除的病块处涂 0.5% 小檗碱原液。对缩叶病、穿孔病、囊果病的防治，应把握在花瓣未展开时进行。喷施 5°Bé 石硫合剂、1∶1∶100 波尔多液、30% 碱式硫酸铜悬浮剂或 70% 代森锰锌可湿性粉剂 500 倍液。

8.1.4 落花后的防治

防治的主要对象是象鼻虫、桑白蚧、蚜虫、李树红点病等。

象鼻虫和桑白蚧： 可选用 90% 敌百虫原药 1 200 倍液或 10% 阿维·哒螨灵乳油 1 000 倍液。或用 22.4% 螺虫乙酯悬浮剂 2 000 倍液 +30% 噻虫嗪悬浮剂 2 000 倍液。

蚜虫： 可在李树落花后至新梢生长期，用 70% 吡虫啉水分散剂 2 000～3 000 倍液喷雾。还可用 10% 啶虫脒乳油 2 000～2 500 倍液、25% 吡蚜酮可湿性粉剂 2 000～2 500 倍液喷雾，这种交替用药的方法对易产生抗药性的蚜虫效果较好。

李红点病： 每年李开花末期子囊破裂，散发大量的子囊孢子，借风雨传播，可在李树开花末期及叶芽开放时，喷施倍量式波尔多液（硫酸铜∶生石灰∶水 = 1∶2∶200），可大量杀灭子囊孢子。或用 10% 苯醚甲环唑可湿性粉剂 1 200 倍液、80% 代森锰锌可湿性粉剂 800～1 000 倍液、70% 甲基硫菌灵可湿性粉剂 1 000～1 200 倍液或 50% 多菌灵可湿性粉剂 800～1 000 倍液，7～10 d 喷 1 次，连喷 2～3 次，可有效防治李树红点病。

8.2 夏季病虫害防治措施

8.2.1 幼果期病虫害防治

李穿孔病和李红点病： 5 月干旱时，病势较缓，如遇多雨天气，应及时防治，可每隔 10 d 补喷 1 次 70% 代森锰锌可湿性粉剂 500 倍液。

李小食心虫： 5 月初，越冬幼虫羽化前，在树冠下地面，每亩用 50% 辛硫磷乳油 300～500 倍液 0.25～0.5 L，毒杀成虫和幼虫。

蚜虫： 5 月中旬繁殖迅速，为害较大，应在虫口数量猛增之前喷施 1 次 10% 吡虫啉乳油1 500～2 000倍液。

叶螨：5月也是防治的最好时机，当越冬成虫死亡、新生的雌虫尚未产卵之时，正是虫害最易控制的时候。可先喷施5%噻螨酮乳油1 500倍液，以后可选用20%哒螨灵可湿性粉剂2 000倍液、48%联苯肼酯悬浮剂2 000～2 500倍液或40%联肼·乙螨唑悬浮剂2 000～3 000倍液喷雾。

潜叶蛾：对发生的果园，应在成虫盛发前喷施20%甲氰菊酯乳油2 000倍液混以50%敌敌畏乳油1 000倍液或25%灭幼脲可湿性粉剂2 000倍液，防治效果很好，可杀死成虫、蛹和幼虫。或用35%氯虫苯甲酰胺水分散粒剂8 000倍液、5%甲氨基阿维菌素苯甲酸盐水分散粒剂3 000倍液+5%氟铃脲乳油1 000倍液。

8.2.2 硬核期、果实膨大期及成熟期病虫害防治

李小食心虫和桃蛀螟：此期正是第一代成虫发生期，可喷施50%杀螟硫磷乳油1 500倍液、2.5%溴氰菊酯乳油500～800倍液或35%氯虫苯甲酰胺水分散粒剂8 000倍液。

红蜘蛛和蚜虫：此期用药与幼果期相同。

裂果病：喷高脂膜，此方法不仅可以防止裂果发生，对病虫害也有作用。或果实膨大期前喷施高钙微肥、螯合钙以减少裂果产生。

成熟期在采收前1个月可喷施1次40%氟硅唑乳油800倍液+10%四螨嗪可湿性粉剂2 000倍液+高效钙300倍液，以防止病虫害再次发生。

8.3 秋季病虫害防治

8.3.1 防治对象

主要防治对象：红蜘蛛、蚜虫、卷叶虫、李穿孔病、李红点病、褐腐病和炭疽病。

8.3.2 防治方法

一是采前喷药，选用杀菌剂和杀虫剂混用加美林钙的方法，共喷3次，第一次在6月底，第二次在7月初，最后一次在7月底。二是采后用药，采后用药主要以杀菌为主，在包装前或入库前用药，以延长果品耐贮性。可采用异菌脲、咪鲜胺等药液浸果的方法或果面喷施的方法。

8.4 冬季病虫害防治

冬季应以防为主、以治为辅,主要任务有 3 项。

8.4.1 清园

病叶、病虫枝、残枝、杂草都是各种病虫害越冬的场所,因此要彻底地将园内一切落叶枯枝、病果病枝、园内及园周围杂草清除干净,集中烧毁。

8.4.2 树体涂白

在 12 月前完成对主干、主枝的刮除老树皮、涂白工作。在刮除老树皮之后,约 1 周内完成涂白任务。涂白剂的配方:生石灰 10 份,硫磺粉 1 份,盐 0.2 份,水 40 份,黏着剂 0.5 份,也可用石硫合剂的残渣代替硫磺粉。

8.4.3 深翻施药

深翻施药可以杀死在土壤中的越冬害虫,深翻后也能使虫卵、虫蛹暴露出来,便于鸟类啄食。土壤施药以粉剂为主,以触杀性药效果为最好。

9 果实采收、包装与贮藏保鲜

9.1 采收

适时采收,以获得最好的果实品质和较高的商品价值。

9.1.1 采收指标

目前,生产上大多根据李果的色泽变化来决定采收期。红色品种,当果实着色占全果面一半时,为长途运输和加工的采收适期;果面彩色占全果面的 4/5 或全着色时,为鲜食采收期。黄色品种,当果皮由绿色转为黄绿色时,为加工和长途运输的采收期;由黄绿色转为黄色时,为鲜食采收期。同一个品种、同一株树的果实成熟期也不一致,所以要分批采收,这样才能保证品质和产量不受影响,又能延长对市场的供应期。果实成熟与否也可以根据果肉硬度测试结果、果实可溶性固形物含量分析结果来判定。

9.1.2 采收要求

分期分批采收，先熟的先采收，后熟的后采收。

9.1.3 采收时间

选择在天气凉爽的晴天，果面露水干后。一般在 8:00—11:00、15:00—18:00 采收较好。雨天、风天、高温时不宜采收。

9.2 果实分级

按果实的成熟度、大小分级，一般分两级即可。剔除病、虫、伤、残果，按质论价。一级果要求充分成熟，果面光滑，无斑点，果形整齐，有亮度，无碰伤，大果占 90% 以上。

9.3 包装

包装应美观大方，重量轻，以提高商品价值。包果的纸要求质地柔软、光滑、洁净，无异味并具有韧性。包装箱底部应用纸条、刨花、锯末填充，以减少晃动。装好后封严，外面应填写产地、品种、毛重、净重、采收期、质检员等。

包装重量不宜过多、过重，一般以 5～10 kg 为宜。

9.4 运输

9.4.1 运前预冷

采收后及时预冷降温，将果温降至适宜的贮藏温度。预冷可在冷库进行，如冷库采用鼓风冷却系统，则更有利于降低果温。

9.4.2 轻装轻卸

在运输过程中一定要轻装轻卸，尽量减少损伤。

9.4.3 冷链运输

采用冷藏卡车、加冰保温列车和冷藏轮船等现代化的运输工具，能够满足李果在运输过程中对温度、湿度的要求，可以减少运输中的损失。

参考文献

蔡达荣，1992. 李树丰产栽培 [M]. 北京：金盾出版社.

陈杰，2006. 李树整形修剪图解 [M]. 北京：金盾出版社.

刘威生，2004. 李无公害高效栽培 [M]. 北京：金盾出版社.

刘威生，2005. 李树杏树良种引种指导 [M]. 北京：金盾出版社.

吕平会，2003. 李树周年管理新技术 [M]. 杨凌：西北农林科技大学出版社.

吕平会，何佳林，2013. 李周年管理关键技术 [M]. 北京：金盾出版社.

❖ 猕猴桃栽培关键技术

1 产业发展概况

1.1 猕猴桃的营养价值

猕猴桃属于猕猴桃科（Actinidiaceae）猕猴桃属（*Actinidia* Lindl.），俗称羊桃、毛梨、毛羊桃、毛桃子等，为多年生落叶藤本植物。猕猴桃果肉富含膳食纤维、维生素、有机酸、多糖、蛋白质、氨基酸等多种营养成分，其中维生素C含量一般为 50～400 mg/100 g，毛花猕猴桃最高可达 1 500 mg/100 g，是其他水果的数倍至几十倍。猕猴桃果肉还含有多种人体必需的中微量元素，如钙、硒、锌、锗等，因此具有降血脂、抗脂质过氧化、清除活性氧自由基、抑制肿瘤细胞等方面的生物活性。猕猴桃因含有丰富的营养，果肉酸甜适中，美味多汁而深受消费者的喜爱。

1.2 猕猴桃产业的发展动态

据联合国粮食及农业组织（FAO）统计，世界上共有 23 个国家生产猕猴桃，遍布五大洲，亚洲是主产区，收获面积和产量均占世界总量的 60% 以上。至 2019 年年底，世界猕猴桃总收获面积大约有 26.88 万 hm²，其中中国 18.26 万 hm²、意大利 2.51 万 hm²、新西兰 1.49 万 hm²，分别占世界总收获面积的 67.93%、9.34%、5.54%。世界猕猴桃总产量为 434.8 万 t，其中中国 216.67 万 t、新西兰 55.82 万 t、意大利 52.45 万 t。世界猕猴桃平均单产为 16.18 t/hm²，新西兰平均单产为 37.41 t/hm²，远超其他国家。

猕猴桃早期的种植品种是以海沃德、布鲁诺为主的美味猕猴桃品种，至今，除新西兰外，意大利、希腊、伊朗等国仍主要是以海沃德为代表的美味绿肉品种。新西兰黄肉品种面积占 42%，绿肉品种面积占 58%，而绿肉品种产量略低于黄肉品种。意大利除绿肉和黄肉品种外，还有少量的红心品种。中国猕猴桃品种呈现多样化，目前猕猴桃绿肉品种的种植面积约占总种植面积的 40%，黄肉和红肉品种各占 30% 左右。综合世界主要生产国的种植情况来看，绿肉品种仍是主栽品种，占 55% 左右，黄肉品种约占 25%，红肉品种约占 20%。

1.3 猕猴桃的适生区域

大多数猕猴桃种类适宜种植在亚热带或暖温带湿润和半湿润气候地区，海拔 800~1 800 m 均可。在海拔 1 000~1 600 m，年均气温 11.3~16.9 ℃，极端高温 42.6 ℃，极端低温 -20.3 ℃，≥10 ℃有效积温 4 500~5 200 ℃，无霜期 160~270 d，年降水量 800 mm 以上，雨量分布均匀，相对湿度 70% 以上是猕猴桃生长的最适区域。幼苗期喜阴凉，成年结果树则需要充足的光照，要求日照时数为 1 300~2 600 h，喜漫射光，光照强度以 40%~45% 为宜。猕猴桃喜土层深厚、疏松透气、有机质含量高的砂质土壤，土壤 pH 值 5.0~7.9。猕猴桃需要微风，在猕猴桃建园时，选择背风向阳的南坡、东南坡和西南坡，坡度一般不超过 30°，既要设防护林，以防大风，又要注意通风透光。

2 主要栽培品种及特性

中国猕猴桃品种丰富多样，生产上应用的猕猴桃主要种类有中华猕猴桃、美味猕猴桃、毛花猕猴桃和软枣猕猴桃等，另外还有授粉雄株品种。

2.1 中华猕猴桃品种

2.1.1 红阳

适合海拔 1 300 m 以下、年均气温 13~16 ℃、年降水量 1 000~1 500 mm 的地区。该品种坐果率高，早产、丰产性强。果实中大、整齐，长圆柱形兼倒卵形，一般平均单果重 60~90 g，果皮绿色或绿褐色，果肉黄绿色，子房鲜红色，沿果心呈放射状红色条纹。果汁特多，肉质细腻，纯甜，清香爽口有香味；鲜食、加工俱佳。果实 8 月中下旬成熟，较耐贮藏，后熟期 10~15 d。不耐夏季高温，在高温高湿环境下影响果肉红色的形成。同时，抗病性和树体的抗药性较差。

2.1.2 东红

果实长圆柱形，平均单果重 70~75 g，果实绿褐色，果肉金黄色，果心四周红色鲜艳。肉质细腻，果汁中等多，风味浓甜，香气浓郁，果实含钙量较高，耐贮性优于红阳。果实采后 30~40 d 以后才开始软熟，果实微软就可以食用，食用期长，在 15 d 以上。果实 9 月中上旬成熟，适应性广，对软腐病和溃疡病的抗性比红阳强。

2.1.3 金艳

果实长圆柱形，果大而均匀，平均单果重101～110 g，最大单果重175 g。果皮黄褐色，果肉金黄，细嫩多汁，味香甜。该品种早果性好，丰产性好。果实成熟期9月下旬至10月中旬。果实特耐贮藏，常温下可贮藏3个月。在低温条件下贮藏期可达120～160 d。

2.1.4 金桃

果实长圆柱形，大小均匀，平均单果重90 g，最大单果重160 g。果皮黄褐色，成熟时果面光洁无毛，果肉黄绿色，后转为金黄色，果心小而软，果肉质地脆，多汁，酸甜适中，品质上等。果实耐贮藏，后熟期可达25 d，贮藏后的风味更佳。该品种耐热性好。

2.1.5 华优

果实椭圆形，平均单果重80～110 g，果皮褐色或黄褐色，果肉黄色或黄绿色，肉质细腻，果汁多，香气浓，风味甜，品质上等；丰产性好，进入盛果期产量可达2 t/亩。果实在室温条件下，后熟期15～20 d，货架期30 d左右，在0 ℃条件下，可贮藏5个月左右。该品种对溃疡病有一定的抗性。

2.1.6 翠玉

果实卵圆形，平均单果重85～95 g，最大单果重129 g。果皮绿褐色，果肉绿色，果肉致密，细嫩多汁，风味浓甜，果肉营养丰富，品质上等。果实较耐贮藏，常温下可贮藏30 d以上，低温条件下，其贮藏期可达4～6个月。该品种结果早，丰产性好，且抗逆性强，抗高温干旱，抗风力也强。

2.1.7 赣猕3号

果实椭圆或长椭圆形，果形端正，平均单果重81.8～107.3 g，最大单果重163 g。果皮褐色或黄褐色，果肉黄色，质细多汁，酸甜适口，微香，品质中上等。果实后熟期约15 d，耐贮性好，室温下可贮藏40 d。抗风，耐高温干旱的能力很强，适应性广，是鲜食和加工兼用的优良晚熟品种。

2.1.8 武植3号

果实椭圆形，平均单果重80～90 g，最大单果重156 g。果皮薄，暗绿色，果肉绿色，肉质细腻，质细汁多，味浓而具清香，果实耐贮藏，采后常温下可贮藏 20 d 以上。该品种耐热性好，丰产稳产，且年年高产。很少发生病虫害，抗旱性强，是一个综合性状好的优良品种。

2.2 美味猕猴桃品种

2.2.1 师宗 1 号

果实扁椭圆形，平均单果重 81 g，最大单果重 150 g；果皮棕褐色，果肉黄绿色，肉质细嫩，汁多，酸甜适口，微香，品质上等。果实后熟期 15～20 d。在室温 12 ℃，相对湿度 75% 的条件下，可贮藏 36 d，在 1～3 ℃的冷藏条件下，可贮藏 5～6 个月。该品种丰产、稳产，盛产期每亩产量可达 3 000 kg 以上。对干旱有较强的抗性。

2.2.2 海沃德

果实椭圆形，平均单果重 80～110 g，最大单果重 165 g。果皮绿褐色，果肉绿色，肉汁多，酸甜，在果肉尚未软化时也可食用，味稍淡，但香气浓，果实极耐贮藏且货架期长。果实的后熟期长，在室温下可贮藏 30 d 左右。

2.2.3 秦美

果实椭圆形，平均单果重 100 g，最大单果重 115 g，果皮绿褐色，果肉淡绿色，肉质细腻，汁多味香，酸甜可口，品质上等。耐贮性中等，常温条件下可存放 15～20 d。丰产性好，盛产期亩产可达到 3 000 kg。

2.2.4 徐香

果实圆柱形，平均单果重 60～70 g，最大单果重 137 g，果皮黄绿色，果肉绿色，汁液多，肉质细腻，具草莓等多种果香味，酸甜适口。果实后熟期 15～20 d，货架期 15～25 d，室内常温下可存放 30 d 左右，在 0～2 ℃冷库中可存放 3 个月以上。

2.2.5 贵长

果实长圆柱形，平均单果重 84.9 g，最大单果重 120 g。果皮褐色，果肉淡绿色，肉质细腻、脆，汁液较多，酸甜适度，清香可口，品质上等，是鲜食与加工兼用的品种。结果早，丰产性能好，抗逆性强，抗缺素症（黄化病）强，抗病虫害能力强，抗低温、干旱和裂果。

2.2.6 翠香

果实卵形，平均单果重 92 g，最大单果重 130 g，果皮黄褐色，果肉翠绿色，质细多汁，甜酸爽口，风味浓，有芳香，品质上等。采后的后熟期为 12～15 d，冷藏条件下可达 100～120 d。该品种丰产性好，早果性较强，一般定植后第三年投产，盛果期亩产可达 2～3 t。适应性较广，抗寒、抗溃疡病的能力较强。

2.2.7 华美 2 号

果实长圆锥形，平均单果重 112 g，最大单果重 205 g。果皮黄褐色，果肉黄绿色，肉质细腻、汁液多，酸甜适口，富有芳香味。果实耐贮藏，在常温下可贮藏 30 d。该品种丰产稳产，抗旱性、抗病性均强。

2.3 毛花猕猴桃品种

华特

果实长圆柱形，平均单果重 82～94 g，最大单果重 132.2 g。果皮褐绿色，果肉绿色，肉质细腻，略酸，维生素 C 含量 628.37 mg/100 g，品质上等。耐贮性好，在常温条件下可贮藏 3 个月。该品种适应性强，耐瘠薄、耐干旱，结果能力强，丰产稳产。

2.4 软枣猕猴桃品种

2.4.1 魁绿

果实扁卵圆形，平均单果重 18.1 g，最大单果重 32 g。果皮绿色光滑，果肉绿色，肉质细腻，多汁，酸甜适口，风味浓郁，品质上等，是鲜食与加工兼用品种。该品种丰产稳产，抗逆性强，在绝对低温 -38 ℃地区无冻害，适应在无霜期 120 d 以上，≥10 ℃年积温 2 500 ℃以上地区栽培。

2.4.2 丰绿

果实圆形，平均单果重 8.5 g。果皮绿色光滑，果肉绿色，肉质细腻，多汁，酸甜适口，品质上等。是鲜食与加工兼用品种。该品种丰产稳产，适应在无霜期 120 d 以上，≥10 ℃年积温 2 500 ℃以上地区栽培。

2.5 授粉雄株品种

2.5.1 磨山 4 号

该品种植株紧凑，每花序常有花 5 朵，最多达 8 朵，花蕾期约 35 d，花期 13～21 d，花粉量大，花粉发芽率 75%。该品系花期长，可作为早熟、中熟，乃至晚熟中华猕猴桃和美味猕猴桃品种的授粉树。

2.5.2 马图阿

该品种树势较弱，定植第 2 年即可开花。花期早，花量多，花粉量大，花粉发芽率 64%，花期长，15～20 d。可用作早、中花期开花的授粉品种。

2.5.3 汤姆利

该品种花期较晚，花量大，花粉发芽率 62%，花期集中，5～10 d，主要作为开花时间较晚品种的授粉树。

3 栽培管理技术

3.1 园区规划建设

3.1.1 园区规划

园区选择应考虑猕猴桃生长对于环境条件的需求和交通条件的便捷性。园区的规划要根据园区的交通、水源、坡度、气候等条件进行，包括道路系统、排灌系统、功能分区授粉树配置等规划。

3.1.2 道路与排灌系统

主干道宽 5～7 m，可通行汽车，便于运输。干道宽 3～5 m，通行小型车辆和农机具，也是园内作业区的分界线。作业道 1～1.5 m，主要为人行道。

排水系统一般与道路系统结合设置，构筑成整个果园的排水网络系统。主干道边的排水沟，为果园内的总排水沟；干道边的排水沟，为各作业区的排水沟。

3.1.3 功能分区

园内建筑物包括休息室、工具间、果品分级包装间、生产资料库、看护房、配药池和化粪池等，应建在交通方便、便于全园管理和操作的地带。作业区大小依地形而定,最好长方形。平地果园每区15～20亩较合适,山地果园作业区可小一些。每区应设置一个贮粪坑、堆肥场和小型蓄水池等。滴灌属于节水灌溉技术,同时能结合施肥,可优先考虑,其次还有沟灌、喷灌、渗灌等灌溉方法。

3.1.4 授粉树配置

猕猴桃为雌雄异株植物,必须配置授粉树。选择授粉树的原则:雄株品种的花期与雌株品种相同或略早于雌株品种1～2 d开花,花量大,花粉多,花粉萌芽率高,亲和性好。雌雄植株的配置比例为(5～8):1。

3.1.5 栽培模式选择

根据当地的天气条件、地势地形、品种特性等选择栽培模式。

生产常采用的架式有"T"形架和棚架。两者基本结构大致相同,均由立柱、横梁、架面等结构构成,不同点在于"T"形架在立柱的顶端形成了一个"T"形的支架,通常只留2个沿行向呈"T"形生长的主蔓,棚架则留3个以上主蔓,在架面上均匀分布。棚架多用于平地果园,"T"形架多用于山地果园。也有两者结合使用的情况。

3.2 定植

定植一般在落叶后至翌年树液流动前进行。定植坑的规格为长、宽、深各 0.6～0.8 m,回填时每个定植坑施入有机肥20 kg、过磷酸钙1 kg,先回填表土后回填底土。选择健壮、无病虫害、无机械损伤的苗木,按规划进行定植,定植时根不能直接与肥料接触,保证根颈部略高出地面,栽好苗后,以苗为中心,培一个直径 50～60 cm、四周高、中间略低的树盘,灌透定根水,再覆盖一层细土,用塑料薄膜或稻草等将树盘覆盖起来,减少水分的蒸发。定植后,当苗木长到一定高度时,进行定干。沿茎干从地面向上数到第3～4个饱满芽处时,剪断茎干,剪口下的第1个芽必须是饱满芽。苗木发芽成活后,要及时检查、补苗。幼苗期需遮阴,可少量多次追施 0.8%～1%氮肥和复合肥,当幼苗萌发新梢后,应及时在幼树旁立竹竿或木棍,将主干绑缚于竹竿或木棍上牵引到棚(篱)架上。

3.3 整形修剪

3.3.1 树体结构与整形

树体结构由主干、主蔓、侧蔓(结果母枝)、结果枝或结果枝组组成。苗木定植后，从定干后萌发的新梢中，选择1条直立向上的健壮新梢作为主干，其余芽全部抹去，待新梢长到距离架面20～30 cm高度时剪梢，促进萌发新梢。选留方向适合的2～4个枝作为永久性主蔓培养。如果选择棚架种植，则选留3～4个枝均匀分布到架面上不同方向，作为永久性主蔓培养；如果是选择"T"形架种植，则选留2个主蔓，作为永久性主蔓培养，分别从架面中央沿行向交叉反向引缚，呈"T"形分布。第二年春季在主蔓上每间隔30～50 cm选留1个枝蔓，作为结果母枝培养，其余的摘心或疏除，侧蔓与主蔓尽量垂直。在侧蔓上间隔20 cm左右选留1个结果枝或者结果枝组用于着果。

3.3.2　冬季修剪

一般在落叶后到第二年的伤流期前进行。主要利用短截、缩剪、疏剪等技术手法。将细弱枝、枯死枝、病虫枝、过密的大枝蔓、交叉枝、重叠枝、竞争枝及下部的无利用价值生长不充实的发育枝等全部疏除，结果母枝间相距30～50 cm，均匀地分布于架面上，形成良好的结果体系，平均1 m^2的架面上有结果母枝3～4条，每个结果母枝留芽10～15个。

3.3.3　夏季修剪

在萌芽后至落叶前均可进行。修剪方法包括抹芽、摘心、疏枝、疏花、疏蕾、新梢绑缚等。分4次进行，第一次在萌芽期进行；第二次在开花后1周完成，雌株对生长旺盛的结果枝从最末1朵花起7～8片叶处摘心，营养枝留10～12片叶摘心，基部徒长枝除去；第三次在花后1月左右，同时还需进行疏果工作；第四次在果实的迅速生长期进行，同时疏去结果过密的果实和畸形果、病虫果等。夏剪完成后，单株的叶果比保持在（6～7）：1以上，叶面积指数控制在3.5～4.0。

3.3.4　雄株修剪

夏剪在雄株谢花后及时进行，以疏剪、长放为主，按架式整理树形，使枝蔓分布均匀合理。冬剪主要是疏除枯枝和纤细枝，剪除发育枝顶部纤细部分，以及病虫枝、过密枝和不必要的徒长枝。

3.4　水肥管理

在进行猕猴桃优质高效栽培过程中，应坚持以有机肥为主、化肥为辅的原则。禁止使用硝态氮肥，限制使用含氯化肥和含氯复合肥。幼苗期以氮肥为主，根据树龄每株施有机肥5～10 kg、复合肥20～50 g，可少量多次地用0.8%～1%氮肥和复合肥进行追肥。

进入结果期的树，采收后及时施入基肥，最好在10—11月，宜早不宜晚，基肥以有机肥为主，应经过充分发酵后施用，作为基肥用量保证做到"斤果斤肥"，根据结果量，株施有机肥20～60 kg。生长期进行适时追肥，一般以速效肥料为主，萌芽前10～15 d施花前肥，以氮肥为主。5月施壮果肥，以钾肥为主，配合氮、磷肥以及微量肥料。果实成熟前25 d施优果肥，以磷、钾肥为主。施肥方式多采用条沟施肥、穴施和灌溉式施肥，施肥后及时灌水，以提高肥料利用率。根据树体和果实的生长情况，还可进行叶面追肥，特别是微肥的施用，如在盛花期可喷施硼砂350倍液，提高受精能力；从果实膨大期开始，每隔15 d左右喷1次磷酸二氢钾300～350倍液，连喷2次，以提高果实含糖量和品质。

猕猴桃叶片大，生长量大，喜水但怕积水，生长季节每3～4 d进行土壤含水量观测，当土壤含水量低于田间持水量的60%时需要及时灌水。萌芽前、花前和花后根据土壤含水情况适时灌水，花期则需要控水。果实膨大期为需水关键期，可进行2～3次灌水，有条件可以进行行间喷雾，以增加空气湿度。果实成熟前15 d左右，停止灌水，休眠前灌1次水。一次的灌水量应使土壤水分含量达到田间持水量的85%以上，浸润深度达到40 cm以上。

3.5 主要病虫害防控

猕猴桃的主要病害有溃疡病、花腐病、根腐病、果腐病等；主要虫害有金龟子、卷叶蛾、介壳虫、叶蝉等。日常管理中应注意使用农业防治、物理防治、生物防治、化学防治等措施进行综合防治。常见病虫害及防治方法如表2-2和表2-3所示。

表2-2 猕猴桃常见病害及防治方法

病害名称	防治时期	防治方法	其他防治措施
溃疡病	早春发病期	1～2年生发病枝条全部剪除；发病大枝及时刮除病斑，用30%氢氧化铜悬浮剂500倍液、20%噻唑锌悬浮剂200倍液涂抹伤口	冬季彻底清园，树干涂白，结合修剪清除病蔓，并烧毁； 选择立地条件好的地块，采用避雨栽培； 刀剪工具注意用75%乙醇消毒，可用两把枝剪交替消毒修剪
	萌芽后至花前	选用1.5%噻霉酮水乳剂600～800倍液、77%氢氧化铜可湿性粉剂600～800倍液或80%代森锰锌可湿性粉剂800倍液等防治，连喷2～3次	
	5—8月	及时全园巡查，去除病部翘皮，对病斑涂抹防腐油等	

（续表）

病害名称	防治时期	防治方法	其他防治措施
溃疡病	采果后至落叶前	选用3%中生菌素可湿性粉剂600倍液进行防治，10～15 d喷1次，连喷3～4次	
	落叶后	清园后地面喷施1次EM（高效微生物菌群）菌剂或15%四霉素母药800倍液	
	冬剪后至萌芽前	均匀喷施3～5°Bé石硫合剂1次，喷施1.5%噻霉酮水乳剂500～600倍液或80%代森锰锌可湿性粉剂800倍液1～2次	
花腐病	萌芽期	3～5°Bé石硫合剂全园喷雾	改善花蕾部的通风透光条件；加强园地肥水管理；摘除病蕾病花
	花期、展叶期	可用65%代森锌可湿性粉剂500倍液或50%退菌特（福美双、福美锌、福美甲胂）可湿性粉剂800倍液或0.3°Bé石硫合剂喷洒全树，10～15 d喷1次	
果腐病	萌芽前	喷施3～5°Bé石硫合剂	冬季彻底清园，消灭病菌载体；加强果园管理，重施基肥，增强树势；改善园区通风透光条件，减少荫蔽
	谢花后2周至果实膨大期	喷施80%甲基硫菌灵可湿性粉剂1 000倍液或80%代森锰锌可湿性粉剂1 000倍液。根外喷施0.2%～0.3%钙肥2～3次	
根腐病	3月和5月下旬	用58%甲霜·锰锌可湿性粉剂500倍液灌根	雨季做好开沟排水，定植不宜过深，有机肥要充分腐熟；土壤黏重可以掺沙改土

表2-3　猕猴桃常见虫害及防治方法

虫害名称	防治时期	使用药剂	其他防治措施
金龟子	①出土时；②5—7月	用50%辛硫磷乳油800～1 000倍液灌杀土中幼虫2～3次；用20%氰戊菊酯乳油1 500倍液，在开花前2～3 d和谢花后喷雾防治2～3次	可用黑光灯或杀虫灯诱杀；敲打树枝震落并捕杀；糖醋液加敌百虫诱杀；冬季清园翻土，挖除越冬幼虫
叶蝉	5月中下旬至7月中下旬	若虫期用70%吡虫啉水分散粒剂4 000～5 000倍液喷雾防治；成虫期用1.5%除虫菊素水乳剂600～1 000倍液喷雾	清理果园，刮除卵块烧毁
果实蝇	幼虫期	用0.1%阿维菌素浓饵剂对植株进行点喷	及时摘除、捡拾病果，集中处理；进行果实套袋，性诱剂诱杀等
	成虫期羽化盛期	地面喷50%辛硫磷乳油700倍液；深翻土壤	
	成虫产卵前	喷90%晶体敌百虫原药+3%～5%糖水，隔5 d喷1次，连续喷2～3次	

（续表）

虫害名称	防治时期	使用药剂	其他防治措施
小薪甲	果实膨大期	喷 522.5 g/L 氯氰菊酯·毒死蜱乳油 1 500～2 000 倍液或 2.5% 高效氯氟氰菊酯乳油 2 000 倍液	消灭越冬虫源；注意控制园内杂草，减少害虫藏匿场所；合理留果，避免果实间过于重叠挤压
吸果夜蛾类	6月中下旬至10月中下旬	喷 1% 苦皮藤素水乳剂 1 000～1 500 倍液	黑光灯或糖醋液诱杀
透刺蛾类	3—10月	喷 20% 甲氰菊酯乳油 1 000～1 500 倍液	发现有蛀孔或虫粪时用小铁丝刺死幼虫
介壳虫	2—11月	喷 25% 噻嗪酮乳油 1 000～1 200 倍液或 48% 毒死蜱乳油 800～1 000 倍液	人工消灭越冬代雌成虫及幼虫；采用阻隔法减少上树虫口；用石硫合剂涂抹介壳虫发生枝干；冬季收集树干周围土壤中的虫卵，集中灭杀

3.6 采收

猕猴桃果实的采收期受多方面因素影响。目前使用较多的是根据可溶性固形物含量判定采收期。一般可溶性固形物含量 6.5%～8.0% 即认为可以采收了。采后要进行果品分级，有人工分级和机械分级两种方式。人工分级可使用选果板，根据果实横径的不同进行分级，一般将横径相差 5 mm 作为等级之间的划分。机械分级是按被选产品的重量与预先设定的重量进行分级。

参考文献

陈东元，黄建民，2004. 猕猴桃无公害高效栽培 [M]. 北京：金盾出版社.
黄宏文，2001. 猕猴桃高效栽培 [M]. 北京：金盾出版社.
黄宏文，钟彩虹，李大卫，等，2013. 中国猕猴桃种质资源 [M]. 北京：中国林业出版社.
郎彬彬，朱博，谢敏，等，2016. 野生毛花猕猴桃种质资源主要数量性状变异分析及评价指标探讨 [J]. 果树学报，33（1）：8-15.
刘兰泉，王东，2016. 猕猴桃栽培及病虫害防治 [M]. 北京：中国农业出版社.
钟彩虹，黄宏文，2018. 中国猕猴桃科研与产业四十年 [M]. 合肥：中国科学技术大学出版社.
钟彩虹，黄文俊，李大卫，等，2021. 世界猕猴桃产业发展及鲜果贸易动态分析 [J]. 中国果树（7）：101-108.

人参果栽培关键技术

1 产业发展概况

人参果（*Solanum muricatum* Aiton）为茄科（Solanaceae Juss.）茄属（*Solanum* L.）多年生草本植物，学名为香瓜茄，原产南美洲安第斯山区。20世纪80年代引入我国栽培，现主要分布在云南、青海、甘肃、贵州、四川等省份，为新兴特色水果。云南从20世纪90年代初引入栽培，云南独特的地理位置、充足的光照、适宜的温度，为云南人参果产业发展提供了良好的气候和生态环境条件，使全省不同区域生产的人参果可实现周年供应，多年来产量稳居全国第一。

据统计，2022年中国人参果栽培面积约32万亩，其中云南27万亩、甘肃2万亩、其他省份3万亩。云南人参果正季栽培主要集中在昆明市石林县，面积20万亩；曲靖市、玉溪市、楚雄州、大理州、保山市有零星栽培，面积约2万亩；成规模的反季栽培始于2017年，主要分布在红河州、文山州、楚雄州、德宏州、普洱市、玉溪市、临沧市等，面积约5万亩，产品上市时间为12月至翌年5月，正值水果淡季，价格高，经济效益好，发展势头强劲，预计5年之内，面积将达25万亩，与正季栽培面积持平。

云南拥有人参果种植、流通龙头企业27家，人参果专业种植合作社45家，其中石林县38家，"西街口人参果"成功注册国家地理标志证明商标。

2 主要栽培品种及特性

目前，石林人参果栽培品种有圆果1号、圆果2号和大紫，主栽品种以圆果2号为主，占全县栽培面积的95%，圆果2号为水果鲜食型品种，口感好，营养价值较高，深受消费者喜爱。大紫为果菜兼用型品种，产量高，抗病性强，但市场占有量小。

2.1 圆果1号

植物学特征：总状花序，花萼星形，中间紫色带白色花边；雄蕊5枚，黄色，环抱雌蕊；雌蕊1枚，柱头高出雄蕊，绿色；果实圆形或心形，果顶圆或略尖，幼果白色，成熟后果面黄色，带紫色条纹或花纹，果肉黄色，有果香，果肉略带腥味。叶片披针形，深绿色，幼嫩叶柄带紫色；茎干圆柱形，幼嫩茎干带紫色，逐渐变绿色；湿度大时易发生根突。耐热性差，超过28 ℃花粉活性较低，不能完成自花授粉，单性结实率低，易出现僵果，成熟期延长；结果期适宜温度白天18～25 ℃，夜晚12～18 ℃，空气相对湿度50%～75%。

生产表现：移栽定植后110～120 d采收。

种植区域：主要集中在昆明市石林县。

2.2 圆果2号

植物学特征：石林称圆果，鲜食水果型。无限生长型，植株生长旺盛，新梢萌发力特别强，抹后即可无限不断萌发。前期不易坐果，结果期推迟。果圆形或扁圆形中等，平均单果重100 g左右，最大单果重260 g。幼果绿白色，长大后有紫色条纹，充分成熟时果皮橘黄色，紫色条纹转为紫黑色，果肉金黄，清香爽口，甘甜多汁，风味纯正。

耐热性一般，30 ℃以上不利于结果，单性结实率低，易形成僵果，仅部分能正常成熟，昼夜温差大有利于提高好品质，未成熟果带酸味，以成熟果上市为主，果肉金黄，果香浓，成熟果可溶性固形物含量10%～13%，品质较好。

生产表现：移栽定植后110 d左右采收。

种植区域：正季栽培主要集中在昆明市石林县，曲靖市、玉溪市、楚雄州、大理市有零星栽培；成规模的反季栽培主要分布在红河州、文山州、楚雄州、德宏州、普洱市、玉溪市、临沧市等。

2.3 大紫

植物学特征：又叫菜果。总状花序，花萼星形，淡紫色带白色花边；雄蕊5枚，黄色，环抱雌蕊；萼片5枚，绿色；果实长椭圆形，果顶尖，果较大，平均单果重250 g左右，最大单果重可达800 g，幼果白色或带淡绿色条纹，成熟后果面带紫色条纹或花纹，充分成熟后果面呈淡黄色带紫色条纹或片状花纹；果肉金黄色，有果香，果肉带腥味；开花结果温度18～30 ℃，结实率高；植株生长较旺盛，无限生长型，发枝力极强；叶片披针形，绿色；茎干圆柱形绿色，湿度大时易发生根突，耐热性、耐肥性和抗寒性较强。以菜用为主，成熟果可溶性固形物含量9%左右。

生产表现：移栽定植后95～100 d采收，挂果率高，抗疫病能力强。

种植区域：主要以甘肃保护地栽培为主，石林县市场销售价格相对较低，栽培面积逐年减少。

3 栽培管理技术

3.1 产地环境要求

3.1.1 自然环境

人参果易受气候环境条件、栽培技术等的影响。资料显示，南美洲人参果原产地年均气温 15.2 ℃，最热月平均气温 19.6 ℃，极端最高气温 32.4 ℃，极端最低气温 -0.1 ℃，年降水量 1 242 mm，年蒸发量 762 mm，年日照时数 2 094 h。由此可知，人参果需要凉爽、光照充足、干而不旱的气候环境。适宜的环境条件是获得高品质人参果产品的重要前提。根据人参果对环境条件的要求，种植地块应光照良好，给排水通畅，年降水量 1 000 mm 左右，白天温度 20～28 ℃，夜间温度宜在 15～20 ℃。春季种植为防止低温和晚霜，应保证夜间温度在 8 ℃以上并且连续 10 d 以上，方可开始移栽。冬季无霜地区，可选择立秋种植。

3.1.2 土壤要求

选择排灌方便、透气性好、有机质含量高、质地疏松、浇水后不易结块的土壤，人参果根系发达，主要分布在 0～30 cm 土层内，培养良好的根系，是提高人参果果实品质的关键。疏松的土壤有利于根系持续生长。圆果 1 号、圆果 2 号耐肥性差，对氮肥特别敏感，氮肥过多容易出现营养生长过旺、花量少、不易结果等问题。

3.1.3 灌溉水要求

灌溉水按照《农田灌溉水质标准》（GB 5084—2021）执行，即基本水质应满足表 2-4 要求。

表 2-4　农田灌溉水质基本控制项目限值

序号	项目类别		作物种类：蔬菜	
			加工、烹调及去皮蔬菜	生食类蔬菜、瓜类和草本水果
1	pH 值		5.5～8.5	
2	水温 /℃	≤	35	
3	悬浮物 /（mg/L）	≤	60	15
4	五日生化需氧量（BOD_5）/（mg/L）	≤	40	15

（续表）

序号	项目类别		作物种类：蔬菜	
			加工、烹调及去皮蔬菜	生食类蔬菜、瓜类和草本水果
5	化学需氧量（COD_{Cr}）/（mg/L）	≤	100	60
6	阴离子表面活性剂 /（mg/L）	≤	5	
7	氯化物（以 Cl^- 计）/（mg/L）	≤	350	
8	硫化物（以 S^{2-} 计）/（mg/L）	≤	1	
9	全盐量 /（mg/L）	≤	1 000（非盐碱地区），2 000（盐碱地区）	
10	总铅 /（mg/L）	≤	0.2	
11	总镉 /（mg/L）	≤	0.01	
12	铬（六价）/（mg/L）	≤	0.1	
13	总汞 /（mg/L）	≤	0.001	
14	总砷 /（mg/L）	≤	0.05	
15	粪大肠菌群数 /（MPN/L）	≤	20 000	10 000
16	蛔虫卵数 /（个 /10 L）	≤	20	10

3.2 栽培技术

3.2.1 品种选择

圆果 1 号、圆果 2 号和大紫等品种。

3.2.2 育苗

3.2.2.1 组培脱毒苗

挑选当年栽培、长势好的优良单株，在晴天下午剪切 2～3 cm 的幼嫩枝条作为备用材料。将枝条去除叶片，切取 1.5 cm 长的茎尖放入 250 mL 干净烧杯中，加入 1 滴蔬菜洗洁精，加入 150 mL 自来水搅拌浸泡 5 min，然后用流动自来水漂洗 10 min，转至超净工作台上进行消毒处理。用 70% 乙醇处理 30 s，无菌水冲洗 3 次，然后放入已消毒的 100 mL 烧杯中，用 0.1% 升汞浸泡 8 min，浸泡时不间断振荡瓶子，以确保枝条得到全面彻底的消毒，最后用无菌水反复冲洗 5 次。

将消毒好的材料在体视显微镜下剥取小于 0.2 mm 茎尖生长点，接种在 MS+KT0.1+IAA0.1+GA0.1+3% 蔗糖 +5 g/L 琼脂的诱导培养基上进行诱导产生无菌芽苗。当无菌芽苗高 5～6 cm 时，进行病毒检测，检测后，将无病毒的无菌苗进行切割转移到继代培养基上进行培养。继代培养基采用不加激素的 MS 培养基即可，培养温度 20～25 ℃，光照强度 2 000～3 000 lx，每 25～30 d 进行 1 次继代培养；人参果苗极易生根，生根培养基采用不加激素的 MS 培养基，切取 2 cm 茎尖接种到生根培养基，每瓶 20 株，10 d 开始生根，15～18 d 即可出瓶炼苗。

采用 98 孔育苗盆，栽培基质采用草炭：珍珠岩 =9∶1 的配方基质，瓶苗栽培后即通过喷雾浇透水，盖上小拱棚，遮光 70%，温度控制在 20～30 ℃，相对湿度 85%～90%，1 周以后，每 3 d 喷施 1 次 0.1% 的叶面肥，15 d 以后逐步增加光照，30～40 d、苗高 15～20 cm 时可移栽大田。

3.2.2.2　扦插苗

人参果侧枝多，扦插简单，易成活。扦插育苗是最简单的繁殖方式，也是目前人参果育苗最主要的繁殖方式。

母本园建立： 选择隔离条件好，远离人参果种植区建立母本园，作扦插采穗用。母本园保护地栽培，使用脱毒组培苗作为母本，栽培密度大约每平方米 8 株，亩栽 5 000 株左右，栽培 70～90 d 后即可分批采芽条进行扦插。

采用 98 孔育苗盆，扦插基质采用草炭：珍珠岩 =9∶1 的配方基质，剪取 8～12 cm 的健壮侧枝作为扦插枝条，在扦插前要对剪下的枝条进行修剪，枝条顶端留 1～2 片叶，其余叶片都剪去，并把枝条剪成长 6 cm 左右的茎尖，每个茎尖带 1～2 节，扦插深度为 2～3 cm，每孔插 1 个茎段，每扦插好一盘后，整齐放入苗床，喷洒水，使基质达到最大持水量。盖上小拱棚，遮光 70%，温度控制在 25～32 ℃，相对湿度 85%～90%，10 d 以后，每 3 d 喷施 1 次 0.1% 的叶面肥，15 d 以后逐步增加光照，40 d 后，苗高 20 cm，健壮敦实，即可作商品苗出售。

3.2.3　栽培

可按照《地理标志证明商标　西街口人参果》(T/YGIIA 001—2023) 标准执行。

3.2.3.1　栽前准备

合理轮作，避免与茄科作物连作。整地做到墒平土细，每亩施腐熟农家肥 1 000～1 500 kg、平衡复合肥 10～20 kg，穴施作底肥与土壤充分拌匀。开墒 1.2 m，墒高 20～30 cm，墒沟宽 50 cm。

3.2.3.2　种植密度

每墒 1 行，单株种植，株距 60 cm，行距 120 cm，每亩定植 800～1 000 株。

3.2.3.3　春季（正季）移栽

由于海拔不同，移栽定植时间稍有不同，原则上在当地霜期结束后开始栽种，早栽可以延长采果期，获得更高的产量。正季栽培时，一般低海拔地区（1 700 m 左右）4月初开始定植，高海拔地区（2 100 m 左右）在4月15日开始定植。定植后浇足定根水再覆膜，覆盖黑膜或银色地膜。地膜有抑制杂草生长、提高地温、保持水分的作用，有利于植株生长。但应选用质量好、使用期较长的地膜，这样可降低后续的管理投入，同时为植株健康生长创造良好的生长环境。边栽边盖膜，盖膜后掏出小苗，用细土将膜掏苗口及四周压实，达到保温、保湿效果。

3.2.3.4 秋季（反季）移栽

反季栽培在冬季无霜的温暖地区，露地栽培，适宜区域主要分布在红河州、文山州、玉溪市、楚雄州、普洱市、临沧市及德宏州海拔600～1 200 m 的地区，规模栽培，宜选有灌溉条件、交通便利的山地或半山地，一般在8—10月定植，其他同春季（正季）栽培。

3.2.3.5 掏苗

栽培覆盖地膜后，掏苗需快速及时，特别是在天气晴朗气候条件下，如果掏苗不及时，就会使膜下温度过高，造成脱水烧苗。轻则，烧伤部分功能叶，苗恢复时间较长，影响产量；重则，因苗移栽脱水死亡。故应边盖膜边掏苗，且掏苗与封口应同时进行，否则掏苗口高温气体外窜，同样会造成烧苗，另外，掏苗口不能太小，一般直径应在10 cm 左右，以便后期灌肥及雨水渗透。

3.2.4 搭架引蔓栽培

3.2.4.1 栽前准备

同 3.2.3.1。

3.2.4.2 种植密度

每墒1行，单株种植，株距30 cm，行距120 cm，每亩定植2 000株左右。

3.2.4.3 春季（正季）移栽定植

同 3.2.3.3。

3.2.4.4 秋季（反季）移栽定植

同 3.2.3.4。

3.2.4.5 搭引蔓支架

人参果枝条萌发能力强，植株直立性较差，挂果后负重大，难以直立生长，进行修枝搭架引蔓可以有效减少植株倒伏、遮蔽及病虫害发生，增加通风透光，促进花芽分化和开花结果。搭架引蔓栽培能显著提高人参果植株挂果率和果实品质。

方法是用水泥柱、木桩或竹竿作为立柱，用铁线搭成与垄平行的架子，高

1.5～1.8 m，一垄一架，苗高 30 cm 左右时要及时吊蔓，每株选取 1～2 条健壮的枝条，用挂线夹夹住枝条，用线缠绕枝条并牵引上架，其余枝条全部修剪掉。此方法搭出的引蔓架稳固性和支撑性均较强，但人力和物力成本相对较高。或可直接以 3 根竹竿搭建成三角形作为每丛人参果植株的立柱，或仅以单根竹竿为立柱，用黏性较强的胶带直接将人参果枝条黏在立柱上，以上方法虽简便易行，但稳固性较差，易被风吹倒。

3.2.4.6 引蔓

当植株生长高度达到 30～35 cm（即开始倾斜）时，可开始引蔓，引蔓线可采用尼龙线、棉线等。随着秧蔓的伸长，每隔 10～15 d，将吊线呈"S"形缠绕枝条引蔓上架。引蔓方式与传统番茄引蔓相似，吊线不能一次收得太紧（植株秆径基本站直即可），因为随着植株生长，还要不断绕线引蔓，同时，线收得太紧，会对植株秆皮造成伤害，从而导致病菌感染，引起病害。另外，结扣应打活扣，以便线长不够时，放线引蔓。引蔓栽培有利于通风透光，减少病虫害获得更高的产量。

3.2.5 植株修剪

3.2.5.1 普通栽培修剪

人参果侧枝萌发力较强，栽培 60 d 后，植株长至 20 cm 左右，根据土地肥力情况，一般保留 4～6 个健壮结果枝，遵循土地肥力好则适当多留、肥力差则少留的原则；结果枝均匀分布，有利于通风透光。栽培过程中，根据实际情况及时打掉下部老叶、病叶，为人参果的发育提供充足的光照，良好的通风条件。

3.2.5.2 引蔓栽培修剪

春季修剪：当植株长至 20 cm 左右，人参果主要枝干保留 2～3 个健壮结果枝。其他同常规栽培。

疏果：待果实坐稳后进行疏果，选留果形好的大果，疏除小果、畸形果及病果，每个花序留 2～3 个果。

3.2.6 除草

以人工或机械为主适时清除杂草，尽量少用或不用化学除草剂，以免土壤累积残留物破坏生态和土壤结构。地膜未覆盖的区域和垄沟内的杂草应及时人工铲除或拔除。除墒面杂草用人工清除外，对路面及地边杂草可采用低毒、残留小、残留期短的除草剂进行除草。

3.3 水分管理

定植初期4月中旬至6月底雨季前，根据墒情，及时浇水，以保证植株生长对水分的正常需求。进入雨季后，雨水多时，做好排水工作，防止积水。进入10月，雨季结束，视墒情及时补充水分。

3.4 施肥管理

肥料使用按照《绿色食品　肥料使用准则》（NY/T 394—2023）标准执行。施肥遵循"有机无机结合、氮磷钾不同时期配合施用、适当增施钾肥、及时补充中微量元素"原则，提倡水肥一体化技术。每次采果后及时补充肥水，宜采用含氨基酸的高钾水溶肥。在人参果生产中所使用的肥料应对环境无不良影响，有利于保护生态环境，保持或提高土壤肥力及土壤生物活性，且应对人参果的营养、口感、品质和植物抗性不产生不良后果。在保障人参果营养有效供给的基础上减少化肥用量，肥料种类的选取以有机肥料为主，化学肥料为辅。

人参果整个生育期内追肥次数、追肥种类及追肥量根据土壤肥力、底肥施用情况及植株长势等确定，一般整个生育期内追肥4次左右。第一次追肥在人参果苗定植成活后1个月进行，结合浇水追肥1次，每亩施用氮磷钾（15∶15∶15）复合肥10～15 kg，兑水浇施；开花结果期，每亩追施高钾复合肥10 kg或硫酸钾8 kg，一般每采收1～2次果进行1次追肥。在坐果期也可以适当喷施叶面肥，主要以磷酸二氢钾和中微量元素肥为主。

3.5 病虫害综合防控

3.5.1 病虫害种类

人参果种植过程中主要病害有疫病、煤污病、黑斑病等；主要虫害有蚜虫、红蜘蛛、白粉虱、蓟马等。

3.5.2 防控原则

按照"预防为主，综合防治"方针，以农业防治为基础，综合运用化学防治、物理防治和生物防治措施，创造不利于病虫害发生的环境条件，减少各类病虫害所造成的损失。

3.5.3 农业防治

避免在低洼地和黏重土壤的地块种植，定植前耕翻土壤，高温晒垡 15～20 d；选用抗性品种及脱毒优质种苗，降低病虫初侵染源；合理密植，适时修枝整形、搭架，摘除老叶、病虫叶，改善通风透光条件，降低田间湿度；加强肥水管理，多施有机肥，合理配合施用氮、磷、钾肥，特别在植株生长的中后期少施氮肥，防止植株徒长；及时清除田间病株，集中烧毁，减少田间病菌量，控制病害扩散蔓延。

3.5.4 生物防治

在人参果整个生育期内，蚜茧蜂释放时间以大田内蚜虫处于点片发生时，每亩释放 1 000～1 200 头。

3.5.5 物理防治

利用昆虫的趋性，园内使用黄蓝板、性激素、杀虫灯等诱杀害虫。黄蓝板每亩放置 30～35 块，黄蓝板比例以 4∶1 为宜。

3.5.6 化学防治

农药使用按照 GB/T 8321 系列标准执行。

在采用农业、生物、物理防治仍不能达到防治效果的情况下，选用化学防治。人参果主要病虫害化学防治措施如表 2-5 所示。

表 2-5 人参果主要病虫害化学防治措施

种类		防治措施
病害	疫病	选用 80% 代森锰锌可湿性粉剂 500 倍液、64% 噁霉灵可湿性粉剂 400 倍液、58% 甲霜·锰锌可湿性粉剂 500 倍液、25% 嘧菌酯悬浮剂 2 000～3 000 倍液、687.5 g/L 氟菌·霜霉威悬浮剂 900～1 000 倍液喷雾均匀喷施，视情况间隔 7～10 d 防治 1 次，连续防治 2～3 次
	病毒病	目前病毒病防控有效策略为"病毒病药剂＋营养叶面肥＋传毒害虫杀虫剂"。苗期流行病株，用 20% 琥铜·吗啉胍可湿性粉剂 1 000 倍液、30% 毒氟磷可湿性粉剂 +5% 氨基寡糖素水剂＋叶面锌肥 600～800 倍液、60% 吗胍·乙酸铜可湿性粉剂 +2% 香菇多糖水剂 800～1 000 倍液等均匀喷施。间隔 1 周，喷施 1 次，持续用 3～4 次
	煤污病	多由刺吸式害虫蚜虫、白粉虱、介壳虫等害虫为害分泌蜜露所致，治虫防病是关键，可选用 10% 吡虫啉可湿性粉剂、25% 噻虫嗪可湿性粉剂、30% 啶虫脒可湿性粉剂等药剂喷施防控虫害；在煤污病点片发生阶段，选用 43% 甲硫·戊唑醇悬浮剂 1 000～1 500 倍液、26% 嘧胺·乙霉威悬浮剂 500～800 倍液、40% 多菌灵悬浮剂 600 倍液、50% 苯菌灵可湿性粉剂 1 500 倍液、50% 多霉灵可湿性粉剂 1 500 倍液、65% 甲霉灵（硫菌·霉威）可湿性粉剂 1 500～2 000 倍液，施用杀菌剂时最好与前面所述的杀虫剂一起混用，均匀喷施，间隔 10～15 d 喷雾 1 次，连续防治 2～3 次

（续表）

种类		防治措施
病害	黑斑病	发病初期，用75%百菌清可湿粉剂500～600倍液、50%异菌脲可湿性粉剂1 000倍液、25%吡唑醚菌酯悬浮剂1 500倍液、60%唑醚·代森联水分散粒剂1 000～1 500倍液、50%克菌丹可湿粉剂400倍液或70%代森锰锌可湿性粉剂400倍液，交替使用，均匀喷施，隔7～10 d防治1次，连续防治2～3次
	灰霉病	选用50%异菌脲可湿性粉剂1 000倍液、50%腐霉利可湿性粉剂1 500倍液、40%嘧霉胺悬浮剂1 200倍液、70%甲基硫菌灵可湿性粉剂700倍液，交替使用，均匀喷施，每隔7～10 d防治1次，连续防治2～3次
虫害	蚜虫	选用10%吡虫啉可湿性粉剂1 000～1 500倍液、20%啶虫脒可溶液剂3 000倍液、50%吡蚜酮可湿性粉剂2 500～5 000倍液喷雾，均匀喷施，根据虫口数量和防治效果，可连续用药1～2次
	红蜘蛛	发病初期，选用34%螺螨酯悬浮剂4 000倍液+43%联苯肼酯悬浮剂1 800～2 500倍液，螺螨酯可以杀卵、若螨和成螨，无内吸作用，应喷洒均匀，持效期30 d。可间隔7 d再用药1次 发病较重时，连续喷药2～3次彻底消灭卵、若螨和成螨。可用20%丁氟螨酯乳油2 000倍液+5%高氯·甲维盐乳油1 500倍液、20%阿维·联苯肼悬浮剂2 000～2 500倍液，间隔7 d再轮换用药1次，然后根据防治效果再酌情用药。丁氟螨酯可以杀卵、若螨和成螨，喷布均匀效果很好
	蓟马	使用25%噻虫嗪水分散粒剂800倍液喷雾，同时可与微乳剂类的阿维菌素混合使用。还可使用25%多杀霉素水分散粒剂2 500～3 000倍液、20%甲氰菊酯乳油3 000～5 000倍液或2.5%溴氰菊酯乳油2 500～3 000倍液，喷雾防治
	白粉虱	使用25%噻嗪酮可湿性粉剂1 500倍液喷施，配用2.5%联苯菊酯乳油2 000～3 000倍液；或用25%噻虫嗪水分散粒剂3 000～5 000倍液喷施，配用10%吡丙醚水乳剂1 000倍液或10%虱螨脲悬浮剂2 500～3 000倍液。在喷施药剂的同时，结合用22%敌敌畏烟剂进行熏烟处理
	茶黄螨	使用45%联苯·乙螨唑悬浮剂8 000倍液、20%阿维·联苯肼悬浮剂3 000～4 000倍液或5%噻螨酮乳油1 500～2 000倍液喷雾防治

3.6 采收

适时采收，人参果一般于定植后110～120 d开始采收，定植后140 d左右进入采收盛期。反季栽培采收时间一般为翌年1—5月。每穗花序中不同果实由于坐果时间不同，生长过程中营养分配不平衡，导致其成熟期不一致。当果实表面呈现明显的紫色条纹，果皮、果肉变成淡黄色时，即为采收适期。根据需要采收，需长期贮存或长途贩运的可适当提前采收，用于短期贮存、短距离贩运和就地销售的在完全成熟时采收。采收时戴上棉布手套，轻轻托起果实，用剪刀剪下，按大小进行分级，每个果实套上包装网，装箱。人参果生育期不断开花、不断结果，应分期成熟、分次采收。

参考文献

陈新党，2023. A级绿色食品人参果的种植技术[J]. 农业科技通讯（4）：223-225.

高凡娥，2019. 浅谈日光温室人参果栽培技术[J]. 农业开发与装备（1）：171.

黄兴龙，王丽芳，陈华兴，等，2022. 人参果搭架引蔓栽培果品品质提高关键技术[J]. 长江蔬菜（24）：59-61.

晏存柱，2019. 乐都区沟岔温室人参果的栽培技术[J]. 青海农技推广（4）：14-15.

云南省地理标志产业协会，2023. 西街口人参果引蔓栽培技术规程：T/YGIIA 003-2023[S]. 云南：云南省地理标志产业协会.

张丽芳，2022. 人参果露地搭架引蔓高效优质栽培技术[J]. 中国果树（6）：84-87，82.

张丽芳，黄兴龙，陈新党，等，2021. 石林县人参果产业现状、存在问题及发展对策[J]. 农业科技通讯（6）：47-50.

香蕉栽培关键技术

1 产业发展概况

香蕉（*Musa nana* Lour.）是热带亚热带地区重要的经济作物和粮食作物，是世界鲜果中产量、贸易量和贸易额最大的水果。世界香蕉种植的核心产区分布在亚洲、美洲和非洲。2020 年，世界香蕉种植面积 520.35 万 hm^2，产量 11 983.37 万 t，主要生产国为印度、中国、印度尼西亚、巴西和厄瓜多尔。据国家统计局的数据，2019 年，我国香蕉总种植面积 33.03 万 hm^2，总产量为 1 165.57 万 t，仅次于印度，是香蕉生产的大国。同时，我国也是香蕉进口的大国，2019 年，我国进口香蕉 194 万 t，仅次于美国，居世界第 2 位。2002—2019 年，我国香蕉的种植呈现"西迁"与"南移"的趋势，目前，我国香蕉的主产区主要分布于福建、广东、广西、海南、重庆、四川、贵州和云南 8 个省（区、市）。作为中国香蕉主产区之一的云南省，香蕉产业是云南高原特色现代农业的重要组成部分。2002—2013 年，云南省香蕉产业发展进入"快车道"；2014—2020 年，受到香蕉枯萎病、寒潮等因素的影响，云南省香蕉产业经历了一定的波动。截至 2020 年，云南省香蕉种植面积 8.24 万 hm^2，占全国种植总面积的 25%，全省香蕉产量为 197.64 万 t，占全国总产量的 18%，分别居全国第 2、第 3 位，其种植面积和产值均居云南省内各水果种类的第 1 位。香蕉产业已成为边疆少数民族农民增收致富的重要途径，在全省尤其是在边疆民族地区巩固脱贫攻坚成果、推进乡村振兴战略实施中起到关键作用，具有重要的政治、社会和经济意义。

2 种植条件

2.1 温度

香蕉是热带果树，生长发育要求较高温度，低温是限制香蕉高产的主要因子。抽蕾期和幼果期温度低于 13 ℃会使幼果受到冻害，1～2 ℃叶片会受到冻害。香蕉尤其害怕霜害，最适生长温度为 27 ℃，最高温不宜超过 38 ℃。

2.2 水分

香蕉属于大型草本植物，水分含量高，叶片宽大，蒸腾量大，因此需要充足的降水量来满足其水分需求。通常，每年的降水量应在 1 000～2 500 mm。

2.3 土壤

香蕉的根系是由球茎抽生出的肉质须根系，分布较浅，而且香蕉生育期短，需要在很短的时间内完成大量生物量的积累，因此要求土壤富含有机质，具有良好的排水性，以保持适度的土壤湿度和氧气供应。避免因过度积水导致根部窒息和病害的发生；土壤为中性至微酸性，通常 pH 值在 5.5～7.0；土壤深度为 60 cm 以上，地下水位以在 1 m 以下为宜。

2.4 种植区域

香蕉生长适宜区域为年均气温≥10 ℃，年积温 6 500 ℃以上，海拔为 90～600 m 的干热河谷和湿热地区。

云南的六大水系（伊洛瓦底江、怒江、澜沧江、元江、金沙江、南盘江）流域的河谷地区，约 8.1 万 km²，基本上都适宜香蕉的生产。目前，云南省内 14 个地州（市）共 60 个县均种植香蕉，其中，红河州、西双版纳州、文山州和普洱市 4 地的种植面积和产量分别占全省总面积、总产量的 86% 和 87% 以上，是云南省香蕉的主产区。

3 主要栽培品种及特性

巴西蕉和威廉斯 2 个品种在云南省的栽培面积占全省总栽培面积的 80% 以上。

3.1 巴西蕉

巴西蕉原品种名为 Nanicao，1989 年由广东省湛江市农业生物技术研究中心通过澳大利亚昆士兰引进试管苗，属于矮把香芽蕉的变异种。植株高大（图 2-1），假茎高 240～330 cm，主干基部粗 81.7 cm，中部粗 54.5 cm，叶序浓密，叶片开张，剑形，长 210～212 cm，宽 95～102 cm；花序由紧密排列的花朵组成，花序通常位于植株的顶部；果穗长 85.5 cm，果梳间较稀疏（图 2-2），果指长 20～23 cm，单株产量为 18～34 kg，果实可溶性固形物含量 18%～21%，香气浓郁。由于单株产量高，果实品质好，适应性强，具有较强的耐寒性和抗旱性，在云南省广泛栽培，对肥水管理要求严格。

图 2-1　巴西蕉植株形态

图 2-2　巴西蕉果穗形态

3.2　威廉斯

据报道，威廉斯是裴济在 19 世纪初从中国台湾引入北蕉后，从其突变单株中选育出的品种，后引种到澳大利亚成为当家品种，20 世纪 80 年代后引入云南省。威廉斯属于中秆偏高的类型，植株高大（图 2-3），抗风力弱，其假茎高 250~300 cm，但秆较细，茎围 80~90 cm。叶子呈大型扇形，叶片青绿而宽阔，直立；果穗较长，梳间较稀，果数不多，但排列紧贴，果形较直，果实总糖 18%~21%，香味较浓；单株产量 20~30 kg，高达 43 kg，果指长 20~24 cm，呈黄色，果肉柔软，口感香甜，果穗整齐，形态美观，适应性强。

图 2-3　威廉斯香蕉植株形态

3.3 桂蕉 6 号

桂蕉 6 号是 1987 年从澳大利亚以试管苗方式引进的香蕉品种（品系）材料，经在国内组培快繁，筛选组培变异株而育成的优良品种，属香芽蕉类，基因型为 AAA。该品种春植、夏植、秋植、冬植生育期分别为 300～420 d、360～400 d、360～400 d、330～390 d。组培苗第一代假茎高 220～260 cm，假茎基部围径 75～95 cm，假茎中部围径 48～65 cm，茎形比为 3.7～4.3。每穗果梳 7～14 梳，每梳果指 16～32 个，每穗果实重 20～30 cm，每亩产量 2 400～3 500 kg。果穗整齐美观，稳产高产，品质优良，适应性强。抗风力中等，不耐霜冻，易受尖孢镰刀菌古巴专化型 4 号生理小种侵染而感染香蕉镰刀菌枯萎病，植株易感香蕉花叶心腐病、香蕉束顶病。

3.4 金粉 1 号

金粉 1 号是广西壮族自治区农业科学院和广西兴旺组培苗公司自主选育的第一个粉蕉品种。植株高大，假茎有光泽，分布少量褐色斑点，黄绿色，高 460～500 cm，基部围径 94～101 cm，中部围径 61～67 cm，茎形比 7.6～8.2，生果皮浅绿色，部分果指有少量不均匀蜡粉，生果肉白色。果穗 10～15 梳，每梳果指 18～20 个，每梳果实两排排列。单株产量 20～30 kg，最高可达 45 kg，每亩产量 2 000～3 000 kg。果实成熟后为金黄色，果肉白色，口感软滑香甜；抗寒抗风能力强，耐轻霜，不抗重霜雪；易感镰刀菌、枯萎病，抗黑星病、束顶病和花叶心腐病，易招引卷叶虫和香蕉象甲，较抗旱但不耐涝；果皮薄，长时间长距离贮运不利于保持商品性。

4 栽培管理技术

4.1 种植规划与园区建设

4.1.1 园址选择

选择正常年份全年无霜冻，空气流通较好，地势开阔，背北向南、背风，坡度 <15°，具有防风林，交通良好的适宜区种植香蕉。1 000 亩以上的大型蕉园，根据不同小环境的地形、地质和蕉类的品种特性划分种植区，并建立品种母本园和种苗基地。

4.1.2 园区道路规划

园区内规划主干道和分支道。主干道用于大型机械通过，路基应宽 3～4 m，根据香蕉品种的秆高和机械类型适当调整主干道宽度，使得机械不会碰触两侧的香蕉植株。规划分支道使小型机械可以顺利通过，可根据园内所拥有小型机械的尺寸进行规划，路基宽度以机械不会擦碰两侧植株为宜。

4.1.3 排灌系统

云南省干、湿季节分明，冬、春季干旱，夏、秋季多雨，蕉园建设必须有排灌的水利设施，一般3～5亩的园地，要有一个容积10 m³的蓄水池和不渗漏的沤粪池。梯地蕉园在园地上方挖一条宽70 cm、深50 cm的截洪沟，每个坡面顺坡开宽60 cm、深40 cm排灌兼用的水沟。平地蕉园沿道路两旁修建互相垂直的纵沟和支沟；有条件的园地，可建设滴灌或微喷灌系统。面积较大的园地，应配建主混肥池和分混肥池，主混肥池建于主灌溉泵旁，方便与水混合灌溉，分混肥池建于管理岗位附近，方便混水肥与沤肥操作。

4.1.4 索道

坡台地可架设香蕉采收拱架式索道和灌溉管网，利用索道的龙骨管作为灌溉主管道，可节省成本。索道可以安装成上下坡的单（双）轨运输线。

4.1.5 整地

在云南省，香蕉种植园多为坡度≥5°的坡地，多数种植区地形地貌复杂，高度差大多超过200 m。因此，种植前，坡度＜25°的坡地，需要环山等高开出台地，台面宽1～1.6 m，向内倾斜5°，向外抬高15～20 cm，边沿筑成宽20～25 cm的小埂，沿台面开宽60～70 cm、深50～60 cm的种植沟或种植穴。株距以台面宽度做适当调整，以每亩种植100～130株为宜。平地园区旱地蕉园采用沟+膜下滴灌栽培方式。种植沟宽1～2 m、深40～50 cm，起垄。

4.1.6 定植

云南省香蕉种植，一般在春季或秋季进行，其中春季种植更佳。春季种植香蕉可在2月初至4月中旬定植，9月至翌年春季抽蕾，2—6月收获。秋季种植在8—10月进行，翌年8—12月收获，可赶在中秋节或国庆节上市。定植时，土壤不能过干也不能过湿，要选择在阴天或16:00以后进行，栽种时不要伤到根系。蕉苗定植后，浇足够的定根水。待蕉苗成活长出1～2片新叶后，喷施0.1%～0.3%磷酸二氢钾、氨基酸、海藻酸等水溶性有机肥。

定植后，应经常检查蕉苗存活情况，及时去除死苗并补种。巡查虫害发生情况，发现斜纹夜蛾、交脉蚜为害要及时喷药防控。蕉苗定植成活初期，定期进行人工除草，也可进行行间覆膜，以抑制杂草生长。

除芽与留芽：为保障母株的健硕，每个植株只能保留两个接班芽，其余萌发的芽，从地面彻底根除，并挖去中间的生长点，以免再次生长。4—8月，每半月除芽1次，其他季节视芽生长情况进行除芽。

4.2 水肥管理

香蕉是速生高产草本植物，生长量大，即便在肥沃的土地上种植，也需要合理施肥，才能保证果实的产量和质量。

4.2.1 水分管理

整个生长季，保持园地土壤湿润，苗期至花芽分化期土壤湿度为土壤田间持水量的65%～75%，挂果期为土壤田间持水量的50%～60%。抽蕾期需要水量急剧增大，应适当增加浇水的次数。挂果中后期适当控制土壤湿度，采收前7～10 d不再浇水。

4.2.2 施肥管理

香蕉生长发育需要的关键矿质元素包括钾、氮、磷、钙、镁、硫、氯，需要的微量元素包括锰、铁、锌、硼、铜、钼。钾对香蕉的产量和质量至关重要，因此对钾肥的需求量大，氮：磷：钾（N：P_2O_5：K_2O）为1：（0.2～0.5）：（2～4）；此外，香蕉对氮、钙、镁吸收比为1：0.69：0.2，对钙、镁肥的需求量也大。

4.2.2.1 基肥

用作改良土壤理化结构、改善土壤的通气性，从而为香蕉整个生长期打下良好的基石。基肥多用有机肥，以腐熟猪牛鸡粪或饼肥为主，辅以复合肥、过磷酸钙等，每株施5 kg农家肥、0.1～0.5 kg钙镁磷肥和0.2～0.5 kg复合肥；或者也可以施1～2 kg生物有机肥、0.3 kg复合肥。施基肥时，要将所用的肥料与土壤充分混合后回填种植沟或种植穴。

4.2.2.2 花芽分化肥

吸芽苗20片叶龄、组培苗28片叶龄时，即假茎高度为1.5～2.0 m时，每株施农家肥5～7 kg＋复合肥0.2 kg＋钾肥0.2～0.5 kg。利用撒施的方式进行追肥，每20 d，每株撒施复合肥0.1 kg、钾肥0.05 kg。抽蕾前撒生石灰4～5次，每亩约15 kg。抽蕾后，每隔20 d，每株可撒施复合肥（氮：磷：钾=15：15：15）0.1 kg，撒施后淋溶。

4.2.2.3 挂果期施肥

挂果期蕉株根系老化,此时采用暖性的水溶性有机肥进行叶面喷施,不宜喷施氮肥,以补充钾肥和微量元素肥为主。

4.3 主要病虫害防控

4.3.1 香蕉枯萎病

香蕉枯萎病是尖孢镰刀菌古巴专化型4号生理小种热带型和1号生理小种引起的一种真菌性病害,又称香蕉巴拿马病、香蕉镰刀菌枯萎病,是全球香蕉产业发生最严重的病害。该病目前不能通过单一的技术进行防控,需采用多位一体的综合防控手段进行防控。

症状:病原菌在土壤中生存并侵入香蕉根系,引起根部病变。感染后叶片出现黄化,并逐渐蔓延到整个植株,导致植株黄化。由于病原菌侵入导致木质部组织受损,假茎和叶鞘维管束组织呈现连续的褐变。随着病情的发展,假茎开裂或腐烂,整株枯死。

发病规律:香蕉枯萎病的潜伏期长,可持续数个月至1年以上;一般全年都有可能发生,10月至翌年2月较为严重。初期感染,病原菌侵入根系后,通过木质部组织向上蔓延;其后病原菌在茎内不断繁殖和扩散,导致茎部组织的衰败和根系失去功能。病害会逐渐扩散到整个植株,导致植株的全面枯萎。病原菌可产生大量的大分生孢子、小分生孢子和厚垣孢子,厚垣孢子在土壤中可存活几年至十几年,病原菌可以通过受感染的土壤、农具、水源和害虫等媒介传播到健康植株上。

防控方法:预防为主,综合防治。①选择抗病性较强的香蕉品种进行种植是预防香蕉枯萎病的关键。种植前检测土壤中尖孢镰刀菌的分生孢子数量,每克土壤孢子数量在10^6个以上时,现有品种基本丧失抗病性,不能再种植香蕉。根据市场情况,选择合适的抗病品种进行种植,目前常见的抗性品种有南天黄、宝岛蕉、桂蕉2号、桂蕉9号、中蕉8号等。②种植园和采后处理厂实行分区管理。蕉园车辆出入口设置消毒池(0.3%高锰酸钾液),人员通道铺设消毒后的地毯。③禁止到疫区调苗,定植时选用无土基质或消毒基质育的种苗或一级组培袋装苗可有效减少发病率。④在已发病的园区种植香蕉,在种植前,采用热处理、化学消毒剂处理等方法对土壤进行消毒处理,以减少病原菌的存在。⑤有条件的产区,避免在同一地块连续种植香蕉,进行轮作休耕;对已经感染的植株进行隔离,并及时清除和销毁受感染的植株,以防止病害的进一步传播。

4.3.2 香蕉灰纹病

香蕉灰纹病是由真菌暗双孢属香蕉暗双孢菌 [*Cordana musae*(Zimm.) Hohn.] 引起的病害。发病初期，下部叶片开始出现暗褐色或灰褐色病斑，半圆形、椭圆形或沿叶缘呈不规则形，大小不一，周缘浸润状；后扩展成两端稍尖的长椭圆形大斑，轮纹状，中央灰褐色至灰色，边缘深褐色且有明显的亮黄色晕圈。随着病程的发展，病斑逐渐向上部叶片扩展，湿度大的时候，叶背面的叶斑有灰褐色霉状物，即病原菌分生孢子梗和分生孢子。香蕉灰纹病菌可通过风、雨水和农具等途径传播。高温和高湿的环境条件，有利于病原菌的繁殖和传播，6—9月雨水季节处于生长期和抽蕾期的香蕉受为害较严重。对于该病害的防治，常用药剂包括三唑类（丙环唑、苯醚甲环唑和戊唑醇等）、嘧菌酯、吡唑醚菌酯、多菌灵、百菌清和代森锰锌等。此外，也可以使用特定的有益微生物，如拮抗菌（如拟青霉菌）或寄生性真菌（如果胶孢）来抑制病原菌的生长和繁殖。

4.3.3 香蕉炭疽病

香蕉炭疽病是由真菌（*Colletotrichum musae*）引起的采后病害。病原菌感病初期，叶片上出现长椭圆形病斑，后期叶片逐渐布满小黑点。青果受害时，果皮上出现长椭圆形的黑褐色病斑，病斑上分布有许多小黑点，但青果受害的情况并不多见，多为熟果受害，病斑黑褐色，呈梭形，中央纵列，部分品种病斑较小，褐色。香蕉炭疽病主要通过风、雨水途径传播。高湿、高温的环境条件有利于病原菌的生长和传播。

防治香蕉炭疽病，可在果实套袋前防治2～3次，间隔5～10 d，可选择的药剂有多菌灵、甲基硫菌灵、咪鲜胺和嘧菌酯等。药液着重喷在果实和附近的叶片上，以喷湿不滴水为宜。

4.3.4 香蕉黑星病

香蕉黑星病病原为香蕉大茎点霉 [*Macrophoma musae*(Cooke)]，属半知菌亚门真菌。主要为害叶片和果实，一般从叶片开始。叶片：在叶片和中脉处产生许多散生突起粗糙的小黑粒，后期小黑粒周围呈淡黄色，黑星布满整片叶，叶片变黄而枯萎。老叶比新叶较易感病，病叶提早凋谢。果实：当病菌为害果实时，病斑由果轴起向果柄发展至果实内弯，因而内排果比外排果严重。起初在果皮出现许多小黑粒，到发病后期，病斑遍布整个果实。果实成熟时，病斑周围形成褐色圆形小斑，中部组织腐烂下陷，其上的小黑粒突起。

高湿高温的条件有利于病原菌的生长和传播。6—9月雨水季节，抽蕾后的植株受害较严重。对于该病害的防治，常用药剂包括三唑类（丙环唑、苯醚甲环唑和戊唑醇等）、嘧菌酯、啶氧菌酯、氟硅唑、吡唑醚菌酯、多菌灵、百菌清和代森锰锌等。

5 采收与采后处理

5.1 采收

5.1.1 适时采收

根据果实棱角的变化，确定成熟度（表2-6），成熟度70%~75%时采收最佳。也可根据抽蕾的时间确定果实的采收时间，春、夏季抽穗的植株，抽穗后80~90 d采收；秋末、冬季抽穗的植株，抽穗后120~150 d后采收。采收时，避免在高温和正午时段，最好在阴天或晴天上午进行采收。

表2-6 成熟度判断方法

	采收成熟度	果指特征
夏、秋季	70%~<75%	接近饱满，棱角非常明显
	75%~<80%	基本饱满，棱角明显
	80%~<90%	圆且饱满，但尚见棱角
冬、春季	75%~<80%	基本饱满，棱角明显
	80%~<90%	圆且饱满但尚见棱角
	90%以上	圆且饱满，基本无棱角

5.1.2 砍蕉

多人配合采果，一人在假茎中部以下将假茎砍倒，另一人或两人托起果穗，在离头梳果顶25~30 cm处砍断果轴，将果穗装放到专用架，运输到蕉园索道或轨道运输线或道路边。

5.1.3 田间运输

有条件的果园一般采用索道运输，人力将香蕉运到索道旁后，用绳子在轴上套牢，然后将果穗挂到潜车挂钩上。果梳间用隔杆分隔开，5~10个果穗形成一组，通过索道统一运输至果园果实处理场所。也有的果园使用拖车运输，在车厢底部垫上一层海绵保护，单层装上果穗运输到果实处理场所。

5.1.4 落梳

香蕉运输至采后商品化处理场所后，将果穗放在落梳架上去轴落梳。落梳期间摘除果指残存花器，用专用下把刀将蕉梳从果穗轴上切下，放入清洗池。

5.1.5 清洗和修把

将蕉梳放入含有 0.1%～0.2% 的明矾溶液（除臭、杀菌和止蕉汁的作用）或清水低压清洗池中循环冲洗，清洗掉果指表面的污垢和蕉汁。清洗的过程中，将切口修整平滑，清除烂果、裂果、反梳果等不合格果。

5.2 分级称重

按照客户要求，参考《香蕉》（GB 9827—1988）、《香蕉等级规格》（NY/T 3193—2018），以及云南省地方标准《地理标志产品 河口香蕉》（DB53/T 912—2019）进行蕉果分级称重。

5.3 保鲜处理

长距离运输的商品，需要预先进行保鲜处理。将蕉梳进行悬挂喷洒或浸泡（10～30 s）保鲜液处理。严禁滥用保鲜剂，选择安全、高效、低毒、低残留的专用防腐杀菌药剂，如咪鲜胺、异菌脲或噻菌灵等杀菌剂。蕉梳放入保鲜池后，药液应浸过果面。保鲜剂要求即配即用，48 h 更换 1 次。为减少运输、贮藏期间炭疽病和冠瘿病的发生，保鲜处理后需要风干。

5.4 包装

风干后，蕉梳之间用吸水纸或珍珠棉隔开，根据客户的要求贴上标签、将蕉梳装入塑料袋（塑料袋预先放入纸箱内）—抽真空扎袋口检查是否漏气—盖上纸盖—封箱—贴商标—分批送入气调房或冷库预贮藏。外包装及包装材料应符合国家和行业等相关标准的要求，产品包装标签应标注产品名称、净含量、保质期、产品执行标准号、包装日期、产地、生产单位名称等。

5.5 长距离运输

在云南省,香蕉的运输主要靠汽车,装车的车厢和四周要垫上软垫,夏季要防高温,冬季要防冻。如有气调运输的条件,把温度调节在 13～14 ℃,相对湿度以 90%～95% 为宜。

参考文献

林欣源,姜雯欣,徐绍荣,2022. 中国香蕉种植面积的时空变迁及影响因素 [J]. 江西农业学报,34(7):218-223.

邹冬梅,范琼,2022. 世界香蕉生产、贸易现状与产业展望 [J]. 广东农业科学,49(7):131-140.

邹瑜,林贵美,牟海飞,等,2011. 粉蕉新品种"金粉 1 号"的选育 [J]. 中国南方果树,40(1):47-48.

杧果栽培关键技术

1 产业发展概况

我国从 1986 年开始规模化种植杧果，30 年时间杧果迅速发展成为仅次于荔枝、龙眼和香蕉的第四大热带水果。主要种植在广西、云南、海南、四川、台湾、贵州、广东、福建等省（区）。2021 年全国杧果种植面积 561.9 万亩，产量 395.8 万 t，分别同比增长 9.1% 和 19.5%。

1.1 云南省情况

1.1.1 自然分布情况

云南省具有生产优质杧果的生态自然优势，杧果种植面积居全国第 1 位，同时也是全国杧果上市期供应最长的省份，供应期为 5—11 月，长达 7 个月。2021 年全省种植面积 172.1 万亩，产量 132.9 万 t，分别同比增长 13% 和 38%。

云南省是我国杧果自然分布最广、种类最多的一个省份，我国有 6 个种杧果，而云南省就有 5 个种，即杧果（*Mangifera indica* L.）、泰国杧（*Mangifera simensis* Warby. ex Craib）、林生杧（尼泊尔杧）（*Mangifera sylvatica* Roxb.）、长梗杧（*Mangifera longipes* Griff.）、桃叶杧（*Mangifera persiciforma* C.Y. Wu et T. L. Ming），分布最广的是杧果（*Mangifera indica* L.），自然分布在云南省 9 个地州（市）46 个县（市）区域。

1.1.2 种植情况

云南省杧果种植分布广泛，从东部文山州至西南的德宏州、滇东北的巧家县至最南边的西双版纳州，97°51′～105°38′ E，21°29′～26°53′ N，海拔在 76～1 700 m 的广阔热带亚热带地区都有杧果种植。云南省下辖 16 个地州（市）129 个县（市、区），目前杧果种植分布有 15 个地州（市）91 个县（市）区域。主栽品种有台农 1 号、贵妃、圣德隆、大赛黄心芒、帕拉英达、椰香、南逗芒 4 号、金煌、凯特、圣心和红象牙等。

2 主要栽培品种及特性

2.1 帕拉英达（PaLa Hin Tha）

2.1.1 品种来源

缅甸。

2.1.2 品种特点

中熟。

2.1.3 品种特性

树势中等，树冠为圆头形。果实平均单果重 260.2 g。果实近似象牙形，成熟果皮黄色，果粉中等，果实成熟期在 6 月下旬至 7 月下旬，果实发育期 90～100 d，果实收获期 30 d，果实耐贮性 15 d，味香甜，食用品质好。

2.1.4 分布区域及适宜范围

主要分布在隆阳区、元江县，最适宜在海拔 800～1 200 m 的区域种植。

2.1.5 优缺点

优点：①高产稳产；②外形美观；③耐贮运，货架期长；④耐旱。
缺点：易受蓟马侵害。

2.2 金煌（Jin Huang）

2.2.1 品种来源

金煌由中国台湾的黄金煌先生选育而成，是白象牙和凯特的杂交后代。

2.2.2 品种特点

中熟。

2.2.3 品种特性

树势直立，长势强壮。果实硕大，果实象牙形，平均单果重650～1 500 g，最大单果重2 400 g。果肩较小，向阳面有红晕，果顶圆钝，果柄粗长，果皮光滑，果点稀少，果皮薄，果窝微凸，果肉黄色，肉质致密、嫩滑、多汁，果肉纤维细、少，味甜，风味浓郁。在云南省每年12月开始花芽分化，2月下旬进入盛花期，7月下旬成熟。

2.2.4 分布区域及适宜范围

全省皆有种植，总体面积不大，最适宜在海拔800～1 200 m 的区域种植。

2.2.5 优缺点

优点：①外观漂亮；②果甜汁多；③可食率高；④早结丰产；⑤货架期长；⑥对蒂腐病、白粉病、炭疽病抗性较强。

缺点：①不耐霜冻；②易感水泡病。

2.3 台农1号（Tainong No.1）

2.3.1 品种来源

台农1号杧果原产中国台湾，由台湾省农业试验所凤山热带园艺试验分所用海顿（Haden）与爱文（Erwin）杂交选育而成。

2.3.2 品种特点

早熟。

2.3.3 品种特性

树冠圆头形，生长旺盛；两性花比例46%，有二次开花习性；果实宽卵形，平均单果重250～300 g；果肩稍斜平，果腹凸起，果背弯斜，果窝浅，果喙大而钝；果皮绿底泛红色，完熟的果实黄色，果肩半部带红色。果肉深黄色至橙黄色，肉质嫩滑，纤维极少，甜香味特浓，品质极佳。果实采收期7月上中旬。在云南保山抽蕾期为1月中旬，高接换种的树一年可抽生5～6次梢，采果修剪后也能抽生1～2次秋梢；开花期为2月下旬至3月中旬，采收期为7月上旬，从谢花到果实成熟需要90～110 d。

2.3.4 分布区域及适宜范围

全省皆有种植,主要分布在元江县、红河县,最适宜在海拔 800～1 200 m 的区域种植。

2.3.5 优缺点

优点:①植株矮小,管理方便;②外形美观,品质好;③高产、稳产,耐贮运,抗炭疽病。

缺点:对白粉病抵抗力弱。

2.4 圣德隆(Seintalone)

2.4.1 品种来源

缅甸。

2.4.2 品种特点

早熟。

2.4.3 品种特性

树姿开张,树形伞形,主干光滑、灰褐色。果实扁球形,成熟果皮黄色,果粉中等,果肉金黄色。果实成熟期在6月中旬至7月中旬,果实收获期30 d,果实耐贮性8 d,味香甜,食用品质好。

2.4.4 分布区域及适宜范围

主要分布在隆阳区、永德县、元江县、红河县,最适宜在海拔1 000 m以下的区域种植。

2.4.5 优缺点

优点:①外形美观;②香气浓、品质好;③稳产,耐贮运,抗炭疽病。
缺点:对细菌性黑斑病抗性较弱。

2.5 南逗芒 4 号（Nam Doc Mai No.4）

2.5.1 品种来源

泰国。

2.5.2 品种特点

中熟。

2.5.3 品种特性

是泰国经过实生树筛选得到的优良品种，树势较强，树冠圆头形，枝条粗壮、生长快。花期为 11 月中下旬至翌年 2 月上旬，具有反季开花特性，花序塔形至圆锥形。果实发育期 100~120 d，果实成熟期为 7 月上旬至 8 月上旬；平均单果重 300~400 g，果实成熟时果皮橙黄色，蜡粉层薄，味甜汁多。

2.5.4 分布区域及适宜范围

零散分布在隆阳区、永德县、元江县，最适宜在海拔 800~1 200 m 的区域种植。

2.5.5 优缺点

优点：①高产稳产；②外观漂亮；③果肉细腻、多汁、香甜可口。
缺点：易感白粉病、细菌性黑斑病。

2.6 贵妃（Gui Fei）

2.6.1 品种来源

中国台湾果农张铭显用爱文 × 白象牙杂交选育品种。

2.6.2 品种特点

早熟。

2.6.3 品种特性

贵妃又名红金龙，树势健旺，树冠圆头形。果实 4—5 月成熟，通常平均单果重 400~800 g。果实长卵形，果形指数（长:宽）约 1.74。果实纺锤形，无果肩，果腹凸出，果背平，果喙点状。果皮极薄，成熟后深红色，光滑，色泽艳丽。果肉金黄色，肉质细腻，多汁，纤维少，味清甜可口，风味纯正。果实发育期为 100～110 d，花期为 2—3 月，果实成熟期 6 月上旬至 7 月中旬，集中在 6 月下旬。

2.6.4　品种分布及适宜区域

主要分布于元江县、红河县，最适宜在海拔 800 m 以下的区域种植。

2.6.5　优缺点

优点：①果形与色泽美观；②香气浓郁，味甜多汁；③抗寒性较强；④抗疮痂病。
缺点：①皮薄，耐贮性差；②花期遇低温易产生无胚果。

2.7　大赛黄心芒（Dasai yellow flesh）

2.7.1　品种来源

从怒江流域大树野生杧果资源中选育而来，在正常的栽培管理下，从播种到挂果只需 3 年时间，该品种因高产、稳产、果肉黄而得名。

2.7.2　品种特性

树势中等，树冠圆头形，主干粗糙、灰褐色。平均单果重 135～265 g，果实近肾形，成熟果皮金黄色，果粉中等，果皮厚 0.131 cm；果肉金黄色，种仁椭圆形，可食率 80.56%；花期 2 月上旬至 3 月中旬，果实成熟期在 5 月中旬至 7 月上旬，果实收获期 30 d，果实耐贮性 4 d，味香，风味好，品质中上等。具有早产、早熟特性，有明显的大小年结果现象，不耐贮运。嫩叶和花易感白粉病和炭疽病，果实易感果实蝇。

2.7.3　品种分布及适宜区域

主要分布于景谷县、隆阳区，最适宜在海拔 800～1 200 m 的区域种植。

2.7.4　优缺点

优点：①早产、早熟特性；②果肉纤维多、长；③味酸甜，适宜加工；④香气浓，复合香味。
缺点：果皮薄，耐贮性差；易吸引果实蝇，易感炭疽病。

3 栽培管理技术

3.1 园区建设

3.1.1 园地选择

选择向阳，土层深厚、地下水位低，排水良好，土壤为微酸性至中性的壤土和砂壤土，花芽分化期天气适当干旱，水源充足，排灌方便。规模化的杧果园最重要的还是必须建立在适宜的生态气候区，且生产和运销条件好。以投资省、花工少、收效快、收益大为原则。

3.1.2 园地建设

杧果园地规划包括果园分区、梯田、防护林、排水系统及道路的规划设计。园地可按气候和地形特点分区，一般小区以防护林为界。因地形地势不同，每小区以15~30亩为宜。平地可用"十"字定标法测量规划，按"井"字形以2∶1至4∶1的长方形设计。在坡地，长边应沿等高线走向，可采用基线定标法进行等高测量规划，以利于耕作及防止冲刷。在平缓地带，最好东西长、南北短（风大或台风多的地区向风面短）。总之，园地的规划要因地制宜，根据地形、地势、土壤条件等来规划设计。

3.1.3 园地开垦

建园时，把杂草、树木清除干净后，进行全垦，充分犁耙，全面垦耕2次，犁深40 cm，使土壤细碎松散，翻晒土地后栽种。

3.2 整形修剪

3.2.1 采后修剪

疏去过密枝、弱枝、下垂枝、衰老枝、交叉枝，回缩短截先端的衰退枝、过长的营养枝。

3.2.2 秋季修剪

疏去内膛枝、交叉枝、弱枝及过密枝；疏去树冠中上部无效枝。

3.3 套袋

3.3.1 套袋前准备

疏果：套袋前疏除病虫果、畸形果等，确保每个花穗留果3～5个。

病虫害预防：套袋前做好细菌性黑斑病及蓟马的预防。

3.3.2 材质与规格

套袋材料：泡沫网袋、尼龙网袋。

袋子规格：泡沫网袋为12 cm×7 cm或14 cm×7 cm，尼龙网袋为45 cm×30 cm。

3.3.3 套袋时间

二次生理落果结束1周内，选择晴天进行。

3.3.4 方法

首先用泡沫网袋把单个果实完全包裹住，然后用尼龙网袋把整穗果实完全包裹，袋口绑紧。

3.4 施肥管理

3.4.1 幼树管理

3.4.1.1 施肥

每年施2～3次复合肥（15∶15∶15），每株施复合肥0.8～1.5 kg，定植后翌年开始增施有机肥，每株施5～10 kg，采用开沟（环状沟或者条状沟）施肥，施肥后覆土。

3.4.1.2 树形培养

培养成圆头形树冠。剪除过密枝，疏去弱枝；一年抽梢3～4次，通过摘心、抹芽、短截、拉枝整形等修剪方式，及早形成早结丰产树冠。

3.4.1.3 田间管理

建立生态型果园，控制杂草高度<30 cm，清除枯枝落叶。

3.4.2 结果树管理

花芽分化期：每年11—12月为花芽分化期，每株施氮钾混配肥1～2 kg，混配肥比例（氮∶钾）为1∶1。

果实膨大期：3—4月为果实膨大期，每株施混配肥0.3～0.5 kg，混配肥比例（氮∶钾）为1∶1.5。

采果肥：果实采收后每株施有机肥20～30 kg、尿素0.8～1.0 kg、氧化钾0.5～1.0 kg、过磷酸钙0.5～1.0 kg。

3.5 主要病虫害防控

3.5.1 主要病害

3.5.1.1 杧果炭疽病（*Colletotrichum gloeosporioides* Penz.）

属半知菌亚门胶孢纲盘长孢状刺盘孢菌。

【为害情况】该病是世界上种植杧果的国家和地区发生最普遍的一种真菌性病害，主要为害杧果叶片、枝条、花序和果实。可引起叶斑、茎斑、花疫、果实糙皮、污斑和腐烂等症状。

【发生规律】主要在杧果树上的病枝、病叶、病果和落地的杧果病残体上存活越冬。高湿的气象条件下，病菌可产生大量的分生孢子，通过风、雨水进行传播，从寄主的伤口、皮孔和气孔等途径侵入，在杧果嫩叶上穿过角质层直接侵入。杧果炭疽病菌再侵染能力较强，在寄主残体上可存活2～3年。杧果炭疽病的发生和流行与气象条件、树势、田间管理、品种抗性、物候期、田间菌源基数及通风透光条件等有着极为密切的关系。

【防治方法】可用多菌灵、代森锰锌、吡唑醚菌酯、咪鲜胺、唑醚·氟酰胺、苯醚甲环唑、嘧菌酯。幼果期禁止使用乳油。施药时间：花期每周喷药1次，药剂选用百菌清、多菌灵、代森锰锌等保护性药剂。结果期每隔半月喷1次，可轮换使用治疗性药剂，如吡唑醚菌酯、嘧菌酯、苯醚甲环唑等。

3.5.1.2 杧果细菌性黑斑病（*Xanthomonas campestris* pv. *mangiferae indicae*）

【为害情况】此病在我国广东、云南、广西、福建等产区均有发生。国外南非、印度、巴基斯坦和巴西等国亦有报道。主要造成早期落叶，果面上形成黑斑。贮运中也可为害。

【发生规律】病菌主要在受侵染的枝梢或病残组织上越冬。翌年当温、湿条件适宜时，病部溢出菌脓，通过雨水或昆虫传播，从寄主的伤口或自然孔口侵入。高温、多雨、潮湿的天气，尤其是暴风雨，有利于该病的发生和传播。种植于风口处或山顶处的杧果园易发生该病。

【防治方法】施药时间：春梢期每隔10～15 d喷1次药，连续喷药3～5次，特别是暴风雨前后要及时喷药。可选用的药剂有波尔多液、春雷·王铜、噻森铜、枯草芽孢杆菌、氢氧化铜、春雷·喹啉铜等。

3.5.1.3 杧果白粉病（*Oidiun mangiferae* Berthet）

【为害情况】此病在我国广东、海南、云南、广西等产区均有发生。国外印度、巴基斯坦、巴西、南非、澳大利亚、斯里兰卡等国亦有报道。主要为害花序、幼果、嫩叶及嫩枝，引起大量落花落果。

【发生规律】低温高湿易引起白粉病的为害，特别是冬、春季阴雨连绵容易暴发此病，主要为害嫩叶、花及幼果。

【防治方法】用70%甲基硫菌灵可湿性粉剂300～500倍液、20%三唑酮乳油1 000～2 000倍液、50%硫磺胶悬剂200～400倍液、70%甲基硫菌灵可湿性粉剂600～800倍液、10%苯醚甲环唑水分散剂800～1 500倍液、12.5%烯唑醇可湿性粉剂2 000倍液、25%吡唑醚菌酯乳油1 500倍液，间隔期15～20 d，不同药剂交替使用。幼果期使用乳油会灼伤果面。

3.5.2 主要虫害

3.5.2.1 蓟马

蓟马是缨翅目昆虫的统称，为害杧果的蓟马种类将近20种，其中比较常见的有茶黄蓟马（*Scirtothrips dorsalis* Hood）等近14种。

【为害情况】为害时间：全年皆可发生，发生有明显的高峰期，发生高峰期与杧果的物候期关系密切，主要在杧果花期、幼果期及嫩梢期为害较大。为害方式：蓟马以锉吸式口器插入杧果组织内，吮吸杧果汁液，成虫和幼虫聚集为害杧果的嫩梢、嫩叶、花期和幼果。为害症状：成虫、若虫为害杧果树小苗叶片，使其呈现无数污黑斑点、叶尖变黑、叶缘卷曲，最后叶片全部落光，整株枯死。

【防治措施】从花穗期至果实第二次生理落果前和每次新梢抽生出3 cm至叶片转绿前，可选用乙基多杀菌素、吡虫啉、啶虫脒、螺虫乙酯或氯氰菊酯等药物进行喷雾，施药间隔为7～10 d，可连续施药2～3次。注意药剂的轮换使用，避免产生抗药性。

3.5.2.2 杧果切叶象甲（*Deporaus marginatus* Pascoe）

【为害情况】为害时间：成虫发生期与杧果抽梢期同步，主要为害杧果的夏梢和秋梢。在云南省通常5月中下旬至7月上中旬为第1个发生为害高峰期；8月中下旬至10月上中旬为第2个发生为害高峰期。老熟幼虫的滞育属短日照滞育类型。在云南省每年11月上旬及翌年3月下旬，光周期为短日照类型，老熟幼虫在土壤里进入滞育状态进行越冬。翌年4月日照逐渐变长，老熟幼虫化蛹羽化。为害部位：主要为害嫩叶。

成虫取食嫩叶的上表皮和叶肉，造成近圆形的取食斑。雌虫将卵产于嫩叶，并从叶片近基部咬断，切口整齐如刀切，带卵部分掉落地面，造成秃梢。为害症状：成虫咬食嫩叶上表皮，留下下表皮，使叶片卷缩、干枯；雌虫在嫩叶上产卵后，在近基部横向咬断，留下刀剪状的叶基部。

【防治措施】可选用阿维菌素、甲氨基阿维菌素苯甲酸盐、高效氯氟氰菊酯、噻虫嗪等药剂。喷药时注意药剂的轮换使用。注意保护猎蝽、蜘蛛等天敌。

3.5.2.3 杧果叶瘿蚊（*Erosomyia mangiferae* Felt）

【为害情况】为害时期：每年抽梢期为害嫩叶。为害部位：叶瘿蚊以幼虫咬破嫩叶表皮钻入其中取食叶肉，被害处叶先见白点后呈褐色斑，穿孔破裂，叶片卷曲，严重时叶片枯萎脱落以至梢枯。为害症状：以幼虫蛀食嫩叶，造成褐斑，穿孔破裂，叶片卷曲，严重时叶片枯萎脱落以至梢枯，致使树冠生长不良。为害高峰期植株和新梢被害率均达100%。

【防治措施】推荐药剂：阿维菌素、吡虫啉、高效氯氟氰菊酯等，每次梢期施药2～3次，施药要均匀，施药间隔时间7～10 d。

3.5.2.4 杧果横线尾夜蛾（*Chlunetia transoersa* Walker）

【为害情况】为害时期：全年各时期为害程度与温度和植株抽梢情况密切相关，气温20 ℃以上时为害较重。一般在4月中旬至5月中旬、5月下旬至6月上旬、8月上旬至9月上旬以及11月上中旬出现4次为害高峰。为害部位：以幼虫蛀食嫩梢和花穗。为害症状：在杧果产区普遍发生，以幼虫蛀食嫩梢和花穗，受害嫩梢或花穗干枯或生长衰弱。

【防治措施】在卵期和幼虫低龄期，采用药物防治。一般在花序和新梢抽出3 cm时，开始喷药，间隔7～8 d喷1次，连喷2～3次。常用药剂：阿维菌素、氰戊菊酯、溴氰菊酯、甲氨基阿维菌素苯甲酸盐、氯虫苯甲酰胺等。

3.6 采收

3.6.1 采收时间

适宜时间为晴天、无露水。

3.6.2 采收标准

果顶颜色由绿色转为黄色，即可采收。

3.6.3 采收方法

剪果时保留果梗0.8～1.5 cm。要轻拿轻放，避免损伤，放置于阴凉干燥处。

参考文献

董玉琛，刘旭，2006.中国作物及其野生近缘植物：果树卷[M].北京：中国农业出版社.

华南农业大学，1989.果树栽培学各论：南方本[M].2版.北京：中国农业出版社.

尼章光，2016.杧果栽培新技术[M].昆明：云南科技出版社.

尼章光，陈于福，解德宏，等，2013.云南芒果产业发展规划研究[J].中国农业资源与区划，34（3）：89-94.

尼章光，张林辉，罗心平，等，2008.怒江低热河谷芒果种质资源调查与分析[J].西南农业学报，21（2）：436-439.

张翠仙，解德宏，陈于福，等，2020.云南芒果产业发展现状[J].中国果树（6）：112-117.

❖ 柑橘栽培关键技术

1 产业发展概况

柑橘是芸香科柑橘属木本植物，属于热带、亚热带常绿果树，喜温暖湿润气候。中国是柑橘的重要原产地之一，柑橘资源丰富，优良品种繁多，有 4 000 多年的栽培历史。我国柑橘经济栽培区主要分布在北纬 16°～37°，海拔 700～1 000 m，最适宜柑橘生长的温度为 23～29 ℃，一年中所需要的日照总量在 1 300～1 400 h，对土壤的适应范围较广，但以疏松、肥沃、排水良好的微酸性至中性土壤最为适宜。柑橘甘甜多汁且营养丰富，与其他水果相比，它具有很高的经济价值，花、果、皮具有较高的药用价值。柑橘类水果所含有的人体保健物质，已分离出 30 余种，其中主要有类黄酮、单萜、香豆素、类胡萝卜素、类丙醇、吖啶酮、甘油糖脂质等。

云南省柑橘种植历史悠久，是全国知名的特早、极晚熟特色柑橘生产基地，也是柑橘起源地之一。柑橘产业是云南省种植面积最广、年产量仅次于香蕉的水果产业。2020 年，云南省柑橘种植面积 146.4 万亩，占全省水果在园面积的 14.67%；产量 135.85 万 t，占全省水果总产量的 14.13%。近年来，云南省柑橘产业快速发展，与 2010 年相比，云南省柑橘种植面积由 51.45 万亩波动增长至 146.4 万亩，增幅约 184.55%，产量从 45.62 万 t 波动增长至 135.85 万 t，增幅约 197.79%。云南省气候类型多样，水、土资源丰富，实现柑橘四季生产、周年供应，在上市时间上与其他国内产区相比具有明显的优势。此外，云南省生态环境优良，有利于柑橘产业的绿色有机生产，为云南省柑橘的品质奠定了优秀的环境基础。

云南省柑橘种植形成以干热河谷及湿热河谷区为主的优势产区，如玉溪市新平县、华宁县、元江县，红河州建水县、弥勒市、河口县、红河县，大理州宾川县、鹤庆县，曲靖市师宗县，昭通市永善县，西双版纳州景洪市、勐海县、勐腊县，丽江市永胜县，普洱市江城县、镇沅县，文山州广南县，德宏州瑞丽市、芒市等。

2 主要栽培品种及特性

2.1 沃柑

沃柑是以色列研发的柑橘品种，系坦普尔（橘橙）与丹西（红橘）杂交种，属于杂柑。该品种属晚熟杂交柑橘品种，生长势强，树冠初期呈自然圆头形，结果后逐步开张。果实中等大小、扁圆形，平均单果重130 g；果皮光滑，橙色或橙红色，易剥离，油胞细密，微凸或与果面平，凹点少；果肉橙红色，果肉细嫩化渣，多汁味甜，可溶性固形物含量13.3%～14.5%；种子数9～20粒，种子棒状，单胚。通常上市时间为1—4月，元江县最早可在11月底至12月初上市，宾川县等部分地区可晚至5—6月采收。

2.2 温州蜜柑类

2.2.1 宫本

宫本由宫川早熟植株上的一个变异枝选育而成。该品种树冠矮，枝梢节间短，有刺；叶片为椭圆形或卵状椭圆形；花单生、白色；果扁圆形，果色橙黄、果皮光滑、油胞平或微凸、易剥皮、平均单果重120 g；果肉橙红色，质地细嫩，汁多、化渣，可溶性固形物含量8.0%～9.0%。果实一般9月中下旬开始采收上市，华宁县一般在7月下旬至9月上旬采收。

2.2.2 大浦5号

大浦5号是从山崎早熟温州蜜柑的枝变中选育而成的特早熟温州蜜柑。大浦5号为常绿小乔木，树冠矮，枝梢节间短，有刺；叶为常绿单身复叶，椭圆形，叶面浓绿，光滑；花单生、白色；果实单性结实，宽皮橘，鲜食，早结性好，生长期短，发育快，着色早；形态扁圆，完熟后果皮橙红色，光滑，味浓、化渣，平均单果重115 g，含糖量8.5%，含酸量0.57 g/100 mL。果实一般9月中旬成熟，华宁县一般在7月下旬至9月上旬采收。

2.2.3 兴津

兴津常绿小乔木，枝梢分布均匀，生长旺盛，有刺；叶片菱形，浓绿色，花较大，单生，白色，花瓣5片；果实高扁圆形，平均单果重180 g；果肉细嫩多汁，可溶性固形物含量 9.0%～10.0%，含酸量 0.7～0.8 g/100 mL。果实 10 月上中旬成熟，华宁县一般在 9 月下旬开始采收上市。

2.3 冰糖橙

冰糖橙原产于湖南省洪江市，因其果实甜脆多汁、化渣、无核或极少核、可溶性固形物含量高、风味好等成为我国特色地方甜橙品种之一。冰糖橙树势健壮，树姿开张，枝梢较粗壮，叶片呈椭圆形，较宽大；果实近圆形，橙红色，果皮光滑；平均单果重 150～170 g，可溶性固形物含量 14.5%，含糖量 12 g/100 mL 以上，含酸量 0.6 g/100 mL，味浓甜带清香，少核。11 月上中旬成熟，元江县等部分地区在国庆节前上市。

2.4 三红蜜柚

三红蜜柚原产于福建省平和县，其果实外皮在相应遮叶下呈现淡粉红色，果皮下的海绵层是粉红色，果肉呈玫瑰红色，得名"三红"。三红蜜柚是琯溪蜜柚芽变成红肉蜜柚后再次芽变的新品种，属于琯溪蜜柚的系列品种。最新品种红皮红肉蜜柚表现为早熟，丰产、品质极优，果大皮薄、瓤肉无籽，汁多柔软，清甜微酸；果形倒卵圆形，平均单果重 1 680 g，果皮红绿色，囊皮粉红色，汁胞红色，可溶性固形物含量 11.55%、总酸 0.74%。一般于 9 月开始上市。

3 栽培管理技术

3.1 种植规划与园区建设

3.1.1 种植规划

品种是柑橘生产的基础。在种植品种的选择上，要根据本地区气候、水利等条件综合选择合适的品种种植。对于优质丰产的新品种应积极种植，但种植前要了解品种是否适合当地栽培，最好进行少量试验种植，观察其是否优质丰产，之后再选择扩种或规模种植。

3.1.2 园区建设

3.1.2.1 园地选择

选择坡度 <25° 的向阳山坡中性或偏酸性的土壤，土层深厚肥沃，有机质含量丰

富，土壤通透性好，果园易排灌，交通便利；生态保持良好，生产柑橘条件优越的区域栽植。

3.1.2.2 改土整地

果园坡度<15°，采取缓坡栽培；坡度在15°～25°，采取坡改梯，台面保持2.5～3 m宽。土壤深翻（深度在0.8 m以上）改土以填埋秸秆、药渣等有机质为主（比例在60%以上）。每亩施打碎的干秸秆、药渣，填埋厚度在20 cm以上，或施专用改土有机肥5 000 kg＋干畜禽粪2 000 kg＋油枯等饼肥500 kg＋生物菌肥1 000 kg（或直接泼撒生物菌），混合覆盖沉实，沿台地开好主沟、背沟，规范平整地面。

3.1.2.3 水利建设

采购智能化、全自动节水性能强的节水灌溉设备。滴灌是利用塑料管带将水通过直径约10 mm毛管的孔口或滴头送到作物根部进行局部灌溉，是目前干旱缺水地区最有效的一种节水灌溉方式，适应不同地形的土壤，肥水精准把控，水肥利用率可达90%以上，具有省水、省肥、省工、省时功效。

3.1.2.4 道路建设

道路建设根据地势间隔200 m修建一条主道，宽3 m；间隔100 m修建一条作业道，宽2 m。因地势不宜修建道路的果园或园内路段，架建单轨道式山地运输车，轨道机房修建在主道路旁，宜于生产物资果实运输，园内道路互通，园区建设向景区发展，便于游客观光体验采摘果实，降低生产成本。

3.1.3 苗木定植

一般在春梢萌动前的2月下旬至3月上旬，或10月下旬至12月上旬。定植密度根据品种特性、地势、土壤、砧木、耕作管理方式等而定。一般早熟温州蜜柑平地每亩40～50株、山地每亩70～80株；中熟温州蜜柑平地每亩40～50株、山地每亩50～70株；甜橙平地每亩20～40株、山地每亩40～60株；椪柑平地每亩50～60株、山地每亩70～80株；柚平地每亩20～40株、山地每亩40～60株。定植前穴内施好定植肥、每穴施入3 kg饼肥、1 kg磷肥，与土充分拌匀，施入穴底层，盖上20 cm细土，放正苗木，舒展根系，填入少量细土后，轻轻向上提苗，使土与根密切结合，填土压实，土齐根颈，每株浇水20 kg，再覆土，露出嫁接口。

3.1.4 定植管理

浇水保湿：苗木定植后约半个月开始成活，土壤比较干燥时，每1～2 d浇1次水，湿润土壤，促进新根生长，苗木成活前只能浇水，不宜追肥。成活后，勤施薄肥，促进根系和枝梢生长。

土壤覆盖：树盘下覆盖稻草、秸秆、绿肥或地膜，保持土壤疏松、湿润，让幼苗安全越冬。

立枝柱防风：新栽苗木根系尚未扎稳，易被大风摇动，应在苗木旁深插一根竹棍，用塑料绳将苗木固定。

摘心：幼树生长旺盛，要及时对长枝摘心，去除主干上和位置不当处抽生的萌芽。

施肥管理和病虫害防治：成活后应勤检查，勤施薄施液肥。发现红蜘蛛、潜叶蛾、蚜虫、炭疽病、溃疡病等应及时防治。

3.2 整形修剪

3.2.1 树形

自然开心形一般为3个大主枝，无中心主干，树干开张而不露干，整形工作分3年进行。第一年：距离地面约50 cm处定干，选配主枝，摘心、抹芽、除萌。第二年：短截主枝延长枝，选配副主枝，摘心、抹芽、除萌，疏除花蕾。第三年：短截主枝延长枝，选配副主枝，摘心、抹芽、除萌、疏花。

自然圆头形是最接近自然的整形树形，定干高40～50 cm，由主干上自然分生2～3个强壮大枝，大枝之间相距10～15 cm，各向一个方向发展。第二年或第三年再留1～2个，上下之间不重叠，各主枝基角约45°，斜向四方发展，共有主枝3～5个，根据其空间，再留1～3个副主枝，各骨干枝上再留大、中、小型枝组，3年后即可成形。

3.2.2 幼树修剪

幼树长势旺盛，一年可抽梢多次，树冠增长快。修剪时应轻剪，先按树形结构定干，在整形带选留主枝培养，保留一定数量的辅养枝，增加叶片数为树体供给养分，增大树冠，促进早投产。

栽植后前2年，每年可放梢3～4次，夏秋梢要集中放梢。放梢前，先抹除1～2次抽发的零星芽，待发梢整体达1/2左右，停止抹芽，统一放梢。新梢转绿前进行1次疏梢，上部枝梢除强去弱留中庸，中下部枝梢除弱留强。新梢展叶转绿后，对过长的新梢摘心，留10片叶左右，促进枝梢老熟。主枝、副主枝和侧枝的延长枝，在每次放梢前短截先端1/3～1/2，使剪口下的2～3个芽正好是放梢中部的健壮芽。随着树体生长，树冠不断增大，其中下部主干、主枝上的辅养枝视情况分次疏除。幼树主干和主枝上生长的徒长枝，如扰乱了树形，及时剪除。当主干过低或主枝分布不合理，内膛空缺较大时，可利用徒长枝替换主干或主枝，须及时摘心，抽发分枝，促进树体成形。幼树期开花坐果会消耗养分，影响树冠扩大，要及时摘除花蕾。花蕾未摘除干净，坐果后的幼果要及时摘除，减少养分消耗。

3.2.3 初结果树修剪

初结果树需继续扩大树冠完成整形，修剪应在保持树冠进一步扩大的前提下，做到早投产、早丰产。初结果树树势旺，营养生长与生殖生长矛盾较突出，采用疏春梢、控夏梢、放秋梢、抹冬梢来协调两者间的矛盾，注重保花保果，平衡树势，力求早挂果、早丰产。柑橘幼树结果母枝主要为秋梢，结果过多，会导致放出的秋梢质量变差。放秋梢前 10～15 d，摘除中上部外围的大顶果，或连萼片一并剪除，促进抽发、培养高质量秋梢，为来年结果打下良好基础。坐果率低的品种，落花落果严重。落花落果枝，一般冬季修剪，但为减少养分消耗，建议夏季放秋梢前进行短剪，以促发秋梢，从而形成优良的结果母枝。秋梢生长好的年份，翌年花量大，易形成结果大年。因此，当年冬剪时，要对 1/3 左右的秋梢进行重短截，疏除弱梢，减少翌年开花量，防止翌年形成结果大年。如末级枝为夏梢，也进行短截控制。枯枝、病虫枝、密生枝、细弱枝、交叉枝、徒长枝、下垂接地枝等无用枝可分次处理。对树体生长影响较大的病虫枝、徒长枝，在生长期剪除，其余的在冬季修剪时一并剪除。

3.2.4 成年结果树修剪

进入丰产期的成年结果树，整形修剪的重点是培养结果母枝，更新结果枝组，协调好营养枝和结果枝的比例，防止出现结果大小年，达到高产、稳产，延长丰产年限的目的。丰产期树冠上部生长旺，易造成上部密生郁闭。修剪时，疏剪大枝以增加内部光照，培养大枝少，小枝多，上空不下空，外空不内空，上下能透光，立体结果的凹凸树形。每年将约 1/3 的衰弱枝进行重短截，更新结果枝组，促发春梢。夏梢留少量，经过摘心和抹芽放梢，使其抽生更多的结果母枝。每年轮换更新结果枝组，可平稳树势，延长丰产年限。进入丰产期后，春梢代替秋梢成为主要的结果母枝，春梢的质量直接影响来年的产量。可通过结果枝组更新促发春梢，也可将无须作延长枝的夏、秋梢，短截至春梢基枝，促发高质量的春梢。谢花后至 7 月上中旬，对落花、落果枝组进行回缩，促进抽发强健的早秋梢，形成优质结果母枝。柑橘丰产期树体较为高大，易封行郁闭，要及时处理和利用好内膛枝，不然会严重影响内膛果的产量和品质。

3.3 水肥管理

水肥管理是柑橘种植的重要工作，有效的柑橘水肥管理可以保证柑橘在整个种植过程中都有着足够的营养，充分的营养可以保证柑橘在种植过程中不会出现果实萎缩的现象，有效保障柑橘的质量。

在旱季要特别注重对柑橘的水分补充，这时可以采用人工灌溉的方式，实际灌溉应用滴灌技术或者微喷灌溉技术，可以有效保证柑橘种植过程中不会出现水分流失的现象，滴灌技术的应用还可以在很大程度上减少对水资源的浪费，实现资源的合理使用。最好可以使用带状的灌溉方式，避免全部的柑橘种植土壤被湿透，柑橘的种植对水分有着一定的要求，水分过多会导致柑橘出现腐烂的情况，直接影响柑橘的正常生长，最终影响柑橘的产量和质量。

施肥应根据不同柑橘品种、砧木、土壤类型、气候条件、肥料种类、树势和种植密度等，合理经济施肥。柑橘结果树施肥主要有花期肥、稳果肥、壮果肥、采后肥。花期肥以速效化肥为主，配合施用有机肥；稳果肥以氮肥为主，配合施用磷、钾肥；壮果肥以速效化肥为主，配合施用有机肥；采后肥以有机肥为主，配合施用速效化肥。

3.4 主要病虫害防控

3.4.1 主要病害

3.4.1.1 黄龙病

农业防治：实施苗木检疫，防止带病苗木、接穗进入无病区；及时挖除病树，对每年春、夏、秋3个梢期，认真逐株检查，发现病株和可疑病株，立即挖除集中烧毁，在挖除病树前喷施1次柑橘木虱药剂。

化学防治：种植管理中要及时防治柑橘木虱，春、秋季，田间零星发现虫害时选用吡虫啉、噻虫嗪、啶虫脒、噻嗪酮、呋虫胺、氟啶虫酰胺、噻虫嗪+吡丙醚等药剂及时喷雾防治；加强栽培管理，保持树势健壮，提高耐病能力。

3.4.1.2 溃疡病

农业防治：实施苗木检疫，防止带病苗木、接穗进入无病区；冬季清园，剪除病枝，清除落叶落果，集中烧毁，减少越冬病源；加强水肥管理，促使新梢整齐抽发，做好潜叶蛾等害虫防治；营造防风林，减少风害。

化学防治：选用100亿芽孢/g枯草芽孢杆菌可湿性粉剂700~800倍液、30%噻唑锌悬浮剂500~750倍液、46%氢氧化铜水分散粒剂1 000~2 000倍液、30%

噻森铜悬浮剂 750～1 000 倍液、77% 硫酸铜钙可湿性粉剂 400～600 倍液、47% 春雷·王铜可湿性粉剂 380～470 倍液或其他铜制剂类及其复配剂进行喷雾防治，药剂轮换使用。

3.4.1.3 炭疽病

农业防治：加强栽培管理，增强树势，提高抗病能力；做好水肥管理和防虫、防日灼等，避免机械损伤，剪除病虫枝和徒长枝，清除地面落叶并集中烧毁。

化学防治：选用 30% 唑醚·戊唑醇悬浮剂 1 500～2 000 倍液、40% 唑醚·咪鲜胺水乳剂 3 000～4 000 倍液、60% 唑醚·代森联水分散粒剂 1 000～1 500 倍液、15% 肟菌·戊唑醇悬浮剂 2 000～2 500 倍液、10% 苯醚甲环唑水分散粒剂 2 000～2 500 倍液、70% 丙森锌可湿性粉剂 600～700 倍液、75% 代森锰锌水分散粒剂 370～470 倍液等及其复配剂进行喷雾防治，药剂轮换使用。

3.4.1.4 日灼病

农业防治：选用发生日灼病较少的品种；提倡园内生草法管理，以调节果园小气候；适当密植，或幼龄结果树在生理落果结束时促放夏梢，以梢遮果减轻日灼程度。对树冠外围的果实通过粘贴纸片、黄色胶带、套袋或涂抹石灰浆的方式防治；保持土壤水分，改善园区小气候；促夏梢推迟放，借枝梢遮挡果实。

3.4.2 主要虫害

3.4.2.1 木虱

化学防治：选用 21% 噻虫嗪悬浮剂 3 360～4 200 倍液、5% 啶虫脒微乳剂 4 000～5 000 倍液、10% 联苯菊酯乳油 1 666～3 333 倍液、2.5% 高效氯氟氰菊酯水乳剂 3 000～4 000 倍液等及其复配剂乙基多杀菌素+甲维·吡丙醚 2 000 倍液、30% 吡丙·虫螨腈悬浮剂 2 000～2 500 倍液；2% 氯氟·噻虫胺颗粒剂+10% 吡丙·吡虫啉悬浮剂 1 000～2 000 倍液、15% 唑虫酰胺悬浮剂 1 500 倍液+5% 甲氨基阿维菌素苯甲酸盐悬浮剂 3 000 倍液进行喷雾防治，药剂轮换使用。

3.4.2.2 红蜘蛛

农业防治：合理修剪，使果园通风、透光；实施生草栽培为捕食螨（红蜘蛛天敌）提供生存环境。

化学防治：冬季清园时可选用 99% 矿物油乳油 150～200 倍液、24.5% 阿维·矿物油乳油 1 000～2 000 倍液喷雾。当 100 片叶平均虫口在 1～2 头时进行全面防治，选用 43% 联苯肼酯悬浮剂 1 900～2 400 倍液、34% 螺螨酯悬浮剂 2 000～3 000 倍液、110 g/L 乙螨唑悬浮剂 2 000～2 500 倍液、5% 唑螨酯悬浮剂 1 000～1 500 倍液、22%

阿维·螺螨酯悬浮剂 4 000～6 000 倍液等及其复配剂喷雾防治，药剂交替使用。

3.4.2.3　潜叶蛾

农业防治：晚秋、早夏梢剪除有幼虫或蛹的枝条；加强水肥管理，彻底抹除零星抽发的早夏梢，集中放梢，切断害虫的食物链。

物理防治：用黑光灯或频振式杀虫灯诱杀。

化学防治：选用 10% 吡虫啉乳油 2 000～2 500 倍液、40% 氯虫·噻虫嗪水分散粒剂 2 000 倍液、20% 虫酰肼悬浮剂 1 500～2 000 倍液或 10% 虫螨腈悬浮剂 800～1 200 倍液及其复配剂 1% 甲氨基阿维菌素苯甲酸乳油 1 500～2 000 倍液、20% 甲氰菊酯乳油 1 500～2 000 倍液、1.8% 阿维菌素乳油 1 500～2 000 倍液、400 亿个孢子/g 球孢白僵菌可湿性粉剂 1 600～2 000 倍液、16 000 IU/mg 苏云金杆菌可湿性粉剂 200～300 倍液喷雾防治，药剂轮换使用。

3.4.2.4　小食蝇

农业防治：捡拾果园的虫果、落果，集中深埋处理；冬季清园时翻土，破坏其越冬环境，减少虫源。

物理防治：设置小食蝇粘胶板诱捕；或用甲基丁香酚性诱剂诱杀成虫。

化学防治：每亩使用 1% 噻虫嗪饵剂 80～100 g、1% 吡虫啉饵剂 70～12 g、1% 阿维菌素浓饵剂 180～270 mL 进行投饵。

3.5　适时采收

3.5.1　成熟度指标

柑橘采收时间要因鲜销或加工、贮藏等用途不同而有所区别。鲜销：以选黄留青、分批采收为原则，采收的成熟度应在 9 成以上，直至完熟到该品种成熟时固有指标。加工或贮藏：一般要求适当早采，通常果实成熟度达到 8 成左右即可采收，果面颜色占品种固有色彩的比例应 >2/3（柚除外）。采收成熟度指标为总可溶性固形物含量：脐橙 ≥11%、冰糖橙 ≥16%、大红甜橙 ≥8.5%、早熟蜜柑 ≥8.5%、中熟蜜柑 ≥12%、椪柑 ≥10%、香柚 ≥12%。

3.5.2　采收技术

宜选择晴天、果实表面水分干后进行采收。使用橘凳自上而下、从外到内的顺序采摘。采取复剪法采果，第 1 剪离果蒂 1 cm 处附近剪下，再齐果蒂剪第 2 剪，做到果蒂平整，保持萼片完整。果实放入采果篓或从采果篓转入果箱（或箩筐），必须轻拿轻放，采果箱（或箩筐）中只能装至 9 成满。采下的果实要防止雨淋日晒，也不宜露天过夜。伤果、病虫为害果应分开放置。

参考文献

邓幸福，2009. 柑橘高标准建园 [J]. 安徽林业，163（5）：34.

黄兰，2020. 海拔和坡向对柑橘土壤养分及果实品质的影响 [D]. 长沙：湖南农业大学.

李树举，龙桂友，杨鸿，等，2010. 大浦5号特早熟温州蜜柑的特征特性及栽培技术 [J]. 湖南农业科学（3）：101-102，106.

卢虹伏，卢前成，2022. 柑橘高标准建园栽培技术 [J]. 现代农业研究，28（8）：122-124.

彭诗怡，2016. 特早熟温州蜜柑生物学性状评价与分析 [D]. 长沙：湖南农业大学.

沈平，沈超，2018. 浅谈柑橘整形修剪技术 [J]. 种子科技，36（11）：76，78.

舒小勇，肖典雄，2014. 柑橘的采收与贮藏 [J]. 湖南农业（10）：32.

宋伟，汪军，李华雄，等，2022. 川南丘区柑橘整形修剪技术初探 [J]. 四川农业科技（5）：25-28.

吴兴明，2012. 特早熟温州蜜柑大浦5号特征特性与栽培技术要点 [J]. 中国果业信息，29（6）：62-63.

佚名，2003. 日南1号品种简介 [J]. 江西园艺（4）：39.

❖ 葡萄栽培关键技术

1 产业发展概况

葡萄作为世界四大水果之一，栽培历史悠久，栽培区域广泛，目前已知的葡萄品种约有 1.1 万个。葡萄栽培管理现代化水平高、栽培方式多种多样，产业链完善、产品丰富，既可鲜食，又可酿酒、制干、制汁及制果酱、果醋等。葡萄以其诱人的鲜食品质和丰富的加工产品而独树一帜。随着农业产业结构调整，葡萄产业以其见效快、收益高等特点，逐渐成为云南省部分地区特色经作产业发展的主导产业。而独特的地理位置、充足的光照、适宜的温度，为云南省葡萄产业发展创造了良好的气候和生态环境条件，使全省不同区域生产的鲜食葡萄可实现周年供应。云南省已成为我国鲜食葡萄的重要产区，葡萄产业已成为云南省巩固脱贫攻坚、实现乡村振兴的重要产业之一。

1.1 生产情况

2010 年以来，云南省葡萄种植面积呈稳步增长趋势，2019 年种植面积居全国第 5 位，产量居全国第 4 位。云南省绿色食品发展中心数据显示，2020 年，云南省葡萄种植面积占全省水果总面积的 5.28%，位于柑橘、杧果、苹果、香蕉、梨和桃之后，居第 7 位，产量占全省水果总产量的 10.21%，位于香蕉、柑橘、梨和苹果之后，居第 5 位。

1.2 区域布局

云南省 112 个县（市、区）均有葡萄种植，种植面积超过 10 万亩的有红河州和大理州，1 万～10 万亩的有楚雄州、曲靖市、昆明市、迪庆州、文山州、昭通市、丽江市和玉溪市，其余州（市）均不足 1 万亩。产量超过 10 万 t 的有大理州、红河州和楚雄州，1 万～10 万 t 的有曲靖市、昆明市、昭通市、文山州、玉溪市和丽江市，其余州（市）产量均不足 1 万 t。

宾川县是云南省葡萄生产规模最大的县，弥勒市次之，宾川县和弥勒市也是云南省仅有的两个栽培面积超过 10 万亩的县（市、区）。宾川县主要以栽培鲜食葡萄为主，弥勒市则主要以栽培酿酒葡萄为主。栽培面积 5 万～10 万亩的有建水县，1 万～5 万亩的有元谋县、陆良县、蒙自市、德钦县、丘北县、永胜县。

1.3 经营主体

云南省拥有省级及以上葡萄种植、加工龙头企业20家，葡萄专业种植合作社37家，共有"七彩云秘""果先锋""楼铁源"等葡萄品牌20余个，其中"七彩云秘""果先锋"均连续3年被评为云南省"十大名品"。"弥勒葡萄"已登记为农产品地理标志，"宾川红提葡萄"成功注册为"中国地理标志证明商标"。

2 主要栽培品种及特性

目前，云南省现有葡萄栽培品种达50余个。鲜食葡萄种植面积占85%以上，主要分布在红河州、大理州、楚雄州、曲靖市等州（市），主栽品种为红地球、夏黑、克伦森、无核白鸡心、巨峰等。近年来，随着鲜食葡萄市场的影响，红地球、夏黑等传统主栽品种面积有所下降，建水县、元谋县、宾川县等通过高接换头，国外引入的部分优良品种栽培面积扩张较快。酿酒葡萄种植面积占15%左右，主要分布在迪庆州的德钦县、维西县，红河州的弥勒市和文山州的丘北县，主栽品种除兼用型的水晶外，还有赤霞珠、红玫瑰、威代尔、法国野等。

2.1 红地球

植物学特征：又称美国红提、晚红、大红球，属欧亚种。植株早春嫩梢呈紫红色，幼叶浅紫红色，叶面光滑，叶背有稀疏茸毛。成叶中等大小，心脏形，5裂，上裂刻深，下裂刻浅，叶正背两面均无茸毛，叶缘钝锯齿形，叶柄较长，呈浅红色，叶柄拱形，两性花。自花坐果率较高，果穗呈紫红色，长圆锥形，果粒大小整齐均匀，为圆形或卵圆形，果粉中厚，果皮为红色或深红色，果柄长，与果实结合紧密，不易裂口；果刷柔软，着生极牢固，耐拉力极强，不脱粒。果皮中厚，果肉硬脆，可削成薄片，刀切无汁，味甜爽口，品质优良，可溶性固形物含量17%以上。

生产表现：鲜食品种。全省可实现周年上市。树势中等，结实力较强。其果穗、果粒极大，高产。耐盐碱，但抗病性比较弱，容易患白粉病、霜霉病、灰霉病、酸腐病等各种真菌性病害。耐贮运，果柄细长，果粒与果柄结合牢固，耐压力、耐拉力均强，不易脱粒。

种植区域：云南省内均有种植，宾川县、陆良县、元谋县种植面积较大。

2.2 夏黑

植物学特征：欧美杂交种，早熟品种。嫩梢绿黄色，有少量茸毛。幼叶浅绿色，带浅紫色晕，叶片表面有光泽，叶背密被丝状绒。成龄叶片大，近圆形，叶片中间稍凹，边缘凸起。多为4裂，裂刻深，叶缘锯齿钝。叶柄洼失形。新梢生长直立，一年生成熟枝条红褐色。两性花。果穗圆锥形，间或有双歧肩。果粒近圆形，紫黑色或蓝黑色，着生紧密，果粉厚，果皮较厚，肉质硬脆，浓甜爽口，有浓郁草莓香味。果实充分成熟时可溶性固形物含量17%～24%，低酸，无核。

生产表现：鲜食品种。全省可实现周年上市。长势好，对土壤要求不严，较耐旱，花芽容易形成，丰产性好。对白粉病、霜霉病的抗性比一般巨峰类品种弱。不裂果，不脱粒，耐贮运。

种植区域：云南省内均有种植，建水县、蒙自市、宜良县、元谋县种植面积较大。

2.3 水晶

植物学特征：多年生蔓黑褐色，一年生蔓浅褐色、新绿色，有茸毛。卷须绿色，少许紫褐色。新梢上端叶片古铜色，有光泽，背面被白茸毛，随着叶龄变老而渐变为灰白色。叶片厚，心形，浓绿，掌状5裂，裂刻中深，叶缘锯齿大而锐，叶背主脉和侧脉凸出。果穗圆锥形、多数无歧肩，结果紧密，果穗重约200 g，最大穗重300 g。果粒圆形，均匀，单粒重4～5 g，果皮绿色，充分熟透后呈黄绿色，果粉厚且晶莹有光泽，果肉草绿色，细嫩，清香甜蜜，可溶性固形物含量14%。

生产表现：酿酒、鲜食兼用品种。云南省5—9月成熟上市。树势强健，根系发达，极性较缓和，抗逆性较强，较耐贫瘠，适应性特广。抗病，易栽培。丰产稳产，品质优良。

种植区域：云南省内弥勒市种植面积最大，其次是丘北县。

2.4 赤霞珠

植物学特征：欧亚种，晚熟品种。原产法国。嫩梢绿色，带有紫色条纹，无茸毛，一年生成熟枝条褐色，节间短而粗。幼叶橙红色，叶表光滑，叶背密生茸毛，成龄叶片小，心脏形，深5裂，叶表面有稀疏茸毛，叶缘锯齿钝，呈圆顶形，叶向上卷，叶柄洼闭合呈圆形，两性花。果穗小，平均穗重165 g，圆锥形。果粒着生中等密度，平均粒重1.9 g，圆形，紫黑色，有青草味。可溶性固形物含量16.3%～17.4%，含酸量0.456%，含糖量高。

生产表现： 酿酒红葡萄。云南省 6—10 月成熟上市。生长势中等，结实力强，易丰产，适应性强，抗病力极强，较抗寒，喜肥水。产量稳定，迪庆州 8 月下旬至 9 月上旬成熟。

种植区域： 云南省内主要种植在弥勒市、德钦县等地。

2.5 克瑞森

植物学特征： 别名克伦生无核、淑女红无核。欧亚种，晚熟品种。嫩梢红绿色，有光泽，无茸毛，幼叶紫红色，叶缘绿色。成龄叶中等大，绿色，深 5 裂。锯齿中等，两侧凸。叶柄长。叶柄洼闭合呈椭圆形或圆形。两性花。果穗圆锥形，单穗重 500～700 g，果粒椭圆形，着生中等紧密或较紧密，平均单粒重 5～6 g，果粒鲜玫红色，果皮中等厚，具白色果粉。果粒椭圆形，果梗长度中等；果肉黄绿色、半透明，肉质细脆、硬度适中，味甜，果皮和果肉不易分离。有香气，无核。果刷长，不易落粒。可溶性固形物含量 18%～22%，含酸量 0.6%。

生产表现： 鲜食葡萄。在云南省 5—12 月成熟上市。植株生长旺盛，萌芽力、成枝力均较强。抗病性稍强，但易感染白腐病、霜霉病、白粉病。采前不落果、采后不落粒，极耐贮运。

种植区域： 云南省内主要种植在宾川县、弥勒市、安宁市等地。

2.6 无核白鸡心

植物学特征： 别名世纪无核。欧亚种，早熟品种。嫩梢绿色，有稀疏茸毛。幼叶微红，有稀疏茸毛，成叶大，心脏形，5 裂，裂刻极深，上裂刻呈封闭状，叶缘锯齿大而锐，叶片正反面均无茸毛，叶柄洼开张呈拱形。同穗果粒成熟期一致，果穗圆锥形，果粒长卵圆形，似鸡心状，着生紧密，未充分成熟时略带涩味，偏酸；充分成熟时果粒黄绿色，味甜，可溶性固形物含量 17.0%～20.6%。无核。

生产表现： 鲜食葡萄。在云南省 3—7 月、10—12 月成熟上市。长势好，自然坐果率高，高产。长势好，较抗旱，对白粉病、霜霉病和黑痘病的抗性中等。果粒过熟容易落粒，耐贮运性中等。

种植区域： 云南省内主要种植在元谋县、宾川县、石林县等地。

3 栽培管理技术

3.1 种植规划与园区建设

葡萄建园应选择无污染、生态环境良好、pH 值 8.2～8.5 的土壤，耕作层 80 cm 以上，地下水位 2 m 以下。优选质地良好、疏松肥沃、有机质含量在 1.5% 以上的土地。根据园区的规模规划设计机耕道路、沟渠、水池、守护房、储物间、包装间等设施。围篱优选铁丝网。按行距 2.5～3 m、株距 0.8～1.2 m 定植。

3.2 整形修剪

采用大"Y"形架设计，立柱高 2.6～3 m，入土 50 cm，架高 2 m。1.2 m 高度使用 40 cm 横档双铁线布置，1.5 m 高度使用 80 cm 横档双铁线布置，1.7～1.8 m 高度使用实心竹或钢管连贯立柱，在立柱两侧 1.1 m 处从横梁下吊一根铁线作第三台架线。定干高度 1.2～1.4 m，"V"形上口宽 2.0～2.4 m。要求立柱高 2.5 m，入土 0.5 m，架高 2.0 m。用实心竹或钢管连贯立柱为横梁，在立柱 1.1 m 处到横梁上一侧的 1.1～1.2 m 处用实心竹或钢管连接，两边对称处理形成"Y"形架面。"Y"形架三带整形，1.2 m 以下为通风带，1.2～1.5 m 为挂果带，1.5～1.8 m 为生产带。长势强、花芽分化节位低的夏黑选择"一"字形或"H"形整形。长势中等、花芽分化节位高的红提、克瑞森选择"双扇"形整形以便结果枝组更换，防止枝梢弱长，影响翌年产量。

冬季修剪。"一"字形整形，冬剪长短结合，预留芽量为设计穗数的 3 倍。冬剪时距主干最近的两个母枝留 5～7 个芽进行修剪，其余母枝留 2～3 个芽进行修剪，亩留芽量为 5 000～7 000 个芽，选 2 000～2 500 个挂果母枝。"双扇"形整形，冬季修剪，单株留 5～6 个结果母枝，亩留 2 000～2 500 个结果母枝，结果母枝留 4 个芽短剪。

3.3 水肥管理

优先采用膜下滴灌技术。浇透促萌水，开花前适当控水，幼果期保持土壤湿润，转色期干湿交替，注意排涝。

秋季每亩施厩肥 3～5 t 或精制有机肥 1 000～1 500 kg + 平衡型控释配方肥 80～100 kg、大粒硼 600 g、大粒锌 500 g。开花前每亩施 3 kg 尿素、3～5 kg 高钙镁水溶肥，谢花后每亩施 5 kg 高钙镁水溶肥。幼果期（硬核期）每亩施高钾型控释配方肥 80～120 kg。转色期每亩施高钾水溶肥 3～5 kg，7 d 施 1 次，共 3 次。其他时期根据苗情用水溶肥适时调节。叶面补充：于采果后、3叶期、开花前、谢花后进行4次叶面喷硼肥，开花前、谢花后进行2次叶面喷锌肥，开花前进行1次叶面喷铁肥，幼果期到成熟叶面多次喷施钙肥、镁肥、磷酸二氢钾、氨基酸等叶面肥。看叶果施肥：要达到树相平衡必须观察枝、叶、果的生长情况，当开花前出现节间过长、叶片大即长势旺的情况可以采取控肥水的方法调节；反之长势弱时稍加强肥水的使用量，达到基础树相，即节间长8～10 cm、叶柄8～10 cm、叶片张开度12～15 cm；转色期（二次膨大）基本停梢。

3.4 主要病虫害防控

葡萄常见病虫害有霜霉病、白粉病、灰霉病、蚜虫等。以农业防治、物理防治和生物防治为主，科学合理使用化学防治。葡萄常见病虫害及防治方法详见表2-7。

表2-7 葡萄常见病虫害及防治方法

防治对象	发病症状	防治时期	农药名称	使用剂量	施药方法	安全间隔期天数（d）
霜霉病	受害叶片首先呈现半透明、边缘不清晰的油渍状小病斑，以后逐渐扩展为黄褐色、多角形病斑，并能相互连接形成大病斑。潮湿时，叶背面产生一层白色霉状物，病斑最后变褐干枯，叶片早落。嫩梢、卷须、穗轴发病，开始为油渍状半透明斑点，以后逐渐变为稍凹陷黄色至褐色病斑，潮湿时，表面产生白色霉状物。枝梢受害，生长停滞、扭曲，甚至枯死。花及幼果感病时呈现深褐色，并生出白色霜状霉层，不久就干缩脱落。果实长到豌豆粒大时受病，最初呈现红褐色斑，后期僵化开裂	发病前或初期时葡萄花期前后	80%波尔多液可湿性粉剂	300～400倍液	喷雾	—
			25%嘧菌酯悬浮剂	1 000～2 000倍液	喷雾	7
			72.2%霜霉威水剂	600～800倍液		
			90%三乙膦酸铝可湿性粉剂、10%氰霜唑悬浮剂	2 000～2 500倍液		
白粉病	叶片染病时，开始在表面上长出灰白色病斑，后生面粉状的霉，严重时叶片焦枯。果实受害后，斑块上面出现黑色网状花纹，上面覆盖一层白粉；果实停止生长，有时变畸形，小而味酸。果实长大后，在多雨时感病，病处纵向裂开后易被腐生菌感染而腐烂。果梗、新梢及穗轴受害时，初期表面呈现灰白色粉斑，后期粉斑下面形成雪花状或不规则的褐色斑，使穗轴、果梗变脆，枝蔓也不能很好成熟	病菌侵染初期	29%石硫合剂水剂	6～9倍液	喷雾	15
			25%三唑酮乳油	1 000～1 500倍液		
			25%苯醚甲环唑乳油	2 000～2 500倍液		
			40%硫磺悬浮剂	400～500倍液		
			40%氟硅唑乳油	8 000倍液		

（续表）

防治对象	发病症状	防治时期	农药名称	使用剂量	施药方法	安全间隔期天数（d）
灰霉病	花序和幼果受害时，初出现似热水烫过的水浸状、淡褐色病斑，后变为暗褐色、软腐，天气干燥时，受害花序和幼果萎蔫干枯，极易脱落；空气潮湿时，受害花序及幼果上长出灰色霉层。果实上浆后感病，发生褐色凹陷病斑，扩展至整个果实软腐，其上长出鼠灰色的霉层	葡萄花期前后	50%异菌脲可湿性粉剂	500～800倍液	喷雾	14
		发病初期	50%嘧菌环胺水分散粒剂	700～1 000倍液	喷雾	7
			50%啶酰菌胺水分散粒剂	1 000～1 500倍液		
		发病前或初期	43%腐霉利悬浮剂	600～1 000倍液	喷雾	14
蚜虫	主要为害嫩梢及嫩叶，严重时造成叶片皱缩畸形，嫩梢尖部生长受阻	发病初期	1.5%苦参碱可溶液剂	3 000～4 000倍液	喷雾	10
			25%吡蚜酮悬浮剂	2 000～2 500倍液		
			70%吡虫啉水分散粒剂	2 000～3 000倍液		

3.5 采收与采后处理

在果实正常成熟、表现出本品种固有的品质特征（色泽、香味、风味、口感等）时进行采收，着色品种单穗的着色果粒应在80%以上，可溶性固形物含量应达到葡萄等级标准规定。采摘时间宜在早上果面露水已干时开始，气温过高时停采。根据成熟度分批采收，采摘时留3～4 cm长的果梗，轻拿轻放，避免太阳暴晒。清除着色不良果、病虫果、机械损伤果和烂果。

需进行长时间贮藏的葡萄，采收后立即进行预冷。为实现快速预冷，应在葡萄入贮前3 d开机，空库降温至-1 ℃。葡萄果品的最佳贮藏温度为-1～0 ℃，要保持库温稳定，波动幅度不超过0.5 ℃；测温仪器精度要求±0.2 ℃。湿度：保持在90%～95%；测湿仪器精度要求±5%。采收并检验合格的产品应按照产品特征进行产品的分级处理。包装应符合《绿色食品 包装通用准则》（NY/T 658—2015）的规定，纸制箱选用钙塑双瓦楞纸箱，技术指标应符合《运输包装用单瓦楞纸箱和双瓦楞纸箱》（GB/T 6543—2008）的规定；塑料箱技术指标应符合《包装容器 复合式中型散装容器》（GB/T 19161—2016）的规定。

参考文献

云南省打造世界一流"绿色食品牌"工作领导小组办公室,2021.云南省"绿色食品牌"重点产业2020年度发展报告[R].2021.7.

中华人民共和国国家统计局.中国统计年鉴(2011—2020)[M].北京:中国统计出版社.

中华人民共和国农业农村部.中国农业统计资料(2011—2020)[M].北京:中国农业出版社.

❖ 菠萝蜜栽培关键技术

1 营养价值

菠萝蜜（学名木菠萝）果香肉厚，清甜多汁，与香蕉、杧果、菠萝和番木瓜等比较，菠萝蜜淀粉、蛋白质、脂肪酸、钙、铁、维生素 B_1 等营养物质含量较多。菠萝蜜果肉富含钙、镁、锌、铁、钠、锰等有益元素和胡萝卜素、类黄酮、挥发性酸、固醇、单宁等多种植物活性成分。每 100 g 菠萝蜜果肉中还含有碳水化合物 16.0%～25.4%、蛋白质 1.2%～1.9%、脂肪 0.1%～0.4%、纤维素 1.0%～1.5%。

菠萝蜜种子是淀粉和膳食纤维的优质食物来源，种子中含有抗性淀粉、蛋白质、皂苷、生物碱、有机酸、氨基酸、微量元素等功能活性成分和人体必需的脂肪酸。每 100 g 种子中含碳水化合物 25.8%～38.4%、蛋白质 6.6%～7.04%、脂肪 0.4%～0.43%、纤维素 1.0%～1.5%。

2 生长习性

菠萝蜜是典型的热带果树，喜欢温暖湿润的热带气候，不耐寒冷，要求种植地的年平均气温≥21 ℃，最冷月平均气温≥13 ℃，绝对最低温度≥0 ℃。相关研究表明，幼树对低温敏感，0 ℃与 -1 ℃条件分别会对叶片、枝条产生冻害，在 -3～-2 ℃条件下树体可被冻死；成年结果树耐寒能力较强，可忍受短期 -3.89～-3.33 ℃低温，在低于 -6.67 ℃的条件下植株在很短时间内会死亡；最适于菠萝蜜生长的温度为年均气温 27～31 ℃。

菠萝蜜花序着生于树干或枝条上，雌雄同株异花，其果实为雌花花序经过授粉发育而成的聚花果，椭圆形。菠萝蜜一年开花 1 次，部分品种可一年开花多次，一般 2—4 月开花，6—8 月成熟，果实发育期 120～150 d。定植后 3～5 年可收获，5～6 年进入盛果期，平均单株结果 30～100 个，平均单果重 10～20 kg，小的 1～2 kg，最大可达 40 kg 以上，亩产可达 4 500 kg。

菠萝蜜在阳光充足、降水充沛（年降水量 1 200 mm 以上）且季节性分布均匀、土层深厚、富含有机质、土壤 pH 值 6.0～7.5、排水良好、海拔 600 m 以下的低丘陵地或者平地的环境中栽培最佳。

3 产业发展概况

印度是最早栽培菠萝蜜的国家，而后广泛传播到世界热带亚热带地区，目前盛产于中国、印度、印度尼西亚、菲律宾、泰国、孟加拉国和巴西等地。据不完全统计，菠萝蜜在全球的栽培面积约390万亩，年产量超362万t。中国栽培菠萝蜜已有1 400多年的历史，早期主要作为庭院树和行道树栽培，管理粗放，树体高大，株产量差异大，果实品质低，没有形成规模种植。近年来，随着我国旅游业的发展及人民生活水平的提高，大众对各种名优稀特水果的需求日益增加。菠萝蜜因果肉芳香味甜、营养丰富，深受消费者喜爱。鲜果及加工产品的销量逐年增加。同时，地方产业发展对小型特色热带水果的政策扶持力度也在逐步加大，种植、加工、物流等产业链不断趋于成熟，中国菠萝蜜产业也迎来更蓬勃的发展。种植面积每年以20%左右的速度增长，在云南、海南、广东、广西等优势产区出现了规模化的商业种植。据不完全统计，截至2021年年底，我国菠萝蜜种植面积为50余万亩，年产量达60余万t，年产值90多亿元。目前，我国海南、广东、广西、福建、云南和四川的热带亚热带地区均有栽培，以海南、云南、广东栽培最多。截至2021年年底，云南省菠萝蜜种植面积为10余万亩，已成为云南省热区大力发展种植的特色水果之一，成为热区增加收入、实现乡村振兴的富民产业。

4 主要栽培品种及特性

4.1 国内主要栽培品种及特性

马来西亚1号（琼引1号）：果实长椭圆形，平均纵径47.63 cm，横径25.06 cm，果形指数1.9，平均单果重20～30 kg，果苞金黄色，可溶性固形物含量16.0%。

琼引8号：果实椭圆形，纵径和横径分别为39.75 cm和24.22 cm，果形指数1.65，平均单果重13.70 kg，果苞金黄色至橙黄色，可溶性固形物含量25.5%，可食率38.6%。

香蜜17号：果实椭圆形，平均纵径35.5 cm，平均横径23.7 cm，果形指数1.5，平均单果重4.5～18.5 kg，果苞橙红色，可溶性固形物含量23.5%～27.8%，可食率45.5%。

海大2号：果苞长圆形，平均单果重7.35 kg，果肉黄色、爽脆，风味浓郁，可溶性固形物含量 21.5%，可食率 58.0%。

红肉菠萝蜜：果实长椭圆形，平均单果重 9.5 kg，果肉橙红色，可溶性固形物含量 18.87%，可食率 78%。

常有菠萝蜜：果实椭圆形，平均单果重 4～6 kg，熟果无胶，果肉金黄色，肉厚爽脆，蜜香浓郁，可溶性固形物含量 26.88%～28.13%，可食率 75%。

秋红：果实长圆形，平均单果重 25 kg 以上，成熟后乳胶少，果肉粉红色，果皮有六角形瘤头突起，可溶性固形物含量达 18% 以上。

4.2 国外主要栽培品种及特性

J-30：马来西亚选育，果肉深橙黄色，质地硬，风味浓甜，清香，可食率 38%。

J-31：马来西亚选育，果肉深黄色，质地硬，风味甜，具有浓郁的香味，可食率 36%。

NS1：马来西亚选育，平均单果重 4～5.5 kg，单株产量 90 kg，5—6 月成熟；果肉深橙色，干苞，皮刺尖，可食率 34%，每颗果实种子数 63 粒，占比 5%，香味浓郁。

Chompa Gob：泰国选育，曾是泰国最好的品种。果肉橙黄色至深黄色，质脆，味香甜，果胶少而易于食用，可食率 30%。

Cheena：澳大利亚选育，是菠萝蜜和尖蜜拉（Champedak）的自然杂交种，果肉橙黄色，质地软，化渣，有少许纤维感，品质优，香气浓郁，可食率 33%。

Lemon Gold：澳大利亚昆士兰选育，果肉柠檬黄色，干苞，皮刺尖，风味甜香，可食率 37%。

Honey Gold：澳大利亚昆士兰选育，果肉深黄色至橙黄色，果肉厚，干苞，皮刺尖，有浓郁的甜香味，可食率 36%。

5 栽培管理技术

5.1 种植规划与园区建设

5.1.1 园地选择

选择年平均气温 ≥21 ℃，最冷月平均气温 ≥15 ℃，绝对最低温度 >5 ℃，年降水量 ≥1 200 mm，坡度 ≤45°，开阔向阳、避风、土层深厚、有机质丰富、结构良好、易于排水、地下水位在 1 m 以下的地区建园。园地环境条件应选择在生态环境良好、远离污染源、具有可持续生产能力的农业生产区域，符合《无公害农产品 种植业产地环境条件》（NY/T 5010—2016）的规定。

5.1.2 园地规划

园地的规划，一般应综合考虑园地规模、坡度、地形、小气候和机械化程度等诸多因素。建设必要的道路（主干道、支道和田间小道）、排灌和蓄水等设施，营造防护林带。防护林应选择速生抗风且不与菠萝蜜存在相同主要病虫害的树种（台湾相思树等）或本土速生抗风树种。以道路、防护林带等将园地划为若干小区，根据园地地形安排小区大小。园址选在山坡地带，应在坡顶挖掘环山的防洪沟，标准是沟面宽 1～1.2 m、沟深 0.6～0.8 m、沟底宽 0.6～1 m，设计时应控制其出口，尽量同排水通道相连接，这样可以减少水土流失。

5.1.3 园地开垦

定植前 4 个月内，未开垦的园地需垦荒深耕，深度 0.4～0.5 m，改善土壤结构，使土壤熟化。垦前应先规划防护林带的位置，选定后给予保留，缺少的部分进行补植、移栽等，清除防护林外的其他林木。园地深耕后即可进行平整筑埂，用以保持水土。一般坡度 <5° 的地块等高种植，同时注意每 4～6 行筑一埂即可；坡度处于 5°～45° 的园地应等高开垦，修筑宽 2～3 m 的水平梯田或环山行，台面向内稍倾斜 4°～5°，单行种植。梯田或环山行间的间距为 5～6 m。

5.1.4 挖定植穴

定植前 60 d 内挖定植穴，定植穴规格为穴面宽 0.8 m、深 0.7 m、穴底宽 0.6 m。挖掘过程中应将表土、底土分开放置，便于日后晒白风化后回土。坡度 <5° 的地块按株距 6 m、行距 7 m 的种植密度挖定植穴，坡度处于 5°～45° 的坡地，在梯田或环山带的梯、带面距离内沿 2/3 的位置按株距 6 m 的种植密度挖定植穴，上下梯田或环山带呈"品"字形挖定植穴，一般每亩 18～20 株。定植穴挖好后，暴晒 20～30 d 后回土，每穴施用腐熟的有机肥 20～30 kg、三元复合肥 1～2 kg、钙镁磷肥 3～5 kg，将肥料同底土混合拌匀。先于穴底回入 0.2～0.25 m 厚的表土，再回入土肥混合物，表层用表土覆盖。注意，回土时土面一般要高出地面 0.2～0.3 m，以备定植。

5.1.5 定植

定植前，对种苗所有叶片进行修剪，去除每片叶片的 2/3，留下 1/3，刨松定植穴，脱去营养袋，轻轻放在定植穴正中间，开始覆土，覆土时尽量用细土，边回边用手轻轻将土压实，注意不要挤散土球，覆土距离芽接口 2～3 cm 的位置即可，以苗为中心，做 50 cm 左右的积水盘，积水盘上覆盖杂草或透水防草布，并及时进行浇水，须浇足浇透，间隔 2 d 左右再浇 1 次透水。定植时，若遇太阳直晒，需进行遮阴保护，待植株恢复后撤除。

5.2　整形修剪

5.2.1　幼树修剪

幼树高 1.5～2 m 时，选择晴天中午打顶，截其主干，促其侧枝生长及分枝。萌发枝芽按东西南北各选留生长健壮、分布均匀、芽体饱满、距离地面 1.2～1.5 m 以上、与树干呈 45°～60° 生长的枝条 3～4 条作为一级分枝，多余的枝芽全部抹去。当一级分枝长至 0.8 m 时，再去顶，促发副主枝，如此培养各级分枝，使其形成枝条分布均匀、合理、通风透光的矮化树冠。

5.2.2　成龄树修剪

开花前 2～3 个月，剪去内部徒长枝、枯枝、残枝、病虫枝、弱枝、过密枝，保持树体通风透光。果实采收后，剪除残留于树干、大枝上的结果枝和雄花枝，剪短过长枝条，剪去交叉枝、下垂枝、徒长枝、枯枝、残枝、病虫枝、弱枝、过密枝及所有不利于发育的枝条。植株高度控制在 4 m 以下。

5.3　肥水管理

5.3.1　施肥原则

应贯彻勤施、薄施、生长旺季多施肥的原则。肥料种类以有机肥为主，适量施用无机肥。

5.3.2　肥料种类

推荐使用的农家肥料和化学肥料按照《绿色食品　肥料使用准则》（NY/T 394—2023）的规定执行。常用有机肥：畜禽粪、畜粪尿、塘泥、饼肥和绿肥等。畜粪尿、饼肥一般沤制成水肥；畜禽粪一般发酵后与表土或塘泥沤制成干肥。常用无机肥包括尿素、复合肥、氯化钾和钙镁磷肥等。

5.3.3 施肥方法与数量

5.3.3.1 幼树的施肥管理

一年生幼树，每株每次施尿素 50 g 或平衡型三元复合肥 100 g，每 3 个月施肥 1 次。10—12 月，增施有机肥 10～20 kg、钙镁磷肥 0.5 kg。2～3 年生幼树，每株每次施尿素 100 g 或平衡型三元复合肥 150 g，每 3 个月施肥 1 次。10—12 月，增施有机肥 15～25 kg、钙镁磷肥 1～1.5 kg。第一年在距离主干基部 0.3～0.4 m 处施用，第二年后在树冠滴水线外围施用。

5.3.3.2 成龄树的施肥管理

花前肥：在花芽分化前 2～3 个月开始施用。推荐每株施有机肥 15～25 kg，高钾三元复合肥 1～1.5 kg。

促花肥：集中抽花蕾时开始施用。以施用速效肥为主，每株施尿素 0.5 kg、氯化钾 0.5 kg，叶面喷施 0.3% 磷酸二氢钾及氨基酸、黄腐殖酸 2～3 次，于阴天或晴天 16:00 至傍晚进行。

壮果肥：在定果以后施用。每株施高钾三元复合肥 1～1.5 kg、钙镁磷肥 1.5～2 kg。

果后肥：采果后即可施用，目的是补充植株营养、恢复植株生势，避免果树早衰。以农家肥为主，少施化肥。每株果树可施用有机肥 20～25 kg、平衡型三元复合肥 2 kg 左右。

以上施肥标准为一般标准，在具体栽培过程中应当积极结合现实生产中的具体情况（如品种、树龄、肥料种类、土壤肥力、土质、类型等）进行综合分析后施用。

5.3.4 水分管理

旱季：开花期及果实发育期及时灌水，保持土壤湿润，灌溉应在上午及傍晚进行，避免中午高温。

雨季：应疏通排水沟，填平凹地，及时排出积水。

5.4 主要病虫害防控

5.4.1 主要病虫害种类

病害主要有炭疽病、花果软腐病、蒂腐病、根腐病；虫害主要有天牛、黄翅绢野螟、金龟子、绿鳞象甲、绿刺蛾。

5.4.2 防治原则

贯彻"预防为主，综合防治"的植保方针，以改善果园生态环境，加强栽培管理为基础，综合应用各种防治措施，优先采用农业防治、生物防治和物理防治等方法。科学合理使用化学防治，其使用药剂防治应符合《农药安全使用规范　总则》（NY/T 1276—2007）中有关的农药使用准则和规定。

5.4.3 防治措施

5.4.3.1 农业防治

选择适应性强、抗病虫能力强的优良品种。加强栽培管理，促进植株生产健壮，提高植株抗病能力。合理修枝，使树冠通风透光。冬季清洁园地，把枯枝、病虫枝叶等集中烧毁，减少传染病。

5.4.3.2 物理防治

使用太阳能诱虫灯，诱杀夜间活动的害虫，每20～30亩的范围内布置1台太阳能杀虫灯。利用害虫的假死性，通过摇树进行人工捕杀。使用防虫板进行防治，将防虫板布置在园地外围。

5.4.3.3 生物防治

果园周围和行间种植蜜源植物，营造有利于天敌繁衍的果园生态环境。繁殖、释放和助迁主要害虫天敌，如捕食性瓢虫、捕食螨等。

5.4.3.4 化学防治

主要病虫害的化学防治参照表2-8或参照《木菠萝栽培技术规程》）（NY/T 3008—2016）的化学防治方法开展防治。

表2-8　主要病虫害的化学防治

防治对象	推荐药剂及使用方法
炭疽病	在发病初期，可选用50%多菌灵可湿性粉剂600～800倍液、75%百菌清可湿性粉剂500～800倍液、450 g/L咪鲜胺水乳剂800～1 500倍液或250 g/L嘧菌酯悬浮剂800～1 500倍液喷雾。隔7～10 d喷施1次，连续喷施2～3次
花果软腐病	在开花期、幼果期喷药护花护果，可选用77%氢氧化铜可湿性粉剂600～800倍液、80%波尔多液可湿性粉剂500～600倍液、25%咪鲜胺乳油1 000倍液、20%噻菌铜悬浮剂500～600倍液或47%春雷·王铜可湿性粉剂500～700倍液喷雾。隔7～10 d喷施1次，连续喷施2～3次
蒂腐病	在开花期、幼果期喷药护花护果，可选用70%甲基硫菌灵可湿性粉剂800倍液、50%多菌灵可湿性粉剂500～600倍液或25%咪鲜胺乳油1 000倍液喷雾。隔7～10 d喷施1次，连续喷施2～3次

（续表）

防治对象	推荐药剂及使用方法
根腐病	在发病初期，可选用70%甲基硫菌灵可湿性粉剂，用量为0.1～0.25 kg/株拌土25～50 kg撒施根际周围；或用80%代森锰锌可湿性粉剂400倍液、50%福美双可湿性粉剂500倍液灌根。隔7～10 d用药1次，连续用药2～3次
天牛	在主干发现新排粪孔时，可用钢丝钩杀蛀入枝干的幼虫，并用蘸有40%辛硫磷乳油20倍液的棉花堵塞洞口，然后用泥封堵蛀孔熏杀、毒杀幼虫；或选用40%噻虫啉悬浮剂3 000～4 000倍液对树干喷雾；或人工捕杀成虫，刮除树皮缝中或皮下的虫卵
黄翅绢野螟	在幼芽抽生期、开花期和幼果期，可选用10%高效氯氟氰菊酯乳油1 000～1 500倍液、1.8%阿维菌素乳油1 000～1 500倍液或8 000 IU/μL苏云金杆菌菌剂800～1 000倍液喷雾。隔7～10 d用药1次，连续用药2～3次
金龟子	在虫害发生初期，可选用5%氯虫苯甲酰胺悬浮剂2 000～2 500倍液、15%茚虫威乳油1 000～1 500倍液或5%甲氨基阿维菌素苯甲酸盐微乳剂2 000～3 000倍液喷雾。隔7～10 d用药1次，连续用药2～3次
绿刺蛾	在虫害发生初期，可选用10%高效氯氟氰菊酯乳油1 000～1 500倍液或1.8%阿维菌素乳油1 000～1 500倍液喷雾。隔7～10 d用药1次，连续用药2～3次
绿鳞象甲	在虫害发生初期，可选用5%氯虫苯甲酰胺悬浮剂2 000～2 500倍液或15%茚虫威乳油1 000～1 500倍液喷雾。隔7～10 d用药1次，连续用药2～3次

5.5 采收与采后处理

5.5.1 果实成熟的标志

果柄呈黄色，或离果柄最近的叶片变黄脱落；用手或木棒拍打果实，发出"噗、噗"混浊音；果皮为黄色或黄褐色，皮刺变钝；用小刀刺果，流出的乳汁清淡。

5.5.2 采收

根据用途和市场需求，将果实按成熟度进行分期分批采收。采收宜选阴天或晴天上午、傍晚进行，中午烈日或雨天不宜采收，整个采收过程中应轻采、轻放、轻运，避免机械损伤。采后果实集中存放于阴凉干燥处，避免暴晒。按《木菠萝》（NY/T 489—2002）的规定条件贮存。

参考文献

黄雄峰，熊月明，苏潮云，2020.菠萝蜜优良单株'秋红'特征特性及栽培技术要点[J].东南园艺，8（3）：36-37.

苏兰茜，白亭玉，吴刚，等，2019.菠萝蜜栽培研究现状及发展趋势[J].热带农业学，

39（1）：10-15，41.

谭乐和，2017.菠萝蜜、面包果、尖蜜拉栽培与加工［M］.北京：中国农业出版社.

王泽槐，潘达富，邓振权，等，2012.菠萝蜜新品种：'红肉波罗蜜'的选育［J］.果树学报，29（3）：518-519，312.

颜彩缤，胡福初，赵亚，等，2023.菠萝蜜优良新品种琼引8号的选育［J］.果树学报，40（3）：600-603.

袁海华，钟慧杰，敖新宇，2018.菠萝蜜种子的营养成分测定及分析［J］.食品研究与开发，39（24）：169-173.

张涛，潘永贵，2013.菠萝蜜营养成分及药理作用研究进展［J］.广东农业科学，40（4）：88-90，103.

西番莲（百香果）栽培关键技术

1 产业发展概况

西番莲（*Passiflora caerulea* Linnaeus）属于西番莲科西番莲属的多年生藤本植物，在北回归线至南回归线的热带、南亚热带地区广泛种植，其生长迅速，具有当年种植当年投产的优点。近年来，中国西番莲种植面积不断增加，2018年全国西番莲种植面积已达 472 km^2，产量 63.5 万 t，主要分布在广西、福建、广东、海南、云南、贵州等热带和亚热带地区，在鲜食、观赏、工业配料和药用等领域得到广泛应用。

西番莲果实成熟后，可散发草莓、番石榴、香蕉、菠萝、柠檬等多种水果的浓郁香味，有"果汁之王"的美誉。果汁占鲜果重的 30%～40%，营养价值很高，全果富含各类有机酸、膳食纤维、抗坏血酸、糖分、维生素、矿物质和果胶等，果汁中含有人体所需的糖、脂肪、蛋白质、维生素、氨基酸、钙、钾、铁、磷等营养物质及元素，食用价值高。

由于品种遗传背景的差异，西番莲对栽培环境的要求及反应也会有差异。①紫果型，能适应南回归线以北至热带气候区之间的亚热带地区及北回归线以南至热带北部间的亚热带地区，能适应一定的冬季冷凉天气，能短期忍受 0 ℃低温和轻霜冻，寒冷天过后易恢复生长。该类型喜旱湿交替的季风型气候，在年降水量 800～2 000 mm 的地区可生长结果，若将其种在常年高温多雨的热带低地，则植株藤蔓会疯长，开花结果少。②黄果型，原产于热带低海拔地区，喜湿热，能适应热带高温、多雨、高湿的环境条件，植株在一年中生长、开花、结果多次交叉进行，一年能收 2～3 造果，但对低温霜冻比较敏感，不能忍受较长时间的 0 ℃以下低温，遇 -2 ℃低温时间较长易被冻死，若遇 35 ℃以上高温、干旱或灌溉不及时，叶色会变黄，甚至干枯脱落。

西番莲的生长结果要求气温在 20～33 ℃，气温在 8～15 ℃生长缓慢，在 25～30 ℃生长最快，越冬最低气温在 5 ℃以上，遇 0 ℃以下低温霜冻会受冷冻害，花期 25～30 ℃为最适温度；年日照时数 1 800～2 200 h，方可满足其优质丰产的需要；年降水量 1 200～2 000 mm，且水量分布均匀较为理想，不积水，怕涝；尽量选择 8 级以上大风较少的地方，避免强风吹倒棚架吹断藤蔓及花果；土壤选择疏松肥沃、pH 值 5.5～6.5 的壤土、砂壤土。

2 主要栽培品种及特性

2.1 主要种类

2.1.1 紫果西番莲（*Passiflora edulis* Sims）

果实成熟后呈深紫色，适应于北回归线以南至热带之间的南亚热带气候区及南回归线以北至热带地区以南的亚热带气候条件，耐寒力稍强。

紫果西番莲生长势强，茎、卷须和叶纯绿色。叶片3裂，叶缘具细锯齿，叶茎心脏形，叶长10～18 cm。花稍小，直径约4.5 cm，新梢每节可长出1朵花。每朵花具5片发白的花瓣和萼片，2轮线状花冠，其基部为暗淡紫色，边缘处为白色，雄蕊5枚，顶部着生1个大花药，子房位于花中部，花柱分生为3个柱头。果实圆形或卵形，直径4～5 cm，平均单果重40～60 g，果汁香味浓、甜度较高，适合鲜食。平均果汁率30%。果实发育期为60～80 d，较耐寒，开花期不耐酷热天气。午夜开花，次日中午前关闭。

2.1.2 黄果西番莲（*Passiflora edulis* Sims f. *flavicarpa* Deg.）

果实成熟后呈黄色，要求热带生态条件。

黄果西番莲生长势较紫果型更强，茎、叶和卷须带有特征性的红色、粉红色或紫色，叶片与紫果型相似，但较大。花较大，直径约6 cm，花丝基部为明亮的深紫色。果形与紫果型相似，果较大，直径约6 cm，平均单果重60～90 g，成熟时果皮呈深黄色或亮黄色，果外表皮的星状斑点较明显。果汁含量可达45%。果肉含酸量比紫果型高，种子呈深褐色。在中午前后开花，21:00—22:00关闭。由于黄果型和紫果型的开花时间正好错开，它们彼此的天然授粉很少发生。花期从春天至晚秋均可出现，初夏有一短暂间歇，故成熟果实从初夏至冬季陆续出现。播种后约10个月开花，果实发育期约70 d；每年可获两季果实。

黄果西番莲生长旺、开花多、产量高、抗病力较强。但该种类不耐寒，遇严重的低温霜冻较易被冻伤、冻死。其果汁酸度大，一般用作工业原料加工果汁，不适合鲜食。黄果西番莲多数品种、品系自交不亲和，需要注意在建园时配置亲缘关系较远的黄果品种植株。

此外，还有大果西番莲、香蕉西番莲、甜果西番莲、樟叶西番莲、苹果状西番莲等，仅有局部地方栽培。

2.2 主要品种和品系

2.2.1 台农1号西番莲

台农1号西番莲是由中国台湾西番莲专家林莹达先生于1981年在台湾凤山热带园艺试验分所以紫果型为母本、黄果型为父本进行杂交所获得的第一代优株，果实鲜红色，圆形，果皮无斑点，略光滑，平均单果重62.8 g，该品种对环境的适应性较强，生长势旺，自交亲和，坐果率高，果较大，产量高，出汁率高，品质优，在北热带气候区的开花期很长，为3月中下旬至11月下旬，一年可收2造或3造果，缺点是不耐病毒病。

2.2.2 黄果选5-1-1

黄果选5-1-1果实鲜黄、果大、产量高，果实出汁率高，果汁色鲜、味浓香，含酸量较高，品质优，适于加工，需要进行异株或异品种（品系）人工辅助授粉才能结果丰产，果圆形，未熟果绿色，熟果鲜黄色，平均单果重68.4 g，自然坐果率46.7%，其加工品质及鲜食品质均较好，出汁率为39.3%，果汁含可溶性固形物15.1%，每100 mL含酸量2.4 g、维生素C 30.32 mg、还原糖14.3 g。

2.2.3 吉龙1号

该品种适应性广，抗逆性强，较耐寒，病虫害发生轻，早结丰产。果紫色或紫红色，果实鸭蛋形或球形，生长快，当年种植当年开花，自花授粉坐果率80%以上，不需进行人工授粉。丰产性好，平均单果重61.28 g，果实品质好，含糖量为15.4%～21%，维生素C含量丰富，每100 g果肉含维生素C 49 mg，香气浓郁。

3 栽培管理技术

3.1 种植规划与园区建设

3.1.1 园址选择

根据种植区域的西番莲鲜果销量以及加工生产线原料需求的数量和质量要求，调查预测以确定建设的规模和目标，选址应注意以下几点。

（1）应在西番莲的生态适宜区内选种植基地，避免基地遭受严重的低温、高温、大风等自然灾害。

（2）交通方便，无须自建道路与已有交通连通，只需要规划好基地内道路，便于生产物资及产品的运输。

（3）选择低丘陵坡地，向阳通风、土层深厚，pH 值 5.5～6.5，土壤有机质含量 1%～1.5%，疏松的壤土或砂壤土，宜修建梯田和排灌系统。

（4）靠近水源，确保干旱时可引水灌溉，雨水太多时又易于排涝，方便引灌水渠建设和滴灌设施建设。

（5）园区与葫芦科、茄科蔬菜基地，烟草基地，瓜果基地有 2 km 以上间隔距离，以防止相同病原的病毒病互相传染。

（6）预先配备好种苗基地，培育足量生产用健壮苗木。

3.1.2　园区规划

（1）按地形地貌、生产生活习惯以及交通路网配置合理规划办公区、职工宿舍区、生产物资仓库、农机具仓库、鲜果分级保鲜包装场、临时周转仓等场地位置。

（2）根据排灌管理等进行园区生产片区划分。

（3）合理布局符合片区生产运输需求的主路、辅路、人行道路路网。

（4）方便管理的排水、灌溉系统管网规划。

（5）设置防风林带。

3.1.3　山坡地果园建园

（1）山坡地果园要按等高线建设梯田。梯田面宽不小于 2 m，梯面向内侧倾斜 3°左右。梯田修建中要先把表土分段堆放，然后再开挖底土，底土叠筑成梯田外壁。外壁的上端修筑高于梯面 20 cm、宽 25 cm 的防冲刷土堤。梯田内侧修宽 25 cm、深 20 cm 的竹节沟，降雨少时可蓄水防旱，降雨多时可及时排水。外侧离土堤 15 cm 处翻挖宽 1 m、深 20 cm 的土面。

（2）按株距挖长、宽、深各 60 cm 的定植穴，每穴施用 10～20 kg 优质有机肥，氮、磷、钾含量均为 15% 的复合肥 1 kg，绿肥禾草碎 20 kg 与表土充分混合 1～2 个月腐熟后待用。

（3）按设计搭架，除坡度特别缓和、梯面特别宽的梯田外，一般只设 1 列架，特别宽阔的梯面才设 2～3 列架。

（4）西番莲棚架通常有单柱单线筒式架、单柱双线"T"形架、双柱三线"人"字架3种，结构简单，有利于攀缘展开，叶片受光面积大，可充分利用光能，生长迅速，有利于促花壮果，管理操作较方便，搭建成本较低。

3.2 整形修剪

西番莲是藤本攀爬植物，在主蔓、侧蔓节间、叶腋附近长出1条卷须。在自然条件下，依靠卷须随机缠绕前方可以缠绕的物件，扩大占领空间，增加叶片受光面积，提高光合作用量，以便获得更多的有机物质。合理修剪整形，可通过人工束缚捆扎、剪截引导，使植株更迅速地占领尚空余的空间，更加合理地利用生长空间。

3.2.1 幼树整形修剪

种苗定植成活后，每3～5 d巡检1次，抹除主蔓上萌生的侧芽。拔除杂草，有苗木倒伏的可插小竹枝捆扎扶正，当主蔓长至50～60 cm高时，用塑料绳或幼株的卷须将幼株固定在攀爬架上，引领幼株向上、向架顶攀长。当主蔓长到100 cm时，将其顶芽摘除，每条主蔓促生2～3条一级侧蔓，继续抹除其他侧芽，当一级侧蔓生长高过架顶20 cm时，将一级侧蔓分向棚架线两侧或分3个方向（如"T"形架）扎紧，将一级侧蔓的顶芽摘除，每条一级侧蔓促生2～3条二级侧蔓。当二级侧蔓长至50～60 cm时，摘去二级侧蔓的顶芽，每条二级侧蔓促生三级侧蔓2～3条。

主蔓和一级侧蔓是主要营养枝，二级至四级侧蔓是挂果时的主要结果蔓。因此，培养好生长健壮、充实、数量充足的二级至四级侧蔓，是当年取得丰产的基础条件。

3.2.2 结果树整形修剪

当年第1造果收果过半时，对结过果的蔓进行回缩修剪，选择较健壮的结果蔓，在其分枝基部前有3～4片完好叶片，离节眼上方1 cm处截去已结完果的部分，促其抽出2～3条四级侧蔓，作第2造果的结果蔓。至于生长势较弱的结果蔓暂不进行处理，只剪掉病虫枝、生长过密的弱枝。以保障秋冬果的产量。

3.2.3 越冬前整形修剪

在入冬前需要对西番莲进行1次清理性的修剪，主要是剪除病虫枝蔓，回缩剪除严重跨株生长的枝蔓，过密部分疏除较弱的枝蔓，让全株可以获得充足光照，减少各植株之间的空间争夺和互相干扰。但修剪也不宜过重，过重易出现一些修剪后枯死的情况。若原先的主蔓领导枝枯萎，则用二次梢代替原领导枝蔓。

3.3 水肥管理

丰产优质的果园应重视施用有机肥，实行有机肥与无机肥相结合的综合培肥措施，每年计划施用的氮、磷、钾适宜施用量中，由有机肥提供的量应占全年的40%左右，从定植起，每年每株都要施优质有机肥20～40 kg，这样的施肥措施能促进果园的土壤熟化，其土壤中有机养分与无机养分含量均保持较高水平。

各种植区域的环境条件差异会直接导致西番莲在各区域的物候期存在差异，可根据各地西番莲的物候期指导施肥，依据西番莲在热带、南亚热带栽培表现，在相同管理的同一品种同一果园同一植株同一时间中，都有处于不同物候态的现象出现，导致施肥时间不好确定。可根据有60%左右枝蔓在同一物候态出现时（如营养枝蔓、花蕾期、盛花期、小果迅速增大期等）定为该物候施肥期的时段，在施肥元素组合及其配比上也要考虑，除了照顾主物候期的需求，也要注意不能对同时出现的其他物候阶段植物器官产生伤害或阻碍（如为了抽春蔓而大量施用氮肥，会使枝蔓营养生长势过旺，引起落花、落果）。在适当增加氮元素比例时，配以适量的钾元素和磷元素，配足其他微量元素，充分满足各方面需要，才能取得最好的效果。

根据紫果西番莲在南亚热带气候的物候期，将南亚热带气候带下正常结果阶段的紫果西番莲园一年中的施肥时段列出以供参考：

第一次，1月中旬至1月下旬每株施尿素150 g、复合肥100 g、硫酸钾100 g，促春芽肥；

第二次，3月上旬至3月中旬每株施复合肥100 g、硫酸钾150 g，促春蕾壮花肥；

第三次，4月中旬至4月下旬每株施尿素50 g、复合肥50 g、硫酸钾200 g，促果大优质肥；

第四次，8月中旬至8月下旬每株施尿素60 g、复合肥50 g、硫酸钾100 g，促秋芽秋蕾肥；

第五次，11月下旬至12月中旬每株施过磷酸钙220 g，施有机肥改土壮树为下年丰产打基础。

根据年度物候实际情况结合品种特性进行具体施肥期和施肥量的调整。有滴灌条件的果园，尽量将水溶性肥料预先在混肥池中按安全浓度调校好，查验合格后随灌溉一起滴施，提高工作效率，有利于及时、安全、高效地补充肥料吸收利用。

3.4 主要病虫害防控

果园病虫害防控坚持以防为主、防治结合、统防统治为指导原则，坚持建设生态健康果园,降低病虫害的发生风险,减少病虫害为害带来的经济损失。西番莲主要病虫害有茎基腐病、病毒病、炭疽病、实蝇、茶翅蝽、稻棘缘蝽、豆芫菁、红火蚁等。

3.4.1 茎基腐病

茎基腐病病原为茄镰刀菌,是西番莲的一种毁灭性病害,主要为害离地面 5~10 cm 的植株茎基部。染病初期在植株茎基部出现深褐色病斑,之后病变皮层产生裂痕,变软腐烂,易脱离木质部。湿度大时,发病部位表面常出现粉红色菌落。发病后期上部叶片和枝蔓出现黄化、凋萎,然后整株枯萎死亡。

防治方法:加强管理,做好果园排水,清除发病枝条或植株后撒施生石灰,用 35 g/L 精甲·咯菌腈悬浮种衣剂、47% 春雷·王铜可湿性粉剂,喷离地 30 cm 以内的主茎,杀除表面病原菌,并灌根。

3.4.2 病毒病

病毒病也叫花叶病,在种植区普遍发生,发病率通常在 30%~40%,症状主要表现为叶片环斑、皱缩、花叶、环斑花叶、死顶和果实木质化等。西番莲多采用无性扦插繁殖,扦插过程中的枝剪没及时消毒及种苗本身带病毒,是该病快速扩散传播的主要途径。

防治方法:清除发病植株,杀灭可能传播病毒的媒介昆虫,选用无毒种苗或种植抗病种苗,合理密植,缩短单株挂果周期,1~2 年更新果园 1 次。也可选用氨基寡糖素和香菇多糖对植株进行喷雾,降低发病症状。

3.4.3 实蝇

目前常见为害严重的实蝇主要有橘小实蝇、瓜实蝇、南瓜实蝇、具条实蝇 4 种,成虫在果实皮下产卵,幼虫潜食,部分蛀空果实,严重时导致大量落果。

防治方法:农业综合防控、理化监测及诱控以及化学防治相结合,可选择性诱剂、食诱剂类绿色防控或触杀性(菊酯类、乙基多杀菌素、阿维菌素)和胃毒性(噻虫啉和吡虫啉)药剂来防治。

3.4.4 茶翅蝽

茶翅蝽为害西番莲叶和梢后症状不明显,果实受害后为害处木栓化,变硬,发育停止而下陷,严重时形成疙瘩或畸形果,失去经济价值。为害部位为叶片、花蕾、嫩梢、果实。同时也可能传播、携带西番莲病毒病。

防治方法： 重点防治春暖开始交尾的越冬代成虫和第 1 代若虫，这时对药剂较敏感，是防治最佳时期，可用 4.5% 高效氯氰菊酯乳油 1 000 倍液、25% 噻虫嗪水分散粒剂 1 500 倍液和 5% 甲氨基阿维菌素苯甲酸盐微乳剂 3 000 倍液等喷洒 1～2 次。

3.5 采收与采后处理

3.5.1 采收

果实成熟后，果实与果柄产生离层，会自动掉落。果实掉落之时起 3 d 内，是西番莲品质最佳、风味最好的时段。

长途运输销售的西番莲鲜果，要尽可能减少果实的外伤。若是为加工提供原料的，为减少用工、降低成本，大多在果熟盛期每隔 1～2 d 巡检果园，捡拾自动落地的熟果，及时集中用水洗净泥沙，晾干水分，剔除烂果，经预冷降温后在较低温度和较高空气湿度下贮藏备用。

禁采未成熟青果并将其混入成熟果中运销、加工、贮藏，因为未熟果含剧毒的氰化物，误食对人体有害。

3.5.2 采后处理

西番莲成熟果实属于典型的呼吸跃变类型，果体内自行释放乙烯较多，因此成熟果实是不耐贮存的。

采收下来的西番莲，放置在室温下的阴凉通风处，只能贮放 3～5 d。若用塑料薄膜小袋包装后贮放在气温 6.5～8 ℃、相对湿度为 80%～95% 的条件下，可以贮藏 20～30 d。因此，控制贮藏环境的温湿度可延长保鲜期；在保鲜袋内放置乙烯吸收剂，可至少延长西番莲保鲜期 6 d。

参考文献

蔡昭艳，董龙，王葫青，等，2023. 百香果花不同发育阶段花粉活力、柱头可授性及其对坐果的影响 [J]. 果树学报，40（5）：969-977.

甘廉生，廖永林，陈晓胜，2020. 百香果优质丰产栽培彩色图说 [M]. 广州：广东科技出版社.

郭艳峰，李晓璐，杨得坡，2019. 不同百香果的挥发性成分分析研究 [J]. 中国南方果树，48（6）：59-63，71.

黄楚韶，吴楚彬，1998. 黄果西番莲新品系 5-1-1 的选育 [J]. 广东农业科学（5）：18-20.

黄修莲，冯锦清，零东宁，等，2022. 百香果保鲜与加工研究进展 [J]. 轻工科技，38（6）：1-4.

李丹萍，寸待泽，李晶，等，2022. 不同氮钾肥用量对百香果生长、品质及产量的影响 [J]. 中国土壤与肥料（12）：123-132.

易籽林，徐卫清，2018. 百香果在海南引种栽培现状与展望 [J]. 热带农业科学，38（7）：25-28.

余东，熊丙全，袁军，等，2005. 西番莲种质资源概况及其应用研究现状 [J]. 中国南方果树，34（1）：36-37.

张蕊，陈媚，冯红玉，等，2022. 百香果病虫害发生规律及综合防治技术研究进展 [J]. 现代农业科技（22）：94-98.

张中润，肖丽燕，冯红玉，等，2022. 海南百香果病虫害种类调查鉴定及其发生危害 [J]. 热带作物学报，43（10）：2114-2121.

周红玲，郑云云，郑家祯，2015. 百香果优良品种及配套栽培技术 [J]. 中国南方果树，44（2）：121-124.

周玉娟，谈锋，邓君，2010. 西番莲属植物的研究进展 [J]. 中国中药杂志，33（5）：1789-1792.

❖ 西瓜栽培关键技术

西瓜［*Citrullus lanatus*（Thunb.）Matsum. et Nakai］是一年生蔓生藤本植物。西瓜在中国栽培范围较广，品种甚多，外果皮、果肉及种子形式多样，以新疆、甘肃兰州、山东德州、江苏东台等地最为有名。西瓜堪称"盛夏之王"，清爽解渴，味道甘甜多汁，是盛夏佳果。西瓜除不含脂肪和胆固醇外，含有大量葡萄糖、苹果酸、果糖、蛋白氨基酸、番茄素及维生素 C 等物质。果肉味甜，能降温消暑；种子含油，可作休闲食品；果皮药用，有清热、利尿、降血压之效，深受广大消费者的喜爱。因其拥有在食用性上清爽解渴、味甜多汁，在种植性上生产适应性强、栽培周期短，在效益性上市场需求量大、增收效果显著等特点，已经成为我国重要的高效园艺作物之一。

1 生长条件要求

1.1 温度

西瓜喜热、不耐低温，适宜生长温度为 18～32 ℃，低于 10 ℃幼苗停止生长，5 ℃受到冻害。生长期需要一定的昼夜温差，适宜的昼夜温差有利于果实中糖分的积累，西瓜含糖量越高品质越好，营养生长的温度相对较低，结果及果实生长需要较高的温度。

1.2 水分

西瓜耐旱、不耐湿，阴雨天多湿度过大，易感病，产量低，品质差。苗期、伸蔓期、果实膨大期需要吸收较多的水分，但浇水过多容易造成烂根。

1.3 光照

西瓜喜光照，日照时长以 10～12 h 为宜，光照充足有利于糖分积累。

1.4 土壤

西瓜对土壤要求不严，以土层深厚肥沃、土质疏松的砂质壤土为佳。忌连作，需与其他非葫芦科作物轮作，其中水旱轮作田需要轮作3～5年、旱地5～7年。

2 品种选择

根据西瓜种植区域的土壤条件、气候条件、水资源状况等自然因素，以及消费市场对西瓜的消费需求选择适合的品种，并且尽量选择产量高、质量好、环境适应性强的品种。

在砧木选择上，应根据生产季节的需要，选择亲和力和共生性强，抗逆性强的南瓜、葫芦或野生西瓜杂交种作砧木。设施栽培或夏秋栽培可选择南瓜杂交种或冬瓜作砧木。

3 栽培管理技术

3.1 整地

3.1.1 耕地

整地前需清除前茬残留物，冬季前深翻30 cm以上，秋冬反季种植应晒垡15 d左右，定植前15 d细碎土壤、平整土地。

3.1.2 理墒

采用一墒单行同向地爬栽培模式，墒宽120 cm，墒与墒间隔130 cm，排水沟深度为25 cm；采用一墒双行两边对爬栽培模式，墒宽300 cm，墒与墒间隔50 cm，沟深25 cm。

3.1.3 基肥

在墒上面每亩施腐熟的农家肥或商品有机肥1 500～2 000 kg、过磷酸钙50 kg、磷酸二铵（氮∶磷=18∶46）30 kg、硫酸钾10 kg和腐熟饼肥50 kg。缺乏微量元素的地块，补充相应肥料1～2 kg。施肥后用旋耕机将肥料拌入土中。

3.1.4 覆膜

一般覆膜栽培，起到保温保湿防草的作用。先铺滴灌管，后选用厚度 >0.01 mm、宽 120 cm 的地膜覆盖墒面，地膜四周用土压实。

3.2 育苗

3.2.1 设施要求

3.2.1.1 棚室

育苗一般在塑料大棚或温室进行，棚内应配有给排水设施、灯光设备，冬春季育苗应配套必要的加温和保温设施，以保证幼苗生长。

3.2.1.2 苗床规格

工厂化育苗一般苗床长度不超过 4 000 cm，宽度分为 3 个系列（165 cm、180 cm、185 cm），高度为 810 cm，可根据地面状况微调；苗床主要结构材料使用热镀锌、铝合金两种，床面的材料为铁丝网、潮汐板、塑料网，具体根据当地的情况进行调整；苗床必须配套给排水设施。

3.2.2 直播育苗

3.2.2.1 基质选择

一般选用 32 孔或 50 孔穴盘，穴盘用洁净的自来水反复冲洗、晾干。重复使用的穴盘，清洗污垢后采用 2% 次氯酸钠水溶液消毒处理。基质采用无污染草炭、蛭石和珍珠岩（草炭∶蛭石∶珍珠岩 =3∶1∶1）配置，1 m³ 基质加氮磷钾平衡复合肥 1.2 kg、50% 多菌灵可湿性粉剂 25 g，充分拌匀放置 2～3 d 后待用。重复使用的基质要进行消毒处理。

3.2.2.2 种子处理

一般未包衣种子需进行种子消毒处理。①晒种：选择晴天将种子晾晒 4～6 h。②浸种：用 55 ℃的温水浸泡种子 20 min，种子冷却至常温时，反复搓洗、搅拌，清除附着在种子上的黏液。③药剂消毒：可用 0.1%～0.2% 高锰酸钾、0.1% 多菌灵溶液、0.1% 硫酸铜溶液或 10% 磷酸三钠溶液等药剂浸种 15 min，后用清水冲洗 4～5 次。包衣种子不需要处理。

3.2.2.3 破壳

（1）有籽西瓜不需要破壳，可直接播种。

（2）无籽西瓜种子的中胚发育不完全，种壳、种皮、种脐较厚，发芽温度较高，需要在育苗之前通过人工破壳，用钳子等工具将种子从脐部缝合线处磕裂1条相当于种子长度1/3的小缝，并创造高温环境进行催芽，以提高出芽率。

3.2.2.4 催芽播种

将消毒后的种子用湿毛巾包裹，置于33～35 ℃的恒温箱中催芽，相对湿度控制在90%～95%，待75%以上的种子"露白"后，将出芽的种子挑选出来，剩下的种子在该环境下持续催芽，48 h后基本完成出芽，淘汰其他没有出芽的种子。

3.2.2.5 播种

每穴1粒，将种子平放，随后在种子上覆盖1 cm厚的基质，播完种后均匀浇透水。幼苗出土时，及时摘除夹在子叶上的种皮，防止子叶因不能及时展开而发黄。

3.2.2.6 苗期管理

萌芽期温度保持在30 ℃，促进种子及幼苗生长发育，破心期（从子叶微展到第一片真叶展出）以及炼苗期采用较低温度防止幼苗徒长，并锻炼幼苗耐逆性。出芽后白天温度保持在25～28 ℃，夜间温度保持在15～20 ℃，空气相对湿度保持在75%左右，基质相对湿度在50%～60%。根据种苗生长情况进行施肥，选择晴天下午喷施0.2%磷酸二氢钾溶液1～2次，每次施肥间隔7～10 d。

3.2.3 嫁接育苗

3.2.3.1 砧木和接穗播种

嫁接育苗在定植前33～40 d播种。采用顶插接法时，需要先播种砧木，砧木播种5 d后播种接穗；采用靠接法时，需要先播种接穗5 d后再播种砧木。嫁接育苗可选择穴盘，也可选择高10 cm、直径8 cm的营养钵，营养钵装土深度7～8 cm。

3.2.3.2 嫁接时间

采用顶插接法，当砧木生长到1叶1心、子叶初展平露心时嫁接；采用靠接法，当西瓜生长到1叶1心至2叶1心、砧木子叶初展平露心时嫁接。

3.2.3.3 嫁接前准备

嫁接操作应在适当遮光的棚内进行，并提前准备好嫁接竹签、刀片、消毒药剂等嫁接所需的工具；嫁接操作工人和嫁接工具均需要用75%乙醇溶液或高锰酸钾溶液消毒；嫁接前1 d对砧木和接穗喷施1次50%百菌清可湿性粉剂1 000倍液+3%中生菌素可湿性粉剂1 000倍液混合液消毒；嫁接前苗床浇透水；嫁接时温度宜控制在25～30 ℃。

3.2.3.4 嫁接方法

（1）顶插接法。先将砧木的生长点剔除，用竹签从生长点以向下45°的方式插入一根竹签，将接穗从子叶下1 cm处切下，下端切成0.5 cm长的楔形面，拔出竹签，立即将接穗插入到砧木中，子叶采用"十"字形的对接方式，保证接穗与砧木的四周完全贴合，同时要有一部分留在外面，最后用嫁接夹固定好。

（2）靠接法。先用嫁接刀挖去砧木的生长点，在砧木子叶下0.5～1 cm处斜向下切，刀口与胚轴的夹角约35°，深及胚轴2/3处，然后在接穗苗与砧木刀口相对方向，在子叶下约1.5 cm处斜向上切，刀口和胚轴夹角30°左右，刀口长度与砧木相等，深度在胚轴的3/4，最后把砧木和接穗的两个舌形刀口对齐嵌合，立即用嫁接夹固定，嫁接后立即把苗栽到营养钵内。

3.2.3.5 嫁接后管理

嫁接后3 d内苗床应密闭、遮阴，保持空气相对湿度90%以上、遮阴率为70%～80%，白天温度控制在25～28 ℃，夜间温度控制在18～20 ℃。3 d后早晚见光、适当通风，嫁接8～10 d后恢复正常管理，在此期间及时除去砧木萌芽。

3.2.4 壮苗

在定植前5～7 d控制浇水，保持基质相对湿度在60%左右，以秧苗不萎蔫为宜，加强棚内通风、透光，适当降温。西瓜秧苗长到3～4片真叶，且节间短，茎粗达到3.5～4.5 mm，叶片浓绿，穴中根系缠绕基质并形成完整的根坨，没有机械损伤，无病虫害，嫁接苗接穗不徒长。

3.3 定植

3.3.1 定植条件

最低气温稳定在5 ℃以上、10 cm深土壤温度稳定在15 ℃以上定植。

3.3.2 定植密度

根据品种特性、栽培季节和整枝方式确定定植密度。露地栽培时，早熟品种株距40 cm左右，每亩定植约660株，中、晚熟品种株距45 cm左右，每亩定植约530株；设施栽培时，早中熟型品种株距30 cm左右，每亩定植1 200～1 400株，晚熟品种株距约45 cm，每亩定植530株。

3.3.3 定植方法

定植时应保证幼苗茎叶和根系所带营养土块完整，定植深度以营养土块表面与畦面相平为宜；嫁接苗接口应高于地面 $1\sim2$ cm；定植后浇透水；根据需要覆盖地膜并用细土压封定植苗地膜口四周。

3.4 水肥管理

3.4.1 浇水

定植后浇足定根水；定植成活后浇 1 次缓苗水；抽蔓时浇 1 次；开花前浇 1 次坐瓜水；坐瓜后 $3\sim4$ d 浇 1 次水，促进幼瓜膨大；采收前 $7\sim10$ d 停止浇水。

3.4.2 追肥

（1）抽蔓初期，以氮肥为主、钾肥为辅，每亩施尿素 $10\sim15$ kg、硫酸钾 $4\sim5$ kg。

（2）果实膨大初期，以钾肥、氮肥为主。每亩施尿素 $20\sim25$ kg、硫酸钾 $10\sim15$ kg，叶面喷施 $0.3\%\sim0.5\%$ 磷酸二氢钾 + 0.1% 硼砂和 0.5% 硫酸亚铁水溶液。

（3）果实膨大中期，每亩施三元复合肥（氮：磷：钾=15：15：15）$10\sim15$ kg。果实膨大初期和中期，叶面喷施 $0.3\%\sim0.5\%$ 磷酸二氢钾 + 0.1% 硼砂和 0.5% 硫酸亚铁水溶液。

（4）第一批西瓜采收后。每亩施三元复合肥（氮：磷：钾=15：15：15）15 kg，促苗恢复坐第二批瓜；第二批瓜果实长至直径 $3\sim5$ cm 时，每亩施三元复合肥（氮：磷：钾=15：15：15）20 kg。

（5）第二批果实膨大中期，每亩施三元复合肥（氮：磷：钾=15：15：15）15 kg。可叶面喷施 $0.3\%\sim0.5\%$ 磷酸二氢钾 + 0.1% 硼砂和 0.5% 硫酸亚铁水溶液。

3.4.3 整枝

整枝宜在坐果期进行，整枝方法包括单蔓整枝、双蔓整枝和三蔓整枝。

3.4.3.1 单蔓整枝

只留 1 条主蔓，并将其余侧蔓去除，单蔓只留 1 个瓜。

3.4.3.2 双蔓整枝

留主蔓，当侧蔓长至 20 cm 左右时，从中选留 1 条健壮的侧蔓。双蔓枝条一般只让主蔓结 1 个瓜，坐瓜前，要将侧蔓全部去掉，坐瓜后长出的侧蔓可以保留，其长势较弱，不会影响果实生长。单蔓整枝或双蔓整枝适用于早熟品种。

3.4.3.3 三蔓整枝

保留主蔓,在侧蔓长至 20 cm 左右时,从中选留 2 条健壮侧蔓。这种方式适应大果型晚熟品种,低密度栽培。只在主蔓留瓜,两侧蔓同样起到营养作用。

3.4.4 理蔓和压蔓

当西瓜蔓长至 50～60 cm 时,应朝同一方向理蔓。选择晴天下午,在蔓上每隔 4～5 节用土块压 1 次,共压蔓 2～3 次,起到固定作用,这样可防止大风将瓜蔓刮乱,避免西瓜幼果受损。

3.4.5 打杈

在西瓜坐瓜前要及时打掉主蔓和侧蔓上的杈枝,防止营养供应不均衡,进而影响西瓜坐瓜和果实膨大。但坐瓜后应减少打杈次数。

3.4.6 留瓜

在蔓长 130～160 cm 处,选主蔓子房大而正、瓜柄直而粗,第二或第三雌花上的瓜果;采用单蔓整枝、双蔓整枝、三蔓整枝时,每株留 1 个瓜;采用多蔓整枝时,每株可留 1～2 个瓜;第一茬瓜采收结束或接近成熟时可再留 1 个瓜,二茬瓜用侧蔓留瓜。

3.4.7 授粉

可采用蜜蜂授粉,以此提高授粉效率。除此之外还可采用人工授粉,每天 7:00—9:00,采当天开放的雄花,剥去花冠,将雄花的花蕊在雌花的柱头上轻轻涂抹,让雄花花粉均匀散落到雌蕊柱头上。

3.4.8 翻瓜

翻瓜是为了保证西瓜着色均匀,具体做法是在果实膨大定型后顺着一个方向转动,每 2～3 d 翻 1 次,每次转动角度不宜超过 30°,每个瓜翻 2～3 次。若遇阴雨天要增加翻瓜次数。

4 病虫害防治

4.1 病害防治

4.1.1 枯萎病

4.1.1.1 为害症状

瓜农叫"死藤",发病时白天凋萎,早晚恢复,经 4～5 d 后死亡,植株基部皮层纵裂,流出胶质,为病菌侵染维管束所致,是西瓜产区的主要病害之一。

4.1.1.2 防治方法

在发病初期用药预防,一旦病害蔓延,可选用以下药剂灌根,50% 甲基硫菌灵可湿性粉剂 600 倍液、10% 咯菌腈悬浮剂 1 000 倍液、10 亿 CFU/g 解淀粉芽孢杆菌可湿性粉剂和 15% 噁霉灵水剂 300 倍液,每株灌药液 300～500 mL,每隔 7～10 d 灌 1 次,可连续灌 2～3 次,严重发病地块,在定植前进行土壤消毒。

4.1.2 蔓枯病

4.1.2.1 为害症状

主要侵染茎蔓,也侵染叶片和果实。叶片染病,呈现圆形或不规则形黑褐色病斑,病斑上生小黑点;湿度大时,病斑迅速扩及全叶,致叶片变黑而枯死。瓜蔓染病,节附近产生灰白色椭圆形至不整齐形病斑,斑上密生小黑点,发病严重时,病斑环绕茎及分权处。果实染病,初产生水渍状病斑,后中央变为褐色枯死斑,呈星状开裂,内部呈木栓状干腐,稍发黑后腐烂。

4.1.2.2 防治方法

发病初期可用 80% 代森锌可湿性粉剂 700～800 倍液、40% 苯甲·吡唑酯悬浮剂 2 000 倍液和 30% 苯甲·嘧菌酯悬浮剂 1 200 倍液,每 7 d 喷 1 次,连续 3～4 次。也可用 50% 甲基硫菌灵可湿性粉剂 500 倍液、40% 甲醛溶液 100 倍液涂抹病部,能起到很好的效果。

4.1.3 疫病

4.1.3.1 为害症状

主要侵害幼苗、叶、蔓和果实,受害部位先呈水浸状,而后干枯死亡。多雨潮湿及积水条件下发病严重；气候干燥、雨水少，则发病轻或不发病。

4.1.3.2 防治方法

（1）农业防治。加强田间排水，雨季瓜田无积水，防止病菌随水滴飞溅而传染。

（2）化学防治。用25%甲霜灵可湿性粉剂1 000倍液、72%霜霉威盐酸盐水剂600倍液和60%唑醚·代森联水分散粒剂600倍液喷雾防治。

4.1.4 炭疽病

4.1.4.1 为害症状

在高温高湿条件下易发病。发病时茎叶及果实表面形成圆形斑点或斑痕，呈水浸状，后变褐变黑，出现同心轮纹，而后干枯死亡。

4.1.4.2 防治方法

发病初期选用80%代森锰锌可湿性粉剂500倍液、50%嘧菌酯水分散粒剂2 000倍液、325 g/L苯甲·嘧菌酯悬浮剂1 500倍液喷雾防治，每7～10 d防治1次，连续防治3次。

4.1.5 白粉病

4.1.5.1 为害症状

发病初期，叶片上出现白色圆形小粉斑，病斑扩大后相连成片；发病后期白粉病呈灰白色，其上有时可见黄褐色的小斑点，严重时叶片正反面、茎蔓上布满一层白粉，病叶枯萎、发脆、卷缩。

4.1.5.2 防治方法

发病初期，用10%苯醚甲环唑可湿性粉剂1 000倍液、70%甲基硫菌灵可湿性粉剂800倍液和40%苯甲·嘧菌酯悬浮剂1 500倍液喷雾防治，每隔7 d喷雾1次，连续2～3次。

4.1.6 病毒病

4.1.6.1 为害症状

又称小叶病、花叶病，叶面出现黄绿镶嵌花斑，茎叶皱缩扭曲，叶面凸凹不平，出现花叶及瓜蔓茎尖叶片变小而卷曲等。

4.1.6.2 防治方法

防治此病重点是防治蚜虫、红蜘蛛等，用吡虫啉、啶虫脒和噻虫嗪等药剂防治，另外要严防接触传染。每年5月底至6月初是露地西瓜病毒病的大发生时期，可用1%

香菇多糖水剂 2 000 倍液、1.5% 烷醇·硫酸铜可湿性粉剂、20% 吗胍·乙酸铜可湿性粉剂 500 倍液喷雾预防。

4.2 虫害防治

4.2.1 蚜虫

4.2.1.1 为害症状

蚜虫会在瓜叶背面或幼嫩茎芽上群集，吸食汁液，被害叶片卷缩，严重时卷曲成团，甚至生长停滞，在苗期甚至会萎缩死亡。其分泌物会覆盖在叶片上，诱发煤污病，影响西瓜叶片的光合作用。蚜虫还是传播病毒病的媒介，会使西瓜植株出现花叶、畸形等症状，严重影响西瓜的产量和品质。

4.2.1.2 防治方法

采用悬挂粘虫板、色膜驱避、通风口处增挂防虫网等方式。当蚜虫普遍发生时，用 25% 噻虫嗪水分散粒剂 5 000 倍液、3% 啶虫脒微乳剂 2 000 倍液、10% 吡虫啉可湿性粉剂 1 000 倍液等喷雾防治。

4.2.2 蓟马

4.2.2.1 为害症状

被害嫩叶、嫩梢缩扭变硬，叶不展，蔓不伸，不坐瓜，或瓜皮粗糙布满锈斑，或瓜僵而不膨大，或为畸形果、裂果。最危险的是蓟马传毒，幼苗受害感染病毒病。

4.2.2.2 防治方法

悬挂蓝色粘虫板进行防治。药剂防治同蚜虫防治。

4.2.3 白粉虱

4.2.3.1 为害症状

成虫和若虫吸食西瓜植株汁液，受到为害的西瓜叶片褪绿、变黄、萎蔫甚至全株枯死。

4.2.3.2 防治方法

（1）棚室内设黄色粘虫板诱杀成虫。

（2）在西瓜定植前，用 80% 敌敌畏乳油 1 000 倍液全棚喷施并连续 3～5 d 密闭棚室；也可用 10% 异丙威烟雾剂烟熏。

（3）用 25% 噻虫嗪水分散粒剂 6 000 倍液、25% 噻嗪酮乳油 1 000 倍液、20% 啶虫脒乳油 2 500 倍液喷雾防治。

4.2.4 红蜘蛛

4.2.4.1 为害症状

西瓜受到红蜘蛛为害后，初期症状有叶片长出黄褐色小斑点，量大时红蜘蛛会在植株表面拉丝爬行，叶片背面出现红色斑块且比较大，后期症状为叶片卷缩、枯黄、脱落等，整株叶片枯黄泛白。

4.2.4.2 防治方法

适时合理追肥、灌水，促进西瓜生长，以增加抵抗力，同时，田间湿度较大，也不利于红蜘蛛发生。药剂可用阿维菌素、哒螨灵、炔螨特等防治。

4.2.5 蛴螬

4.2.5.1 为害症状

幼虫可咬断西瓜幼苗的根、茎，使全株死亡，造成缺苗断垄；或啃食根、茎，使西瓜苗生长衰弱直接影响产量和品质。

4.2.5.2 防治方法

（1）灯光诱杀。在成虫盛发期，用灯光诱杀，可极大地压缩虫口，减轻为害。

（2）土壤处理。每亩用2.5%敌百虫颗粒剂1.5～2 kg，拌细土撒于苗床底层或定植穴。

（3）化学防治。用90%敌百虫可溶粉剂800倍液、25%噻虫嗪水分散粒剂500倍液灌根，或用0.05%甲氨基阿维菌素苯甲酸盐颗粒剂35～45 kg/亩穴施。

5 采收

西瓜采收时应根据品种成熟特性和上市销售情况，按照果实的成熟度分批采收。对采收成熟度要求不严的品种可适当早收，对采收成熟度要求严的品种应达到所需成熟度才能采收。供应当地市场可采摘9成熟的瓜，运往外地的要采7.5～8成熟的瓜。

5.1 采收时间

春季栽培西瓜一般于4月下旬开始采收第1批瓜，或开花后40 d左右采收第1批瓜。第2批以后在西瓜花后28～32 d采收。早熟品种需28～30 d，中熟品种需

30～35 d，晚熟品种需35～45 d。采收时宜选择在早晨和傍晚，避免阴雨天采瓜，以防裂瓜增多。

5.2 成熟度判断

西瓜成熟后一般表皮坚硬光滑具有光泽，果面花纹清晰，果实脐部和果蒂部位向里收缩、凹陷，果柄处刚毛稀疏不显，轻拍有清脆的回声，即为熟瓜，应及时采收。

5.3 采收方法

采收时保留瓜柄和一段瓜蔓，准备贮藏较久的西瓜，最好连同一段瓜蔓割下。采收后应防止日晒、雨淋，及时运送出售，暂时不能装运的，要放到地头或路边阴凉处散去田间热，并轻拿轻放，瓜下垫一些瓜蔓或草。

6 采后处理

6.1 预冷

预冷最简单的方法是在田间进行。利用夜间较低的气温冷却一夜，在清晨气温回升之前装车或入库，有条件的地方可采用机械风冷法预冷。

6.2 分级

西瓜一般按品质、颜色、个体大小、重量、新鲜程度、有无虫病伤等方面分级。

特级：具有本品种的典型形状和色泽，大小一致，并且包装排列得整齐，允许有4%的误差（数目或重量）。

一级：具有本品种的典型形状、色泽和风味，允许呈现某些缺陷，但不影响外貌和保存品质。

6.3 包装

西瓜可根据果形的大小和商品价值采用相应材料进行包装或散装。一般多为散装，少数用条筐、板箱、塑料网袋等包装。

6.4 贮藏

暂时不鲜销的，可以进行贮藏，贮藏的库房应阴凉并具有通风条件。常温贮藏主要依靠通风降温，一般是白天关闭门窗，夜间通风，温度保持在15～20 ℃，空气相对湿度70%～80%。果皮坚硬、果肉致密、不易倒瓤的品种较耐贮，无籽西瓜较耐贮。果实采收后应及时进行药剂处理，发现病果、烂果、表皮受伤果，要及时剔除。

参考文献

冯翠，刘慧颖，田鹏飞，等，2022. 早春西瓜—夏丝瓜—秋花椰菜绿色高效周年栽培模式［J］. 中国蔬菜（6）：128-130.

马江黎，徐红，孙兴祥，2022. 不同处理对连作土壤及西瓜产量和品质的影响［J］. 中国瓜菜，35（7）：50-55.

孟佳丽，吴绍军，王夏雯，等，2020. 稻前茬西瓜栽培技术［J］. 中国瓜菜，33（1）：79-81.

王驰，杨瑜斌，林怡，等，2021. 设施西瓜多种接茬种植模式与关键栽培技术［J］. 中国蔬菜（11）：117-121.

杨洛滨，王海英，张梅志，2020. 露地盐碱地富硒西瓜栽培技术［J］. 中国瓜菜，33（3）：79-80.

第三章

蔬　菜

番茄栽培关键技术

番茄（*Lycopersicon esculentum* Mill.）是茄科（Solanaceae）番茄属（*Lycopersicon*）一年生或多年生草本植物，别名：西红柿、番柿、柿子等。番茄原产南美洲的秘鲁、厄瓜多尔、玻利维亚。番茄具有适应性强、栽培容易、产量高、营养丰富、用途广泛等优点，中国南北方广泛栽培。2021年，中国番茄栽培面积为 1 144 821 hm²，占世界总栽培面积的 22.2%。番茄的果实营养丰富，具特殊风味，可以生食、煮食、加工（番茄酱、番茄汁）或整果罐藏。

1 生长条件

番茄喜温、喜光、怕霜、怕热、耐肥、半耐旱，温度、光照、水分、土壤等因素影响番茄生长发育。

1.1 温度

番茄对温度的适应范围为 15～33 ℃，一般在 20～25 ℃下生长发育良好，番茄对温度的反应因生长阶段和发育不同而有所差异。

1.2 光照

番茄是喜光的作物，光照不足或连续阴雨天气常引起落花落果。光照强度对番茄的生长、发育有较大的影响。

1.3 水分

番茄根系比较发达，吸水力较强，因此对水分的要求表现为半耐旱。既怕旱又怕涝，要求土壤排水要好、地下水位低、水分均匀供给，不同的生长发育时期对水分要求不相同。

1.4 土壤

番茄对土壤条件要求不严格，以土层较厚、排水良好、富含有机质的肥沃壤土为适，pH 值以 6～7 为宜。

2 类型和品种

2.1 类型

番茄品种类型繁多，在园艺学上大体可以分为以下几种类型：按植物生长习性分为无限生长型、有限生长型；按叶型分为普通叶型、薯叶型、皱缩型；按果实大小分为大果型（150 g 以上）、中果型（100～149 g）、小果型（100 g 以下）；按颜色分为红色果、粉红色果、黄色果等。

2.2 品种

生产上选用丰产性、耐热性、抗病性均适应当季栽培、商品性适宜本地市场销售或外销、耐贮运的番茄品种。春夏栽培应选择耐低温弱光的早中熟品种，夏秋栽培应选择抗病毒病、耐热的品种。近年来，国内番茄黄化曲叶病毒病时有发生，建议选择抗病品种，减少损失。

3 栽培季节和方式

根据当地气候环境和市场需求合理安排播种期。华南部分地区及海南省、云南省部分地区终年无霜，可以露地越冬栽培，其他地区露地栽培均在无霜期内栽培。可进行露地栽培和设施栽培，云南可实现周年栽培。

4 栽培管理技术

4.1 播种育苗

育苗是蔬菜集约化生产的重要环节，培育和应用适龄幼苗是蔬菜早熟、优质、高产的重要保证。目前生产上常用的育苗方法有穴盘育苗、苗床育苗、营养钵育苗 3 种。其中，穴盘育苗有利于工厂化育苗（方便管理、运输等），除少数偏远地区外，蔬菜大面积生产地区都采用穴盘育苗。本节介绍穴盘育苗的方法。

4.1.1 播种前准备

云南省中部及气候相似地区于 1—2 月在塑料大棚内育苗，目前生产上番茄育苗多采用 50~72 穴的育苗盘。育苗场地应选择排灌良好、前茬未种过茄果类作物（如番茄、茄子、辣椒）的田块。穴盘育苗不仅可减少土传病害的发生，而且出苗整齐、粗壮，一次成苗，无须分苗等操作。放置苗盘的地块提前平整，避免地块出现高低不平，影响出苗的质量，有条件的地方可采用育苗架。育苗基质可用蔬菜育苗基质、烟草育苗基质，也可以自配。自配可用腐熟猪粪和田园土按 1∶1 混合，每 1 m^3 加入 100 g 复合肥、10 g 多菌灵拌匀，田园土可选用 1~2 年内未种过茄果类蔬菜的地块 15 cm 以内的表层土，或用未种过蔬菜的红土。播种前一天把育苗盘装满育苗基质或育苗土，用小木板刮平后用花洒把育苗盘浇足底水待用。育苗盘 2 盘一组摆放整齐，以便工人播种、间苗操作。

冬季无霜地区越冬栽培于 8 月育苗，此茬番茄苗期温度高，可用遮阳网覆盖等措施降温。

4.1.2 浸种催芽

番茄育苗的播种量一般为每亩 10~15 g，为达到早出苗、出齐苗、育壮苗的目的，在播种前一般进行种子处理，主要进行浸种和消毒（包衣种子不需要处理）。常用的方法是温汤浸种，具体方法：先将种子装入尼龙网袋或纱布袋内放入常温水中浸 15 min，然后转入 55~60 ℃的热水中浸泡 15 min，水量为种子体积的 10 倍。15 min 后让其自然冷却，在常温环境下泡种 4~6 h。也可将温汤浸种过的种子在 50% 多菌灵可湿性粉剂 1 000 倍液或 10% 磷酸三钠溶液中浸泡 20 min，以钝化种子携带的病毒，捞出后用清水冲洗干净，捞出后用纱布或湿布包住催芽。

将处理过的种子用湿毛巾包好，放在盆钵中。种子袋上面再盖几层湿毛巾，以保持湿度，然后放在 25~30 ℃条件下 3~5 d，有 50% 种子"露白"后即可播种。催芽过程中，每天翻动种子 1 次，使种子处于松散状态。如发现种子发黏，应立即用清水清洗毛巾。一般每天清洗 1 次，清洗后控出多余水分，以免种子吸水过多，造成种子缺氧，发芽不良。若无条件催芽，可浸种后直接播种。

4.1.3 播种

播种前一天把育苗盘或育苗床浇透水，育苗盘一般每穴播种 1 粒。播后盖 1 cm 厚的细土，不宜太薄，也不宜太厚，太薄浇水后种子容易露出或易出"戴帽苗"，太厚影响出苗。浇水后出现种子露出的情况,要及时覆土。播种后的苗盘上覆盖地膜保湿,有幼苗出土后,再将地膜揭去。低温度时大棚内育苗最好加盖小拱棚,或在苗床上用竹片搭小拱棚覆盖的薄膜保温,夜间有霜冻的地方还应覆盖草席保温,利于出苗。

4.1.4 苗床管理

从播种到出苗,要求床土水分充足、通气良好,保持苗床较高的温度,促使种子尽快发芽出苗。当幼苗出土,子叶张开时,白天要及时揭去草席薄膜透风,降低床温,防止徒长,但夜间仍应覆盖薄膜草席保温。若苗徒长,小棚应早揭晚盖,并适当控水。在育苗过程中,要适当追肥,一般都结合浇水,追施0.1%～0.2%的尿素。浇水要用细网眼喷壶或喷雾器喷洒,避免用瓢浇泼,以免土壤湿度过大,苗期病害严重。苗期主要病害是猝倒病,一经发现病株,马上拔除,并在病株周围撒多菌灵或百菌清原粉以防止病情蔓延。秧苗定植前10 d开始炼苗,加强通风,减少浇水,控制生长量,让幼苗尽快适应露地环境条件,缩短缓苗时间。

苗床浇水一般在9:00—11:00进行,每1～2 d浇水1次。苗期如果出现叶片发黄、叶小、茎细等生长不良的缺肥症状,可以采用根外追肥的办法,用0.3%磷酸二氢钾＋0.2%尿素溶液叶面喷施。

4.2 整地理墒

选择土质疏松、肥沃、排水良好,2～3年未种过茄科作物的砂壤土或壤土地块,水旱轮作最佳。

在前茬作物采收后及时深耕晒垡。每亩施腐熟农家肥2 500～3 000 kg、复合肥50 kg、过磷酸钙40 kg和钾肥15 kg,耙平地面作宽1.2 m的畦(包沟),深沟高畦,沟深30 cm左右。

4.3 适期定植

一般春季露地栽培的番茄定植均在当地晚霜终止后即可定植。番茄栽植的密度要根据品种特征特性、整枝方式、气候、栽培方法和土壤肥力水平等因素来决定。一般早熟品种比晚熟品种密度大;株型紧凑的品种比植株开展度大的品种密度大;土壤肥力低地块比土壤肥力高地块的密度大。

番茄苗5～7片真叶时定植。一般株行距(30～40)cm×60 cm,每畦种2行。用小锄头挖穴,放入苗四周盖上土压实即可,栽培深度以到子叶节处为好。定植后浇足定根水,以利缓苗。一般阴天定植,忌在雨天湿土定植,湿土定植难发新根,缓苗慢。

番茄壮苗标准为子叶完好，苗高 15～20 cm，有 5～7 片真叶，叶片肥厚、宽大，叶色深绿；茎粗壮，节间短；根系发达须根多，白嫩，无病虫害。

4.4 田间管理

4.4.1 水分管理

定植缓苗后适当控水，控制地上部植株生长，促进根系向纵深发展，调节营养生长和生殖生长的平衡，以积累更多养分供给果实的发育。到第一穗果坐果后，开始增加浇水量，5～7 d 浇水 1 次，保持土壤湿润。雨季做好排水工作。

4.4.2 施肥

番茄是连续生长结果的蔬菜，产量高，需肥量大。在施足基肥的基础上，提倡追肥早施、勤施。根据不同的生长期需求，适时、适量分期追肥。定植后 10 d 左右施提苗肥，可追施适量尿素 5～7 kg/亩。从第一穗果膨大开始，每隔 10～15 d 追肥 1 次，每亩交替施尿素 10 kg、复合肥 15 kg。同时，结果期除适当的土壤追肥外，还应结合叶面施肥，以弥补根部吸肥能力的不足，可用 0.2% 尿素和 0.3% 磷酸二氢钾混合液，每 15 d 喷施 1 次，叶面喷肥应选择晴天进行，喷施后应保证 24 h 无雨。喷施时，最好在傍晚或早晨尚有露水时进行，不宜在中午进行，以免因气温高而加快药肥液浓缩速度而造成肥害。

4.4.3 中耕除草、培土

一般搭架前要及时中耕除草，并结合培土，可保证土壤疏松透气，从而促进新根的发生，增强根系的吸收能力。一般从定植开始，每隔 10～15 d 中耕 1 次。中耕宜在土壤半干湿时进行。开花结果期中耕，应结合追肥和培土进行。搭架后植株封行，以清沟培土为主，结合除草。

4.4.4 整枝摘叶、立支架

番茄植株高 40 cm 左右要采用吊架或搭架方式进行吊蔓或绑蔓，采用竹竿支架绑蔓，多采用"人"字形，高 1.8～2.0 m。要及时绑蔓、整枝，整枝一般采用单干或双干式整枝（单干式整枝只留主茎，所有侧枝完全摘除；双干式整枝，除主枝外，再留第一花序至下叶腋所生的一条侧枝，其他侧枝全部摘除），侧枝生长到 7.5～10 cm 时剪除。无限型品种多采用单干整枝，整枝以晴天下午进行为宜，伤口容易愈合。当植株长到支架顶部，留2～3片叶摘心，集中养分，提高上层花序坐果率。生长后期应及时摘除基部老叶、病叶，有利于通风透光，减少养分消耗和病虫害发生。摘除的老、病、残叶，及时清理出园并进行深埋或烧毁。

4.4.5 保花保果

造成番茄落花的原因主要是光照不足，温度过高（夜温25 ℃以上或日温35 ℃以上）或过低（15 ℃以下）。营养不良、水分不足、施肥不当，植株茎叶徒长、病虫为害等都会引起落花落果。在生产上，除加强水肥管理、整枝摘叶外，还可用药剂进行保花保果，促进果实快速膨大。一般用2,4-滴10～15 mg/L或对氯苯氧乙酸纳水剂20～25 mg/L浸花或涂抹花柄，每隔3～4 d用1次。注意在配好的药水中加入有色墨水做标记，以避免重复点花而造成药害。2,4-滴对嫩叶和生长点会产生药害，注意小心使用。

4.4.6 疏花疏果

坐果后，及时疏花疏果，即人工去除多余的花，保证营养供给，发育成为有商品价值的果实，避免果实大小不均匀，畸形果多，商品率低。一般大番茄品种每花穗选留3～4个形状好的果实，中番茄品种每花穗选留4～5个果实，小番茄品种每花穗选留8～15个果实。

5 病虫害防治

以"预防为主，综合防治"的方针，采用以农业防治、物理防治、生物防治为主，化学防治为辅的防治原则。

5.1 农业防治

选用抗病品种，实施轮作；培育壮苗；采用深沟高厢栽培；合理密植；及时整枝搭架；拔除病重株；清洁田园；深翻炕土，减少病虫源；科学施肥。

5.2 物理防治

利用黄色粘虫板、蓝色粘虫板、频振式杀虫灯诱杀成虫；田间铺银灰膜或悬挂银灰膜条趋避蚜虫；大棚使用防虫网防止害虫进入，人工摘除害虫卵块和捕杀害虫等。

5.3 生物防治

保护有益生物。选用植物源农药、微生物、农用抗生素等生物制剂防治病虫害。

5.4 化学防治

5.4.1 主要病害及防治

番茄主要病害有病毒病、晚疫病、早疫病、青枯病、灰霉病、白粉病等。

5.4.1.1 病毒病

（1）症状。由病毒引起的病害，在田间主要表现的症状是花叶、蕨叶、条纹、丛生、卷叶、褪绿黄化等。该病害在秋播番茄生长中发生较为严重。

（2）防治方法。在发病前或发病初期，可选用5%氨基寡糖素水剂2 000倍液、1%香菇多糖水剂600倍液、20%吗胍·乙酸铜可湿性粉剂1 000倍液、5.9%辛菌·吗啉胍水剂300倍液等药剂喷雾防治，每隔7～10 d喷1次，连续2～3次。

5.4.1.2 晚疫病

（1）症状。主要为害叶片和果实，也为害茎和叶柄。病斑大多先从叶尖或叶缘开始，初为水浸状褪绿斑，后逐渐扩大，可扩大叶的大半以至全叶。天气干旱时病斑干枯呈褐色，叶背无白霉，质脆易裂，扩展慢。茎部皮层形成长短不一的褐色条斑，湿度大时病部表面产生白色霉层。果实染病多在青果期，发病部位可为果柄、萼片和果实，发病初期为油浸状浅褐色斑，发病部位多从近果柄处开始，逐渐蔓延，引起萼片发病，并向果实四周扩展呈云雾状不规则病斑，病斑边缘没有明显界限，发病果实的病部表面粗糙，果实质地坚硬，扩展后病斑呈暗棕褐色，湿度大时病斑边缘长出稀疏白色霉层。

（2）防治方法。叶柄和茎秆发病初期，可选用72%霜脲·锰锌可湿性粉剂600倍液、68%精甲霜·锰锌可分散粒剂600倍液、75%百菌清可湿性粉剂500倍液等药剂，每隔5～7 d防治1次，连续喷2～3次。

5.4.1.3 早疫病

（1）症状。主要为害叶、茎和果实等部位，叶片初期呈深褐色或黑色，圆形至椭圆形的小斑点，逐渐扩大后成为直径1～2 cm的病斑，病斑边缘深褐色，中央灰褐色，具有明显的同心轮纹，有的边缘可见黄色晕圈。潮湿时病斑表面生有黑色霉层。病害常从植株下部叶片发生，逐渐向上蔓延，严重时病斑相互连接形成不规则的大病斑，病株下部叶片枯死、脱落。茎部病斑多在茎部分枝处发生，病斑灰褐色、椭圆形、稍凹陷，具有同心轮纹，发病严重时病枝折断。果实上病斑多发生在蒂部附近和有裂缝处，圆形或近圆形，黑褐色，稍凹陷，具有同心轮纹，为害严重时，病果常提早脱落。在潮湿条件下，各受害部位均可长出黑色霉状物。

（2）防治方法。对连年发病的温室大棚，在定植前密闭棚室后，每亩用45%百菌清或10%腐霉利烟剂11～13 g熏烟。露地栽培可选择喷洒72%霜脲·锰锌可湿性粉剂600倍液、58%甲霜·锰锌可湿性粉剂800倍液、10%苯醚甲环唑水分散粒剂1 000倍液、70%甲基硫菌灵可性粉剂700倍液、50%异菌脲可湿性粉剂1 000倍液等，7～10 d喷药1次，注意轮换交替使用农药。

5.4.1.4 青枯病

（1）症状。该病是细菌性维管束组织病害。初始叶片顶部嫩叶先表现症状，特别是中午萎蔫下垂，傍晚恢复正常，后期很快扩展至整株萎蔫，并不再恢复而死亡。剖开病茎，维管束变褐色，横切后用手挤压可见乳白色黏液渗出。

（2）防治方法。目前对番茄青枯病尚无理想的防治药剂，防治上应采用综合措施，初果期开始加强田间巡查，一旦发现病株随即拔除，收集烧毁，病穴及附近植株淋灌77%氢氧化铜悬浮剂800倍液，或50%琥胶肥酸铜可湿性粉剂400倍液，每10 d灌根1次，连续灌根3～4次以上，间隔5～7 d，淋透淋足（200～500 mL/株）。

5.4.1.5 灰霉病

（1）症状。在整个生育期间，植株的各个部位均可感染。番茄灰霉病主要引起叶片及果实腐烂，一般先从较衰弱的子叶及真叶的边缘开始，叶片变软下垂后在病处产生大量的灰色霉层，最后病株折倒。严重时，田间幼苗成片腐烂；成株期发病，可为害地上的各个部位。番茄叶片染病多从叶尖及叶缘开始，初为水渍状，后颜色变淡，呈淡褐色，稍有深浅相间的轮纹，叶片病斑多呈"V"形，扩大后呈不规则形或圆形轮纹斑，边缘明显，叶面产生灰色霉层，有时病斑破裂；病斑往往不受叶脉限制继续向全叶扩展，致使叶片最后干枯死亡。茎部染病初呈水渍状小点，后病斑扩大，湿度大时病斑上产生灰色霉层，严重时引起植株枯死。病菌多从花瓣或柱头处染病，致使花腐烂，长出淡灰褐色的霉层，并引起落花。果实被害可造成烂果或外缘白色、中央绿色的圆形斑，即"花脸斑"。

（2）防治方法。番茄蘸（喷）花时加防治灰霉病的药剂，如在配好的对氯苯氧乙酸钠或2,4-滴稀释液中，加入50%异菌脲可湿性粉剂，或50%腐霉利可湿性粉剂，或50%多菌灵可湿性粉剂稀释液，然后进行蘸（喷）花。发病初期，可选用50%腐霉利可湿性粉剂800倍液，或40%嘧霉·啶酰菌悬浮剂2 000倍液，或65%甲霉灵可湿性粉剂800倍液，或25%腐霉·福美双可湿性粉剂800倍液等药剂喷雾防治2～3次，轮换交替使用农药，间隔期为7～10 d。

5.4.1.6 白粉病

（1）症状。番茄叶片、叶柄、茎和果实均可染病，一般下部叶片先发病，逐渐向上部发展。发病初期，叶面出现褪绿小点，扩大后呈近圆形或不规则形的病斑，表面生有白色粉状物，开始时白色粉层比较稀疏，以后逐渐加厚，并向四周扩展，严重时整个叶片布满白粉，抹去白粉可见褪绿的叶组织，最终病叶变黄褐并逐渐枯死。其他部位染病时也可产生白粉状病斑。

（2）防治方法。合理密植，及时摘除植株下部老叶。发病初期，可选用40%氟硅唑乳油6 000倍液，或15%三唑酮可湿性粉剂500倍液，或70%甲基硫菌灵可湿性粉剂600倍液等轮换交替喷施，连续防治2～3次，间隔7～10 d。

5.4.2 虫害防治

主要害虫有棉铃虫、白粉虱、蚜虫、蓟马等。

5.4.2.1 棉铃虫

（1）为害特点。主要以幼虫蛀食花蕾、花、果等，并且食害嫩茎、叶和芽。花蕾受害后，苞叶张开，变成黄绿色，2～3 d后脱落，花蕾和幼果常被吃空引起腐烂而脱落，成果期受害引起落果造成减产。

（2）防治方法。药剂防治在幼虫孵化盛期，可选用2%甲氨基阿维菌素苯甲酸盐乳油2 000倍液、32 000 IU/mg苏云金杆菌可湿性粉剂300倍液、600亿PIB/g棉铃虫核型多角体病毒水分散粒剂3 g/亩、14%氯虫·高氯氟微胶囊悬浮剂2 000～3 000倍液等喷雾防治，每隔3～5 d喷1次，连喷2～3次。

5.4.2.2 白粉虱

（1）为害特点。白粉虱为杂食性害虫，繁殖速度快，种群数量在秋季达高峰。以成虫、若虫群集在叶片背面，用刺吸式口器吸吮汁液，被害叶片褪绿、变黄，使植株生长缓慢、萎蔫，甚至死亡。还能分泌大量的蜜露，引发煤污病。亦可传播病毒病。

（2）防治方法。发生初期，可选用25%吡虫啉可湿性粉剂3 000倍液、70%啶虫脒水分散粒剂5 000倍液、25 g/L联苯菊酯乳油2 000倍液、1.8%阿维菌素乳油2 000倍液、25%噻嗪酮乳油2 500倍液、20%甲氰菊酯乳油2 000倍液等，每隔3～5 d喷1次，连喷2～3次。

5.4.2.3 蚜虫

（1）为害特点。以成虫或若虫群聚在叶片背面、嫩叶、嫩茎、花苞及近地面叶上，以刺吸口器吸食植株汁液和养分，分泌蜜露，常造成植株严重失水和营养不良。幼叶被害，卷曲皱缩，轻者褪绿，有斑点，叶片发黄；重者叶片卷缩变形枯萎，甚至枯死。

蚜虫还能传播病毒病。

（2）防治方法。发生初期，可选用25%吡虫啉可湿性粉剂3 000倍液、20%氰戊菊酯乳油2 000倍液、20%甲氰菊酯乳油1 500倍液、28%阿维·螺虫酯悬浮剂3 000倍液、14%氯虫·高氯氟微胶囊悬浮剂4 000倍液等喷雾，每隔3～5 d喷1次，连喷2～3次。

5.4.2.4 蓟马

（1）为害特点。成虫和若虫在嫩叶叶背毛丛和花中吸吮汁液，被害叶片背面茸毛及叶肉呈灰褐色，变硬老化，花蕾受伤脱落。受害植株生长缓慢，节间缩短。蓟马繁殖速度极快，一年可发生10～20代，世代重叠，终年繁殖，5—9月为发生高峰期，以秋季最严重，成虫活跃善飞，怕光，若虫落入表土化蛹。

（2）防治方法。可选用25%吡虫啉可湿性粉剂3 000倍液、25%噻虫嗪悬浮剂3 000倍液、28%阿维·螺虫酯悬浮剂4 000倍液等喷雾防治，每隔3～5 d喷1次，连喷3次。喷药重点是植株的生长点、嫩叶背面、花蕾等部位。

6 适时采收

番茄从开花到果实成熟，早熟品种40～50 d，中晚熟品种50～60 d。应根据需要适时采收和贮果催熟。

番茄采收期应根据番茄产品销售市场来确定，番茄成熟有绿熟、变色、成熟、完熟4个时期，长途运输在绿熟期（果顶及果面变白）采收；短途运输可在变色期（果实的1/3变红）采摘；就地出售或自食应在成熟期（即果实1/3以上变红）采摘。采收时最好不要扭伤果柄，用番茄剪沿果柄根部轻轻剪下，果柄不要露出果面，轻摘轻放，避免机械伤害。采收后进行分级包装，贮存或运输销售。

参考文献

林鉴荣，刘士亚，丘漫宇，等，2004. 番茄茄子无公害生产彩色图说 [M]. 广州：广东科技出版社.

全国农业技术推广服务中心，国家大宗蔬菜产业技术体系，2017. 番茄高效栽培与病虫害防治彩色图谱 [M]. 北京：中国农业出版社.

中国农业科学院蔬菜花卉研究所，2010. 中国蔬菜栽培学 [M]. 2版. 北京：中国农业出版社.

❖ 辣椒栽培关键技术

辣椒（*Capsicum* spp.）别名番椒、海椒、辣子，为茄科（Solanaceae）辣椒属（*Capsicum*）一年生或多年生植物，是全世界广泛种植消费的蔬菜和香料作物之一，起源于中南美洲热带、亚热带地区，是人类种植的最古老的农作物之一。早在公元前7000年辣椒有在南美种植记载，16世纪传入欧洲，17世纪由欧洲引入中国。

目前辣椒属中有5个种驯化栽培，分别为一年生辣椒（*C. annuum*）、中华辣椒（*C. chinense*）、浆果状辣椒（*C. baccatum*）、灌木辣椒（*C. frutescens*）和茸毛辣椒（*C. pubescens*），其中一年生辣椒和灌木辣椒在全球范围内广泛种植。

据国家大宗蔬菜产业技术体系统计，近年来我国辣椒年种植面积稳定在3 200万亩以上，占蔬菜种植面积的9.28%，成为中国种植面积最大的蔬菜，辣椒不仅用作蔬菜和调味品，还因含有丰富的辣椒素、辣椒红素、维生素C和维生素E等生物活性物质被作为重要的工业原料广泛应用于工业、医疗、化妆品、军事和航海等领域。

1 生长条件

1.1 温度

辣椒种子发芽适温25～30 ℃，生长发育温度20～30 ℃，坐果温度20～25 ℃；低于15 ℃或高于35 ℃坐果率下降；低于10 ℃停止生长；低于5 ℃易受冷害；低于0 ℃易受冻害。

1.2 光照

辣椒为喜温喜光作物，光补偿点为1 500 lx，光饱和点为30 000 lx，但对光周期不敏感。辣椒在短日照条件下，开花结果较快，故春播可适当促进早熟。

1.3 水分

辣椒是茄果类蔬菜中较耐旱的作物，忌水涝，否则极易引发病害，导致大面积死棵；辣椒苗期对水需求敏感，坐果期需水量增大，土壤含水量为土壤田间持水量的

60%～80%时生长良好，坐果率高。

1.4 土壤

辣椒对土壤的适应性较强，以土层深厚、土质疏松、肥水条件好、土壤pH值6.2～7.2的砂壤土为宜。辣椒根系一般分布在20～40 cm深的土层范围内，疏松、肥沃、排水良好的土壤有利于辣椒根系发育。地下水位较高、排水条件差、土质黏重的地块不适宜辣椒种植。

1.5 肥料

辣椒对氮、磷、钾肥的需求较高。整个生育期对氮的需求最多，占60%；钾次之占25%；磷为第3位，占15%。同时，还需要钙、镁、铁、硼、钼、锌等多种中微量元素。云南省种植的辣椒要适量增施硼、钼肥，对增产和提高果实品质有较好的效果。此外，硝态氮肥更容易被辣椒吸收利用，硝态氮和铵态氮的适宜比例为7∶3。

2 类型和品种

2.1 类型

辣椒品种类型多样，按照消费习惯主要分为鲜食型、加工型和鲜食加工兼用型三大类。鲜食型辣椒包括皱皮椒、甜椒、牛角椒和螺丝椒等；加工型辣椒包括干制辣椒、美人椒和工业辣椒等；线椒、朝天椒和小米辣则既可用作鲜食又可用作加工。

2.2 品种

鲜食型辣椒主产区以保山市为主，栽培品种类型繁多，主要以市场需求为导向；朝天椒在云南各地均有分布，文山州、红河州、普洱市、西双版纳州、临沧市等地为主产区，主要种植单生朝天椒品种。

以丘北辣椒、乐业辣椒为代表的干椒是云南省辣椒特色品种之一。丘北辣椒因其突出的品质驰名海内外，以丘北县为中心分布种植在文山州境内，主栽品种有云干椒系列、丘椒系列和文干椒系列等。乐业辣椒历史悠久，香气口感独具一格，主要种植于会泽县乐业镇。小米辣作为另一个特色品种，以泡椒凤爪、泡椒方便面等加工产品为人熟知，在云南省热区均有种植，红河州、文山州和普洱市是小米辣的优势主产区。

近年来，工业辣椒种植和配套精深加工产业异军突起，在云南省辣椒产业版图中占据了一席之地。目前，工业辣椒品种较为单一，品种抗病性和产量有待进一步提高。

3 栽培管理技术

3.1 种子处理

3.1.1 种子杀菌消毒技术

辣椒播种前对种子进行杀菌消毒处理，杀灭种子携带的有害微生物和病原菌，减少苗期病害，有利于培育壮苗。操作方法如下：第一，60 ℃温水浸泡15 min，其间不断翻动搅拌；第二，1%硫酸铜或50%多菌灵可湿性粉剂500倍液浸泡1 h；第三，10%磷酸三钠溶液浸泡1 h；第四，清水漂洗3～5次，取出晾干即可用作播种育苗。

3.1.2 吸湿—回干种子处理技术

经上述处理过的种子，可进一步进行吸湿—回干技术处理。该技术可增强种子活力，提高发芽率、发芽势，使得出苗快、出苗齐，还可提高幼苗抗旱性，提高移栽成活率和短期干旱的耐受力。

处理方法：①吸湿，将以上消毒后的种子置于25～30 ℃的清水中浸泡吸湿6 h左右。②回干，将吸湿的种子晾干至消毒处理前重量，或以手抓握微松手种子会从指缝滑落为晾干标准。③放置3～5 d后再重复吸湿—回干1次，晾干即可播种。需要注意，应在播种期前7～10 d对种子进行该处理。

3.2 育苗技术

3.2.1 主要育苗方式

当前生产中辣椒主要采用穴盘育苗，穴盘育苗具有基质用量小、秧苗根系发达、移栽时不易伤根、缓苗期短甚至几乎没有缓苗期等优点。

3.2.2 穴盘育苗技术

3.2.2.1 育苗基质与消毒

育苗基质对培育壮苗非常关键，可直接购买商品育苗基质或农家肥堆沤自制育苗基质。农家肥堆沤自制育苗基质配比：农家肥或腐殖土2 t＋鸡粪或羊粪300 kg＋复合肥（氮∶磷∶钾=15∶15∶15）20 kg+尿素 5 kg+过磷酸钙 50 kg。

3.2.2.2 播种

将基质装入穴盘，适度压紧，在基质中心挖一个 1 cm 深的小孔，将辣椒种子放置在小孔内，播种后盖 1 cm 厚的细土，不宜太薄，也不宜太厚，太薄浇水后种子容易露出或形成"戴帽苗"，太厚则影响出苗，然后浇足水。根据播种季节的温度选用塑料薄膜或遮阳网覆盖在穴盘上，保温保湿以利于出苗。穴盘须离地 50 cm 以上放置，避免土传病害的发生。切忌不能将穴盘直接放在水泥地或塑料薄膜上，以免烫伤根系。

3.2.2.3 出苗前的管理

出苗前保持温度 25～30 ℃，5～9 d 即可出苗。当 70% 左右的种子拱土出苗后，要及时撤掉覆盖物，适时浇水保持盖土湿润。出苗后通过揭膜通风等措施保持白天 20～23 ℃，夜间不低于 15 ℃。

3.2.2.4 出苗后的管理

温度保持在白天 27～28 ℃，夜间 18～20 ℃，土壤相对湿度为 70%～80%，空气相对湿度 50%～70%。根据秧苗生长情况每隔 3～5 d 浇施 1 次水溶性配方肥。浇水宜在晴天 10:00 左右，一次浇足，避免少浇勤浇。苗期还可喷 0.2%～0.3% 的磷酸二氢钾溶液 2 次。

定植前 1 周应逐步通风降温、控水控肥进行炼苗，最低夜温降至 10 ℃ 左右。定植前 1 d 浇足起苗水。

3.3 整地与栽培模式

3.3.1 选地整地

宜选择排灌方便、前茬为非茄科作物、富含有机质、土层深厚的地块。每亩施 3 000～4 000 kg 有机肥后翻耕约 30 cm，整平耙细土壤。

3.3.2 栽培模式

起厢栽培生产中使用较为广泛。采用 1.2 m 包沟开厢，沟深 30 cm，每厢 2 行，行距 50 cm，穴距 40 cm，按 50 cm—70 cm—50 cm 的宽窄行方式栽培。

干旱少雨地区可配合旱覆膜集雨栽培技术进行栽培，具体操作方法：做穴塘，覆盖地膜，穴塘中心用直径 5～6 cm 的尖木棒打孔以收集降水，并用少量疏松泥土覆盖孔口。4 月下旬气温回升后，遇有少量雨水即可定植，干制辣椒每穴定植 2 株，注意 2 株并拢不要分开，其他辣椒单株定植。

3.4 定植与生长期管理

3.4.1 定植

苗高 10～15 cm、7～8 片真叶时为最佳定植期。定植后浇足定根水，在植株周围用干细土覆盖地膜，防止地膜烫伤辣椒苗，同时可保持和提高地膜内地温。

3.4.2 生长期管理

3.4.2.1 生长前期管理

定植后到采收前的管理。定植后要提高地温促使尽快缓苗，基质育苗的辣椒苗，缓苗期一般仅需 2～3 d。定植后 1 周缓苗后施 1 次提苗肥，每亩浇施 6～8 kg 的 0.5% 硝酸铵或硝磷酸铵溶液。然后蹲苗 10～15 d，通过控水控肥提高地温，使土壤空气充足促进根系生长。

蹲苗结束后结合施肥浇灌 1 次大水，每亩施 15 kg 硝酸铵，同时叶面喷施 0.3% 磷酸二氢钾或氨基酸肥料 1～2 次。辣椒对硼和锌较敏感，可适当叶面喷施 0.1% 硼砂和 0.05% 硫酸锌微量元素肥料。

3.4.2.2 生长中期管理

开始采收至盛果期的管理。主要目标是促秧攻果，争取盛果期提前到来，并在高温季节来临前植株封行。具体措施：①及时采收，盛果期应每隔 4～6 d 采收 1 次；②加强肥水管理，一般每采收 1～2 次后，追肥灌水 1 次，每亩追施硝酸铵 10 kg + 复合肥 15 kg，并每隔 10 d 左右进行 1 次叶面喷肥。

3.4.2.3 生长后期管理

此期管理的要点是促使植株发新枝，促进第 2 次结果高峰。具体措施如下。①进行整枝，清理病残株。保留有分枝能力的健壮枝条和下部已抽生的枝条，剪除老枝、弱枝等已丧失分枝能力的枝条，清除老病叶。②肥水管理。一般每隔 10～15 d 施 1 次速效肥，每亩施 10～15 kg 硝酸铵，促使新枝生长。③浇水与排涝。高温多雨季节保持土壤湿润，及时浇水降温；在雨后及时排除沟内积水，防止沤根。

4 病虫害防治

病虫害防治应遵循"预防为主，防治结合"的绿色防控原则，采用以农业防治、物理防治、生物防治为主，化学防治为辅的综合防治原则。

4.1 病害防治

4.1.1 猝倒病

4.1.1.1 主要症状

出土前发病引起烂种烂芽；幼苗出土后发病，茎基部呈黄绿色水渍状，后很快转黄褐色并发展至绕茎一周。病部组织腐烂干枯而凹陷，产生缢缩。水渍状自下而上扩展，幼苗倒伏于地。发病初期，苗床上只有少数幼苗发病，几天后，以此为中心逐渐向外扩展蔓延，最后引起幼苗成片倒伏死亡。

4.1.1.2 发病条件

病原菌主要通过灌溉水或雨水传播，高湿条件下发病重，辣椒苗期如遇连续阴雨、光照不足、幼苗生长势弱等易发病。

4.1.1.3 防治方法

（1）农业防治。厩肥或堆肥必须经过高温发酵，充分腐熟；加强通风，防止苗床土过湿。

（2）化学防治。严格进行苗床消毒，发病初期可用噁霉灵、精甲·噁霉灵、霜脲·锰锌喷雾防治，喷药待叶面干后再撒施草木灰。

4.1.2 立枯病

4.1.2.1 主要症状

一般在辣椒真叶出现以后、开花结果以前为害。幼苗白天萎蔫，夜间恢复，反复几天以后，植株枯萎死亡。茎基部生椭圆形、暗褐色病斑，略凹陷，扩大到茎基部周围，病部收缩干枯，叶色变黄凋萎，根变褐腐烂，直至全株死亡。湿度高时，病部生褐色稀疏的蛛网状霉。

4.1.2.2 发病条件

播种过密、间苗不及时、高湿、16～24 ℃容易发病。

4.1.2.3 防治方法

（1）农业防治。选用无病土或基质育苗，增施磷钾肥，防止土壤忽干忽湿。

（2）化学防治。发病前喷百菌清或敌磺钠预防；发病后撒草木灰或拌有噁霉灵、多·福药剂的干细土；也可用38%甲霜·福美双可湿性粉剂800倍液、30%噁霉灵水剂1 500倍液、50%异菌脲可湿性粉剂2～4 g/m^{2}或30%甲霜·噁霉灵水剂2 000倍液泼浇和灌根。

4.1.3 疫病

4.1.3.1 主要症状

茎枝病部开始为暗绿色水渍状，后变为褐色坏死长条斑，病部凹陷缢缩，植株上部萎蔫枯死。叶片受害产生暗绿色水渍状圆形或近圆形的病斑，直径 2～3 cm；湿度大时整叶腐烂，干燥时，病斑淡褐色，病叶易脱落。果实受害始于蒂部，产生暗绿色水渍状病斑，湿度大时变褐软腐，表面长出白色稀疏霉层，干燥时形成僵果残留于枝上。根部受害变褐腐烂，整株萎蔫枯死，但维管束不变色，该症状有别于镰刀菌引起的枯萎病。

4.1.3.2 发病条件

重茬、低洼地、排水不良，氮肥使用偏多、密度过大、植株衰弱均有利于疫病的发生和蔓延。该病害发生适宜温度为 20～30 ℃、土壤相对湿度 95% 以上时，2～3 d 即可完成一个侵染循环，所以灌水量大或遇大暴雨、气温高，易暴发流行。

4.1.3.3 防治方法

（1）农业防治。进行水旱轮作；选用抗病品种；适当控制氮肥，增施磷、钾肥，施用腐熟农家肥；深沟高厢栽培，避免久旱后灌水，严禁大水漫灌。

（2）化学防治。种子处理，种子吸足水后可用 69% 烯酰·锰锌可湿性粉剂 1 000 倍液浸种 5 min 消毒。发病初期，可选用 58% 甲霜·锰锌可湿性粉剂 600 倍液、30% 甲霜·噁霉灵水剂 600～800 倍液、50% 烯酰吗啉可湿性粉剂 1 500 倍液等药剂进行灌根或根茎喷洒，每株灌根 200～250 mL，连续用药 2～3 次，间隔期 7～10 d。

4.1.4 炭疽病

4.1.4.1 主要症状

炭疽病主要为害果实和叶片，也可侵染茎部。叶片染病，初呈水浸状褪色绿斑，后逐渐变为褐色。病斑近圆形，中间灰白色，有轮生黑色小点粒，病斑扩大后呈不规则形，有同心轮纹，叶片易脱落。果实染病，初呈水渍状黄褐色病斑，扩大后呈长圆形或不规则形，病斑凹陷，上有同心轮纹，边缘红褐色，中间灰褐色，轮生黑色点粒，潮湿时，病斑上产生红色黏状物，干燥时呈膜状，易破裂。

4.1.4.2 发病条件

结果盛期天气高湿、早晨结露较重或大雨后遇晴暴晒造成日烧病较重的情况下容易发病。

4.1.4.3 防治方法

（1）农业防治。加强栽培管理，清除病残体，田间发现病果随即摘除带出田外销毁；合理密植，辣椒封行后行间不郁闭，果实不暴露，减少日灼病。

（2）种子消毒。该病主要是种子带菌，种子消毒处理有较好的预防效果。可用55 ℃温水浸种 30 min 后移入冷水中冷却，晾干后播种；或将种子在冷水中浸 10～12 h，再用 1% 硫酸铜水溶液浸种 5 min，冲洗干净后催芽播种；还可用 50% 多菌灵可湿性粉剂 500 倍液浸种 1 h，冲洗干净后催芽播种。

（3）化学防治。发病初期，可选用 25% 咪鲜胺乳油 1 500 倍液，或 10% 苯醚甲环唑水分散粒剂 800 倍液，或 68% 精甲霜·锰锌水分散粒剂 500 倍液，或 75% 肟菌·戊唑醇水分散粒剂 4 000～6 000 倍液，连续用药 2～3 次，间隔期 7～10 d。

4.1.5 白粉病

4.1.5.1 主要症状

主要感染叶片，发病重时感染枝干和茎。发病初期主要在叶面或叶背产生白色圆形霉状物呈粉斑点状，从下部叶片开始向上部叶片发展，严重时叶面会有一层白色霉层，发病后期叶片白色霉层呈灰褐色，叶片发黄坏死。

4.1.5.2 发病条件

温暖环境，阴天及密植、窝风环境易发病，25～28 ℃及稍干燥条件下易于流行。大水漫灌，湿度大，肥力不足，植株生长后期衰弱发病严重。

4.1.5.3 防治方法

（1）农业防治。加强栽培管理，提高辣椒的抗病力，大棚栽种时，注意控制棚室内的温湿度，防止棚室过于干燥、湿度过低。

（2）化学防治。发病初期可用 12% 苯甲·氟酰胺悬浮剂 1 000 倍液，或 20% 咪鲜胺乳油 1 000～1 200 倍液，或 30% 啶氧菌酯·戊唑醇悬浮剂 2 000 倍液，或 10% 苯醚甲环唑水分散粒剂 1 000 倍液喷雾防治，连续用药 2～3 次，间隔期 7～10 d。

4.1.6 疮痂病

4.1.6.1 主要症状

主要为害叶片，初期叶背面生隆起斑点，水浸状，扩大后病斑为不规则形，周缘稍隆起，暗褐色，内部色较淡，稍凹陷，表面粗糙呈疮痂状。病斑融合连在一起可形成较大斑点，引起叶片脱落。果实染病可见果面隆起白色圆点。凸起带轮纹病斑是诊断辣椒疮痂病的典型症状。

4.1.6.2 发病条件

长时间结露和暴雨天气发病严重。

4.1.6.3 防治方法

（1）农业防治。选用抗病品种，与非茄科作物轮作；雨天及时排水。

（2）化学防治。用3%中生菌素可湿性粉剂1 000倍液浸种30 min可预防该病害。发病初期，可选用30%琥胶肥酸铜可湿性粉剂800倍液、2%春雷霉素水剂800倍液、20%噻菌铜悬浮剂700倍液喷雾防治，连续用药2～3次，间隔期7～10 d。

4.1.7 病毒病

4.1.7.1 主要症状

常见的辣椒病毒病症状有以下4种类型。

（1）花叶病。病叶出现明显黄绿相间的花斑、皱缩，或产生褐色坏死斑。

（2）叶片畸形或丛簇型。初期植株心叶叶脉褪绿，逐渐形成深浅不均的斑驳、叶面皱缩、后期病叶增厚，产生黄绿相间的斑驳或大型黄褐色坏死斑，叶缘向上卷曲。幼叶狭窄、严重时呈线状，植株上部节间短缩呈丛簇状。重病果果面有绿色不均的花斑和疣状突起。

（3）条斑型。叶片主脉呈褐色或黑色坏死，沿叶柄扩展到侧枝和主茎，出现系统坏死条斑，常造成早期的落叶、落花、落果，严重时整株枯死。

（4）枯顶病毒病。植株矮缩、黄化、不结实或果实小且僵化不长。

4.1.7.2 发病条件

高温干旱天气，传毒昆虫（蚜虫、蓟马）发生严重促使病害的流行。

4.1.7.3 防治方法

（1）农业防治。选用抗病品种，隔离防虫育苗，减少农事操作造成伤口。

（2）化学防治。清水浸种3～4 h后放入10%磷酸三钠中浸泡40～50 min进行种子消毒；防治蚜虫、蓟马等传毒害虫；选用2%氨基寡糖素水剂300倍液、20%盐酸吗啉胍可湿性粉剂500倍液、20%宁南霉素可湿性粉剂1 000倍液等药剂预防和抑制病害的发生，连续用药2～3次，间隔期5～7 d。

4.2 虫害防治

4.2.1 蓟马

4.2.1.1 为害特点

为害辣椒的蓟马主要分为花蓟马和西花蓟马。成虫、若虫多群集于花内取食为害，花器、花瓣受害后呈白化状，经日晒后变为黑褐色，为害严重的花朵萎蔫。叶片受害后呈现银白色条斑，严重时枯焦萎缩。成虫在枯枝落叶层、土壤表皮层中越冬。

4.2.1.2 防治方法

（1）农业、物理防治。夏季休耕时采取高温闷棚；防虫网阻隔；蓝色粘虫板诱杀等。

（2）化学防治。初期可用吡虫啉、多杀霉素、乙基多杀菌素、溴氰虫酰胺等药剂喷雾，加入有机硅作为渗透剂效果更佳，重点喷植株的上部、嫩叶的背面和嫩茎。

4.2.2 叶螨类

4.2.2.1 为害特点

为害辣椒的叶螨类有茶黄螨、朱砂叶螨（红蜘蛛）和二斑叶螨等。成螨和幼螨集中在植株幼嫩部分刺吸。受害叶片背面呈灰褐色或黄褐色油渍状，叶片边缘向下卷曲；受害嫩茎、嫩枝变黄褐色，扭曲变形，严重时植株顶部干枯；受害果皮变黄褐色。

4.2.2.2 防治方法

（1）农业、生物防治。加强栽培管理，清理地块或棚室周围杂草，释放捕食螨，以螨治螨。

（2）化学防治。发生初期可轮换使用阿维菌素、螺螨酯、虫螨腈、联苯肼酯等喷雾，喷药时上喷下翻，注重喷幼嫩部位。

4.2.3 蚜虫

4.2.3.1 为害特点

为害辣椒的蚜虫主要有桃蚜和瓜蚜。群居在叶背、花梗或嫩茎上吸食植物汁液，分泌蜜露。被害叶片变黄，叶面皱缩卷曲。嫩茎、花梗被害呈弯曲畸形，影响开花结实，植株生长受到抑制，甚至枯萎死亡。蚜虫还可传播病毒病。

4.2.3.2 防治方法

（1）农业、物理防治。选用抗性品种，间作套作，加强田间管理；用黄板诱杀有翅蚜，用银灰膜趋避蚜虫。

（2）生物防治。保护天敌、释放天敌、喷施蚜霉菌。

（3）化学防治。蚜虫发生初期用啶虫脒、吡蚜酮、噻虫嗪、吡虫啉、呋虫胺等药剂喷雾防治，温室还可用异丙威等烟剂熏蒸。

4.2.4 烟粉虱

4.2.4.1 为害特点

烟粉虱是热带和亚热带地区主要害虫之一，被业界公认为超级害虫，在植株上分泌蜜露，产生煤污病，严重影响叶片光合作用。

4.2.4.2 防治方法

（1）农业、物理防治。避免连茬、连作；大棚内使用40目防虫网防治；使用黄板诱杀。

（2）化学防治。发生初期可用吡虫啉、噻虫嗪、啶虫脒、螺虫·噻虫啉、氟吡呋喃酮、烯啶虫胺喷雾防治。

4.2.5 棉铃虫、烟青虫

4.2.5.1 为害特点

以幼虫蛀食蕾、花、果，常在果实蒂部形成蛀孔，造成果实腐烂、脱落。

4.2.5.2 防治方法

（1）农业、物理防治。及时整枝打杈，摘除嫩叶、嫩枝上的卵块；化蛹高峰期灌水灭蛹；清除田间及地边杂草；用杀虫灯诱杀。

（2）化学防治。在低龄幼虫尚未蛀果前，用高效氯氰菊酯、甲氨基阿维菌素苯甲酸盐、苏云金杆菌、棉铃虫核型多角体病毒、氯虫·高氯氟等喷雾防治。施药以上午为宜，重点喷洒植株上部。

5 适时采收

辣椒采收要抢晴避雨，一般不在下雨天和露水未干时采收，以免滋生病菌。根据不同品种类型的商品需求，成熟一批采收一批。采收时要注意辣椒的成熟度，多数辣椒果实发育饱满，果面有光泽并且果实大小不再变化时即可采收。为了保证果实的新鲜，一般在早晨和傍晚采收为宜。采收以后，果实要及时放到阴凉通风处存放，有贮藏条件的要及时入库贮藏。

6 采后处理

辣椒采收后应尽快挑出病果烂果，在完成清捡工序后进行贮藏前处理。预冷处理是新鲜辣椒入库贮藏前的常规处理方法，可降低新鲜辣椒的呼吸速率，延长新鲜辣椒贮藏期；适当温度的热激处理可以减少辣椒鲜果的腐烂，改善果蔬品质；短波紫外线处理是一种无化学污染的物理处理方法，通过辐照诱导提高果蔬抗病性，减少化学保鲜剂的应用，减轻采后腐烂损失，是一条绿色环保的贮藏保鲜途径。

参考文献

邹学校，2002.中国辣椒[M].北京：中国农业出版社.
邹学校，2021.辣椒育种栽培新技术[M].长沙：湖南科学技术出版社.

茄子栽培关键技术

1 概况

茄子（*Solanum melongena* L.）是茄科（Solanaceae）茄属（*Solanum* spp.）中以浆果为产品的一年生草本植物，在热带为多年生植物。原产于东南亚、印度，早在公元4—5世纪就传入中国，一般认为中国是茄子的第二起源地。2021年，中国茄子栽培面积为804 381 hm²，占世界总栽培面积的41.0%。茄子是中国南北各地栽培最广泛的蔬菜之一，其产量高，适应性强，栽培较容易。茄子以煮食、炒食为主，但也可以制作茄干、茄酱或腌渍。

2 生长条件

2.1 温度

茄子喜温，耐热性较强。结果期间的适宜温度为 25～30 ℃。

2.2 光照

茄子为喜光性蔬菜，对光周期的反应不敏感，光照强度影响其光合作用的强度。

2.3 水分

茄子枝叶繁茂，生育期间需水量大，通常以土壤相对湿度 70%～80% 为宜。但在门茄形成之前，需水量较少，不宜多浇水，防止秧苗徒长、根系发育不良和落花率增加。门茄结果以后需水量逐渐增多，对茄收获前后，需水量最多。水分不足会严重影响产量和品质，但土壤过湿会造成土壤通气性不良，引起沤根。

2.4 土壤

茄子对土壤的适应性较强，砂质土壤和黏质土壤均可栽培。由于茄子耐旱性差，喜肥，宜选用土层深厚、保水性强、土壤 pH 值 6.8～7.3 的肥沃壤土或黏质壤土种植，有利于茄子根系发育，形成旺盛根群。地下水位较高、排水不良的地块及耕层浅、土质黏重的土壤，不利于茄子根系发育。

2.5 肥料

茄子在结果期，果实与茎叶同期生长，需肥量大，对矿物质肥的要求以钾最多，氮次之，磷最少。氮对植株生长、花芽分化和果实膨大有重要作用。缺氮时，植株长势弱，分枝减少，花芽发育不良，短柱花多，落花率高，果实生长停顿，皮色不佳。茄子是一种耐肥蔬菜，生长结果期长，所以要多次追肥，才能保证产量的提高。

3 类型和品种

3.1 类型

按植物学分类将茄子栽培种分为圆茄、长茄、矮茄 3 个变种。按成熟期可分为早熟、中早熟、中熟、中晚和晚熟种；按茄子果实的颜色可分为黑紫色、紫色、紫红色、绿色和白色。

3.2 品种

茄子受消费习惯影响大，区域性极强，因此品种选择应根据市场需求、生产目的进行，在大面积种植之前，需要进行引种试验，充分了解品种的特征特性，合理安排生产计划，以达到增产增收的目的，切忌盲目种植。

4 栽培管理技术

4.1 播种育苗

育苗是蔬菜集约化生产的重要环节，培育和应用适龄幼苗是蔬菜早熟、优质、高产的重要保证。目前生产上常用的育苗方法有穴盘育苗、苗床育苗、营养钵育苗 3 种。其中，穴盘育苗有利于工厂化育苗（方便管理、运输等），除少数偏远地区外，蔬菜大面积生产地区都采用穴盘育苗。本节介绍穴盘育苗的方法。

4.1.1 播种前准备

南亚、东南亚国家大部分区域均可全年栽培，最适宜栽培时期为冬、春季。育苗场地应选择排灌良好，前茬未种过茄果类作物（如番茄、茄子、辣椒）的田块。穴盘育苗不仅可减少土传病害的发生，使出苗整齐、粗壮，而且可一次成苗，无须分苗等操作。放置苗盘的地事先平整，避免地块高低不平，影响出苗的质量。有条件的地方，用育苗架更好。育苗基质可用蔬菜育苗基质、烟草育苗基质，也可以自己配制。育苗基质可用腐熟猪粪和田园土按1∶1混合，每立方米加入100 g复合肥、10 g多菌灵拌匀，田园土可从1～2年内没有种过茄果类蔬菜的田园上15 cm以内的表层土选取，或用未种过蔬菜的红土均可。播种前一天把育苗盘装满育苗基质或育苗土，用小木块擀平后用花洒把育苗盘浇足底水待用。育苗盘2盘一组摆放，方便工人播种、间苗时操作。

4.1.2 浸种催芽

茄子育苗的播种量一般为每亩15～20 g，为达到早出苗、出齐苗、育壮苗的目的，在播种前一般进行种子处理，主要进行浸种和消毒（包衣种子不需要处理）。常用的方法是温汤浸种，具体做法：先将种子装入尼龙网袋或纱布袋内，放在常温水中浸15 min，然后将种子放入55～60 ℃的热水中，浸泡15 min，水量为种子体积的10倍。15 min后让其自然冷却，在常温环境下泡种4～6 h。也可将温汤浸种过的种子在50%多菌灵可湿性粉剂1 000倍液或10%磷酸三钠溶液中浸泡20 min，捞出后用清水冲洗干净后用纱布或湿布包裹催芽。

将处理好的种子用湿毛巾包好，放在盆钵中，种子袋上面再盖几层湿毛巾，以保持湿度，然后放在25～30 ℃下3～5 d，有50%种子"露白"后即可播种。催芽过程中，每天翻动种子1次，使种子处于松散状态。如发现种子发黏，应立即用清水清洗毛巾。一般每天清洗1次，清洗后控出多余水分，以免种子吸水过多，造成种子缺氧、发芽不良。若无条件催芽，可浸种后直接播种。

4.1.3 播种

播种前一天把育苗盘或育苗床浇透水，育苗盘一般每穴播种1粒。播后盖1 cm厚的细土，不宜太薄，也不宜太厚，太薄浇水后种子容易露出或易出"戴帽苗"，太厚影响出苗。浇水后出现种子露出的情况，要及时覆土。播种最好选在晴天的上午，有利于播后提高苗床内的温度，播种覆土后用花洒浇透水。

4.1.4 苗床管理

从播种到出苗，要求床土水分充足、通气良好。当幼苗出土，子叶张开时，苗床温度保持在 20～25 ℃，防止徒长。在育苗过程中，要适当追肥，一般都结合浇水，追施 0.1%～0.2% 尿素。浇水要用细网眼喷壶或喷雾器喷洒，避免用瓢浇泼，以免土壤湿度过大，苗期病害严重。苗期主要病害是猝倒病，一经发现病株，马上拔除，并在病株周围撒多菌灵或百菌清原粉以防止病情蔓延。秧苗定植前 10 d 开始炼苗，加强通风，减少浇水，控制生长，让幼苗尽快适应露地环境条件，缩短缓苗时间。

苗床浇水一般在 8:00—10:00 或 17:00—19:00 进行。苗期如果出现叶片发黄、叶小、茎细等生长不良的缺肥症状，可以采用根外追肥的办法，用 0.3% 磷酸二氢钾 + 0.2% 尿素溶液叶面喷施。

4.2 整地理墒

选择土质疏松、肥沃、排水良好的壤土定植茄子，应避免连作。在前茬作物采收后及时深耕晒垡。每亩施腐熟有机肥 3 000～4 000 kg、复合肥 50 kg、过磷酸钙 30 kg，耙平地面作宽 1.3 m 的畦（包沟），沟深 30 cm 左右。

4.3 适期定植

一般春季露地栽培的茄子定植均在当地终霜期、最低气温稳定在 12 ℃时开始定植。茄子栽植的密度要根据品种特征特性、栽培方法和土壤肥力水平等因素来决定。一般早熟品种比晚熟品种密；株型紧凑的品种比植株开展度大的品种密；土壤肥力低的比土壤肥力高的密。一般中早熟品种株行距 30 cm×60 cm；中熟品种株行距 40 cm×60 cm；晚熟品种株行距 50 cm×60 cm，每畦种 2 行。用小锄头挖穴，放入苗四周盖上土压实即可，栽培深度以到子叶节处为好。定植后浇足定根水，以利于苗的成活。要在晴天定植，忌在雨天湿土定植，湿土定植难发新根，缓苗慢。茄子壮苗的标准：子叶完好、宽大；苗高 15～18 cm，有 6～8 片真叶，叶片肥厚、宽大，叶色深绿；茎粗壮，节间短，花芽分化良好，第一花蕾已现，但未开；根系发达颜根多，白嫩，无病虫害。

秧苗从播种至定植的苗龄随播种地点、季节及管理水平而有所不同，在昆明市一般冬春育苗需 60～90 d，夏秋育苗需 30～40 d。

4.4 田间管理

4.4.1 施肥

茄子追肥要掌握"轻施苗肥，稳施花肥，重施果肥"的原则。缓苗后施提苗肥，一般每亩施尿素 10~15 kg。四门斗相继坐果膨大期是茄子需水的高峰期，应在对茄开始膨大时中耕培土，进行追肥，一般每亩施复合肥 25~30 kg，或尿素、钾肥各 15 kg。四门斗果膨大期重施肥，每隔 10 d 追施 1 次。中后期要增施钾肥，少磷肥，因为缺钾植株易感病倒伏，多施磷肥易引起果实僵硬。结果期除适当的土壤追肥外，还应结合叶面施肥，以弥补根部吸肥能力的不足。叶面喷肥应选择晴天进行，用 0.2% 尿素和 0.3% 磷酸二氢钾混合液，每 15 d 喷施 1 次，应保证在 24 h 无雨的天气喷施，此条件下喷施效果好。喷施时，最好在傍晚或早晨尚有露水时进行，不宜在中午进行，以免因气温高而加快药液浓缩速度而造成药害。

4.4.2 水分管理

土壤水分不足，植株生长缓慢，甚至引起落花、果实果皮粗糙、无光泽，导致品质差。在云南各地 3—4 月茄子定植的季节，雨季尚未到来，天气干燥，前期一定要勤浇水，每 2~3 d 就要浇 1 次，到茄子苗成活为止。到门茄开花时适当控水蹲苗。蹲苗的目的主要是适当控制地上部分植株生长，促进继续向纵深发展，调节营养生长和生殖生长的平衡，以积累更多的养分供给果实的发育、膨大，获得丰产。6—8 月雨量充沛，应注意排出积水；门茄瞪眼期结束蹲苗及时浇水。茎叶和果实同时长，生长速度快，需要的水分也显著增多，对茄和四门斗相继坐果膨大时，需水最多。从门茄瞪眼后，视天气和植株生长情况，应每 5~7 d 浇水 1 次，以经常保持土壤湿润，防止忽干忽湿。果前以长茎叶为主，需肥水少，结果后需肥水多。根据不同发育阶段的特点进行合理的追肥，是丰产的主要措施之一。

4.4.3 中耕培土

中耕结合除草进行，早期可以中耕深些，后期浅些。当植株生长到 30 cm 左右，要结合中耕进行培土。中耕培土多结合除草进行，可保证土壤疏松透气，从而促进新根的发生，增强根系的吸收能力。一般从定植开始，每隔 10~15 d 中耕 1 次。中耕宜在土壤半干湿时进行。开花结果期中耕，应结合追肥、培土进行。

4.4.4 整枝摘叶、立支架

茄子开花后，摘除门茄以下侧枝，并及时搭架。若整枝过早，不利于发根，过迟则会引起徒长，不利于提早坐果。整枝要在晴天进行，以防止伤口感染病菌。茄子的整枝有双秆整枝法和四秆整枝法两种。双秆整枝法是只留主枝和门茄下的第一侧枝，其余侧枝全部去掉。四秆整枝法是除了保留主枝和门茄下的第一侧枝外，也保留主枝和门茄下的第一侧枝上长出的侧枝，门茄以下基部的其余侧枝全部摘除。目前生产上多用双秆整枝法。立支架时距植株主秆外 10 cm 处插立杆，每隔 2 m 立 1 根杆，然后在距离地面 30 cm 的高度绑上横杆，将植株主干用细绳或布条绑在架上防倒伏。茄子生长中后期，可以适当摘去门茄以下的黄叶、老叶、病叶，以利于通透光和减轻病虫为害。摘除老、病、残叶，及时清理出园并进行深埋或烧毁。

4.4.5 保花保果

造成茄子落花的主要因素是光照不足、营养不良、温度过高（38 ℃以上）或过低（15 ℃以下）。在生产上，除加强水肥管理、整枝摘叶外，还可用 2,4- 滴 25～30 mL 或对氯苯氧乙酸钠（防落素）40～50 mg/L 浸花或涂抹花柄，每隔 3～4 d 进行 1 次，保花保果，促进果实快速膨大。气温低、湿度高时浓度高些，气温高、湿度低时浓度低些。2,4- 滴对嫩叶和生长点会产生药害，注意小心使用。

4.4.6 疏花疏果

有的茄子品种是单花，不需要疏花疏果；但有的品种是花序，一个花枝有 2～6 朵花，需要疏花疏果，一般只留 1 朵花，人工去除多余的花，保证营养供给，避免果实大小不均匀，畸形果多，商品率低。

5 病虫害防治

以"预防为主，综合防治"为策略，采用以农业防治、物理防治、生物防治为主，化学防治为辅的防治原则。

5.1 农业防治

选用抗病品种，培育壮苗，实施轮作；采用深沟高畦栽培，合理密植；及时整枝搭架；拔除病重株，清洁田园；深翻土壤，减少病虫源；科学施肥。

5.2 物理防治

利用黄色粘虫板、蓝色粘虫板、频振式杀虫灯诱杀成虫；田间铺银灰膜或悬挂银灰膜条趋避蚜虫；大棚使用防虫网防止害虫进入，人工摘除害虫卵块和捕杀害虫；使用性诱剂杀虫等。

5.3 生物防治

选用植物源、微生物源、农用抗生素等生物制剂防治病虫害。

5.4 化学防治

5.4.1 主要病害及防治

茄子主要病害有猝倒病、黄萎病、绵疫病、褐纹病、白粉病等。

5.4.1.1 猝倒病

（1）症状。又叫卡脖子、倒苗、小脚瘟。幼苗出土后染病，在幼茎基部出现淡黄色至黄褐色水浸状病斑，进而病斑绕茎1周，病部收缩成线状，子叶或幼叶尚未凋萎幼苗即倒伏于地，出现猝倒现象。幼苗出土前也可受害，引起烂种。湿度大时，病部附近能见一层白色絮状菌丝。

（2）防治方法。幼苗出土及时喷药防治，发现病株及时拔出，并撒百菌清药粉在病株穴内。发病初期喷洒58%甲霜·锰锌可湿性粉剂600倍液、722 g/L霜霉威水剂600倍液、70%代森锰锌可湿性粉剂600倍液、50%多菌灵可湿性粉剂700～800倍液喷雾等药剂，每隔7～10 d喷1次，连续2～3次。喷药后，可撒干土或草木灰降低苗床土层湿度。晴天要加强通风。

5.4.1.2 黄萎病

（1）症状。多在结果期表现症状，发病初期叶脉间褪绿变黄，逐渐发展到全叶，多数茄子植株半边发病，半边正常，俗称"半边疯"。早期叶片晴天中午萎蔫，早晚或阴雨天尚可恢复，数日后全株萎蔫叶片脱落。横切茎部，可见维管束变褐色。挤压变色部不渗出浑浊液，可区别于青枯病。

（2）防治方法。嫁接栽培是目前生产上用来防治最有效的方法；其次是在定植前塘施10 g甲基硫菌灵药粉，拌匀后再栽苗，或移栽后浇定根水时用30%甲霜·噁霉灵水剂800倍液+96%硫酸铜原药1 000倍液灌根，可以有效防止黄萎病的发生；也可发病初期用50%多菌灵可湿性粉剂500倍液、50%苯菌灵可湿性粉剂1 000倍液灌根，每株灌配好的药液300～500 mL，隔10 d灌1次，连续灌2～3次。

5.4.1.3 绵疫病

(1) 症状。一般为害果实部位,以近地面果实特别是果实顶部接触地面的部位先发病。最初在果上产生水浸状褐色圆形病斑,边缘不明显,病部稍凹陷,扩大后可遍及整个果实。田间湿度大,果肉变黑腐烂,潮湿病部生有白色棉絮状菌丝,病果易脱落或收缩成僵果。茎部发病初期为水浸状,后变褐腐烂,并缢缩以致折断。叶片上有不规则或近圆形水浸状病斑,有轮纹,可迅速扩展。潮湿时病斑边缘不清,生有稀疏白霉,干燥时病斑边缘明显,易干裂。

(2) 防治方法。选用抗病品种,注意轮作,有条件水旱轮作最佳。发病初期用58%甲霜·锰锌可湿性粉剂600倍液、722 g/L霜霉威水剂600倍液、70%代森锰锌可湿性粉剂600倍液、72%霜脲·锰锌可湿性粉剂500～700倍液、50%烯酰吗啉可湿性粉剂600～800倍液、75%百菌清可湿性粉剂500倍液等药剂,每隔7 d喷1次,连续喷2～3次。

5.4.1.4 褐纹病

(1) 症状。主要为害叶、茎、果。染病多出现在下部叶片,初生白色小点,后病斑变为褐色近圆形或不规则形,上有轮纹,生小黑点。茎部染病,最初在茎上出现菱形病斑,边缘深紫褐色,中间灰白色凹陷的干腐状溃疡,上生黑色小点。果实染病,初生浅褐色椭圆形凹陷病斑,后扩大为黑褐色,其上轮纹状生黑色小黑点。

(2) 防治方法。可用75%百菌清可湿性粉剂600倍液、70%代森锰锌可湿性粉剂500倍液预防,门茄膨大期用50%异菌脲可湿性粉剂1 000倍液、25%吡唑醚菌酯悬浮剂1 000～1 500倍液、40%苯醚甲环唑悬浮剂1 000～1 500倍液等交替喷雾防治,每隔7～10 d喷1次,连续喷2～3次。

5.4.1.5 白粉病

(1) 症状。以为害叶片为主。在叶面上生成不规则、大小各异的白粉霉斑扩大到整个叶片,使叶组织变黄后干枯。

(2) 防治方法。合理密植,及时摘除植株下部的老叶。发病初期可用29%吡萘·嘧菌酯悬浮剂1 500倍液、40%氟硅唑乳油6 000～10 000倍液、10%苯醚甲环唑水分散粒剂1 500～2 000倍液、70%甲基硫菌灵可湿性粉剂600倍液等进行防治。

5.4.2 虫害防治

主要虫害有蓟马、白粉虱、蚜虫、红蜘蛛等。

5.4.2.1 蓟马

(1) 为害特点。成虫和若虫藏匿于嫩叶叶背毛丛和花中吸吮汁液,被害叶片背面茸毛及叶肉呈灰褐色，变硬老化，花蕾受伤脱落。受害植株生长缓慢，节间缩短。蓟马繁殖速度极快，一年可发生10～20代，世代重叠，终年繁殖，5—9月为发生高峰期，以秋季最严重，成虫活跃善飞，怕光，若虫落入表土化蛹。

（2）防治方法。采用地膜覆盖能大大降低虫口密度，及时摘除老叶、病叶，清除田间杂草，采收期结束后清除残枝残叶集中烧毁，减少虫源。当每株虫口达3头时即应喷药防治，可用25%吡虫啉可湿性粉剂3 000倍液、25%噻虫嗪水分散粒剂2 500～3 000倍液、25%乙基多杀菌素水分散粒剂2 000～3 000倍液、10%烯啶虫胺水剂1 000～1 500倍液、10%高效氯氰菊酯乳油2 000倍液等喷雾防治，隔3～5 d喷1次，连喷3次。喷药重点是植株的生长点，嫩叶背面、花和花蕾等部位。

5.4.2.2 白粉虱

（1）为害特点。白粉虱为杂食性害虫，繁殖速度快，种群数量在秋季达高峰。以成虫、若虫群集在叶片背面，用刺吸式口器吸吮汁液，被害叶片褪绿、变黄，植株生长缓慢、萎蔫，还能分泌大量蜜露。

（2）防治方法。发生初期，可用10%吡虫啉可湿性粉剂1 000倍液、25%噻嗪酮乳油2 500倍液、20%甲氰菊酯乳油2 000倍液，加入10%吡丙醚乳油1 000倍液或10%虱螨脲悬浮剂2 500～3 000倍液一起喷施，每隔3～5 d喷1次，连喷2～3次。

5.4.2.3 蚜虫

（1）为害特点。常群集在叶片背面和嫩茎上以刺吸口器吸吮植株汁液，分泌蜜露，造成植株严重失水和营养不良。幼叶被害，卷曲皱缩，轻者褪绿，有斑点，叶片发黄；重者叶片卷缩变形，甚至枯萎。蚜虫还能传播病毒病。

（2）药剂防治。用50%抗蚜威可湿性粉剂2 000倍液，10%吡虫啉可湿性粉剂1 000～1 500倍液、20%啶虫脒可溶液剂3 000倍液、50%吡蚜酮可湿性粉剂2 500～5 000倍液、20%氰戊菊酯乳油2 000倍液或20%甲氰菊酯乳油3 000倍液喷雾，视田间虫情每隔7～10 d喷1次，连喷2～3次。

5.4.2.4 红蜘蛛

（1）为害特点。在叶背刺食汁液，并吐丝结网。被害叶片初期出现白色小斑点，后褪绿变为黄白色，严重时变锈褐色呈火烧状，造成叶片早落，果皮变粗，果僵硬不能长大，植株枯死。

（2）防治方法。清除田边杂草及收获后的枯枝烂叶，集中销毁，以减少虫源。发生初期为防治重点。可选用73%炔螨特乳油1 500倍液、34%螺螨酯悬浮剂4 000倍液+43%联苯肼酯悬浮剂1 800～2 500倍液、20%阿维·联苯肼悬浮剂2 000～2 500倍液、5%噻螨酮乳油3 000倍液喷雾，每隔7～10 d喷1次，连喷2～3次。

6 适时采收

茄子授粉后果实迅速膨大，从开花到果实采收一般为 20～25 d，茄子的采收必须掌握适时，过早采收产量低，过晚果实硬、种子多，不堪食用，且消耗养分过多，影响植株开花结果及枝条正常生长。

6.1 采收标准

果实长到品种应有的长度和粗度，基部的颜色较浅，其他部分的颜色变深；用手按压果皮表面，表现出较强的弹性；果皮变亮，有光泽。判断果实采收的适宜时期，可根据萼片果实相连处（称为"茄眼"）的宽度判断，茄眼较宽，表明果实正在迅速生长，不宜采收；茄眼不明显，表明果实生长变慢或已停止生长，可以采收。

6.2 采收方法

茄子正确的采收方法是用刀或枝剪齐果柄根部采下，不带果柄，以避免在装箱运输途中刺伤果皮，影响果实的外观品质。采收的时间以早晨为好，果实新鲜柔嫩，品质佳，贮藏性能好。采收后及时分级包装销售。

参考文献

林鉴荣，刘士亚，丘漫宇，等，2004. 番茄茄子无公害生产彩色图说 [M]. 广州：广东科技出版社．

中国农业科学院蔬菜花卉研究所，2010. 中国蔬菜栽培学 [M]. 2版. 北京：中国农业出版社．

菜豆栽培关键技术

菜豆（*Phaseolus vulgaris* L.）是豆科（Leguminosae）一年生植物，又名四季豆，起源于美洲，后经过数百年的传播和栽培，已成为全球范围内被广泛种植和食用的蔬菜作物，尤其在热带和亚热带地区菜豆的种植和食用非常普遍。菜豆是一种营养丰富、口感清爽的蔬菜，富含蛋白质、碳水化合物、膳食纤维、维生素和矿物质，尤其富含钾、钙、镁、铁等矿物质。

得益于云南省立体气候环境，云南省已实现了菜豆的周年生产，成为冬春菜豆的优势产区，产品销往全国各地。近年来，利用云南省毗邻东南亚国家的区位优势，部分企业和种植户在老挝、缅甸等国种植菜豆返销中国，带动了东南亚国家蔬菜产业的发展。

1 生长条件

1.1 温度

菜豆生育适宜温度为 10～30 ℃，不耐霜冻，种子发芽适宜温度为 20～25 ℃，35 ℃以上、8 ℃以下种子不易发芽。幼苗生长适宜温度为 18～20 ℃，开花结荚适宜温度为 18～25 ℃。

1.2 光照

菜豆生长发育对日照长度的要求不严格，即在较长的日照或较短的日照下均能开花。但菜豆生长、开花结荚需要较强的光照，如果光照不足，则容易发生徒长、落花落荚。

1.3 水分

菜豆根系发达，侧根多，较耐旱而不耐涝。种子发芽，需要吸足水分，但水分过多，土壤缺氧，种子易腐烂。植株生长适宜的土壤相对湿度为土壤田间持水量的60%～70%。开花结荚期对水分要求严格，其适宜的空气相对湿度为65%～80%。结荚期如遇高温、干旱天气，则嫩荚生长缓慢，荚壁中果皮易硬化，内果皮细胞分裂加速，子室间空腔提前发生，内果皮变薄，品质降低。开花时遇大雨，土壤和空气湿度过大也会影响花粉发芽，过多的水分会降低雌蕊柱头上黏液的浓度，使雌蕊不能正常授粉而落花、落荚，而且容易引起病害的发生。

1.4 土壤

菜豆最适宜在土层深厚、松软、腐殖质多且排水良好的砂壤土、壤土中生长，最适pH值为6～7。营养元素的吸收量以氮、钾为多，磷较少。开花和结荚期对氮、钾的吸收量渐增，茎叶中的氮、钾随着生长中心的变化随之转移至果荚中，嫩荚迅速伸长时要吸收大量的钙。

2 品种选择

菜豆品种有蔓生和矮生两种类型。目前蔓生的品种有红花青壳、双青玉豆、春风4号、秋抗6号、大鱼鳅背、泰国架豆王、世纪架豆和绿龙等。矮生的品种有沙克莎、美国供给者、8916矮生豆和法国芸豆等。

3 栽培管理技术

3.1 整地

菜豆是一种对土壤肥力要求较高的作物，适量的底肥可以提高菜豆的生长速度和产量。每亩施腐熟有机肥1 500～2 000 kg、磷酸二氢铵30 kg、硫酸钾10 kg，深耕土壤25～30 cm，将肥料与土壤充分混匀后作畦。一般采用高垄种植，四周设置40 cm深排水沟渠。

3.2 栽培季节

云南可实现菜豆周年生产，一般滇中、滇东北、滇西北3—7月播种；滇西、滇南、干热河谷地区9—11月冬春错峰种植。东南亚的热带和亚热带地区9—11月种植。具体播种可以根据当地的气候和降雨情况进行调整，选择合适的品种和适宜的种植时间。

3.3 种子处理

选择表面光滑、籽粒饱满的种子。播种前晾晒 1～2 d，提高种子内酶的活性，促进发芽整齐，此外对种子处理提高种子发芽率和整齐度。

3.3.1 热水浸种

将种子用 80 ℃左右的热水浸泡 5～8 min，然后迅速冷却，可有效杀灭种子表面的病菌，促进发芽。

3.3.2 药剂浸种

将种子用 0.3% 高锰酸钾溶液，或 1% 硫酸铜溶液浸泡 20～30 min，然后用清水洗净，杀灭种子表面的真菌和细菌，提高种子发芽率和苗期抗病能力。

3.3.3 药剂拌种

用种子重量 0.2% 的 50% 多菌灵可湿性粉剂拌种，防治苗期的立枯病、猝倒病和炭疽病等多种病害。

3.4 播种及出苗后管理

3.4.1 播种

菜豆一般采用直播，每亩用种量 1.0～1.5 kg。播前先浇底水，播种后畦面盖地膜，膜宽 100 cm，每畦播 2 行，畦面宽 80 cm，沟宽 40 cm、深 20 cm，每穴播 3～4 粒种子，穴深 5～6 cm，播后盖土 1～2 cm。

3.4.2 苗期管理

播种后 5 d 左右及时破膜放苗，四周膜边用土压实，有利于保温保湿，防止白天膜下高温蒸气伤害幼苗。出齐苗后及时查苗、补苗，发现缺苗立即补种。2 片真叶时，间除病株、弱株，每穴留 2 株健壮苗。苗期尽量少浇水，促进根系生长，防止幼苗徒长和发生根部病害等。

3.4.3 搭架

植株具有 5 片叶顶端开始甩蔓时进行搭架，架子的材质可以用竹子、木条、铁管等材料制作的垂直支架或横架，以便菜豆攀爬生长。在搭建架时，还需要留出足够的空间，便于采摘和田间管理。

菜豆的攀爬高度和架子的形式可以根据品种和生长情况进行调整。通常选用高 1.8～2.2 m 的架杆，在距离植株根部 10～15 cm 处插入地下，深 15～20 cm，避免伤根。每株插 1 根架杆，畦上相邻两行的每 4 株菜豆搭 1 个"人"字形架，每 4 根架杆成 1 捆，搭架后按逆时针方向引蔓上架。落蔓后及时调整茎蔓朝向，将茎蔓引缠到空缺位置，充分利用空间，尽量使茎蔓相互不缠绕、不堆积、不重叠。部分菜豆品种分枝过多，会造成田间局部郁闭或茎蔓堆积，要及时整理并摘除下部老叶、病叶，以利于通风透光。

3.5 水肥管理

水肥管理遵循"苗期少、抽蔓期控、结荚期促"的管理原则。

3.5.1 苗期

出苗后，视土壤墒情浇一次齐苗水。此后控秧促根，以促进营养生长为主，一般到临近开花前才开始浇水，具体根据土壤墒情灵活掌握。开始开花前适量追施磷、钾肥，以促进花芽的形成和开花结荚，提高产量。落花落荚是影响菜豆高产的一个重要因素，而缺硼则是导致落花落荚的主要原因之一，补硼能提高菜豆开花坐荚率，因此可以在幼苗期补充硼肥，每亩可施硼砂 2 kg。

3.5.2 开花期

花芽形成期和开花期需要大量的磷、钾肥，因此可以在菜豆幼荚坐稳后结合天气状况和植株长势适当增加水肥施用量。开花期如果土壤过于干燥，会造成花粉早衰，不能完成受精；土壤和空气湿度过大，花粉黏着不利于散粉，会出现授粉受精不良，最终造成落花落荚。切忌在开花盛期进行大水浇灌，以免植株营养生长与生殖生长竞争养分，引起重度落花落荚。30 ℃以上的高温和 10 ℃以下的低温直接影响花芽的正常分化或使花器发育不完全而出现败育，间接造成落花落荚，果荚短小或畸形，导致产量降低。因此，高温阶段浇水尽量在傍晚时进行，以降低地温，促进枝叶和豆荚协调生长，同时要注意雨后排涝。

3.5.3 结荚期

结荚期需要大量的氮肥，在菜豆结荚后根据土壤养分状况适量追施氮肥，一般可以采用有机肥和化肥结合的方法，以维持土壤肥力平衡，促进荚果的发育和增大，提高产量。每亩追施复合肥 20 kg，冬季阴冷天气和阴天前尽量不易施肥浇水，以减少病害发生。

3.5.4 采收期

结合浇水，7～10 d 施 1 次肥料，每亩施用三元复合肥（氮：磷：钾 =15：15：15）10 kg 和尿素 5 kg，叶面喷施 0.2% 磷酸二氢钾和硼砂 500 倍液 2～3 次，每 10 d 喷施 1 次。此后每采收 2～3 次追施 1 次三元复合肥（氮：磷：钾 =15：15：15），叶面喷施 0.3% 磷酸二氢钾。

3.5.5 采收后期

根据市场价格调整植株管理方式，若市场价格较高时可通过水肥管理来促进植株翻花，延长采收期，可施用 1～2 次三元复合肥（氮：磷：钾 =15：15：15）促进植株生长，每次施用 10 kg/ 亩。整个生长期应及时拔除病株和杂草，疏去黄叶，增加光照，加强透风，理顺荚条，防止因机械阻碍引起的畸形荚发生。

4 病虫害防治

4.1 病害防治

4.1.1 锈病

4.1.1.1 为害症状

主要为害叶片，初呈黄绿色，后渐为褐色的圆形病斑、黄绿色晕环，病斑有褐色至黑褐色的小粒点，最后褐色部分脱落，形成穿孔。叶片背面，有大量的锈孢子密集在一起，似黄白色至淡黄褐色的粗绒状霉。叶脉、叶柄及蔓茎被侵染后，病斑梭形或近梭形条状，稍隆起，褪绿有水渍状，在蔓茎上有时出现纵裂，中央持有褐色至黑褐色小粒点；病害发生严重时还可侵染茎蔓、豆荚。

4.1.1.2 发生特点

以温差大、雾露天气多的春、秋季易发生、扩展快；菜豆生长中一般到生长中期开始发病；菜豆长势弱或偏施氮肥易发病。

4.1.1.3 防治措施

（1）农业防治。实行轮作倒茬；选择抗病品种；阴雨季节做好菜田清沟排水，防止低洼地积水,降低田间湿度;合理密植,保证通风良好,做好菜园清洁,收获后将病叶清除干净集中烧掉;避免前期氮肥施用过多。

(2)化学防治。底脚叶片零星发病时,可用70%硫磺·锰锌可湿性粉剂、20%苯醚甲环唑微乳剂或60%唑醚·代森联水分散粒剂等药剂喷雾防治;中下部叶片普遍发病时选用戊唑醇+苯甲·醚菌酯、苯甲·嘧菌酯+戊唑醇或已唑醇、吡唑萘菌胺+乙嘧酚+磷酸二氢钾等组合药剂喷雾防治。间隔5～7 d,连续防治3次以上。

4.1.2 炭疽病

4.1.2.1 为害症状

发病时在茎上、豆荚上产生梭形或长条形病斑。初期病斑为紫红色,后色变淡,稍凹陷以至龟裂,病斑上密生大量黑点,以豆荚受害最为严重。雨季发病较多,病部往往因腐生菌的生长而变黑,加速茎组织的崩解。轻者生长停滞,重者植株死亡。

4.1.2.2 防治措施

病菌侵染潜伏期较长,防治应以保护预防为主,进入果荚期和叶片茎蔓零星发病期可用70%甲基硫菌灵可湿性粉剂、70%代森锰锌可湿性粉剂、25%溴菌腈微乳剂或50%异菌脲可湿性粉剂等药剂喷雾预防保护;田间出现普遍发病可用25%咪鲜胺乳油、20%硅唑·咪鲜胺水乳剂、咪鲜胺+三唑类或咪鲜胺+吡唑醚菌酯等药剂喷雾防治,间隔7～10 d,连续防治2～3次。

4.1.3 白粉病

4.1.3.1 为害症状

初期叶片背面产生圆形小白斑,后扩大,相互连接,遍布全叶,沿叶脉扩展成粉带,颜色由白色转为灰白色至紫褐色,严重时叶面形成病斑以致叶片枯黄脱落。

4.1.3.2 防治措施

发病初期可用15%三唑酮可湿性粉剂、10%苯醚甲环唑可湿性粉剂、12.5%烯唑醇可湿性粉剂、43%戊唑醇悬浮剂喷雾防治,间隔5～7 d,连续防治3次以上。

4.1.4 根腐病

4.1.4.1 为害症状

植株下部叶片变黄,病部产生点状病斑,由支根蔓延至主根,引起整个根系腐烂或坏死,病株易拔起,纵剖病根,可见维管束呈红褐色,病情扩展后向茎部延伸。主根全部发病后,地上部茎叶萎蔫枯死。湿度大时,病部产生粉红色霉状物,即病菌的分生孢子。

4.1.4.2 防治措施

（1）农业防治。选用抗病品种；水旱轮作，或与非豆科作物实行2年以上轮作；深沟高畦，防止积水，雨后及时排水；加强田间管理，增施磷、钾肥，提高植株抗病力。

（2）化学防治。发病初期，可用甲基硫菌灵或敌磺钠液喷淋根、茎基部；还可用噁霉灵、甲霜·噁霉灵、氰烯菌酯等药液灌根，每隔7 d浇淋1次，连续防治2～3次。

4.1.5 病毒病

4.1.5.1 为害症状

嫩叶出现花叶、明脉、褪绿或畸形等症状，新生叶片上浓绿部位稍突起呈疣状；有的病株产生褐色凹陷条斑，叶肉或叶脉坏死。病株生长不良、矮化、花器变形、结荚少，豆粒上产生黄绿花斑；有的病株生长点枯死或从嫩梢开始坏死。

4.1.5.2 防治措施

（1）农业防治。选用耐病品种；避免与豆类、茄果类、瓜类等蔬菜连作；加强栽培管理，提高植株抗病力。

（2）化学防治。①治虫防病，切断传播媒介，在田间有蚜虫、蓟马、白粉虱、红蜘蛛等虫害发生时，选用吡虫啉、噻嗪酮、阿维菌素、啶虫脒、乙基多杀菌素等药剂防治，切断病毒传播、扩散。②药剂防控，发病前或发病初期，可用宁南霉素、菇类蛋白多糖、吗胍·乙酸铜等＋氨基寡糖素＋芸苔素内酯喷雾防治，间隔期7～10 d，连续防治2～3次。

4.1.6 菜豆细菌性疫病

4.1.6.1 为害症状

叶片边缘先出现水渍状坏死，后扩大，病斑沿叶脉分布，病斑周围有黄色晕圈；发病多从下部底脚叶开始发病，沿同侧叶片向上扩展；低洼积水地块易发病；暴雨、暴晴天气易发病。

4.1.6.2 防治措施

（1）农业防治。选用耐病品种；注意田间排水防涝，避免大水漫灌；加强栽培管理，提高植株抗病力。

（2）化学防治。田间零星植株底脚叶出现病害症状时，选用噻菌铜、喹啉铜、氢氧化铜、春雷霉素、中生菌素等药剂喷雾防治，间隔期 7～10 d，连续防治 2～3 次。

4.2 虫害防治

菜豆主要害虫有豆荚螟、美洲斑潜蝇、红蜘蛛、蚜虫、白粉虱、蓟马、甜菜夜蛾、斜纹夜蛾、烟青虫等。以"预防为主，综合防治"的方针，坚持以农业防治、物理防治、生物防治为基础，化学防治为辅，治少、治小，从而达到经济、安全、有效地控制害虫的目的。

4.2.1 物理防治

4.2.1.1 色板诱杀

根据蚜虫、白粉虱、蓟马、美洲斑潜蝇等成虫对黄/蓝色有较强趋性的特点，在田间悬挂黄/蓝板对其进行诱杀和监测，每亩菜地放置 30 块粘虫板。

4.2.1.2 灯光诱杀

利用棉铃虫、地老虎、斜纹夜蛾、甜菜夜蛾、烟青虫等成虫的趋光性诱集并消灭，15～30 亩菜地安装 1 个太阳能杀虫灯。

4.2.1.3 性诱剂诱杀

利用雌性信息素吸引田间寻求交配的雄蛾，将其诱杀在诱捕器中，使雌虫失去交配的机会，不能有效地繁殖后代。根据不同的害虫，一般每亩设置不同数量的性诱剂诱捕器。

4.2.1.4 毒饵诱杀

针对害虫有营养补充、夜间活动的特性，用糖醋液、麸糠或鲜嫩青草等制作毒饵，投放诱杀蛾类、蟋蟀等害虫。

4.2.2 生物防治

（1）生物天敌。保护和利用瓢虫、草蛉、食蚜蝇、蜘蛛等捕食性天敌和赤眼蜂、丽蚜小蜂等寄生性天敌。利用生物天敌防治害虫，如利用丽蚜小蜂防治温室白粉虱，草蛉防治蚜虫、红蜘蛛等，瓢虫防治蚜虫，赤眼蜂防治大豆食心虫等。

（2）微生物源农药。利用苏云金杆菌防治斜纹夜蛾、甘蓝夜蛾等；利用球孢白僵菌、金龟子绿僵菌防治白粉虱、蚜虫、金龟子等；利用核型多角体病毒防治斜纹夜蛾等。

（3）植物源农药。利用苦参碱防治红蜘蛛、蚜虫、白粉虱等；利用除虫菊素防治菜青虫、红蜘蛛、斑潜蝇等；利用藜芦碱防治白粉虱等。

4.2.3 化学防治

（1）豆荚螟。可用4.5%高效氯氰菊酯乳油、10%虫螨腈悬浮剂、35%氯虫苯甲酰胺水分散粒剂、30%茚虫威水分散粒剂等药剂喷雾防治。

（2）蚜虫。可用3%啶虫脒微乳剂、25%吡蚜酮悬浮剂、10%吡虫啉乳油、10%溴氰虫酰胺可分散油悬浮剂等喷雾防治。

（3）叶螨类。可用15%哒螨灵乳油、43%联苯肼酯悬浮剂、1.8%阿维菌素乳油、24%螺螨酯悬浮剂等喷雾防治。

（4）美洲斑潜蝇。可用1.8%阿维菌素乳油、70%灭蝇胺可湿性粉剂、10%虫螨腈悬浮剂、2.5%溴氰菊酯乳油等药剂喷雾防治。

（5）白粉虱。可用70%吡虫啉水分散粒剂、25%噻虫嗪水分散粒剂、3%啶虫脒微乳剂、10%烯啶虫胺水剂等药剂喷雾防治。若叶片背面多虫态共存可在上述药剂中加入吡丙醚、灭幼脲、氟铃脲等杀卵剂一起喷施。

（6）斜纹夜蛾、甜菜夜蛾。可用5%甲氨基阿维菌素苯甲酸盐微乳剂、5%氯虫苯甲酰胺悬浮剂、2.5%溴氰菊酯乳油、4.5%高效氯氰菊酯乳油等药剂喷雾防治。

5 适时采收

菜豆开花结荚盛期，下部花序已结荚，中上部花序又相继开花，需要消耗大量养分，如果植株负担得过重，容易因养分失调而引起落花落荚。因此，除在盛花初期注意施肥外，还应及时采摘嫩荚，以减少植株营养负担。矮生菜豆播种后50～60 d，蔓生菜豆播种后65～80 d开始采收，可连续采收30～60 d或更长时间。落花后10～15 d为采收适期，采收过早影响产量，过晚则降低品质。盛荚期3 d左右采收1次。采摘后的菜豆不宜长时间存放，应保持通风干燥，否则易发霉变质。采收后应尽快分拣包装运输，以免影响品质。

参考文献

代程，何玉华，包世英，等，2017.云南蔓生型普通菜豆资源形态学遗传多样性分析[J].西南农业学报，30（2）：256-261.

秦伟，陈昆，刘颖颖，等，2017.菜豆日光温室高产高效栽培技术[J].安徽农学通报，23（13）：61-62.

瞿云明，郑仕华，马瑞芳，等，2021.菜豆化肥农药减施栽培技术规程[J].中国瓜菜，34（2）：92-94.

田如霞，2020.春播露地菜豆栽培技术改进[J].中国蔬菜（6）：111-112.

汪宝根，董君旸，汪颖，等，2022.浙江省地方菜豆种质资源鉴定与遗传多样性分析[J].浙江农业学报，34（11）：2416-2427.

汪骞，陶婧，袁艺，等，2018.老挝北部山区水稻—菜豆水旱轮作栽培模式[J].热带农业科学，38（4）：27-30.

中国农业科学院蔬菜花卉研究所，2010.中国蔬菜栽培学[M].2版.北京：中国农业出版社.

速生叶菜栽培关键技术

速生叶菜一般指种子萌发后，生长周期相对较短的叶菜类蔬菜，通常在种植后4～6周内就能够生长到成熟并开始收获。速生叶菜生长快速，适合在短时间内获得丰收，同时也有利于在短暂的生长季节内快速增加产量，提高土地资源利用率和复种指数。常见的速生叶菜包括芥蓝、菜心、青梗菜、散叶生菜（意大利生菜）、苋菜、油麦菜等。随着市场需求的扩大，速生叶菜已成为人类日常饮食中不可或缺的重要组成部分。

云南充分利用良好的自然环境和独特的立体气候条件，建立常年蔬菜、夏秋蔬菜、冬春蔬菜三大优势蔬菜产区，实现了不同类型蔬菜周年生产周年供应的发展模式。其中，周年蔬菜产区以速生叶菜为主，形成了以陆良、嵩明、晋宁、沾益等为代表的速生叶菜优势产区，这些区域属于北亚热带高原季风气候类型，年均温14～18 ℃，海拔1 800～2 000 m，适宜速生叶菜塑料大棚周年栽培。

1 常见速生叶菜

1.1 芥蓝

1.1.1 营养价值

芥蓝（*Brassica oleracea* var. *alboglabra*）是十字花科芸薹属甘蓝种芥蓝亚种植物，又名白花芥蓝、绿叶甘蓝、芥蓝菜等，为一年生草本植物。每100 g芥蓝花薹鲜样中含水量90%左右、还原糖0.74～1.00 g、蛋白质1.60～2.08 g、纤维素1.20 g、β-胡萝卜素0.96～2.0 mg、维生素C 81～101 mg，同时还含有其他微量元素。芥蓝菜薹柔嫩、鲜脆、清甜、味鲜美，可炒食、煮汤食，也可作配菜。

1.1.2 品种与分布

芥蓝起源于中国，原产地在中国的南部。芥蓝栽培地区分布在广东、广西、福建、江苏、浙江、云南等地，其中产地主要集中在华南地区。在国外地中海沿岸也有芥蓝分布。主栽品种有幼叶早芥蓝、柳叶早芥蓝、福建黄花芥蓝、抗热芥蓝等。

1.2 菜心

1.2.1 营养价值

菜心（*Brassica rapa* var. *parachinensis*）为十字花科芸薹属白菜亚种的一个变种，为一、二年生草本蔬菜植物，别名菜薹、绿菜薹、菜尖等。菜心每千克可食用部分含蛋白质 13～16 g、脂肪 1～3 g、碳水化合物 22～42 g、钙 410～1 350 mg、磷 270 mg、铁 13 mg，以及胡萝卜素 1.0～13.6 mg。菜心具有一种芸薹属蔬菜的特殊清香，质地柔嫩，其丰富的纤维素可促进肠壁蠕动，帮助消化，防止大便干燥，稀释肠道毒素，广东、港澳等地居民把菜心作为不可缺少的上等菜。

1.2.2 品种与分布

菜心主要分布在中国广东、广西、台湾、香港、澳门等地。20世纪后期在日本引种成功。目前，菜心是广东栽培面积最大的蔬菜之一，年种植面积达 27 万亩以上，产量占蔬菜年上市量的 30%。近年来菜心在宁夏、四川、甘肃、云南等地均有栽培，被列为一种名优特色蔬菜。一般按照菜心生长期的长短和对栽培季节的适应性可分为早熟、中熟和晚熟 3 种类型，生产中的主栽品种有四九菜心、四九菜心 19 号、桂林柳叶早菜心、宝青 40 天、青梗柳叶中菜心、三月青菜心、特青迟心 4 号等。

1.3 青梗菜

1.3.1 营养价值

青梗菜（*Brassica rapa* L. ssp. *chinensis*），俗称不结球白菜，是十字花科芸薹属植物，又叫小白菜、上海青，具有耐热、耐寒、适应性强、经济效益高等优点。青梗菜富含维生素、蛋白质、胡萝卜素以及钙、磷、铁、钾、钠、镁、氯等矿物质，据测定，其所含的维生素 C 是大白菜的 3 倍多。

1.3.2 品种与分布

青梗菜原产于中国，在我国有 1 000 多年的栽培历史，福建更是青梗菜的主要产区。青梗菜品种以常规种为主，主栽品种有夏帝、华冠、金品 1 夏、金品 552、苏州青、四月慢、五月慢等。

1.4 散叶生菜

1.4.1 营养价值

散叶生菜（*Lactuca sativa* Crispa）又叫意大利生菜，是菊科莴苣属中的一、二年生草本植物，是以叶为产品器官的叶用莴苣。散叶生菜是奶油生菜、紫叶生菜、玻璃生菜、散叶生菜、直立生菜、野生生菜、花叶生菜等多种不结球生菜的总称。散叶生菜营养含量丰富，含有大量 β- 胡萝卜素、抗氧化物、维生素 B_1、维生素 B_6、维生素 E、维生素 C，还有大量的膳食纤维素和中微量元素如镁、钙，以及少量的铁、铜、锌。味道鲜嫩爽口、有较高的营养价值、经济价值和独特的保健功能。

1.4.2 品种与分布

散叶生菜属半耐寒性蔬菜，喜冷凉，忌高温，最适宜生长温度为白天 16～22 ℃、夜间 10～12 ℃，能耐 30 ℃高温和 -1 ℃的低温。品种间对温度要求差异很大，玻璃生菜、奶油生菜、紫叶生菜、野生生菜等不耐高温，适宜在较低的温度下生长。美国加州大速生菜、意大利生菜耐热又抗寒，适应性广。散叶生菜可周年种植，喜光照忌荫蔽；适宜湿润的环境，不耐干旱；要求有充足的氮肥，同时配合磷肥、钾肥和微量元素肥料；适宜在土层深厚、疏松、富含有机质的土壤种植。主栽品种有益农四季抗热耐抽薹生菜、美国大速生菜、罗莎生菜、罗马生菜、散叶生菜等。

1.5 油麦菜

1.5.1 营养价值

油麦菜（*Lactuca sativa* Longifolia）是菊科一、二年生草本植物，以叶为主要食用器官，含多种营养成分。油麦菜含有丰富的维生素 A，居蔬菜前列；油麦菜还富含抗坏血酸和叶酸。抗坏血酸能刺激人体的造血机能，促进血中胆固醇转化，使血脂下降；叶酸能保护心血管。油麦菜比较适合素炒或凉拌，食之均嫩脆、清香可口。

1.5.2 品种与分布

我国是油麦菜的主要生产国之一，油麦菜在我国南方地区广泛种植，主要以地方品种和耐热散叶品种栽培为主，如北散生、紫玉、萨琳娜斯 88、云翠、红帆紫叶生菜、绿裙生菜等。

2 主要种植模式

速生叶菜主要采用露地、小拱棚、钢架大棚、连栋大棚等种植,以设施种植为主。由于速生叶菜生产具有生长周期短、成熟速度快等特点,不同地区对速生叶菜周年优质高效生产茬口进行了探索研究,摸索出适宜不同区域的高效种植模式,以实现速生叶菜周年生产、均衡供应的目标。

2.1 青梗菜周年蔬菜模式

以江苏苏中地区为例,探索建立了青梗周年种植模式,以上海青为主栽品种,在盐城、南通等地全年栽种6茬,夏秋40 d采收,冬春60 d采收。生产中联合运用机械翻耕、播种、收获、水肥一体化等核心技术,配套运用诱虫色板、频振式杀虫灯等绿色防控技术,提高产品产量和品质。

2.2 上海速生叶菜周年生产模式

以上海地区为例,经过多年的生产实践证明,茼蒿—芥蓝—菜心—油麦菜—芥蓝栽培模式,该模式中茼蒿于1月中旬播种,3月上旬采收;第1茬芥蓝于3月下旬播种,5月下旬采收;菜心于6月上旬播种,7月中旬采收;油麦菜于7月底播种,10月上旬采收;第2茬芥蓝于10月中旬播种,翌年1月上旬采收。该模式可以有效提高土地利用率,实现合理的轮作,产品质量有保障。

2.3 云南速生叶菜周年生产模式

以云南省为例,通过轮作换茬减少病虫害发生,形成了"油麦菜—黄白菜—青梗菜—意大利生菜—黄白菜—油麦菜"的栽培模式。根据以上4种速生叶菜生长特性和市场需求等因素,建立了1年6茬速生叶菜栽培模式,既提高了土地利用率又产生了较好的经济效益,利用轮作换茬、土壤高温消毒技术降低病虫害发生率,减少农药的使用量,提高蔬菜品质。

3 栽培管理技术

3.1 土壤消毒

选择大棚空闲期开展消毒工作,在消毒前,要彻底清除棚内残留物及其他蔬菜、杂草，并结合整地作业施入适量的农家肥，通过灌水将土壤相对湿度维持在60%左右，接下来用地膜覆盖大棚，高温闷棚13 d左右，确保土壤升温至60 ℃，杀灭土壤内部病原菌及寄生虫，进而达到改良土壤的目的。

3.2 育苗

为了培育壮苗和提高土地利用率，速生叶菜采用育苗移栽，可以缩短幼苗在大田20～25 d生长期，通过集约化育苗实现种苗病虫害集中防控和营养集中供应，确保移栽后大田植株的整齐度。一般生产中采用商品基质穴盘育苗，每穴播种1粒，播种深度0.5～1 cm。一般播种后4～5 d出苗，夏、秋季中午高温时覆盖遮阳网，同时增加棚内通风降低温度，出苗后7 d施0.1%复合肥水溶液（氮：磷：钾=15：15：15，下同），15 d施0.2%复合肥水溶液，冬季育苗期施肥2～3次。夏、秋季苗龄20～25 d，冬、春季苗龄30 d左右。移栽前进行炼苗，即控制浇水次数，增大通风和光照，使幼苗尽量适应外界环境，增加移栽成活率。

3.3 整地理墒

速生叶菜植株较小、生长快，一般适合宽墒密植。整地前亩施入2 000 kg腐熟农家肥或商品有机肥、30 kg复合肥和200 kg微生物菌肥。松翻土壤后耙平种植墒面，将整个大棚作为1个墒面，不单独预留操作道。黄白菜株行距30 cm×30 cm，青梗菜株行距15 cm×15 cm，意大利生菜株行距25 cm×25 cm，油麦菜株行距15 cm×20 cm，菜心株行距10 cm×10 cm，芥蓝株行距10 cm×10 cm。冬季覆膜栽培。

3.4 水肥管理

选择阴天或晴天下午栽苗，防止中午高温灼烧叶片，移栽后及时浇足定根水，提高苗的成活率，缓苗期适当增加浇水次数，速生叶菜一般采用喷灌浇水，根据天气情况和植株生长状况进行适宜浇水。由于速生叶菜生长周期短，肥料以底肥为主，生长期追施1～2次肥料即可满足其生长。移栽缓苗后追施0.3%尿素水溶液提苗，黄白菜移栽20 d追施第2次肥料，每亩追施复合肥20 kg；青梗菜移栽后15 d进行第2次追肥，每亩追施复合肥15 kg或喷灌喷施浓度为0.5%水溶肥；意大利生菜移栽后20 d开始第2次追肥，每亩追施复合肥15 kg；油麦菜移栽后15 d进行2次追肥，每亩追施复合肥15 kg或喷灌喷施0.5%的水溶肥；菜心和芥蓝在定植后10 d和20 d，按每亩追施复合肥15 kg。速生叶菜可根据植株生长状况，冬季适当增加追肥次数，同时，结合喷打农药，叶面喷施1～2次叶面肥。

4 病虫害防治

速生叶菜主要病害有霜霉病、软腐病、黑腐病、黑斑病、菌核病、灰霉病、炭疽病、病毒病等；主要害虫有黄曲条跳甲、蚜虫、猿叶虫、小菜蛾、菜青虫、斜纹夜蛾、甜菜夜蛾等。

4.1 农业综合防治

种植前对地块进行全面清理，即清除杂草、植株残体，集中回收废弃物等；蔬菜生长期间及时摘除病虫为害的叶片，拔除病虫残株；蔬菜收获后彻底清理菜地，将病株残体带出田外集中烧毁、深埋或放于粪窖内，以减轻病虫害的发生和蔓延；利用换茬间隙，冬季深翻冻融，夏季晒垡炕地、翻犁地前撒施生石灰消毒杀灭栖息土壤中的病虫；适当减少中耕除草，防止人为造成的伤口，预防病菌侵入；避免同一类蔬菜在同一地块多年种植，尽可能减少土传病害传播。

4.2 微生物农药防治

在蔬菜药剂防治上应优先选用生物农药，如用3%中生菌素可湿性粉剂、2%春雷霉素水剂等防治软腐病、黑腐病等；用8%宁南霉素水剂、2%氨基寡糖素水剂等预防病毒病；用10亿活孢子/g枯草芽孢杆菌、1.5亿活孢子/g木霉菌防治炭疽病、枯萎病等。用2%阿维·苏云菌可湿性粉剂防治菜青虫、小菜蛾等；用斜纹夜蛾核型多角体病毒防治斜纹夜蛾等；用80亿孢子/g金龟子绿僵菌防治金针虫、蛴螬、地老虎等。

4.3 植物源农药防治

用0.5%大黄素甲醚水剂500～600倍液防治白粉病、霜霉病、灰霉病、炭疽病等；用4%小檗碱水剂300～500倍液防治白粉病、霜霉病等；用1%蛇床子素水乳剂800～1 000倍液防治白粉病等。用0.3%苦参碱水剂500～700倍液防治蚜虫、白粉虱等；用0.5%藜芦碱醇溶液400～600倍液防治菜青虫、蚜虫等；用3%除虫菊素乳油50～80倍液等防治菜青虫、小菜蛾、蚜虫等。

4.4 化学药剂防治病害

4.4.1 霜霉病

底脚叶初见病斑时可用80%代森锌可湿性粉剂、722 g/L霜霉威水剂、70%丙森锌可湿性粉剂、70%氟菌·霜霉威悬浮剂、10%氰霜唑悬浮剂等喷雾防治。

4.4.2 软腐病、黑腐病

发病初期可用2%春雷霉素水剂、1%中生菌素可湿性粉剂、40%噻唑锌悬浮剂、20%噻菌铜悬浮剂、47%春雷·王铜可湿性粉剂或50%氯溴异氰尿酸可溶粉剂等喷雾防治。

4.4.3 黑斑病

发病初期可用50%异菌脲可湿性粉剂、70%丙森锌可湿性粉剂、10%苯醚甲环唑水分散粒剂、25%嘧菌酯悬浮剂、45%咪鲜胺水乳剂或25%吡唑醚菌酯悬浮剂喷雾防治。

4.4.4 炭疽病

发病初期可用10%苯醚甲环唑水分散粒剂1 500倍液、50%异菌脲可湿性粉剂1 000倍液或45%咪鲜胺水乳剂2 000倍液喷雾防治。

4.4.5 病毒病

首先，防控好传毒媒介害虫蚜虫、白粉虱、蓟马、红蜘蛛等，切断病毒病传播、扩散途径；其次，采取药剂防治，发病前或生长前期及时喷施1.5%烷醇·硫酸铜可湿性粉剂、20%吗胍·乙酸铜可湿性粉剂或8%宁南霉素水剂等。每隔7～10 d喷1次，连续防治2～3次，注意农药交替使用。

4.4.6 菌核病

田间零星出现发病点或中心病塘时，选用50%腐霉利可湿性粉剂、50%异菌脲可湿性粉剂、40%菌核净可湿性粉剂、70%甲基硫菌灵可湿性粉剂等药剂喷雾防治，间隔5～10 d喷1次，连续防治2～3次，注意农药交替使用。

4.5 虫害防治

4.5.1 物理防治

4.5.1.1 防虫网防治

单独使用时，应选择银灰色或黑色防虫网；与遮阳网配合使用时，以选择白色40～60目为宜。夏秋栽培一般采用防虫网全棚覆盖。夏秋栽培的速生叶菜类蔬菜，因其生育期短，采收相对集中，可用小拱棚覆盖栽培。覆盖防虫网之前及时杀灭棚内自生虫源。覆盖防虫网后，有些病害虫还可通过土壤、种子和种苗等进行传播，因此，栽植时种苗最好进行杀虫杀菌处理，并挑选无病虫害的健壮植株。

4.5.1.2 诱虫板防治

黄板可诱杀蚜虫、白粉虱、烟粉虱、飞虱、叶蝉、斑潜蝇等，蓝板可诱杀种蝇、蓟马等。诱虫板应选用材质较好，可双面诱杀，无毒、抗日晒、耐雨水冲刷的产品。每亩设施大棚悬挂诱虫板 30~40 张（长 30 cm、宽 25 cm），悬挂高度应高于植物的生长点 15~20 cm，以达到较好的防治效果。随着植株生长的不断增加，黄板高度也应随之调节。当害虫布满诱虫板时要及时更换。

4.5.1.3 灯光诱杀防治

蔬菜生产上常用频振式杀虫灯。主要防治甘蓝夜蛾、小地老虎、斜纹夜蛾、甜菜夜蛾、烟青虫、菜螟、小菜蛾、叶甲等害虫。吊挂高度一般为 1.5 m，在田中呈棋盘状分布，灯距 100~150 m，每盏灯的控制面积为 20~40 亩。注意及时清理捕杀袋中的害虫尸体。

4.5.1.4 性引诱剂防治

可采用性引诱剂诱杀斜纹夜蛾、甜菜夜蛾、小菜蛾、小地老虎等。小菜蛾性引诱剂在 5—6 月和 7—9 月使用，每亩放 1 个诱捕器，内置 3 个诱芯；斜纹夜蛾和甜菜夜蛾性引诱剂在 7—10 月使用，每亩放 1 个诱捕器，内置 1 个诱芯。诱捕器间距 30 m，将诱捕器挂在棚架或木棍上，高出蔬菜 30 cm，春、秋季每 30 d 更换 1 次诱芯，夏季每 20 d 更换 1 次诱芯。

4.5.1.5 生物天敌防治

粉虱类害虫可用丽蚜小蜂防治，定植 7~10 d 后，发现粉虱类害虫即可释放，释放量 2 000 头 / 亩，7~10 d 释放 1 次，连续释放 3~5 次；蓟马类害虫可用小花蝽类天敌防治，释放量 500 头 / 亩，7~10 d 释放 1 次，连续释放 2~4 次；螨类害虫可用智利小植绥螨防治，叶部撒施智利小植绥螨 5~10 头 /m²，点片发生时中心株释放 30 头 /m²，每 2 周释放 1 次，共释放 3 次；蚜虫类害虫可用瓢虫防治，发现害虫后按 2 000 头 / 亩释放瓢虫（卵）、蚜茧蜂，7~10 d 释放 1 次，连续释放 2~3 次；鳞翅目害虫如小菜蛾、甜菜夜蛾、棉铃虫、斜纹夜蛾等可用赤眼蜂、绒茧蜂等防治，按 2 万头 / 亩释放，5~7 d 释放 1 次，连续释放 3 次，即能获得较好的防治效果。

4.5.1.6 化学防治

蚜虫、白粉虱和蓟马可用 3% 啶虫脒微乳剂、10% 吡虫啉乳油、25% 噻虫嗪水分散粒剂或 60 g/L 乙基多杀菌素悬浮剂等药剂喷雾防治；小菜蛾和菜青虫可用 5% 甲氨基阿维菌素苯甲酸盐微乳剂、5%氯虫苯甲酰胺悬浮剂、3%阿维菌素乳油或2.5%溴氰菊酯乳油等药剂喷雾防治；斜纹夜蛾和甜菜夜蛾可用2%甲氨基阿维菌素苯甲酸盐微乳剂、10%虫螨腈悬浮剂或30%茚虫威水分散粒剂等药剂喷雾防治；潜叶蝇可用5%阿维菌素微乳剂、70%灭蝇胺可湿性粉剂或10%虫螨腈悬浮剂等药剂喷雾防治。每隔7~10 d喷施1次，连喷2~3次，农药交替使用。

5 采后处理

5.1 采收标准

根据速生叶菜的品种特性及市场需求及时采收。一般早晨和下午采收，避开中午采收，露地种植速生叶菜应在露水干后采收，采用塑料周转箱存放菜，采收完毕及时运输到冷库。

5.2 贮藏保鲜

有条件的地方采用真空预冷20~40 min，去除田间热，保持新鲜状态。真空预冷后的叶菜，放入3~5 ℃冷库中贮藏。可在冷库中进行蔬菜初加工分级，分级标准根据目标市场要求制定。

5.3 包装运输

初加工分级后的速生叶菜一般采用泡沫箱包装，泡沫箱内放包装用的保鲜膜，在泡沫箱对角线放2个冰瓶，冰瓶外包裹1层报纸防冻。有条件的地方可冷链车运输，采用挂车运输时，在车包裹保温棉被和封车薄膜，降低运输过程中蔬菜损耗。

参考文献

汪骞，孔令明，陶婧，等，2017.老挝北部雨季速生叶菜栽培技术[J].中国蔬菜（3）：96-98.

汪骞，念红艳，袁艺，等，2019.云南速生叶菜设施一年多茬栽培技术应用[J].农业工程技术，39（4）：72-74.

中国农业科学院蔬菜花卉研究所，2010.中国蔬菜栽培学[M].2版.北京：中国农业出版社.

❖ 南瓜栽培关键技术

南瓜是葫芦科（Cucurbitaceae）南瓜属（Cucurbita）一年生蔓生草本植物。起源于美洲，于16世纪初传入中国，目前中国是世界南瓜第一生产国和消费国。南瓜主要有5个栽培种：中国南瓜（Cucurbita moschata Duch.）、美洲南瓜（Cucurbita Pepo L.）、印度南瓜（Cucurbita maxima Duch.）、黑籽南瓜（Cucurbita ficifolia B.）、灰籽南瓜（Cucurbita argyrosperma Huber.）。本节主要介绍中国南瓜栽培技术。

中国南瓜又称麦瓜、番瓜、倭瓜、饭瓜、北瓜等，除果实外，茎、叶、花、种子均可食用，其可食部分达到80%以上。果实皮薄肉厚、组织细密、风味甜美，且营养价值较高，富含淀粉、脂肪、还原糖、氨基酸、维生素、矿物质和纤维素等。据报道，每100 g南瓜中，含有蛋白质0.7～2.0 g、脂肪0.1～0.5 g、糖类3.3～11 g、维生素A 0.34～0.78 mg、矿物质6～48 mg。此外，研究表明南瓜中的多糖类、类胡萝卜素等物质还可降低血脂、调节血糖，提高机体免疫能力。

1 生长条件

1.1 温度

南瓜种子发芽所需温度为28～30 ℃，一般7～10 d子叶即可展平；幼苗期白天适宜温度为25～30 ℃，开花结果期白天适宜温度为25～28 ℃、夜间适宜温度为13～18 ℃。生长期温度高于35 ℃或低于13 ℃，均会导致植株生长不良，花期温度高于35 ℃则会导致花器官发育不良，花粉败育。

1.2 水分

南瓜根系发达，抗旱能力较强，其对水分的要求不高，土壤相对湿度保持在55%～70%即可。一般来说，幼苗期需水量较小，浇足定根水即可；伸蔓期至开花结果期，当地块出现明显干旱时，浇1次水即可，若水分过多，则容易徒长，植株生长过旺，导致后期不易坐果；果实膨大期，应保证水分充足，缺水容易引起果实畸形，严重时产量降低。空气湿度过低易导致植株萎蔫，过高易造成落花落果，过低过高交替出现，易造成白粉病暴发，且不利于开花授粉。因此，栽培过程中根据土壤墒情确定浇水。

1.3 光照

南瓜属于喜光的短日照植物。在 8 h 短日照下，促进雌花分化，当光照超过 12 h，雌花分化率降低。其在光照充足的条件下生长健壮，在弱光下植株细弱、易徒长，坐瓜率下降，甚至引起化瓜。

若进行人工遮光，将光照控制在 8 h，可降低雌花节位，促进雌花分化，增加坐瓜率，提高产量。

1.4 土壤养分

南瓜喜温耐旱，适合在土层深厚、疏松透气、肥力较高的中性或微酸性壤土或砂壤土地块种植。南瓜抽蔓期前需肥量较少；坐果期养分需求量大；果实膨大期，对磷、钾吸收量增多，对氮吸收量减少。

2 品种选择

根据当地气候、栽培目的和市场需求，选择优质、丰产、抗逆性强、商品性好的优良品种，如密本南瓜、云南本地姜柄瓜等。

3 整地施肥

播种或移栽前 1 个月，将土壤深翻、暴晒；播种或移栽前 2～3 d，施入充分腐熟的有机肥 3 000 kg/亩和三元复合肥 30 kg/亩，并混匀作畦，畦面宽 3.0～4.0 m（包沟）。在节约肥料和提高肥料利用率方面，可在打塘后，每塘放入 1 kg 有机肥和 50 g 三元复合肥，覆土 2～3 cm 后拌匀备用。

4 播种育苗

4.1 种子处理

选择饱满、粒大的新种子，剔除瘪粒、霉粒，为保证出苗率，播种前须做发芽试验。播种前将种子放入 50～55 ℃温水浸泡 15 min，其间要不断搅拌，然后在常温下浸种3~4 h后放入28 ℃恒温箱中催芽，2~3 d即可"露白"。

4.2 直播

直播期一般选择在当地无霜后进行，如在云南1 800~2 000 m海拔的地区，一般在4月中旬至4月底播种。打塘播种，播种前先在塘内浇水至渗出，每塘播2~3粒，种子之间要分开，并盖上2~3 cm的细土。播种深度过深，不易出苗，播种深度过浅，苗易戴帽出土。一般爬地栽培的行距3~4 m、株距0.6~0.8 m；搭架栽培的行距1.0~1.5 m、株距0.6~1.0 m。当秧苗2~3片真叶时，分两次进行定苗，第一次将有机械损伤或病虫为害的幼苗拔除，第二次选择一株生长最为健壮的苗留下。

4.3 育苗移栽

将铺满基质的32孔穴盘浇透水，种子平播于土上或芽尖朝下，每个穴盘孔中播1粒种子，覆土2 cm。保持苗床温度25~28 ℃，5~7 d可出苗。出苗后，白天保持20~25 ℃，夜间13~15 ℃。定植前7 d开始通风、降温，白天温度降至20~22 ℃，夜间温度逐渐降至10~12 ℃。苗床相对湿度保持在60%~70%，缺水时及时喷水。基质育苗应在真叶出现后适当补充水溶性肥料，第一片真叶出现后，喷施0.2%尿素或磷酸二氢钾溶液。

壮苗标准：苗龄20~25 d，具有3~4片真叶，苗高10~15 cm，叶片肥厚，叶色深绿，茎基部节间短而粗，无病虫害的幼苗。

定植前1周开始炼苗，保持温度在15~25 ℃，加强通风，适当控水。移栽时，于晴天下午选取壮苗带土定植，定植深度以土壤盖住土坨为宜，切忌让根部直接接触肥料，移栽后及时浇足定根水。

4.4 苗期管理

早春育苗时，浇水过多会导致地温下降，不利于幼苗生长；温度高、光照不足时，浇水过多会导致幼苗徒长，且容易造成苗期病害的发生，如霜霉病，因此，苗期应小水勤浇，保证土壤见干见湿。

5 田间管理

5.1 肥水管理

伸蔓期应控制水肥，促进发根，少浇或不浇水，以防水肥过足，导致植株生长过旺。开花结果期，第一瓜坐稳开始膨大后，浇 1 次水，促进瓜的膨大，并随水追 1 次肥，一般采用三元复合肥（氮∶磷∶钾=15∶15∶15）或高钾水溶肥 10～15 kg/ 亩。注意在开花结果期，切忌大量追施氮肥，容易造成植株徒长、落花落果等现象。若采收嫩瓜，每采收一批瓜后，待新梢长出，随水追肥 1 次。

5.2 整枝、留瓜

一般每株只留 1 个主蔓，其余侧蔓全部摘除。食用老瓜，则 1 株只留 1 个瓜，待瓜坐稳后，在距离瓜 40～50 cm 处打顶，留 6 片以上功能叶。食用嫩瓜，一般根据结瓜习性，可留 4～6 个嫩瓜，边采收边结瓜的方式，待最后一个瓜坐稳后，去除生长点。

若采用搭架栽培，其间需打掉根部老叶、病叶，并根据主蔓长度进行适当降蔓。若爬地栽培，在瓜苗伸蔓前于叶节处进行培土，瓜蔓长 1 m 时第一次压蔓，之后每长 1 m 左右进行 1 次压蔓，瓜后 2 节处压 1 次。压蔓方法：在蔓边开出 1 条浅沟，将蔓轻轻移到沟内，用土压住。压蔓时间一般中午或下午温度高时进行。

5.3 中耕除草

由于栽培行距较大，杂草容易蔓延，所以要进行中耕除草。除草时要注意伤到根系。第一次中耕除草应在缓苗后，中耕深度为 3～5 cm；第二次中耕应在伸蔓时，结合引蔓进行，后期根据杂草生长情况进行中耕。

5.4 人工授粉

为提高坐果率，须进行辅助授粉。一是昆虫辅助授粉：通过在瓜地里种植花期长、颜色鲜艳的花卉，吸引昆虫或人工养殖的蜜蜂进行授粉。二是人工辅助授粉：若花期遇低温阴雨天气或设施内无昆虫授粉的情况，需人工辅助授粉。一般于开花前 1 d 进行套袋，第二天 10:00 以前结束授粉。人工辅助授粉方法：去除雄花花瓣，将雄蕊对准雌花，轻轻涂抹在雌花的柱头上即可。注意授粉时不要碰到小瓜，易造成病菌感染引起化瓜。授粉后，用绳子轻轻捆绑花瓣，防止雨水将花粉冲掉。

6 缺素症状及防治

6.1 缺氮

症状：植株叶片小，新叶淡绿色，从下到上慢慢变黄。先是叶脉间发黄，之后花落后坐果量少，果实膨大缓慢。

防治方法：根据南瓜对氮、磷、钾三要素和微肥的需求，施用腐熟的有机肥，防止缺氮。低温条件下可施用硝态氮，田间出现缺氮症状时，应立即在根部增施氮肥，也可叶面喷施。

6.2　缺钾

症状：主要从植株下部的老叶上出现不正常叶色或黄色斑点，叶缘开始变黄焦枯，如"镶金边"，接着叶缘出现坏死。

防治方法：应追施钾肥或叶面喷 0.2% 硫酸钾或磷酸二氢钾溶液。

6.3　缺镁

症状：首先表现在老叶上，老叶的叶脉间出现黄化，由叶片中间向叶缘，继而整个叶片发黄或者发白。

防治方法：可叶面喷施 0.2% 硫酸镁溶液。

6.4　缺锌

症状：表现在叶片和果实上，植株生长受到抑制，尤其是在节间生长严重受阻，并表现出叶片脉间失绿或白化。

防治方法：不过量施用磷肥，而有选择地施用酸性肥料来降低土壤 pH 值，田间可喷施 0.2% 亚硫酸锌溶液。

6.5　缺钙

症状：主要表现在新叶、顶芽、果实等生长旺盛而幼嫩的部位。缺钙时，叶片尖端部分弯曲黄化或者白化，叶缘向上或者向下皱褶（降落伞形）；果实易裂开，根系常常变黑腐烂，发病的特点为果实脐部或植株顶部。

防治方法：叶面喷施钙肥，如叶面喷施 0.3% 氯化钙溶液 +0.2% 磷酸二氢钾溶液，或追施钙镁磷钾肥 10 kg/ 亩。

6.6　缺铁

症状：最先表现在幼叶上，缺铁与缺镁相似，不同的是缺铁先从新叶的叶脉间出现黄化，叶脉仍为绿色，继而整个叶片发黄或者发白。

防治方法：土壤 pH 值应在 6～6.5；防止土壤过干或过湿；叶面喷施 0.1%～0.2%

硫酸亚铁水溶液。

6.7 缺硼

症状：缺硼易造成花而不实的现象。生长点附近的节间显著缩短，生长点停止生长，根系不发生；上位叶的叶缘向上卷曲，叶缘部分变为褐色，叶片展开得慢，同时可见上位叶的叶脉有萎缩现象；果实上有污点，果实表面出现木质化。

防治方法：增施有机肥，在合理增施有机肥的基础上，控制氮肥，叶面喷施 0.1% 硼砂水溶液。

7 病虫害防治

7.1 病害防治

南瓜生产中的主要病害有白粉病、病毒病、霜霉病、疫病等。

7.1.1 白粉病

白粉病是南瓜常见且严重的病害之一，一般在夏季高温不透风的环境下暴发，南瓜白粉病致病菌主要是二孢白粉病菌（*Erysiphe cichoracearum*）和单囊壳菌（*Podosphaera xanthii*），为专性寄生菌，很难离体保存。

7.1.1.1 为害症状

主要为害叶片，其次是茎和叶柄，严重时也会为害果实表面。发病初期叶片上有白色小霉点，逐渐扩大成圆形斑点，直至整个叶面覆满白色粉状物，严重时叶背、茎秆也有白色斑块，从而导致叶片干枯。

7.1.1.2 防治方法

（1）农业防治。选用抗病品种；与其他科作物轮作，忌与瓜类连作；合理施肥，避免偏施氮肥，增施有机肥和磷、钾肥；加强植株管理，及时摘除基部病叶、老叶，修剪枝杈，加强植株通风透气性。

（2）化学防治。一旦发现南瓜发生白粉病，应及时采取防治措施，喷药时着重喷施叶背面，发病初期可选用15%三唑酮可湿性粉剂1 500倍液、25%嘧菌酯悬浮剂1 000～2 000倍液、25%乙嘧酚磺酸酯乳油1 500倍液、250 g/L吡唑醚菌酯乳油1 500倍液等药剂喷雾防治，每7～10 d喷施1次，连续用药2～3次。

7.1.2 病毒病

病毒病发病症状复杂多样，易与缺素症、虫害症状混淆，在实际生产中，病毒病可通过农事操作、传毒昆虫、植株间摩擦等交叉接触感染，传染迅速，寄主范围广。

7.1.2.1 为害症状

病毒病典型症状是叶片和瓜果不规则形褪绿或呈现浓绿与淡绿相间斑驳，植株叶片受侵害后先产生淡黄色不明显的斑驳，后期呈现浓淡不均浅黄绿镶嵌状花叶，叶片会变小，叶缘向叶背卷曲变硬发脆。有的植株枝杈顶端生长点部位的幼嫩叶片变褐坏死成顶枯；有的植株节间变短，枝叶丛生呈丛簇状。瓜果表面上形成褪绿斑纹或突起。为害严重时病叶和病瓜畸形皱缩，植株生长缓慢或矮化，结小瓜。

7.1.2.2 防治方法

（1）农业防治。选用抗病品种；与非葫芦科作物轮作，忌与瓜类连作；合理施肥；及时清除田间杂草；发现病株立即拔除带出田外深埋或烧毁。

（2）化学防治。可选用5%氨基寡糖素水剂2 000倍液、1%香菇多糖水剂600倍液、80%吗胍·乙酸铜可湿性粉剂1 000倍液、5.9%辛菌·吗啉胍水剂300倍液等药剂喷雾防治，间隔5~7 d防治1次，连续2~3次。

7.1.3 霜霉病

南瓜霜霉病是由古巴假霜霉菌［*Pseudoperonospora cubensis*（Berk. et Curt）Rostov］引起的侵染性病害，其孢子囊常附着在叶片表皮毛上，可借助气流、雨水、机械和农事传播。

7.1.3.1 为害症状

在苗期和成株期均可发生，主要为害叶片，发病初期在叶片背面形成水浸状小点，病斑边缘不明显，叶片正面形成黄化病斑，后期扩大呈不规则黄褐色病斑，一般由植株下部逐渐向上部发展，最后导致植株发黄枯死。

7.1.3.2 防治方法

（1）农业防治。选用抗病品种；与其他科作物轮作，忌与瓜类连作；及时摘除病叶；采用搭架栽培，及时修剪枝杈，增加透气性；设施内增加二氧化碳浓度。

（2）化学防治。可选用0.5%几丁聚糖可湿性粉剂600倍液喷雾预防；发病初期可选用58%甲霜·锰锌可湿性粉剂1 000倍液、75%百菌清可湿性粉剂600倍液、72%霜脲·锰锌可湿性粉剂800倍液、80%烯酰吗啉水分散粒剂2 000倍液等药剂喷雾防治，间隔5~7 d防治1次，连续2~3次。

7.1.4 疫病

南瓜疫病由疫霉菌（*phytophthora* spp.）引起的植物病害。当气温25～30 ℃、空气相对湿度高于85%，或通风透光差的地块发病严重。雨季或大雨后天气突然转晴，气温急剧上升，病害易暴发。

7.1.4.1 为害症状

整个生育期发生，根、茎、叶、果均可发病，主要为害南瓜的茎和果，造成茎和果实腐烂，叶片发黄，以致全株萎蔫枯死。茎部发病部初呈水渍状，淡褐色，后渐渐变褐色，湿腐，病部有粉状的白色小点；叶片发病初期出现圆形暗色水浸状斑，软腐，下垂，干燥时呈灰褐色，易脆裂；果实主要是爬地栽培的南瓜易受害，病斑初呈暗绿色水渍状，后渐湿腐，病部表面有粉状的白色小点。

7.1.4.2 防治方法

（1）农业防治。选用早熟抗逆性强的品种；与禾本科等非寄主作物进行4～5年轮作；防止栽培过密，以利通风透光，降低土壤湿度，减少发病机会；及时中耕除草、整枝压蔓、翻瓜；对发病植株和病瓜要及时清出田外，集中深埋或烧毁。

（2）化学防治。发病初期，可选用722 g/L霜霉威盐酸盐水剂600倍液、58%甲霜·锰锌可湿性粉剂600倍液、30%甲霜·噁霉灵水剂600～800倍液、50%烯酰吗啉可湿性粉剂1 500倍液等喷雾防治，连续用药2～3次，间隔期7～10 d。

7.2 虫害防治

南瓜生产中主要的害虫有白粉虱、蚜虫、地老虎、蓟马等。

7.2.1 白粉虱

7.2.1.1 为害特点

白粉虱以吸取南瓜植株汁液为害，吸食时会把病毒注进植物体内；分泌的蜜露会影响植株光合作用，成虫的排泄物会阻碍植株的呼吸，甚至诱发病害，为害果实和叶片。

7.2.1.2 防治方法

（1）农业防治。清洁田园，及时清除杂草；加强整枝打杈，清除老叶、病叶。

（2）物理防治。利用黄板诱杀。

（3）化学防治。在白粉虱发生初期，可选用10%吡虫啉悬浮剂1 000倍液、25%噻嗪酮可湿性粉剂2 000倍液喷雾防治。白粉虱飞行能力较强，药剂不易喷施到虫体上，所以一般先喷水打湿其翅膀再喷施药剂。每周1次，连续喷药3次。

7.2.2 蚜虫

7.2.2.1 为害特点

蚜虫在南瓜的整个生育期内都会产生为害，常以成群的成虫或若虫聚集刺吸叶片背面和嫩茎上的汁液，叶片和嫩茎受害后卷曲萎蔫，甚至枯死，影响果实生长。蚜虫还会分泌蜜露污染南瓜的叶片和果实，携带病毒。

7.2.2.2 防治方法

可用3%啶虫脒乳油1 000倍液、10%吡虫啉可湿性粉剂1 500倍液、25%噻虫嗪水分散粒剂4 000倍液喷雾防治。

7.2.3 地老虎

7.2.3.1 为害特点

地老虎又叫土蚕、地蚕等，一般地老虎在1～2龄幼虫取食作物的心叶或嫩叶，形成白斑或小洞，3龄以后从近地面处咬断作物的茎或者叶柄，出现缺苗断垄。

7.2.3.2 防治方法

（1）农业防治。及时清除杂草，防止成虫产卵。

（2）物理防治。用糖∶醋∶酒∶水按照3∶4∶1∶2的比例，加入少量敌百虫配成杀虫液，放置高度为离地面50～100 cm，每亩放置10～15份。

（3）化学防治。可用2.5%溴氰菊酯乳油1 500倍液、40%氯氰菊酯乳油2 000倍液、5%高效氯氟氰菊酯微乳剂2 000倍液喷雾防治。

7.2.4 蓟马

7.2.4.1 为害特点

成虫、若虫以锉吸式口器取食心叶、嫩芽、花器和幼果汁液，嫩叶嫩梢受害，组织变硬缩小，茸毛变灰褐或黑褐色，植株生长缓慢，节间缩短。幼瓜受害，果实硬化，瓜毛变黑，造成落瓜。嫩叶受害后使叶片变薄，叶片中脉两侧出现灰白色或灰褐色条斑，表皮呈灰褐色，出现变形、卷曲，生长势弱。

7.2.4.2 防治方法

（1）农业防治。及时清除杂草，防止成虫产卵；播种前对土壤进行消毒。

（2）物理防治。蓟马有趋蓝性，可在田间放置蓝板。

（3）化学防治。可用25%噻虫嗪水分散粒剂5 000～6 000倍液、20%啶虫脒可湿性粉剂5 000～8 000倍液、28%阿维·螺虫酯悬浮剂4 000倍液等药剂喷雾防治。

8 适时采收

中国南瓜一般食用老熟果实,雌花授粉后60 d左右,当果皮变硬,呈现本品种固有色泽,果粉增多,果柄木质化时可采收;也可食用嫩瓜,一般在授粉后20 d左右,达商品成熟度时,即可采收。采收时,留瓜柄2~3 cm,注意不要损伤瓜皮;采收后,放置于阴凉、通风、干燥处,一般可贮藏2~3个月,最长可贮藏6个月。

黄瓜栽培关键技术

黄瓜（*Cucumis sativus* L.）是葫芦科一年生蔓生或攀援草本植物，是全球性、大众化的重要蔬菜，其栽培面积仅次于番茄、甘蓝和洋葱，名列第四位。黄瓜在亚洲栽培面积最大，约占世界总面积的50%，其次是欧洲、北美洲、中美洲。我国黄瓜栽培面积约占世界栽培面积的28%，居各国之首。云南各地均有不同类型的黄瓜栽培，生产中一般采用塑料大棚栽培。

黄瓜营养丰富，每100 g鲜果含碳水化合物1.6～2.4 g、蛋白质0.4～0.8 g、钙10～19 mg、磷16～58 mg、铁0.2～0.3 mg、维生素C 4～16 mg。黄瓜含有纤维素，有加速体内腐败物排出、降低胆固醇的作用，另外还有清热解毒、利水解渴的作用。近代医学临床实践证明，黄瓜藤有良好的降血压和降低胆固醇的作用。

1 生长条件

1.1 温度

黄瓜喜温，不耐寒，遇霜冻即枯死。生长适宜温度白天为20～25 ℃、夜间为12～16 ℃，<10 ℃或>35 ℃时发生生理障碍，停止生长。

1.2 光照

黄瓜属于短日照植物，喜光，也耐弱光，日照8～11 h有利于提早开花结实，光照不足会影响黄瓜产量和品质。黄瓜的光饱和点为55 000～60 000 lx，光补偿点为15 000 lx，最适光照强度为20 000～60 000 lx，当光照强度低于20 000 lx时，植株生育迟缓。

1.3 水分

黄瓜产量高，需水量大，适宜土壤相对湿度为60%～90%。幼苗期水分不宜过多，土壤相对湿度为60%～70%；结果期应水分充足，土壤相对湿度保持在80%～90%。

1.4 土壤和营养

黄瓜根再生能力弱,吸肥力差,受涝易沤根,因此宜选择疏松、肥沃、透气的砂壤土,土壤pH值以5.5～7.2为宜。黄瓜对肥水要求严格,特别是氮或钾不足时,易引起落花和果实带苦味(含葡萄糖苷)。每生产1 000 kg黄瓜,需氮1.7 kg、磷0.99 kg、钾3.49 kg,而且在结瓜期需肥量占总需肥量的80%以上。

2 品种选择

选择耐低温弱光、抗病性强、主侧蔓均可结瓜、全雌或强雌、单性结实能力强的黄瓜品种。目前,云南省推广的黄瓜品种大多是从省外或国外引进,主要有津杂1号、津杂2号、津杂3号、津春1号、早春2号、津春3号、津研7号、中农5号、中农7号、亮条王、津盾10-12、雅美特2188、春旺F1-A黄瓜王、金皮黄瓜、耐热王、改良型超级(454)无刺黄瓜、莱福13-18无刺小黄瓜、绿优瓜王等。

3 栽培管理技术

3.1 整地作畦

前茬作物收获后,及时翻耕20～25 cm深,临近种植前期每亩施腐熟的农家肥2 000～3 000 kg、过磷酸钙25～30 kg或磷酸二铵10～15 kg,并将土壤与肥料充分混合。定植前翻耕作畦,畦宽80 cm、高20 cm以上,沟宽30 cm,双行种植。开厢起沟完成后,覆盖地膜,将地膜拉紧,使其紧贴厢面,并用土封严。

3.2 播种育苗

3.2.1 常规育苗

3.2.1.1 基质选择

以草炭、蛭石、珍珠岩替代营养土作育苗基质。草炭:蛭石:珍珠岩的配比为3:1:1,1 m³基质加三元复合肥(氮:磷:钾=15:15:15)1.2 kg、50%多菌灵可湿性粉剂25 g,充分拌匀放置2～3 d后待用。

3.2.1.2 种子消毒

黄瓜种子上通常会带有枯萎病、黑星病、疫病、立枯病、褐斑病、角斑病等多种病害的病原菌，在育苗前对种子进行必要的消毒处理能够有效防止由种子带菌引发的病害。黄瓜种子消毒通常采用药剂浸种或药剂拌种的方法。一般选用 0.1% 多菌灵盐酸液浸种 1 h，随后用清水冲洗干净，再用清水浸种 4 h，然后催芽。根据种子带菌情况，预防枯萎病、黑星病选用 40% 甲醛 300 溶液倍液浸种 1.5 h 后捞出洗净；针对病毒病，选用 10% 磷酸三钠溶液浸泡 30 s 后捞出洗净，防病效果较为理想。药剂拌种通常选择克菌丹、敌磺钠、多菌灵等药剂，用药量一般控制在种子重量的 0.3%～0.5%，具体操作过程：将种子放入瓶装器皿内，然后加入药剂，加盖后进行摇动，直至药剂与种子充分混合，药粉均匀黏在种子表面。

3.2.1.3 浸种催芽

种子经过消毒处理后，便可进入浸种催芽环节。具体做法：把经消毒处理的种子捞出洗净后浸种 5～6 h，捞出晾干后用布包好，放在通气、避光处，温度控制在 26 ℃左右，勤翻动，以防温度不均，一般 24 h 后开始出芽。

3.2.1.4 播种

（1）播种时间。根据品种特性、各地气候环境、栽培模式及市场需求等情况决定播种时间。定植前 30 d 为最佳播种育苗时间。

（2）播种方法。每穴 1 粒，将种子平放或芽尖朝下，播种深度为 1～1.5 cm，播后覆盖基质，用木尺、木条等刮平，然后用喷壶浇透水。

3.2.1.5 控温

黄瓜育苗中温度控制至关重要，播种到出苗前，要合理控制苗床温度，通常白天需控制在 25～30 ℃，夜间控制在 22～23 ℃，这样的温度有利于出苗。出土后可适当降低温度，白天温度控制在 20～25 ℃，夜间控制在 14～16 ℃。缓苗后要适当调控温度，防徒长，促花芽分化，特别需注意降低夜间温度，定植前 5～7 d 炼苗，白天 20～23 ℃，夜间 10～12 ℃，经过炼苗增强黄瓜种苗的抗病性，提升种苗定植后对环境的适应能力。

3.2.1.6 壮苗标准

株高 15 cm 左右，3～4 叶 1 心，子叶完好，节间短粗，叶片浓绿、肥厚、有光泽，根系发达、健壮、无病，苗龄 30 d 左右。

3.2.2 嫁接育苗

黄瓜要早于砧木 3～5 d 播种，当砧木两片子叶充分展开，第 1 片真叶展开前，而接穗子叶由黄变绿，充分展开而真叶初露时是嫁接的最适时期。

一般采用靠接法。用刀片将南瓜生长点从子叶处去掉，以免长出侧芽。用刀片在南瓜子叶下 1 cm 处，以 35°～40° 角向下斜切一刀，深度为茎粗的 1/2，然后在黄瓜子叶下 1.5～2 cm 处向上斜切一刀，角度 30° 左右，深度为茎粗的 3/5，把 2 个切口互相嵌入，使黄瓜两片子叶压在南瓜子叶上面，用嫁接夹固定。在嫁接过程中，要避免感染病菌，而且不能损伤瓜苗以免引起瓜苗组织坏死。嫁接时，要使室内保持一定的温湿度和阴暗条件，不要在阳光直射处嫁接，防止接穗蒸腾失水而影响成活率。

嫁接后的管理：温度和空气相对湿度是嫁接苗成活的关键，嫁接好的小苗浇足水，扣上小拱棚，然后 1～4 d 内要求拱棚内相对湿度保持在 95% 左右，白天 25～28 ℃，夜间 15～18 ℃，在高温、高湿条件下促进伤口的愈合，4 d 后逐渐放风、放光，1 周后接口愈合，用刀片断去黄瓜根，去掉嫁接夹并喷施百菌清消毒。一般在接后的 25～35 d 即可定植。

3.3 田间管理

3.3.1 定植

3.3.1.1 定植时间

春茬瓜苗 4～5 片真叶、株高 5～10 cm，在确保定植后不受冻的前提下尽早定植，一般要求夜间最低温高于 5 ℃，0～10 cm 土壤温度高于 12 ℃。秋茬瓜苗 2～3 片真叶、苗龄 20 d 左右定植。

3.3.1.2 定植密度

4 000～4 500 株 / 亩，大小行定植，小行距 40 cm，大行距 80 cm，株距 25～30 cm。

3.3.2 插架

定植后及早插架，插架可采用花架或"人"字形架，架高 1.8～2 m，距离根部 10 cm。

3.3.3 绑蔓

当黄瓜苗长至 30 cm 时采用"8"字方法绑蔓，防止磨伤茎蔓和茎蔓下垂。每 3～4 叶节绑 1 次。操作应在下午进行，上午茎蔓易折断，绑蔓的松紧度应抑强扶弱，对于生长势强的植株适当绑紧，尽量使黄瓜苗的高度保持在同一水平线上。

3.3.4 整枝与掐尖

以主蔓结瓜为主的品种，应及时摘除侧枝。对于主、侧蔓都可结瓜的品种，一般第 1 个瓜（根瓜）下面的侧蔓要尽早除去，上面的侧蔓留 1 个瓜后，瓜前留 2 叶摘心。主蔓爬到架顶时摘心，及时打掉下部老叶、黄叶、病叶、畸形瓜，以节约养分，加强通风透光，减少病虫害发生。

3.3.5 水肥管理

3.3.5.1 浇水

及时浇水与中耕，水量及次数根据天气、生育期而定。从定植到第 1 个瓜坐稳前，管理上应以"控"为主，适当蹲苗，多中耕松土，少浇水，改善生长环境，以达到促根壮秧，使花芽大量分化的目的。进入结瓜期后，要突出"促"字，原则上应掌握"轻、重、轻"的规律，根瓜期浇水量不宜过大，保持畦面见干见湿即可。中部坐瓜时期，温度高、光照足，植株生长旺盛，需水量大。

3.3.5.2 追肥

应符合肥料使用准则，并遵循"前轻后重、少量多次，增施磷钾肥"的原则。一般随水施，1 次清水 1 次带肥水，坐瓜盛期可水带肥。追肥时将有机肥、缓释肥、速效肥配合施用，氮肥施用总量控制在每亩不超过 40 kg 纯氮。每次每亩追施尿素 8～10 kg、硫酸钾 10 kg，配合叶面喷施 0.3% 磷酸二氢钾。

4 病虫害防治

4.1 病害防治

4.1.1 霜霉病

4.1.1.1 为害症状

初期在叶片背面形成水渍状小点，后病斑逐渐扩大，因受叶脉限制，呈多角形、水渍状。潮湿时，病斑上生紫黑色霉层。叶正面病斑初期黄色，边缘不明显，后期变黄褐色。严重时病斑连片，全叶卷缩、干枯，仅留心叶。

4.1.1.2 防治方法

（1）农业防治。选用抗病品种；与非葫芦科作物轮作或水旱轮作。

（2）物理防治。高温闷棚是防治黄瓜霜霉病的有效办法，选晴天上午，浇 1 次大水后封闭棚室，将棚室温度提高到 43～45 ℃，持续 1.5～2 h，然后慢慢加大放风口，缓慢使温度下降。闷棚后加强肥水管理。

（3）化学防治。发病初期可选用80%烯酰吗啉水分散粒剂2 000倍液、68.75%氟菌·霜霉威悬浮剂1 000倍液、75%百菌清可湿性粉剂500倍液、58%甲霜·锰锌可湿性粉剂400倍液等药剂交替轮换喷雾防治，每7 d喷1次，连续防治2～3次。

4.1.2 炭疽病

4.1.2.1 为害症状

幼苗发病，子叶边缘出现褐色半圆形或圆形病斑。茎基部受害，患部缢缩、变色，幼苗猝倒。叶片病斑呈红褐色，外围晕圈黄色。未成熟的果实不易受害，成熟果实受害时先呈现淡绿色水浸状斑点，很快变为黑褐色，并逐渐扩大，凹陷，中部色较深，病部着生黑色小粒点。病果弯曲变形，留种黄瓜发病较多。

4.1.2.2 防治方法

发病初期可选用70%甲基硫菌灵可湿性粉剂1 000～1 500倍液、50%咪鲜胺锰盐可湿性粉剂1 000～2 000倍液、70%甲硫·福美双可湿性粉剂600倍液、35%氟菌·戊唑醇悬浮剂2 000倍液喷雾防治，每7 d喷1次，连续防治2～3次。

4.1.3 白粉病

4.1.3.1 为害症状

主要侵染叶片，亦为害茎部和叶柄，一般不为害果实。发病初期，叶片正面或叶背面产生白色近圆形的小粉斑，以后逐渐扩大成边缘不明显的连片白粉斑。随后许多病斑连在一起布满整个叶面，白粉状物渐变灰白色或红褐色，叶片变成枯黄而发脆，但一般不脱落。

4.1.3.2 防治方法

发病初期可选用250 g/L嘧菌酯悬浮剂600倍液、250 g/L吡唑醚菌酯乳油1 500倍液、15%三唑酮可湿性粉剂1 500倍液、50%甲基硫菌灵可湿性粉剂1 000倍液喷雾防治，每7 d喷1次，连续防治2～3次。

4.1.4 细菌性角斑病

4.1.4.1 为害症状

叶片受害，叶正面病斑淡褐色，背面受叶脉限制呈多角形，初呈水渍状，后期病斑中央组织干枯而脱落。果实及茎上病斑初期呈水渍状，表面可见乳白色细菌菌脓。幼苗也可受害，子叶上初生水渍状圆斑，稍凹陷，后变褐色干枯，如果病部向幼茎蔓延，可引起幼苗软化死亡。

4.1.4.2 防治方法

可用77%氢氧化铜可湿性粉剂500倍液、30%琥胶肥酸铜可湿性粉剂200倍液、47%春雷·王铜可湿性粉剂600倍液、3%中生菌素可湿性粉剂600倍液、14%络氨铜水剂300倍液喷雾防治，每5～7 d喷1次，连续防治2～3次，交替用药。

4.1.5 枯萎病

4.1.5.1 为害症状

该病的典型症状是萎蔫，发病初期植株表现为叶片从下向上逐渐萎蔫，似缺水状，中午更明显，早晚尚能恢复，数日后整株叶片枯萎下垂，不再恢复常态。茎蔓基部稍缢缩，有的病株受害溢出琥珀色胶体物。病株根部褐色腐烂。茎基部常纵裂，在潮湿条件下，病部表面产生白色或粉红色的霉层。幼苗发病，子叶萎蔫或全株枯萎，茎基部变褐缢缩，多呈猝倒状。

4.1.5.2 防治方法

（1）农业防治。与非瓜类作物合理轮作，选择抗性较强的黄瓜品种，培育健壮种苗，采用嫁接苗种植，及时清理感病植株并集中烧毁。

（2）化学防治。发病初期可用80%甲霜·锰锌可湿性粉剂800倍液+80%代森锌可湿性粉剂1 000倍液混合药液灌根和植株喷雾，或用2%春雷霉素可湿性粉剂673～900 g/亩灌根，每7 d喷灌1次，连续防治2～3次。

4.1.6 黑星病

4.1.6.1 为害症状

黄瓜黑星病在整个生育期均可侵染发病，为害部位有叶片、茎、卷须、瓜条及生长点等，以植株幼嫩部分如嫩叶、嫩茎和幼果受害最重，而老叶和老瓜对病菌不敏感。幼苗染病，子叶上产生黄白色圆形斑点，子叶腐烂，严重时幼苗整株腐烂。侵染嫩叶时，起初在叶面呈现近圆形褪绿小斑点，进而扩大为2～5 mm淡黄色病斑，边缘呈星纹状，干枯后呈黄白色，后期形成边缘有黄晕的星状孔洞。嫩茎染病，初为水渍状暗绿色菱形病斑，后变暗色，凹陷龟裂，湿度大时病斑长出灰黑色霉层。生长点染病时，心叶枯萎，形成秃桩。

瓜条感病起初为圆形或椭圆形褪绿小斑，病斑处溢出透明的黄褐色胶状物，凝结成块，以后病斑逐渐扩大、凹陷，胶状物增多，堆积在病斑附近，最后脱落，湿度大时，病部密生黑色霉层。

4.1.6.2 防治方法

（1）农业防治。选用抗病品种；种子消毒处理；水旱轮作或非瓜类作物轮作；黑星病属低温高湿病害，早春大棚及冬季温室经常发生，加强田间管理，栽培时应注意种植密度，升高棚室温度，采取地膜覆盖及滴灌等节水技术，及时放风，以降低棚内湿度，缩短叶片表面结露时间。

（2）化学防治。发病初期，可选用400 g/L氟硅唑乳油4 000倍液、50%多菌灵可湿性粉剂500倍液、75%甲基硫菌灵可湿性粉剂600倍液、20%腈菌·福美双可湿性粉剂500倍液喷雾防治，间隔7 d防治1次，连续防治2～3次。

4.2 虫害防治

4.2.1 蚜虫

4.2.1.1 为害症状

蚜虫以刺吸式口器从植物中吸食大量汁液，使植株长得矮小，叶片卷曲，花蕾不能开放，植株提前老化、早衰。蚜虫刺吸过多的植株汁液排出体外，招引蚂蚁，感染霉菌，诱发煤污病。

4.2.1.2 防治方法

初期设施内可选用70%啶虫脒水分散粒剂7 000～10 000倍液、10%吡虫啉可湿性粉剂1 000～1 500倍液、25%噻虫嗪水分散粒剂3 000～5 000倍液喷雾防治。

4.2.2 茶黄螨

4.2.2.1 为害症状

受害的叶片背面呈现灰褐色或黄褐色，且具有油质光泽或油浸状，叶片僵硬变厚，边缘下卷。受害嫩茎、嫩枝变为黄褐色，扭曲畸形，严重时会导致黄瓜的生长点消失，顶部干枯。

4.2.2.2 防治方法

初期可选用25%噻虫嗪水分散粒剂1 500倍液、10%吡虫啉可湿性粉剂1 000～1 500倍液喷雾防治。

4.2.3 白粉虱

4.2.3.1 为害症状

成虫群集于叶背，吸食叶片汁液，造成叶片褪绿枯萎，叶片褪色、失绿，果实畸形僵化，引起植株早衰。另外，还会分泌大量的蜜液，造成叶片和果实污染，往往会引起大片煤污产生。

4.2.3.2 防治方法

可选用20%呋虫胺可溶粒剂1 500倍液、25%噻虫嗪水分散粒剂7 000倍液、25%噻嗪酮可湿性粉剂2 500倍液、2.5%高效氯氟氰菊酯乳油3 000倍液等药剂喷雾防治。

4.2.4 蓟马

4.2.4.1 为害症状

叶片受害后，失绿后变成黄白色；果实受为害，生长缓慢，而且会形成畸形瓜，受害严重时造成落瓜，严重影响黄瓜的质量和产量。

4.2.4.2 防治方法

可选用20%呋虫胺可溶粒剂1 500倍液、5%吡虫啉乳油1 500～2 000倍液、1.8%阿维菌素乳油4 000～5 000倍液、10%甲氰菊酯乳油1 000～1 500倍液、40%氟啶·吡蚜酮水分散粒剂3 000倍液喷雾防治。

5 适时采收

根据品种特点，达到商品瓜标准时及时采收。结瓜初期每隔3～4 d采收1次，盛瓜期应每天或隔天采收。单瓜重前期100～150 g，中后期150～250 g。采摘时，瓜把处留1 cm长果柄，用剪刀剪断，采收时要轻拿轻放，避免发生机械伤害。采收后，要避免日晒、雨淋，放在阴凉处或冷库中预冷，并及时包装。

采收要及时，过早采收产量低，产品达不到标准，而且风味、品质和色泽也不好。过晚采收不但赘秧，影响产量，而且产品不耐贮藏和运输。采收时保留果梗，采后宜放于阴凉场所或预冷库中预冷散热，避免将产品置于阳光下暴晒，有助于保持产品的品质，防止产品质量下降。

6 采后处理

6.1 分级包装

黄瓜采收后要进行修整，不留果柄并清理果皮上的污物，然后进行分级和包装，按相同品种、相同等级、相同大小规格整齐摆放于箱内或筐内，黄瓜包装箱不宜过高，否则在运输过程中会造成底层黄瓜压伤。包装箱上应标明品名、等级、规格、净重、产地等基础信息。

6.2 运输

运输装卸车要防止机械损伤，产地距离运输采用普通车要注意防晒保湿通风，夏天注意降温，冬天注意防冻，预冷过的黄瓜销售运输时间在10 h内，可用保温车，超过10 h要用冷藏车，冷藏车温度控制在12 ℃。

6.3 贮藏保鲜

黄瓜易腐烂，对贮藏条件要求严格，为了延长供应期，提高经济效益，不能及时出售的应在场地或销售地进行贮藏保鲜。

参考文献

曹荣荣，2023.设施黄瓜栽培技术要点[J].现代农村科技（5）：29-30.

郭海霞，2022.黄瓜的特征特性及栽培技术[J].河南农业（22）：40.

康欣娜，武亚红，周彦伟，等，2023.日光温室黄瓜复合基质栽培技术要点[J].现代农村科技（6）：31-32.

刘艳凤，2022.黄瓜常见病害防治技术[J].现代农村科技（4）：31.

肖运成，2004.无公害黄瓜采后产品标准化处理技术[J].江苏农业科学（4）：97-98.

尹德兴，赵俊杰，李英，等，2023.水果黄瓜压蔓再生越夏轻简化栽培技术[J].长江蔬菜（1）：29-30.

苦瓜栽培关键技术

苦瓜（*Momordica charantia* L.）是葫芦科（Cucurbitaceae）苦瓜属（*Momordica*）的一年生蔓性草本植物。苦瓜又称凉瓜、金荔枝、癞葡萄、锦荔枝、癞瓜、红姑娘等。苦瓜一般是采食嫩瓜，东南亚很多地方的群众还采食嫩梢、叶和花等，也有用其汁作药用的。

苦瓜营养价值高，嫩果中含有丰富的矿物质、氨基酸和多种维生素。苦瓜全身是宝，其根、茎、叶、花、果及种子都有相关药用记载，《本草纲目》中称苦瓜有"去邪热、解劳乏、清心明目"的功效。据报道，苦瓜还具有降血糖、抗菌、消炎、抗病毒、提高人体免疫力的作用。苦瓜不仅富含多种活性成分，如植物胰岛素、苦瓜素、多肽、糖类和氨基酸等，而且含有大量的维生素C、粗蛋白质和可溶性糖。苦瓜肉苦而微甜，鲜嫩而清香，是一种营养价值较高的保健蔬菜，其中有益人体健康的钙、镁、铁等微量元素的含量远远超过了番茄、茄子、辣椒等蔬菜，其维生素C含量在瓜类蔬菜中居首位，分别是黄瓜和冬瓜的14倍和5倍。

随着人们对苦瓜的营养价值和诸多食疗功效的重新认识，近年来我国苦瓜生产发展迅速，栽培面积逐年扩大，现已成为市场畅销的蔬菜。

1 生长条件

1.1 温度

苦瓜生长要求有较高的温度，耐热、不耐寒，适应性较强，10～35 ℃均能适应。种子萌芽的适温为30～35 ℃。苦瓜的种皮虽厚，但在40～50 ℃温水中浸种4～6 h，在适温下催芽48 h开始发芽，60 h便有70%以上萌芽。20 ℃以下发芽缓慢，13 ℃以下发芽困难。在25 ℃发芽快且植株生长健壮，经15～20 d后真叶可以达到4～5片。开花结果的最适温度为25 ℃左右，在20～25 ℃范围内，温度越高越有利于苦瓜植株的生长发育，结果早，产量高。

1.2 光照

苦瓜属于短日作物，喜光不耐阴。春播苦瓜，常遇到低温阴雨，光照不足，导致其幼苗徒长，叶色发黄，植株细弱。开花结果期需要较强的光照，充足的光照有利于光合作用，积累更多有机养分，提高坐果率，增加产量，提高品质。

1.3 水分

苦瓜喜湿而怕涝，在生长期间要求空气相对湿度和土壤相对湿度均达到70%～80%。但如遇较长时间的阴雨连绵天气或排水不良时，植株生长不良，极易发生沤根死苗和暴发病害。

1.4 土壤

苦瓜对土壤的要求不太严格，适应性较广，云南省各地均可栽培，但根系对积水缺氧敏感，所以选择土壤排水良好、通气性好的肥沃壤土或砂壤土栽培有利于植株生长，产量高。

1.5 营养

苦瓜耐肥而不耐瘠薄，对肥料三要素的吸收以钾最多，氮次之，磷最少，生产上注意栽种前施足基肥，苗期适当追肥，促进茎叶生长。开花结果期持续供肥。如果有机肥充足，植株生长粗壮，枝叶繁茂，开花结果多，品质好。特别是生长后期，若肥水不足，则植株衰弱、叶色黄绿、花果少、果实细小、苦味增浓，品质下降，因此及时追肥，特别在结果盛期要求有充足的氮、磷肥。

2 品种选择

根据种植目的、市场需求，以及当地的小气候环境，选择生长势强、分枝能力强、抗热、抗病、高产的苦瓜品种。在云南比较有代表性的地方品种有玉溪大白苦瓜、绿皮苦瓜、小苦瓜；引进品种有夏丰苦瓜、川绿1号苦瓜、农友6号苦瓜、绿宝石苦瓜等，云南热区还有种植野苦瓜的习惯。

3 栽培管理技术

3.1 整地

3.1.1 选地和耕地

苦瓜种植宜选择地势较高、排灌方便、有机质含量高的肥沃地块，前作忌瓜类，翻耕晒垡后整地作畦。合理密植是种植苦瓜获得高产的有效途径，一般采用双行种植，畦宽 1.5～1.8 m（包沟），株距 0.3 m，行距 1～1.3 m。前茬作物收获后，将田间残株杂草清除干净，深翻土地晒垡。

3.1.2 底肥

定植前，每亩施过磷酸钙 50 kg，拌入充分腐熟的有机肥 500～1 000 kg，再加复合肥 40 kg、硼砂 1.5 kg 施于穴中翻匀，然后把苗定植于穴中，苗的根系不能直接接触肥料，浇足定根水。

3.2 育苗

3.2.1 播种期

苦瓜正常生长发育的气温要求在 15～30 ℃。云南立体小气候环境比较独特，各区域栽培时间有明显差异，因此具体播种时间，要根据各地气候环境而定。一般春植苦瓜播种期 3—4 月，秋植苦瓜播种期 7—8 月。

3.2.2 种子处理

苦瓜种子种皮坚硬，播种前可采用人工破壳，即用钳子等工具将苦瓜种子从脐部缝合线处破开一条相当于种子长度 1/3 的小缝。

此外，播种前还需进行种子消毒。种子先用清水浸 3～5 h 后，可用 10% 磷酸三钠溶液浸 20 min 或用 50% 多菌灵可湿性粉剂 500 倍液浸 60 min，然后用清水冲洗干净后，即可放入温度 25～30 ℃的恒温箱中催芽或直接播种。也可以将种子放入 55 ℃水中，要使种子受热均匀，直至水温降至 30 ℃左右时，停止搅拌并浸泡 4～5 h，放到干净湿润的纱布上包好，置于 25～30 ℃的恒温下催芽，3～4 d 后待 75% 种子"露白"时便可播种。

3.2.3 育苗准备

为防止定植过程中对植物根茎造成伤害,降低病害发生率,一般采用营养袋或穴盘进行育苗,营养土可以选择火烧土或没有种植过瓜类蔬菜的菜园土,掺入约30%充分腐熟的有机肥以及适量复合肥后拌匀,并用50%多菌灵可湿性粉剂500倍液、70%甲基硫菌灵可湿性粉剂1 000倍液喷洒营养土。

3.2.4 播种

一般选择晴天播种,播种前浇淋营养土,将底水浇透,之后于营养钵或育苗床中点播种子,覆盖1.5～2 cm厚的细土。播种后覆盖地膜,并扣小拱棚。出苗前确保苗床温度保持在30～35 ℃,为出苗提供有利条件。

3.2.5 苗期管理

出苗后及时揭去地膜,同时适当降低苗床温度,保持白天25～30 ℃,夜间15～20 ℃,在2片真叶期后可用0.3%磷酸二氢钾溶液叶面喷施2～3次,可使幼苗生长健壮,提高幼苗的耐低温能力。定植前7～10 d进行炼苗,白天温度控制在25 ℃左右、夜间温度控制在12～15 ℃,加强通风,逐渐将覆盖的薄膜揭开,使幼苗逐渐适应外界环境条件。苦瓜播种至出苗前一般不浇水,出苗后保持基质见干见湿,补水在晴天午后进行。

3.2.6 嫁接

为了防止发生青枯病和枯萎病,可以采用嫁接育苗,用丝瓜或黑籽南瓜作砧木,当砧木幼苗长到2叶1心,接穗长到1叶1心时,去掉砧木的心叶,从顶端斜削掉1叶,再把接穗的胚轴削为斜面,使砧木的接穗削面接在一起,用嫁接夹固定好。嫁接后喷药消毒,放到小拱棚中,盖上遮阳网,避免阳光直射,确保空气相对湿度在95%以上,温度保持在25～28 ℃。

3.3 田间管理

3.3.1 中耕培土

苦瓜定植后应及时进行中耕、除草和培土,以防土壤板结。一般在定植后10 d左右,田间开始长出杂草、土壤表层出现板结时要进行第一次中耕除草、培土。第二次中耕在第一次中耕后10～15 d进行,中耕时要注意保护新根,宜浅不宜深。

3.3.2 搭架引蔓

幼苗3~4片叶时搭架或吊线引蔓，当植株长至10~12片真叶时，应及时压蔓，以促发不定根，扩大根系吸收范围，促进茎蔓生长。当瓜蔓长到30 cm左右时进行绑蔓，每间隔4~5节绑1道，将瓜蔓引上架。一般早上9:00露水干后绑蔓，绑蔓时要防止断蔓。

3.3.3 植株调整

苦瓜主蔓有较强的分枝能力，若不进行干预，将长出过多的侧枝，对主蔓的开花结果和正常生长造成影响，因此应及时疏除多余侧枝。通常留1~2枝从地表根茎部发出的侧枝，去掉地表节以上到第15节的营养侧枝，带有雌花的侧枝从雌花以上留1~2个叶片后将侧蔓枝去掉，整个生长期要注意去除过密的衰老茎蔓及细弱的侧枝，确保其通风和透光良好。

主蔓50 cm以下不留瓜，应把雌花摘掉以利于植株健壮生长。待主蔓坐稳6~7个瓜后，留5~6片叶打顶。

3.3.4 肥水管理

苦瓜耐肥不耐瘠，喜湿不耐涝。苦瓜耐湿，但需水量较大，生长期间需要保持70%~80%的空气相对湿度和土壤相对湿度。基肥以有机肥为主，每亩施充分腐熟有机肥1 500~2 000 kg。追肥以氮肥为主，春植苦瓜全期追肥6~8次，在高温高湿地区夏秋季苦瓜因生育期短，可减少追肥次数。整个生育期追肥：定植10 d后，每亩用尿素8 kg、氯化钾5 kg，间隔10 d再施1次。当出现雌花时，每亩施45%三元复合肥25 kg。盛果期要及时追肥，每亩可用硫酸钾5 kg、过磷酸钙15 kg、尿素15 kg，一般间隔10 d追施1次，促进蔓叶生长，延长采收期。

3.3.5 人工授粉

苦瓜具有较高的单性结实率，但人工授粉可以提高坐果率，同时为果实的发育提供保障。在授粉过程中要注意在雌雄花开放当天展开。当气温稳定在15~30 ℃，每天6:00—10:30是开花盛期，中午之后基本不开花。通常，1朵雄花可人工授粉4朵雌花，当苦瓜开花后，采摘雄花，在雌花柱头蘸上雄花的花粉，雌花受精后，子房逐渐变大，约20 d便可采果。

4 病虫害防治

苦瓜的主要病害有猝倒病、炭疽病、褐斑病、白粉病和病毒病等，虫害主要有瓜蚜、蓟马、白粉虱和瓜实蝇等。

4.1 病害防治

4.1.1 猝倒病

4.1.1.1 为害症状

猝倒病在幼苗期易发生。首先幼苗茎基部呈水浸状，而后病部变淡褐色，幼苗近地面处明显缢缩，子叶尚未凋萎而倒伏。土壤温度低、湿度大时有利于病菌的生长与繁殖，所以一般夜晚凉爽、白天光照不足、苗床湿度大时发病严重。

4.1.1.2 防治措施

可选用72%霜霉威盐酸盐水剂750倍液、75%代森锰锌可湿性粉剂800倍液、58%甲霜·锰锌可湿性粉剂700倍液、32%甲霜·噁霉灵可湿性粉剂300倍液进行灌根处理。

4.1.2 炭疽病

4.1.2.1 为害症状

瓜条、叶片及茎蔓均可发病。瓜条上病斑圆形或不规则形，初期淡黄褐色，后期变红褐色至淡褐色，稍凹陷。叶片病斑圆形或不规则形，灰褐色至棕褐色，略湿腐。茎蔓上病斑长圆形、褐色、凹陷，严重时出现龟裂。

4.1.2.2 防治措施

（1）农业防治。选用抗病品种并进行种子消毒；加强与非瓜类蔬菜进行轮作；避免偏施、过施氮肥，增施磷、钾肥和中微量元素肥，同时适当控水，雨后要排出积水；及时摘除病叶、病枝和病瓜，并要保持良好的田间通风透光性。

（2）化学防治。发病初期可选用25%咪鲜胺乳油1 500倍液、10%苯醚甲环唑水分散粒剂800倍液、75%肟菌·戊唑醇水分散粒剂4 000～6 000倍液喷雾防治，连续用药2～3次，间隔期7～10 d。

4.1.3 褐斑病

4.1.3.1 为害症状

主要为害叶片。初期叶片上产生褐色圆形小斑点，后逐渐扩展为近圆形或不规则形，黄褐色，周围常有褪绿晕圈，环境条件适宜时，病斑迅速扩大连接成片，最终整叶干枯。

4.1.3.2 防治措施

（1）农业防治。选择抗病品种；选择地势较高、排水良好的地块种植；加强田间管理，改善田间通风透光条件；基肥以有机肥为主；雨后及时排出田间积水，注意控制田间湿度；重病地要实行不同作物轮作或水旱轮作。

（2）化学防治。发病初期可选用25%咪鲜胺乳油1 500倍液、10%苯醚甲环唑水分散粒剂800倍液、75%肟菌·戊唑醇水分散粒剂4 000～6 000倍液、32.5%苯甲·嘧菌酯悬浮剂1 500倍液喷雾防治，连续用药2～3次，间隔期7～10 d。

4.1.4 白粉病

4.1.4.1 为害症状

叶片、叶柄和茎均可发病。叶片发病时，初期叶正背面产生白色小粉点，扩展后为圆形、近圆形稀疏白色粉斑，随病情发展粉斑连片，叶面布满白粉。病重时，叶片逐渐变黄，最后干枯，植株生长及结瓜受阻，生育期缩短。叶柄、茎蔓发病，病部长满稀疏白粉。

4.1.4.2 防治措施

（1）农业防治。选用抗病品种；合理密植，及时搭架、整蔓和摘叶，增加植株通风透光；采用配方施肥，增施中微量元素肥；保持土壤见干见湿。

（2）化学防治。发病前，可选用325 g/L苯甲·嘧菌酯悬浮剂2 000倍液预防；发病初期可选用10%苯醚甲环唑水分散粒剂1 500倍液、80%硫磺悬浮剂600倍液、25%三唑酮可湿性粉剂1 000倍液、250 g/L嘧菌酯悬浮剂600倍液等药剂喷雾防治，连续用药2～3次，间隔期7～10 d。

4.1.5 病毒病

4.1.5.1 为害症状

病株叶片呈现黄绿相间的花叶，植株矮小，顶部幼嫩茎叶症状明显。新叶不舒展，叶面皱缩，产生黄绿斑驳，后期黄斑变坏死斑。早期发病，瓜苗生长不良，节间短缩，从下部叶片往上黄枯。

4.1.5.2 防治措施

（1）农业防治。选用抗病品种并进行种子消毒；实行轮作；培育健壮苗；及时防治蚜虫、蓟马和白粉虱，切断传染源；发现病株及时拔除、烧毁。

（2）化学防治。发病初期，可用2%宁南霉素水剂300倍液、20%吗胍·乙酸铜可湿性粉剂1 500～2 000倍液、1%香菇多糖水剂600倍液等药剂喷雾防治，连续用药2～3次，间隔期5～7 d。

4.2 虫害防治

4.2.1 瓜蚜

4.2.1.1 为害症状

瓜蚜群集在叶背、嫩茎和嫩梢刺吸汁液，引起嫩叶卷缩，生长点枯死，瓜苗萎蔫，严重时瓜苗期造成整株枯死。瓜蚜排泄的蜜露污染叶面，引起煤烟病，影响光合作用。瓜蚜还会传播病毒，引起花叶、畸形、矮化等症状，受害株早衰，造成更大的损失。

4.2.1.2 防治方法

（1）物理防治。保护地提倡采用24～30目、0.18 mm的银灰色防虫网或采用黄板诱杀。

（2）生物防治。瓜蚜发生初期每亩释放1 500头七星瓢虫，控制虫口密度。

（3）化学防治。可选用3%啶虫脒乳油1 500倍液、10%吡虫啉可湿性粉剂2 000倍液、25%噻虫嗪水分散粒剂4 000倍液喷雾防治。

4.2.2 蓟马

4.2.2.1 为害症状

苦瓜被害植株生长点萎缩、变黑而出现丛生现象，心叶不能展开，影响正常坐瓜。受害幼瓜的茸毛变黑，表皮锈褐色，生长缓慢，甚至畸形。受害严重时造成落果，极大影响产量及品质。

4.2.2.2 防治方法

（1）农业防治。及时清除田间杂草集中烧毁或深埋，消灭越冬成虫和若虫，减少虫源，加强水肥管理，促使植株生长健壮。

（2）物理防治。蓟马对蓝色有趋向性，将蓝板悬挂在苦瓜行间监测和诱杀，及时清理板上的害虫。

（3）化学防治。可选用40%吡虫啉可湿性粉剂2 000～3 000倍液、20%呋虫胺可溶粒剂1 500倍液、40%氟啶·吡蚜酮水分散粒剂3 000倍液、18%杀虫双水剂250～400倍液喷雾防治。

4.2.3 白粉虱

4.2.3.1 为害症状

成虫和若虫均栖息在嫩茎、嫩梢或叶背吸取液汁，受害叶片褪色、变黄、卷缩，植株生长受到抑制，并传播病毒。

4.2.3.2 防治方法

（1）物理防治。每亩悬挂30～40块黄板；银灰膜趋避蚜虫；利用频振式杀虫灯、黑光灯、高压汞灯等诱杀害虫。

（2）生物防治。释放丽蚜小蜂防治蚜虫，10 d释放1次，共3～4次。

（3）化学防治。可选用20%呋虫胺可溶粒剂1 500倍液、3%啶虫脒乳油1 500倍液、10%吡虫啉可湿性粉剂2 000倍液、25%噻虫嗪水分散粒剂4 000倍液喷雾防治。

4.2.4 瓜实蝇

4.2.4.1 为害症状

受害瓜先局部变黄，后全瓜腐烂变臭，大量落瓜；刺伤处凝结流胶，畸形下陷，果皮硬实，瓜味苦涩，品质下降。

4.2.4.2 防治方法

（1）农业防治。及时摘除和收集菜园中的受害瓜和落地瓜，集中深埋、水浸或焚烧；避免与瓜类蔬菜连作。

（2）物理防治。利用频振式杀虫灯、黑光灯、高压汞灯等诱杀。

（3）套袋护瓜。发生严重的地区，在瓜果完全谢花、花瓣萎缩时进行套袋，应注意果实套袋前喷1次农药，防治其他的病虫害，确保瓜果套袋后的质量。

（4）化学防治。可选用1.8%阿维菌素乳油2 000倍液、4.5%氯氰菊酯乳油1 000倍液、5.2%阿维·高氯乳油1 000倍液喷雾防治。

5 适时采收

苦瓜生长快，一般在开花后10～12 d果实已发育充分，瓜肉细嫩，微苦带甜，品味俱佳。采收过晚，瓜肉粗纤维增加，且不利于后茬瓜生长。采收时要求果实的条状或瘤状突起比较饱满，果皮有光泽，果顶颜色开始变淡。

6 采后处理

6.1 贮藏

苦瓜采收后，尽快置于阴凉处，有条件的地方可在冷库中预冷，使苦瓜温度接近贮运的适宜温度。苦瓜适宜的贮藏温度为 10～13 ℃，低于 10 ℃会发生冷害。贮藏环境空气相对湿度为 85%～90%。

6.2 运输与包装

根据采收地、集散地或销售地的距离及气候条件等因素确定苦瓜运输和包装的具体要求。如果是随产随销或者中短途运输，一般可以采取常温运输；如果外界天气较炎热或出现连续阴雨天气，要采取遮阳或遮雨设施；严冬季节要在瓜上铺盖棉被或者覆盖稻草等以防止出现苦瓜冻坏现象。另外，如果进行长途运输则应采用低温的方式，可用加内衬的纸箱或竹筐等作为包装。

参考文献

胡庆华，2012. 保健蔬菜：南瓜栽培技术 [M]. 北京：科学技术文献出版社.
梁金兰，1996. 蔬菜病虫实用原色图谱 [M]. 郑州：河南科学技术出版社.
龙荣华，张思竹，2014. 云南保健蔬菜栽培 [M]. 昆明：云南科技出版社.
陆国一，2003. 瓜类蔬菜周年生产技术 [M]. 北京：金盾出版社.
宋曙辉，2002. 14 种名特瓜类蔬菜栽培技术 [M]. 北京：中国农业出版社.
王广印，张建伟，2000. 苦瓜种子发芽特性研究 [J]. 中国农学通报，16（1）：45-46.
王久兴，2009. 瓜类蔬菜病虫害诊断与防治原色图谱 [M]. 北京：金盾出版社.
谢慧媛，黄淑霞，邓胡宁，等，1998. 中药苦瓜化学成分研究 [J]，中药材，21（9）：458-461.

❖ 洋葱栽培关键技术

洋葱（*Allium cepa* L.）是单子叶百合科（Liliflorae）葱属（*Allium*）中以肉质鳞茎为产品的两年生草本植物，别名：葱头、圆葱、球葱等。洋葱原产于中亚和地中海沿岸地区，栽培历史悠久，在我国南北各地已有近百年的栽培历史。洋葱的营养极为丰富，不仅含有较多的蛋白质、维生素，还含有硫、磷、铁等多种矿物质。洋葱耐贮存、耐运输，除内销外，还是出口日本、东南亚和俄罗斯等国家和地区的主要蔬菜之一。洋葱适应性强，虫害较少，可以与粮、棉等作物间作，如与大麦、豌豆间作，对大麦黑穗病和豌豆黑斑病有一定抑制效果，是茄科、葫芦科和十字花科蔬菜轮作、倒茬的理想作物。洋葱除鲜食外，也可加工成洋葱粉、洋葱酱、洋葱油、脱水洋葱和洋葱汁等。洋葱含有植物杀菌素，还含有前列腺素样物质及激活血溶纤维蛋白活性的成分，具有舒张血管、降血压、降血脂的作用。

云南洋葱常年种植 10 万亩，主要种植区域为红河州、楚雄州、昆明市、玉溪市、保山市、大理州等地，其中种植面积最大的区域为红河州和楚雄州，种植面积占全省的 90%。立足于云南多样的气候类型，洋葱从 1 月初建水县上市到 6 月姚安县上市，是中国上市最早、供应时间最长的洋葱产区。红河州主要种植红葱品种，在广东、广西市场存在区位优势。元谋县气候干燥，昼夜温差大，光照充足，冬季少雨，有利于黄皮、白皮洋葱生长，其洋葱水分足口感甜、质量好，出口优势大。总的来说，云南洋葱因为其口感好、新鲜脆爽，深受消费者喜爱，通过剥皮、加工、真空装箱出口日、韩等国。

1 生长条件

1.1 温度

洋葱为半耐寒性蔬菜。种子发芽的最低温度为 4 ℃，最高温度为 33 ℃，适宜温度为 12~25 ℃。幼苗期生长适温为 12~20 ℃，但幼苗抗寒性强，能忍耐 -7~-6 ℃的低温。植株旺长期以 20 ℃左右为宜，如超过 25 ℃则生长不良。根系在低于 5 ℃时基本停止生长，其适温范围比地上部稍低，土壤温度在 26 ℃以上有促使根系老化的作用。

鳞茎肥大生长期所要求的温度差异较大，短日照型早熟品种，鳞茎肥大生长期适宜温度为 15～20 ℃；长日照型中晚熟品种，鳞茎肥大生长期则需 20～26 ℃。

1.2 光照

洋葱种子在发芽过程中不需要光照。鳞茎肥大生长期对日照长度的要求：长日照型品种需 13.5～15 h，短日照型品种则需 11.5～13 h。另外，也有一些品种在形成鳞茎时对日照要求并不十分严格。一般北方品种大多属于长日照型晚熟品种，南方品种大多为短日照型早熟品种，故引种时必须注意。例如，天津的大水桃和荸荠就是长日照型品种，引种到重庆、上海等地常因日照长度不足而减产。洋葱在生育期间适宜中等光照强度，适宜光照度 20 000～40 000 lx。

1.3 水分

洋葱在发芽期、植株旺长期和鳞茎肥大生长期需要充足的水分。秋季定植后应适当控水，但在越冬前要浇足水以利越冬保苗。采收前 1～2 周应停止供水，促使进入生理休眠，提高耐贮性。

1.4 土壤和营养

洋葱对环境条件的适应性很强，对土壤质地要求不严。土壤质地比较黏重对根系生长不利，但鳞茎质地紧密；疏松的砂质土壤有利于根系延伸生长，但保水保肥力弱。早熟栽培宜选用砂质土；洋葱最适 pH 值为 6～8，但幼苗期对盐碱反应敏感。

洋葱育苗应增施磷、钾肥而适当控制氮肥。洋葱形成 1 000 kg 商品需要氮 2.37 kg、磷 0.7 kg、钾 4.1 kg。洋葱栽培需要土壤提供充足的养分，以适应其短期速生的需要，当土壤肥力不足时，应适当追肥。

2 主栽品种

栽培洋葱品种按鳞茎形态和生长特性不同可分为普通洋葱、分蘖洋葱和顶球洋葱。

2.1 普通洋葱

一般每株形成 1 个肥大的鳞茎，个体大，品质好，产量高，耐寒性较强，以种子繁殖，栽培广泛。鳞茎颜色有紫红色、铜黄色、淡黄色及白色；鳞茎形状有扁圆形、圆球形、高圆球形、纺锤形。按其鳞茎膨大对日照条件的要求分为长日照型、短日照型、中日照型 3 个生态型；按其生长成熟期长短不同分为早熟、中熟、晚熟品种。

2.2 分蘖洋葱

分蘖洋葱每株分蘖成多个大小不规则的鳞茎。鳞茎铜黄色，品质差，产量低，耐贮藏。植株抗寒性极强，很少开花结果，多用分蘖小鳞茎繁殖。

2.3 顶球洋葱

顶球洋葱也叫顶生洋葱。此类型洋葱鳞茎能正常抽薹，但通常不开花结实，在花茎上形成 7～8 个至 10 多个气生鳞茎，用气生鳞茎繁殖。耐贮性和耐寒性强，适于严寒地区种植。可供加工腌制。

3 栽培管理技术

3.1 播种育苗

洋葱在云南西部和南部等亚热带地区、干热河谷地区宜秋播，一般于 9—10 月播种，其他区域春季播种。适时播种是洋葱生产中非常重要的环节，播种过早，冬前幼苗过大，翌年春天容易出现早期抽薹，不能正常形成肥大鳞茎，影响洋葱产量；播种过迟，营养生长期短，会导致鳞茎细小产量锐减。育苗分为穴盘育苗和苗床育苗。苗床育苗应选择土壤疏松、肥沃且保水性强，在 2～3 年内没有种植过葱蒜类蔬菜的地块，切忌在低洼易涝处进行育苗。育苗床面积为栽培面积的 1/15 左右，每 100 m^2 苗床施用充分腐熟、细碎的农家肥 300 kg，播种量为 0.6～0.7 kg。

3.2 种子处理

将种子置于 50 ℃温水中浸泡 3～5 h，浸种后在 20～25 ℃条件下催芽，在催芽过程中每天要用清水淘洗 1 次，当种子"露白"时及时播种。

3.3 苗期管理

长出第 1 片真叶后适当控水，2 片真叶时结合浇水追施氮肥，一般 100 m^2 施硫酸铵 3.4～5.1 kg 或尿素 1.7～2.5 kg，叶面喷洒 0.2%～0.4% 磷酸二氢钾 1～2 次。追肥前结合除草进行间苗。

3.4 整地作畦

整地时耕翻深度不低于 20 cm，每亩施入充分腐熟的堆肥、厩肥或其他农家肥 1 500～2 000 kg，过磷酸钙 40 kg。洋葱一般采用高畦栽培，畦宽 1.3～1.5 m，畦间沟深 0.3 m，长度根据地块确定。

3.5 定植

3.5.1 定植期

定植期因气候和品种而异。应在月平均气温 4～5 ℃时、苗高 20～28 cm 时适时定植。春季定植应尽量提早有利于增产。定植时幼苗的大小，与越冬后抽薹率和产量关系很大。苗过大，越冬时容易通过春化阶段，到翌年春季时抽薹早，洋葱小；苗过小，则生长衰弱，虽不易抽薹，但产量低。定植规格株行距 10 cm×20 cm 或 15 cm×20 cm，适度密植，不会影响单一洋葱的大小，但过度密植会降低洋葱的重量。

3.5.2 定植方法

覆盖地膜即按预定株行距用竹签等物穿膜扎孔，按孔插苗后在苗四周用土封严地膜。不覆盖地膜按行距开沟，按株距摆苗。定植时，苗过大者，宜剪去 1/3～1/2 叶片和 1/3 左右的根系。直立栽植，不要倾斜。深度要适宜，栽植过深，生长不良，而且将来鳞茎变长，不呈扁圆球形；栽植过浅，则易受干旱，发根差，而且洋葱肥大后，露出土面过多，较易引起开裂。栽培深度以 3～5 cm 为宜。定植后即浇定根水，使根系与土壤良好地接触，以快速恢复生长。

3.5.3 浇水

定植后缓苗期 7～10 d，需要小水勤浇，直到植株已充分生长将转向鳞茎肥大生长时，要控水蹲苗 10 d 左右，即当外叶深绿、叶面蜡质增多、心叶颜色相应加深时结束蹲苗，要大量浇水。一般从定植到收获共浇水 12～15 次，当田间个别植株开始倒伏、洋葱成熟时除停止浇水外，还应注意排水，促进洋葱成熟充实，以提高其耐贮性。

3.6 施肥

洋葱根系浅，生长期长，除施足基肥外，还应多次追肥。春季定植的在缓苗后，晚秋定植的在返青以后进行第一次追肥。结合灌水每亩追施磷酸二铵 10～15 kg 和硫酸钾 8～10 kg。此后再追 1 次提苗肥，每亩追施硫酸铵 10～15 kg，以保证地上部功能叶生长的需要。在鳞茎开始膨大生长后进行 2~3 次追肥（催头肥），催头肥应以鳞茎肥大生长中期为重点，每亩追施硫酸铵 10~15 kg 和硫酸钾 5~10 kg。

3.7 中耕、培土

如不覆盖地膜应进行中耕，尤其在蹲苗前必须中耕，中耕深度一般在 3~4 cm。结合中耕进行培土则能提高产量。

3.8 防止早期抽薹

栽培条件不良、管理措施失当会导致洋葱早期抽薹，早期抽薹直接影响洋葱的产量、品质和耐贮性。防止洋葱抽薹可选择不易早期抽薹、冬性强的品种，适时播种，培育壮苗，在苗期喷洒 250 mg/L 乙烯利溶液可减少抽薹率。出现早期抽薹的植株，应及时摘除或折断花薹以减少营养损耗。

4 病虫害防治

4.1 病害防治

4.1.1 紫斑病

4.1.1.1 为害症状

紫斑病在整个生长期均可发生。主要为害叶片和花薹，初呈水渍状白色小斑点，继而扩大成圆形或纺锤形凹陷斑。病斑由小到大，呈暗褐色至暗紫色，并产生同心轮纹状黑色霉层。为害严重时，多个病斑连成一片，致全叶、整个花茎和花柄枯萎。夏、秋季气温 20~30 ℃适合该病流行。

4.1.1.2 防治方法

（1）农业防治。实行轮作；清理病残株；增施有机质肥料，特别是多施磷、钾肥，提高植株抗病能力。

（2）化学防治。用 40% 甲醛溶液 300 倍液对种子进行消毒，浸种 3 h 捞出后用清水冲洗备用。发病初期可用 75% 百菌清可湿性粉剂 500~600 倍液、50% 异菌脲可湿性粉剂 1 500 倍液、64% 噁霜·锰锌可湿性粉剂 500 倍液、58% 甲霜·锰锌可湿性粉剂 500 倍液等喷雾，每隔 5~7 d 喷 1 次，连续防治 2~3 次。

4.1.2 霜霉病

4.1.2.1 为害症状

侵染葱叶、花茎和花柄，产生椭圆形淡黄色病斑，边缘不明显，湿度大时表面产生白色霉层，后期变为淡黄色或暗紫色。中下部叶片染病，病部以上渐干枯下垂，严重时枯黄凋萎。病原卵孢子在种子、土壤及病株残体上越冬。

4.1.2.2 防治方法

发病初期喷施 75% 百菌清可湿性粉剂 600 倍液、64% 噁霜·锰锌可湿性粉剂 600~800 倍液、72% 霜脲·锰锌可湿性粉剂 500 倍液等，每隔 7~10 d 喷 1 次，连续防治 2~3 次。

4.1.3 锈病

4.1.3.1 为害症状

锈病在春秋多湿条件下发生，主要侵染叶、花薹和绿色茎部。初期叶呈白色小斑点，病斑凸起，后表皮破裂散出褐色真菌孢子粉末，空气湿度大时发生严重。

4.1.3.2 防治措施

发病初期可喷施 15% 三唑酮可湿性粉剂 2 000~2 500 倍液、25% 已唑醇悬浮剂 4 000 倍液、50% 萎锈灵乳油 700~800 倍液等药剂、70% 代森锰锌可湿性粉剂 500 倍液喷雾防治，每隔 7~10 d 喷 1 次，连续 2~3 次。

4.1.4 病毒病

4.1.4.1 为害症状

感病植株初期无明显症状，其后绿叶初呈淡绿色短条斑，逐渐发展成多道黄绿相间的长条斑。严重时叶身皱褶由圆变扁，新叶叶鞘伸长受阻，叶身短皱，各叶丛生，产量、品质和贮藏性都显著降低。目前尚无有效防治该病毒病的药物。主要采取大区轮作，精选葱秧，及时拔除中心病株，杀灭传毒蚜虫、蓟马等措施预防病毒侵染。

4.1.4.2 防治措施

防治此病重点是防治蚜虫、蓟马等，用吡虫啉、啶虫脒和噻虫嗪等药剂防治。可用 5% 菌毒清水剂 500 倍液、1.5% 十二烷基硫酸钠悬乳剂 1 000 倍液、2% 香菇多糖水剂 1 000 倍液、80% 吗胍·乙酸铜可湿性粉剂 1 000 倍液、5% 氨基寡糖素可溶液剂 600 倍液等药剂喷雾，每隔 5~7 d 喷 1 次，连喷 2~3 次。

4.2 虫害防治

4.2.1 蓟马

4.2.1.1 为害症状

蓟马以成虫、若虫锉吸葱叶和花茎汁液，被刺伤口处产生灰白色条纹或斑点，严重时大量伤斑可连成一片，使叶片干枯。

4.2.1.2 防治措施

（1）农业防治。早春清洁田园、勤浇水和除草可减轻为害。

（2）化学防治。蓟马发生初期用 60 g/L 多杀菌素悬浮剂 1 000 倍液、10% 吡虫啉可湿性粉剂 1 000 倍液、10% 高效氯氰菊酯乳油 1 000 倍液、25% 噻虫嗪水分散粒剂 1 000～1 500 倍液喷雾防治。

4.2.2 斑潜蝇

4.2.2.1 为害症状

斑潜蝇成虫产卵于叶表皮下，孵化出幼虫潜食叶肉，使叶片布满迂回曲折的隧道和窄带状枯斑。虫口密度大时枯斑可连片致叶片枯萎，影响生长，并使叶片丧失食用价值。

4.2.2.2 防治措施

可用 1% 阿维·高氯乳油 1 500 倍液、10% 虫螨腈水剂 1 500 倍液、20% 阿维·杀虫单微乳剂 1 500 倍液喷雾防治。

4.2.3 葱蝇

4.2.3.1 为害症状

葱蝇幼虫潜入土中，为害幼苗假茎基部，导致洋葱植株生长点受伤，可造成鳞茎腐烂，叶片枯黄萎蔫，甚至成片死亡。

4.2.3.2 防治措施

（1）农业防治。有机肥要充分腐熟均匀撒入土中，立即深翻，不要裸露于土面，以免招引成虫产卵。

（2）化学防治。药液灌根，杀灭幼虫，可选用 90% 敌百虫原药 800～1 000 倍液、50% 辛硫磷乳油 800 倍液等浇灌植株根部。防治成虫，90% 敌百虫原药 800 倍液、2.5% 溴氰菊酯乳油 1 000 倍液喷雾防治。

4.2.4 甜菜夜蛾

4.2.4.1 为害症状

甜菜夜蛾是一种多食性害虫，以幼虫蚕食或剥食叶片造成为害，低龄时常群集在心叶中结网为害，然后分散为害叶片。

4.2.4.2 防治方法

可用10%氯氰菊酯乳油2 000～3 000倍液、2%甲氨基阿维菌素苯甲酸盐乳油1 000倍液、20%虫酰肼悬浮剂800倍液或10%虫螨腈微乳剂1 500倍液喷雾防治。

5 采后处理

5.1 采收标准

洋葱叶鞘颈变细、皱缩、软萎是葱球成熟的标志，表明洋葱已充分长大，叶子枯萎时即可采收。休眠期短、耐贮性弱的品种在30%～50%发生倒伏时应及时收获。中、晚熟，耐贮性强的品种可在倒伏率达到70%时采收。洋葱采收后需要放置在通风、干燥的地方晾晒，以便去除表面的水分，减少病虫害的发生。晾晒时间一般为3～5 d，具体时间根据天气情况确定。

5.2 分级

洋葱采收后需要按照其大小、品质进行分级。

5.3 贮藏

洋葱鳞茎在贮藏前必须使管状叶和鳞茎外皮呈干燥状态，因此应在田间晒蔫，晾晒时使后排叶片盖上前排鳞茎，不让洋葱直接暴晒，之后进行编辫或捆扎后继续洋葱朝下晾晒，然后可在室内室外挂贮或腾空堆贮，初期还要经常翻倒。当夜温低于-5 ℃时，必须入室贮藏。应选择阴凉干爽的地方进行贮藏，传统贮藏方法有垛藏、囤藏、堆藏、挂藏等。

参考文献

程玉琴，徐践，2003. 葱洋葱无公害高效栽培［M］. 北京：金盾出版社：71-101.

王久兴，孙成印，2005. 蔬菜病虫害诊治原色图谱：葱蒜类分册［M］. 北京：科技文献出版社：15-35.

中国农业科学院蔬菜花卉研究所，2010. 中国蔬菜栽培学［M］. 2版. 北京：中国农业出版社：394-402.

Key Cultivation Techniques for Main Fruits and Vegetables in the Lancang-Mekong Region (Yunnan)

Chapter One
Overview of the Development of Fruit and Vegetable Industry in Yunnan

Chapter One Overview of the Development of Fruit and Vegetable Industry in Yunnan

Yunnan is located on the southwest border of China, with latitudes ranging from 21°8′ N to 29°15′ N and longitudes ranging from 97°31′04″ E to 106°11′48″ E. It stretches about 877.5 km from north to south and 721.4 km from east to west, with a total area of approximately 394 300 km^2. The main part of Yunnan province belongs to the western part of the Yunnan-Guizhou Plateau, known as the Diandong (Eastern Yunnan) Plateau, while the northwest is the southern extension of the Qinghai-Xizang Plateau, known as the Dianxi (Western Yunnan) longitudinal valley. The terrain generally features a stepped plateau landform that slopes from north to south, featuring alpine valleys, mountains, plateaus, hills, basins, river valleys, and alluvial plains with elevation differences of several hundred to over a thousand meters. The mountainous and plateau areas account for about 94% of the area of Yunnan, and the predominant soil type is mountain red soil.

Overall Yunnan has a subtropical plateau monsoon climate, with an average annual temperature of 14.5 °C. The hottest month (July) has an average temperature of 19–22 °C, while the coldest month (January) has an average temperature of 6–8 °C. The climate is characterized by small annual temperature differences but large daily temperature differences, with relatively mild winters. Yunnan is intersected by six major river systems running mainly from northwest to southeast or from north to south, and most areas receive an annual average precipitation of about 1 100 mm. Although the annual precipitation is moderate, the distribution across seasons is uneven, with distinct dry and wet periods, the dry season lasts from November to April of the following year, and the rainy season from May to October, resulting in warm winters and cool summers. Yunnan's diverse climate characteristics, influenced by its geographical location, atmospheric circulation, and complex geographic environment, result in the presence of six climate types ranging from temperate to tropical as it spans from north to south. This diverse range of climate types endows Yunnan with abundant plant resources and diverse small-scale crop cultivation.

As a typical low-latitude plateau region, Yunnan's unique geographical location and climatic conditions provide ample sunshine, high-quality water sources, clean air, and diverse fertile soils for the production of agricultural products. These natural advantages endow Yunnan's agricultural products with a natural and original ecological attribute of "safe and ecological Yunnan products". Yunnan's main agricultural products include grains, vegetables, fruits, tea, Chinese medicinal herbs, nuts, fresh-cut flowers, and various livestock and poultry products. The export of agricultural products has remained relatively stable for many years, with fruits and vegetables occupying a crucial position in agricultural trade and serving as a barometer for Yunnan's agricultural product exports.

1 Overview of the development of Yunnan's fruit industry

1.1 Planting area and yield of fruit trees in Yunnan

Yunnan boasts a rich variety of fruits, mainly temperate, subtropical, and tropical fruits. It has approximately 133 identified fruit tree species, accounting for about 60% of the nation's total fruit tree species. According to data from the National Bureau of Statistics and the Yunnan Provincial Bureau of Statistics (Figure 1-1), the total area of fruit orchards in Yunnan has shown a steady growth trend since 2010, with the orchard area exceeding 5 million mu[①] in 2011. After 10 years of development, the total area surpassed 10 million mu (approximately 10.626 2 million mu) in 2021, and the production increased from 4.071 4 million t in 2010 to 11.426 0 million t in 2021. From 2019 to 2021, Yunnan's fruit orchard area ranked 7th nationwide for three consecutive years, while the production has been ranked 12th to 13th.

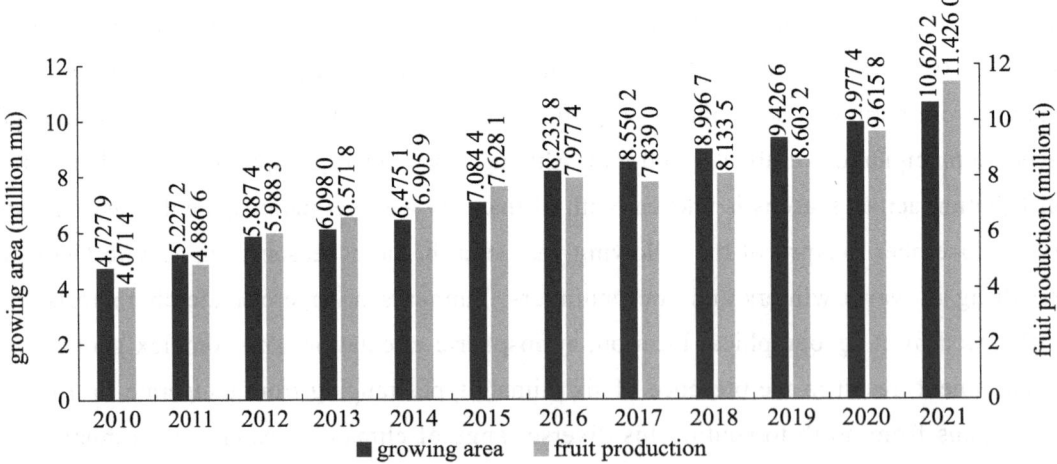

Figure 1-1　Fruit planting area and yield in Yunnan from 2010 to 2021

(Source: National Bureau of Statistics, Yunnan Provincial Bureau of Statistics)

---

① 　1 mu ≈ 667 m^2，15 mu=1 hm^2。

Chapter One Overview of the Development of Fruit and Vegetable Industry in Yunnan

1.2 Distribution

The distribution of fruit cultivation in Yunnan exhibits distinct climate regional characteristics, and Yunnan's unique "three-dimensional climate" creates a time difference in the ripening of most fruits compared to other producing areas, enabling fresh fruits to be produced "four-season production, year-round supply". Some fruits in Yunnan have the characteristic of "earlier early-ripening and later late-ripening". The relatively large-scale fruit categories in terms of planting area and production include banana, citrus, mango, grape, pear, apple, peach, watermelon, pomegranate, and pineapple, while the planting scale of some emerging specialty fruits such as sweet persimmon, loquat, ginseng fruit, kiwifruit, waxberry, and lychee has gradually expanded in recent years. Among them, the area of temperate fruits represented by apples and pears accounts for approximately 38.5%, the area of subtropical fruits represented by citrus fruits and grapes accounts for approximately 28.2%, and the area of tropical fruits represented by mangoes and bananas accounts for approximately 33.3%, forming a preliminary pattern of balanced development.

1.3 Trade

Since 2015, fruit exports in Yunnan have consistently ranked as the province's top agricultural export category, accounting for about 40% of the total agricultural export value. It is the leading province in fruit exports nationwide, with an over 25% share of the country's fruit export value. In 2021, the total export value reached 1.87 billion US dollars, accounting for 43% of the province's total agricultural export value and 25% of total value of the country's fruit export (7.51 billion US dollars). The export volume of citrus fruits and grapes has remained the first place in the country.

1.4 Characteristics of Yunnan's fruit industry

In recent years, the development of distinctive and advantageous fruits, promotion of refined production techniques, establishment of standard demonstration bases, and brand cultivation have significantly enhanced the role of Yunnan's fruit industry in regional economic development and effectively consolidating poverty alleviation achievements while dovetailing them with rural revitalization. A regionalized pattern of fruit production has basically taken shape, such as the apple-producing areas in Northeastern and Northwestern Yunnan, blueberry and pear-producing areas in Eastern Yunnan, grape, banana, pineapple, and pomegranate-producing areas in Honghe Basin, and peach and strawberry-producing areas in Central Yunnan. The branding of the fruit industry has been substantially formed, with a total of 531 certified green food products and 225 organic products in the province by the end of 2020. However, Yunnan's fruit production also faces challenges such as weak basic supporting facilities, low degree of scale development, lagging technology promotion, and outdated post-harvest facilities, which has prevented the full transformation of environmental advantages into quality advantages.

1.5 Development strategies

To address these challenges, it is necessary to increase infrastructure construction such as water, electricity and road supply, increase investment in production facilities for water and fertilizer integration, and support the construction of guaranteed water source projects and basic cold chain storage and transportation facilities. Furthermore, efforts should be made to strengthen the breeding, production, and sales system of improved varieties, and intensify the integrated research and application of standardized key production and management technologies. It is suggested to strengthen the leading and demonstration roles of new business entities such as enterprises, cooperatives, and industry alliances, support and consolidate the leadership role of leading enterprises, and promote the full play of brand benefits such as geographical indication of agricultural products. Lastly, efforts should be made to emphasize scientific and technological coverage of the whole industry chain to promote the improvement and efficiency of the fruit industry and ensure its sustained and healthy development.

2 Overview of the development of Yunnan's vegetable industry

2.1 Planting area and yield of vegetables in Yunnan

Vegetables are the largest economic crop in terms of planting area in Yunnan, a bio-industry with prominent advantages and a solid industrial foundation in the province. It has played an irreplaceable role in the transportation of southern vegetables to the north and western vegetables to the east and vegetable exports in the country.

According to data from the National Bureau of Statistics and the Yunnan Provincial Bureau of Statistics (Figure 1-2), the planting area and output of vegetables in Yunnan have shown a continuous annual growth trend since 2010. The planting area exceeded 15 million mu in 2015, and it reached nearly 20 million mu (about 19.373 7 million mu) in 2021. In 2013, the output exceeded 15 million t, and it began to exceed 20 million t in 2017, reaching nearly 30 million t (about 27.483 6 million t) in 2021. Yunnan's vegetable planting area has consistently ranked 10th in the country, while its output has remained in the 11th place.

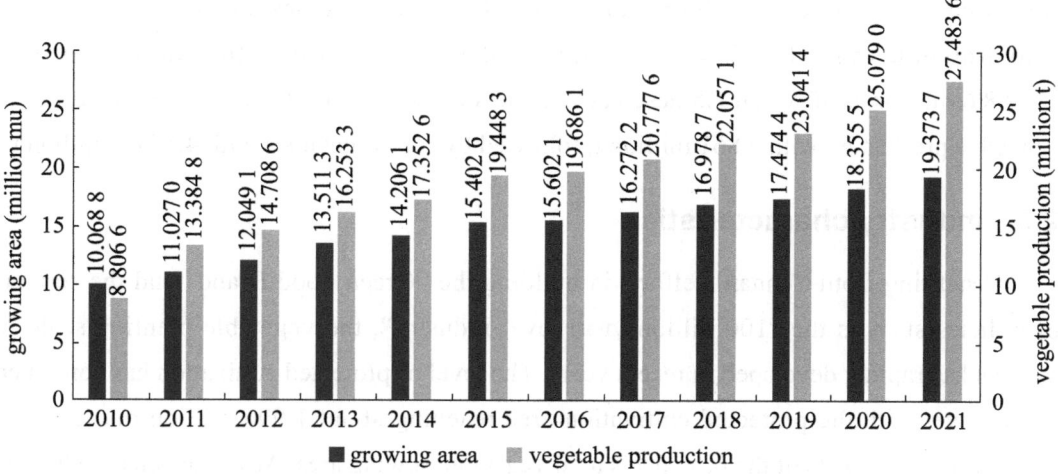

Figure 1.2 Vegetable planting area and output in Yunnan from 2010 to 2021
(Source: National Bureau of Statistics, Yunnan Provincial Bureau of Statistics)

2.2 Distribution and output structure

Relying on its low-latitude three-dimensional climate, giving full play to the advantages of "natural greenhouse" and "natural cool shed", Yunnan has formed advantageous production areas for winter and spring vegetables focusing on Baoshan City, Pu'er City, Xishuangbanna Prefecture, Dehong Prefecture, Lincang City, and the low-temperature valley area in the central and northern parts of the province, mainly producing solanaceous vegetables, onion, bitter gourd, cowpea, and green bean. The summer and autumn vegetable production areas mainly include Zhaotong City in the northeast, Dali Prefecture and Lijiang City, Nujiang Prefecture and Diqing Prefecture in the northwest, and Wenshan Prefecture in the southeast, mainly producing heading cabbage, Chinese cabbage, radish, and other cool-loving vegetables, as well as chili peppers and tomatoes. The perennial vegetable production areas mainly include Kunming City, Qujing City, Yuxi City, Chuxiong Prefecture, and the northern part of Honghe Prefecture, mainly producing common Chinese cabbage, lettuce, and scallions and garlic. In 2019, the planting area of winter and spring, summer and autumn, and perennial vegetable production areas in Yunnan accounted for 25.11%, 23.77%, and 51.12% respectively of the province's total, achieving staggered market availability of vegetable products throughout the year, ensuring year-round balanced supply in the domestic vegetable market.

According to the *Yunnan Statistical Yearbook*, the vegetable production in Yunnan is structured as follows: "Chinese cabbage > root vegetables > leafy vegetables >

solanaceous vegetables > scallion and garlic > legumes > melons and gourds > cabbage > hydroponic vegetables". In 2021, the total vegetable production in Yunnan was 27.488 6 million t, of which Chinese cabbage accounted for 21.32% of the province's total vegetable production with 5.861 million t, followed by root vegetables with 4.379 0 million t.

2.3 Industry characteristics

Benefiting from Yunnan's efforts in building the "Green Food Brand" and the support of policies such as the "100 billion-yuan level industry", the vegetable planting scale in Yunnan has rapidly developed in recent years. The level of protected cultivation has continued to improve, with the protected cultivation area reaching about 1.996 million mu in 2019, accounting for 11.39% of the province's total vegetable sowing area. Advantageous production areas have gradually formed, achieving balanced year-round supply of various types of vegetables. There are a wide variety of vegetables and 23 types of vegetable categories with a sowing area exceeding 200 000 mu. The leading role of leading enterprises and the role of new rural cooperatives and social service organizations continue to be strengthened, with a large number of households actively participating in the development of the vegetable industry. The vegetable sales channels are constantly expanding, and Yunnan has gradually become a well-known "vegetable garden" for the important "transportation of southern vegetables to the north" and "transportation of western vegetables to the east" in the country.

2.4 Challenges and strategies

First, the level of facility equipment is insufficient and needs to be further improved. It is suggested to rely on existing facilities and equipment to strengthen the transformation and upgrading of production machinery for different regions and vegetable varieties, to develop and promote new equipment, so as to better adapt to the needs of modern industrial development and enhance the ability to resist natural disasters.

Second, vegetable processing is mostly limited to primary processes such as sorting and packaging, lacking advanced processing. It is recommended to strengthen the construction of processing facilities and equipment, improve post-harvest treatment capacity and level, and conduct processing technology improvements and production line upgrades based on vegetable varieties and their value. Finally, continuous efforts should be made to increase technical research and product development in new vegetable processing areas.

Third, the capacity of breeding and seed production for innovative vegetable varieties is insufficient. It is suggested to strengthen the breeding research of innovative vegetable varieties, expand the independent innovation ability of vegetable varieties, increase the construction of new variety trial demonstration bases and factory-based seedling facilities, and vigorously promote the production of healthy vegetable seedlings and the construction of factory-based seedling facilities.

Fourth, efforts to build regional vegetable brands, enterprise varieties, and product brands are not sufficient. The "Yun Cai" (Yunnan vegetables) lacks visibility and influence in society. It is recommended to vigorously introduce large vegetable processing and distribution enterprises from home and abroad to settle in Yunnan, cultivate a group of large vegetable enterprise groups, and increase the promotion efforts for vegetable brands and products to enhance the overall social influence of "Yun Cai".

Chapter Two

Fruits

❖ Key Techniques for Apple Cultivation in Low-latitude Plateau

1 Overview of industrial development

Apple has at least two thousand years of cultivation history in China, which is mainly planted in Liaoning, Hebei, Shanxi, Shandong, Shaanxi, Gansu, Sichuan, Yunnan, Xizang, etc.

1.1 Natural distribution

Yunnan and its surrounding areas is one of the regions with the most abundant *Malus* resources in China, accounting for more than 70% of the total *Malus* resources in China. Among the known resources, 9 species are endemic to the region with diverse types. At present, 21 species of *Malus* germplasm resources have been found in Yunnan, 16 of which are wild types and 5 of which are cultivated types.

In 2021, the apple planting area in Yunnan was 1.408 million mu, with a total yield of about 1.347 million t, ranking 12th in China. The fruit quality is excellent, the price advantage of early, medium and late ripening varieties is stable, while the fruit is known for its bright appearance, early ripening, sweet and crisp taste, and excellent internal quality. Therefore, Yunnan has become a high-quality apple production base in South China.

1.2 Planting condition

Apple planting in Yunnan is mainly distributed in the production areas of Northeast (Zhaotong City and Qujing City), Northwest (Lijiang City, Dali Prefecture and Diqing Prefecture), Central region (Kunming City) and Southeast (Honghe Prefecture), with an altitude of 2 000–2 700 m in their impoverished mountainous and semi-mountainous areas, which have cold highland areas with complex topography, 5–18 °C of average annual temperature, 500–800 mm of average annual precipitation, slightly acidic or neutral soil, and excellent drainage and irrigation system. Fuji superior, Golden Delicious superior and Gala superior are the main cultivars.

2 Main cultivars and characteristics

2.1 Late-ripening Red Fuji

【Variety source】Japan.

【Suitable area】It can be cultivated universally in apple planting areas and is suitable for planting in areas with an altitude of 2 000–2 400 m.

【Characteristics】The tree has robust growth potential and semi-open appearance. Saplings or strong branches have a marked habit of axillary flowers and buds bearing. In the early fruiting stage, the long fruiting branches and axillary flowers and buds occupy a certain proportion, but they will soon turn to bear fruit on short fruit branches mainly, which accounts for about 70% in the full bearing stage. It features late-ripening, purplish red flowers, the average fruit weight of 220 g, the maximum fruit weight of 650 g, oblate to approximately round shape, medium-thickness and tough pericarp and bright red appearance, which has red glow and stripes, slices and strips of red, and small flecks on sunlit side. The fruit surface is smooth, no rust, and much fruit powder, while the pulp is creamy yellow, crispy, juicy, moderately sour and sweet, 14.5%–15.5% of soluble solids condent, excellent quality, high yield and excellent storage resistance.

【Production performance】It bears fruit early and produces plentifully. Vigorous rootstock apple trees begin to bear fruit in 4–5 years, while dwarfing rootstock trees begin to bear fruit in 3 years, and enter the full bearing period after 5 years. After strengthening the comprehensive management and adopting the measures of promoting flower-formation, flowers can be seen in the second year of final planting, but the fruit tree is prone to appear on-year and off-year. Under normal pollination conditions, the fruit setting percentage is relatively high, with about 70% of inflorescence fruit setting and 16.2%–40% of flower fruit setting. The bourse shoot is thin, with poor continuous bearing capacity but strong disease resistance. It ripens in early October in Zhaotong City and mid to late October in Lijiang City. The self-flowering and seed setting rate is low, so a certain proportion of pollinating trees is often required for final planting, and the pollination and seed setting rate of different varieties are high. This variety has less physiological fruit dropping and pre-harvest fruit dropping, without dehiscent fruits before ripening. The average yield is 3 000–4 000 kg per mu in the fertile period.

2.2 Yanfu No. 3

【Variety source】It is selected and bred by Yantai Institute of Pomology, Shandong.

【Suitable area】It can be cultivated universally in apple planting areas and is suitable for planting in areas with an altitude of 2 000–2 400 m.

【Characteristics】It features late-ripening, nearly round shape, glossy and small flecks. The fruit surface is smooth, no rust and much powder, the upper and lower, inside and outside of crown are well colored, reddish, and the ratio of all red fruits can reach about 80%. The pericarp is medium-thickness and tough, and the pulp is creamy yellow, crisp, juicy, and moderately sour and sweet, with superior quality. The soluble solids content is 14.8%–15.4%.

【Production performance】The fruit growth and development period is 190 d, with an average single fruit weight of 250–300 g, high yield, disease resistance, and storage resistance. After removing the bag, it reaches full red within 5–7 d, especially in autumn when the temperature is high and the temperature difference between day and night is small. It has a significant coloring advantage compared to other Fuji varieties. It ripens in mid October in Zhaotong City and September in Mengzi City.

2.3 Qinguan

【Variety source】It is selected and bred by Northwest A&F University.

【Suitable area】It can be cultivated universally in apple planting areas and is suitable for planting in areas with an altitude of 1 800–2 200 m.

【Characteristics】It is bred by cross hybridization, with Golden Delicious of high yield and quality as female parent, with Jiguan of early-fruiting, high yield and strong resistance as male parent. It is a late-ripening variety with smooth bark, dark reddish brown perennial branches and brown annual branches, long internodes, large dense and oval lenticels, little fuzzes, red flowers. Fruit is conical and the average single fruit weight of 200–250 g with high yield, whose ground color is yellow-green, dark red glow and intermittent red stripes on sunlit side, often with white rust, and can reach full dark red when ripening. The surface is smooth and waxy with obvious flecks, the pericarp is thick and tough, and the pulp is creamy yellow, crisp and slightly dense, containing 16.5% of soluble solids, 0.19% of acid, 2.31 mg/100 g of vitamin C. It is highly resistant to storage.

【Production performance】It has strong growth potential, robust growth and powerful central leader. This variety has a high germination rate, easy to flower, plump flower buds, not easy to form on-year and off-year, high fruit setting rate, continuous bearing capacity, drought resistance, cold resistance, disease and pest resistance. It ripens in mid-to-late October and early November in Zhaotong City, and leaves begin to fall heavily in late November.

2.4 Golden Delicious

【Variety source】It was the earliest introduced variety in Yunnan, originally from USA.

【Suitable area】It can be cultivated universally in apple planting areas and is suitable for planting in areas with an altitude of 1 800–2 200 m.

【Characteristics】It is a medium-ripening variety with the average single fruit weight of more than 150 g, conical, slight convex at the top. The stalk is slender, and the pericarp is thin, slightly rough, green and yellow in color, becomes golden after a little storage, and occasionally has light red glow on the sunlit side when harvesting late. The pulp is yellowish-white, with a fine and dense texture. When just harvesting, it is crispy and juicy, with rich sweet and fragrant aroma, and excellent quality. The pulp of the late picked fruit is light yellow, the fresh flavor is excellent, but the storability is slightly poor. It has high yield and stable performance.

【Production performance】The tree has moderate growth potential, average growth of branches, strong central leader, high germination rate, and weak branching rate. The branches are relatively open, making it easy to form flowers and not easy to form on-year and off-year. The flowers are white, with a high fruit setting rate, continuous fruit bearing capacity, strong disease resistance, and easy to produce fruit rust. It ripens in late August in Zhaotong City, and the leaves begin to fall heavily in mid November, which are neat and consistent. During the high yield period, It can produce around 4 500 kg per mu.

2.5 Gala

【Variety source】New Zealand.

【Suitable area】It can be cultivated universally in apple planting areas, the Central Yunnan region is better and it is suitable for planting in areas with an altitude of 1 600–1 800 m.

【Characteristics】It belongs to the medium and early ripening variety. The tree has strong growth potential, high germination rate, easy flowering, and strong branching ability. The average length of internodes is 2.7–3.3 cm, which is easy to form short branches. Long, medium, and short branches and axillary buds can all bear fruit, and the proportion of axillary buds fruiting in saplings is high. During the full bearing period, the trees mainly bear fruit on short fruit branches. The fruit is conical, slightly pentagonal, with an average single fruit weight of 170 g, a coloring index of 65%–85%, and cardinal red stripes. The pulp is yellowish-white, crisp, juicy, moderately sweet and sour, and thick in aroma. It contains 14.58% of soluble solids, 11.16% of total sugar, 0.22% of total acid, and 7.0–8.3 kg/cm^2 of hardness. It is resistant to storage, and can be stored for 1 month at room temperature, and for 3 months under refrigeration conditions.

【Production performance】It has strong continuous bearing capacity, high yield, ripens in batches, and is susceptible to powdery mildew. The variety has a high self-flowering and seed setting rate and well-coloring, and can produce high-quality fruits with gorgeous colors without bagging in Qujing City and Lijiang City.

2.6 Red Delicious

【Variety source】It originated in California, USA.

【Suitable area】It can be cultivated universally in apple planting areas and is suitable for planting in areas with an altitude of 1 600–2 200 m.

【Characteristics】The fruit shape is upright with a high pile, and the calyx is pentagonal. The average single fruit weight is 200 g, and the larger ones can reach more than 500 g. The pulp is crisp, juicy, sweet, and aromatic. When the fruit starts coloring, there are obvious intermittent red stripes, followed by red glow. After full coloring, the whole fruit is cardinal red, with obvious purplish red thick stripes on its glossy and bright surface, light brown or grayish-white flecks, and an uneven shoulder. The pulp is yellowish-white, crisp, medium coarse, juicy, sweet, soluble solids content is 14%, with a strong aroma, and the quality is superior.

【Production performance】It has strong continuous bearing capacity and high yield, with a growth and development period of about 145 d. It ripens generally in mid September and is resistant to storage. The best tasting period is after 1–2 months of post-harvest storage.

2.7 Jonagold

【Variety source】It was hybridized with Golden Delicious and Jonathan at the New York State Agricultural Experiment Station, USA.

【Suitable area】It can be cultivated universally in apple planting areas and is suitable for planting in areas with an altitude of 1 600–2 200 m.

【Characteristics】The fruit is conical, with an average single fruit meight of 220–250 g, the background color is green yellow or light yellow with bright red glow and unclear intermittent stripes on most of the sunlit surface, and the surface is smooth and glossy with waxy substance and unclear small flecks. The pulp is creamy yellow, crisp, medium coarse, juicy, sweet and sour, with slight aroma, containing about 14% of soluble solids, with excellent quality and storage resistance.

【Production performance】Seedlings can bear fruit in 3–4 years after planting, and enter the massive fruiting period in 7–8 years with high yield. It is not only a fresh food variety, but also an excellent variety for juice making and processing.

3 Cultivation and management techniques

3.1 Garden plot selection

The agricultural production area should be chosen with excellent ecological conditions, away from pollution sources, sustainable production capacity, and relatively flat hilly areas with good irrigation conditions. If the garden plot is built in mountainous areas, the slope gradient should be lower than 25°. For the purpose of preventing typhoons, conserving water and soil, wind break and sand fixation and other problems, windbreak forests should be built. The species of windbreak trees should be selected with strong adaptability, rapid growth, long lifespan, no same diseases and pests or intermediate hosts as fruit trees, and high economic value. Cypress plants are prohibited within 5.0 km around the newly built apple orchard. The standard for fertile soil in the garden plot is loose and airy with good moisture conservation, an active layer of 0.6 m, and the content of organic matter above 0.8%.

3.2 Efficient cultivation technology pattern

3.2.1 Arborization cultivation technology pattern

It has a history of nearly 80 years in Yunnan, and is currently the main cultivation pattern in Zhaotong City apple production area. According to the main cultivars, the corresponding pollinator tree should be configured. The vegetables, peanuts and other crops can be reasonably interplanted according to the row spacing, mainly legumes, which can not only increase the income of fruit farmers, but also combine planting and improving the soil fertility.

Points to note: Apple arborization cultivation should pay attention to the cultivation density, and the tree shape should be selected according to the density. When growing grass or planting green manure in the orchard, the seeds should be sown 1 m away from the trees in the row, and all the green manure must be dressed or plowed before November. There should be neither green manure growth in the winter, nor the reserved seeds for planting in the apple orchard.

3.2.2 High-density dwarfing cultivation pattern

It is the future direction of apple industry development, which has gradually been recognized by countries around the world due to its small tree form, convenient operation, strong continuous bearing capacity, high-quality, and labor saving. At present, this technology adopts two-year seedling of high-quality dwarf self rootstock M9-T337 for final planting, installing grid system facilities and integration facilities of water and fertilizer, and the row spacing of final planting is 1.0 m×3.5 m, with 192 plants per mu, 1 m stem height, 0.8 m×1.2 m crown width, and 2.5−3.0 m plant height, pruning into slender spindle. With grass in wide rows and micro-drip irrigation in narrow rows, apple orchards adopt vertical-culture technique, and pruning is mainly done by pulling branches every year. Pruning techniques include thinning and non-pruning, and short-cutting is rarely used. The main tree shape is cylindrical with fruiting branch groups directly growing on trunks. The cultivars are Royal Gala, Gala Mitchgla, Red Delicious, Huashuo, Honeycrisp, Pink Lady, Yanfu No. 3, etc.

Through wide-row dense planting and grid cultivation, "four savings, two highs, two earlies" can be achieved, namely saving 60% of water , saving 70% of fertilizer, saving 70% of land, and saving 80% of labor, with the characteristics of high quality, high yield, early fruiting, and early profitability. The adoption of this cultivation pattern can effectively promote the standardization, large scale, industrialization, and mechanization of Yunnan's apple production.

3.3 Pruning techniques

The quick shaping technique of column-free dwarfing and dense planting pruning can be used in cool highland apple orchards. After germination, buds below 60 cm should be removed in time. One strong bud among terminal buds should be selected as the central leading branch, and the 2−3 adjacent buds below it should be removed in time. When the lateral branches grow to 30 cm after germination, they should be pressed flat with toothpicks.

The 1st year winter pruning: In autumn, the young shoots on the main trunk should be pulled to 95°−120°, and winter pruning should be delayed to late February to early March. During pruning, all lateral branches are left with the horse's teeth shaped by re-short cutting, the elongated head of the middle stem is lightly shortened or not shortened, and lateral branches below 80 cm should all be removed.

The 2nd year winter pruning: In autumn, the young shoots on the main trunk should be pulled to 95°−120°, all lateral branches are left with the horse's teeth shaped by re-short cutting in winter, the elongated head of the middle stem is lightly shortened, and the tree height basically reaches about 3 m.

The 3rd year winter pruning: Annual branches with moderate growth potential and large angles should be selected as main branches without cutting. Every year, all lateral branches that are more than 1 m long and the base thickness is more than 1/4 of the trunk thickness should be removed as needed, and lateral branches that are more than 25 cm should be pulled below the level to keep the growth of the tree gentle, so that the tree forms and begins to bear fruit.

3.4 Water and fertilizer management

Germination to flowering stage: Often in March and April, the first peak period of apple root system growth occurs, around the time when the aboveground part begins to grow, and slowing down when the young shoots enter the growth peak period. Top dressing follows the technical requirements of "appropriate, shallow, skillful and even", generally twice a year. The first topdressing should be done before germination, with nitrogen fertilizer as the mainstay and phosphorus fertilizer as a supplement. The second topdressing should be carried out after germination to before flowering, generally based on nitrogen and phosphorus fertilizers, including diammonium phosphate, which should be carried out by hole application combined with watering germination water, with 10 kg urea, 20 kg diammonium phosphate, 5 kg potassium sulfate, and 10 kg amino acid organic fertilizer for drip irrigation per mu.

Flowering to fruit setting stage: 60−70 kg diammonium phosphate should be applied per mu, and soil conditioner, humic acid organic fertilizer or ecological organic potassium, biogas slurry and so on can also be applied as topdressing. In case of drought during this period, watering should be done in time to ensure soil moisture. For trees with a large number of flowers, weak growth potential, and no nitrogen fertilizer before germination, urea should be applied as topdressing. For 12−13-year-old trees, 0.5−1 kg urea should be applied per plant, and for 20-year-old trees, 1−1.5 kg urea should be applied per plant, followed by watering. For trees with strong growth potential, moderate or less flower quantity, watering should be done before flowering without nitrogen fertilizer.

Fruit enlargement period: From early June to early-to-mid July, topdressing of this time is to meet the needs of fruit enlargement, branch and leaf growth, and flower bud differentiation. Therefore, this topdressing is mainly potassium, and should select hole application or shallow application of tic-tac-toe shaping groove, with 40 kg potassium chloride and 5 kg diammonium phosphate applied per mu. For an apple orchard with a yield of 2 000−2 500 kg, 18−23 kg pure nitrogen, 20−25 kg pure phosphorus, and 25−30 kg pure potassium should be applied throughout the year. From August to September, the fruits enlarge rapidly, and nitrogen, phosphorus, and potassium fertilizers are mainly applied at this time. Apply 30 kg urea, 12 kg monoammonium phosphate, 30 kg potassium sulfate, and 20 kg amino acid organic fertilizer for drip irrigation per mu.

Fructescence stage: 20−30 d before fruit ripening, leaf spray should be conducted on cloudy or sunny days before 9: 00 and after 16: 00, so that the fertilizer can maintain a long time of moisture on the leaves and be absorbed for a long time. From the early stage of fruit coloring to the coloring stage, leaf fertilizer should be sprayed every 10–15 d. Medium and late ripening varieties should be sprayed more frequently, and generally stop spraying 15–20 d before fruit harvest.

Post-harvest stage: Post-harvest topdressing can boost root system growth and development, increase tree nutrition accumulation, improve flower bud quality, raise nutrient utilization rate, and enhance growth potential and frost resistance. When fertilizing, organic fertilizer should be the mainstay, with appropriate amounts of nitrogen, phosphorus, potassium compound fertilizers, medium and micro element fertilizers, and root-specific fertilizers. For the annual fertilizer application amount, the application amount of nitrogen should account for about 30% while phosphorus and potassium should account for 60%, which is, 10–15 kg pure nitrogen (N), 10–12 kg pure phosphorus (P_2O_5), and 18–25 kg pure potassium (K_2O). The optimal fertilization time is from early September to mid-to-late October. After harvesting, 0.5% nitrogen fertilizer should be sprayed on the leaf surface again to promote root system growth and tree nutrition storage, which is conducive to high yield in the following year.

3.5 Prevention and control of major diseases

3.5.1 Valsa canker

【 **Damage situation** 】 The main damage is to the branches and fruits of bearing trees. The symptoms of branches are divided into two types: ulcerative and deadwood. The ulcerative type is characterized by smelling like vinasse, small black flecks, and yellowish threads, while the deadwood type occurs mostly on 2–4-year-old twigs, often showing cross tawny and brown wheel pattern flecks. In the early stage of disease, lesions are maroon, and the middle of the lesions appear small black particles in the late stage.

【 **Occurrence rules** 】 The peak of disease occurs in early spring, and ends in May.

【 **Prevention and treatment** 】 The main trunks and branches that have been infected with valsa canker should be treated in time. There are two commonly used treatment methods: conventional scaling and coating method, and no scaling and scratching coating method. No matter which method is adopted, the scar site must be applied with chemicals, and large scars must be conducted bridge grafting. The latest treatment method is to mix the biological bacterial fertilizer and pure soil in a mass ratio of 1∶1, dilute with water and stir into a paste, apply on the scar after scaling with 1 cm, wrap it with plastic film externally, and untie it after 3 months, combined with underground application of biological bacterial fertilizer to strengthen the soil and trees, increase fertilizer efficiency, enhance resistance, and improve the yield and quality.

3.5.2 Powdery mildew

【Damage situation】This disease occurs rapidly and is prone to recurrence. It mainly harms leaves and tender shoots, especially the top leaves of young shoots are prone to infection. An important condition for the occurrence of powdery mildew is humidity, so the continuous rain leads to excessive humidity in the orchard, which provides conditions for the occurrence of powdery mildew. Therefore, rainfall during the flowering stage will definitely aggravate the powdery mildew in the later stage.

【Occurrence rules】Fungal diseases caused by ascomycota can occur twice a year, during the spring germination stage (from April to June) and the autumn shoot growth stage (end of August).

【Prevention and treatment】Prevention should be well conducted before flowering, and prevention measures should be strengthened if it rains during flowering. It is advisable to employ lime sulphur of 3°Bé on the trees before germination, 40% myclobutanil SC (1 : 3 000) or 10% kresoxim-methyl EW (1 : 1 000) before flowering, and 10% hexaconazole SC (1 : 1 500) or of 25% pyraclostrobin SC (1 : 2 000) on the tree 1–2 times after flower dropping.

3.5.3 Rust disease

【Damage situation】It mainly harms young leaves, petioles, young shoots, and young fruits, as well as other green and tender tissues. After initial infection, orange flecks appear on the surface of the leaves, and with the progression of the disease, scallion-root-like spores appear on the back of the affected leaves. After petiole being affected, the affected part becomes orange-yellow, spindle-shaped, and swells up, with yellow flecks and spores also appearing on it. When young fruits are affected, circular, orange-yellow flecks are formed near the stalk cavity, and later the lesions turn brown.

【Occurrence rules】Rust disease is a fungal disease caused by basidiomycetes, and it is prone to occur in early spring when temperature is high and rainfall is abundant.

【Prevention and treatment】Starting from the leaf-expansion stage of apple trees, protective fungicides including mancozeb, chlorothalonil, and carbendazim should be sprayed every 10–15 d for 2–3 consecutive times to protect the leaves from rust fungus infection. From late April to early May, it is advisable to employ 30% difenoconazole SC (1 : 2 000) or 43% tebuconazole SC (1 : 3 000) on the tree. No cypress trees are allowed within 5 km of the orchard.

3.5.4 Apple scab

【Damage situation】It is a fungal disease caused by ascomycetes and mainly harms leaves, fruits, flowers, buds, and tender shoots.

【Occurrence rules】The peak of the disease occurs from June to July. In the initial onset of the disease, nearly round white lesions with a diameter of 12 mm appear on the leaves. With the progression of the disease, the lesions expand, appearing nearly round or radial, with obvious edges. Dark brown to dark green mold layers appear on both sides of the leaves. Several small and close lesions can also be connected into a large lesion, with irregular edges. In the later stage, the lesions thicken, causing uneven leaves, or even twisted. The non-lesion part of infected leaves has symptoms of chlorosis, leading to early leaf dropping. The lesions on the fruits are scabby, which is nearly round and light yellowish-green in the early stage, and gradually expands. The surface of the diseased fruits also has black or dark green mold layers, with slightly darker edges. With the growth of the fruits, they concave and crack, causing fruit deformity.

【Prevention and treatment】In winter, the deciduous leaves and diseased fruits should be thoroughly cleaned up, taken out of the orchard, and buried deeply. When diseased leaves are found, they should be immediately removed, put into plastic bags, taken out of the orchard and buried deeply to prevent secondary spread. In the young fruit stage, using safe chemical agents such as difenoconazole, azoxystrobin and pyraclostrobin, and brassinolide and amino acids can be added to spray each time. In the fruit enlargement stage, it is recommended to utilize difenoconazole + pydiflumetofen and captan + tebuconazole, and spray once 10 d before packing. In order to achieve satisfactory control effect, the following points must be done. Early prevention is vital, and treatment should start early. Prevention and treatment should be continuous, paying attention to spraying once every 5−7 d for 2 consecutive times. Spraying should be done on both the trees and the ground at the same time, without missing any areas. Ensuring the amount of spraying, only meticulous spraying can ensure the effectiveness.

3.6 Prevention and control of major pests

3.6.1 Oriental fruit fly

【Damage situation】It mainly harms fruits. When the imagoes lay eggs, they will cause wounds on the pericarp and lay eggs inside the fruit. When the larvae grow up, they will directly eat the pulp, so the inside of the fruit begins to rot.

【Occurrence rules】The oriental fruit fly can grow in temperatures of 14−34 °C, with the optimal temperature being 25–30 °C. Places with high temperatures and dry conditions in spring and mild and less-rainfall conditions in summer are conducive to the occurrence of the pest.

【Prevention and treatment】The orchard should be cleaned up well, and the rotten and diseased fruits should be cleaned away in time, buried deeply or destroyed after soaking with drugs. If conditions permit, the orchard can be irrigated 2–3 times to eliminate the larvae, pupae, and newly emerged imagoes in the soil. A suitable environment for the survival of oriental fruit fly's natural enemies including pteromalus puparum, rove beetle, ant and so on can be created. Sex attractant or sugar-aceticacid liquid can be used to lure and kill imagoes. When the fruit is ripe, cyhalothrin, fenvalerate, avermectin and spinetoram and so on can be sprayed for prevention and treatment.

3.6.2 Apple aphid

【Damage situation】It harms young leaves and young shoots.

【Occurrence rules】Apple aphids do not transfer hosts, and the climate of that year has a great impact on them. The winter temperature is higher than the average temperature of previous years, the temperature rises quickly in early spring, the young shoots grow vigorously, and less rainfall and drought, which are all conducive to the occurrence of apple aphids. The reproduction rate is extremely fast, with more than 10 generations a year. The eggs generally overwinter in the bud side and the bud axil, and the overwintering eggs begin to hatch during the germination period of the fruit tree. The larvae concentrate on the bud to feed, and climb to the leaflets after the apple leaf-expansion. The infected shoots begin to appear in late May, and the peak period of infection is from June to July. Winged aphids will be produced within the aphid population for migration and spread. In late July, winged aphids migrate to weeds to reproduce in the form of parthenogenesis, and in October, winged aphids fly to the orchard to copulate and lay eggs.

【Prevention and treatment】In the initial stage, it is advisable to employ 75% spirotetramat·pymetrozine WDG (1∶5 000), 22% sulfoxaflor SC (1∶8 000), 70% imidacloprid WDG (1∶5 000) or 15% pymetrozine WP (1∶4 000), which can effectively reduce the number of yellow aphids in the orchard, and has no lasting effect on aphid natural enemies and their progeny.

3.6.3 Woolly apple aphid

【Damage situation】It harms branches, young shoots, leaf axils, fruit navels, and exposed root system, causing wounds and other diseases.

【Occurrence rules】There are 12–18 generations of woolly apple aphids every year, and they overwinter in the scars and cracks of trunks and near-ground roots.

【Prevention and treatment】During April to May, the soil should be dug up to expose the roots, covering with original soil after applying chemical agent. During the growing period, the chemical agent can be sprayed or the barks of branches and trunks can be lightly scraped and applied with the chemical agent.

For fruit trees in dormancy stage, combined with the prevention and control of spider mites and scale insects, spraying a mineral oil emulsion with 5% of oil before the fruit trees germination can kill the overwintering aphid eggs.

Applying the chemical agent to the trunk is particularly suitable for non-fruiting trees with distant water sources and difficulty in accessing water, and wrap with plastic cloth or waste newspaper after applying.

3.6.4 Spider mite

【Damage situation】It harms the leaves and fruits of plants, causing severe thinning, whitening, and even falling off of the leaves, affecting the normal growth of the plants.

【Occurrence rules】There are 7–9 generations of spider mite every year. Winter eggs overwinter on short fruiting branches, bourse shoot and branchlets forking, leaf scars, bud rings, and rough barks which are over two years old. Winter egg hatching is relatively concentrated, normally 10–20 d. Therefore, the peak period of winter egg hatching is the first critical period for chemical control.

【Prevention and treatment】

Limewater: This method is mainly used for woody plants. After the leaves of the plants fall and enter dormancy in late autumn, the dead leaves, weeds, and dead barks of the trunk should be cleaned up in time, and then the trunks should be brushed with limewater to kill the eggs parasitic on the surface of the trunks.

Sulfur powder: In late autumn, the dead leaves, dead branches and weeds on the soil surface should be cleaned up, and then a small amount of sulfur powder should be evenly spread, and the soil should be turned over to make the sulfur evenly distributed in the topsoil.

Lime sulfur: In late autumn, after the plant is dormant, the lime sulfur powder [1 : (300–500)] should be diluted, and then evenly sprayed on the trunk of woody plants and soil surface until it is moistened by 0.5–1 cm.

Spirodiclofen: In early spring, 20% spirodiclofen SC [1 : (4 000–6 000)] should be diluted and evenly sprayed on the front and back of the leaves and branches, Irrigate the medicinal infusion later.

Biological control: In spring and early summer before mite infestation occurs, predatory mites should be released, preferably on cloudy days or evenings (avoiding rainy days). Paper bags can be nailed directly to the trunks of large trees. The effective period of each bag is generally one quarter.

Chemical control: It is advisable to employ 73% propargite EC (1 : 1 500), 34% spirodiclofen SC (1 : 4 000) + 43% bifenazate SC [1 : (1 800−2 500)], 20% avermectin· bifenazate SC [1 : (2 000−2 500)], 2.5% cyhalothrin EC (1 : 4 000), and 5% hexythiazox EC (1 : 3 000), once every 7−10 d for 2−3 consecutive times.

4 Harvest

The appropriate time for harvesting is clear days without dew, and harvesting can be achieved if the fruit surface is more than 60% colored. Fruits should be handled with care to avoid damage and placed in a cool and dry place.

References

JIANG Z W, YU Q, SONG L Q, et al., 2010. Standard integrated management technology of soil, fertilizer and water for mature apple orchards with high quality and high efficiency [J]. Yantai Fruits (3): 32−35.

LI B Z, 2010. Technology of rootstock apple garden construction and sapling form pruning [J]. Northwest Horticulture (10):13−16.

LIU H T, YU H J, LIU P, et al., 2022. Key Techniques for establishment and management of apple orchards with modern cultivation pattern [J]. Deciduous Fruits, 54 (5): 84−85.

WANG Y L, 2010. High quality and efficient cultivation techniques of apple [J]. Hebei Fruits (2): 8−11.

ZHANG J, 2022. Technology of integrated pest and disease management in apple orchard [J]. World Tropical Agriculture Information (1): 49−50.

❖ Key Techniques for Pear Cultivation

1 Overview of industrial development

Pear, a perennial deciduous tree or shrub fruit tree in the Rosaceae family, is rich in nutrition, with crisp and juicy flesh, a delicious sweet and sour taste, and an excellent flavor. Fresh pears are rich in protein, fat, carbohydrates, calcium, phosphorus, iron, carotene, vitamin B_1, vitamin B_2, vitamin B_3, vitamin C, and other nutrients and are known as the "king of fruits". Traditional medicine believes that pears have the functions of nourishing fluids, clearing heat and phlegm, nourishing the heart and lungs, and treating hot coughs, thirst, heat injuries, and constipation. Regular consumption may prevent urinary and digestive system diseases. Due to its rich dietary fiber, low calorie content, and abundant plant nutrients in the peel, pears are known for their high content of flavonoids, which have good antioxidant and anti-inflammatory effects and may have certain preventive effects on diseases such as heart disease, type 2 diabetes, and cancer, earning pears the reputation of being a healthy fruit.

Yunnan is one of the original habitats of pears in China, with a history of pear cultivation in the province for at least 1 200 years. Pear trees are distributed in 127 counties and cities in Yunnan, at altitudes ranging from 450 m to 3 400 m, among which the most suitable planting area is found at elevations between 1 800 m and 2 200 m. As of 2022, the pear cultivation area in Yunnan is approximately 1.02 million mu, with the most extensive distribution and largest area in central Yunnan, followed by the northwest and northeast, and the least in southwest Yunnan. Pears are categorized as red-skinned, brown-skinned, and green-skinned, and all three varieties are cultivated in Yunnan, with red-skinned pears being particularly famous, covering an area of approximately 290 000 mu. Yunnan's red-skinned pears are one of the four major advantaged regions for Chinese characteristic pears. Unique low-latitude highland geographical conditions have created a distinct climate, resulting in red-skinned pears produced in Yunnan being characterized by vibrant colors, rich flavor, moderate sweetness and sourness, and a crisp, juicy texture, making them highly favored by domestic and foreign consumers, suitable for the production of high-end fresh fruit. Yunnan has over 400 pear varieties and types, with 320 being local varieties selected under the influence of local natural conditions and cultivation, some of which have high processing value. Currently, Yunnan's pear processing products include pickled pears, canned pears, pear paste, pear vinegar, pear wine, and pear beverages.

2 Main cultivars and characteristics

2.1 Meiren Su

The average single fruit weight is 275 g, longitudinal diameter of 9.8 cm, and transverse diameter of 9.5 cm. It is inverted oval shape, fresh and bright red skin on the sunny side of the fruit. There are 5 locules in the fruit core. The flesh is pale yellow-white, crisp and juicy with a sweet and slightly astringent taste, and slightly fragrant. It contains soluble solids of 13.2%, with medium to high quality. It can be stored at room temperature for 10 d. The tree has strong vigor, a semi-open canopy, strong budding and branching abilities, and a high yield. One-year-old branches are reddish-brown. The tips of the branches are hairy, with lanceolate leaves that are 9.5 cm long and 7 cm wide, gradually tapering at the tips and wide at the base. Flower buds are pale pink, with an average of 7–9 flowers per inflorescence, averaging 8 flowers. There are 31–34 stamens with an average of 33 stamens. And the diameter of the corolla is 3.8 cm. In the Anning District, the fruit ripens in late August.

2.2 Mantian Hong

The average single fruit weight is 280 g, longitudinal diameter of 8.5 cm, and transverse diameter of 8.7 cm. It is nearly round, bright red skin on the sunny side of the fruit. There are 5 locules in the fruit core. The flesh is milky white, crisp and juicy with a sweet and slightly astringent taste, and slightly fragrant. It contains soluble solids of 13.8%, with medium to high quality. It can be stored at room temperature for 10 d. The tree has strong vigor, semi-open canopy, strong budding ability, moderate branching ability, and high yield. One-year-old branches are reddish-brown, with hairy tips. The lanceolate leaves that are 9.3 cm long and 8.1 cm wide, gradually taper at the tips and wide at the base. The small leaves are purple-red. The flower buds are pale pink, with an average of 7–9 flowers per inflorescence, averaging 8 flowers. There are 30–31 stamens with an average of 30 stamens. The diameter of the corolla is 3.7 cm. In the Anning District, the fruit ripens in late August.

2.3 Zao Bai Mi

The average single fruit weight is 160 g, longitudinal diameter of 8.5 cm, and transverse diameter of 8 cm. It is oval or nearly round, yellowish-green skin with a faint red blush on the sunny side. There are 5 locules in the fruit core. The flesh is milky white, fine texture, crisp and juicy with a sweet, and slightly fragrant taste. It contains soluble solids content of 11.2%, with high quality. It can be stored at room temperature for 10 d. The tree has moderate vigor, semi-open canopy, strong budding ability, moderate branching ability, and high yield. One-year-old branches are greenish-yellow, gray-brown, yellow-brown, red-brown, brown, purple-brown, or black-brown. The ovate leaves that are 9.5 cm long and 6.5 cm wide, gradually taper at the tips and wide at the base. The flower buds are pale pink, with an average of 6–8 flowers per inflorescence, averaging 7 flowers. There are 30–38 stamens with an average of 35 stamens. The diameter of the corolla is 3.8 cm. In the Anning District, the fruit ripens in late June.

2.4 Yunhong Pear No. 1

The average single fruit weigh is 320 g, with a vertical diameter of 6.2 cm and a horizontal diameter of 5.8 cm, and is oblong in shape. The skin is more than 85% red. The fruit has 5 locules and the flesh is white, dense, sweet, and slightly fragrant, with a soluble solids content of 13.6%. The fruit has medium quality and good storage properties, with a shelf life of over 30 d at room temperature. The tree has vigorous growth, an erect posture, strong budding and branching abilities, and high yield. The one-year-old branches are reddish-brown, and the leaves are lanceolate, 11 cm long and 8 cm wide, gradually tapering at the tips and heart-shaped at the base. The flower buds are light pink. There are 6–8 flowers per inflorescence, with an average of 7 flowers. And there are 18–21 stamens, with an average of 20 stamens. The diameter of corolla is 4.1 cm. The fruits mature in early October in the Anning District.

2.5 Caiyun Hong

The tree has moderate vigor, an open posture, strong budding ability, weak branching ability, and high yield. The one-year-old branches are reddish-brown, and the young leaves are light red and oval, with a length of 11.5 cm and a width of 7.3 cm, gradually tapering at the tips and wedge-shaped at the base. The pedicel is hairless. The flower buds are light pink. There are 5–11 flowers per inflorescence, with an average of 8 flowers. The corolla diameter is 3.45 cm. There are 29–31 stamens, with an average of 30 stamens, and purple-red anthers. The average single fruit weight is 240 g, longitudinal diameter is 8.0 cm, and transverse diameter is 8.3 cm. The shape is near circular. Calyx depression is deep and calyx is detached. The peel has yellow background color and orange blush on the sunny side, with a large coloring area, a natural coloring rate of over 80% of the surface, a quick coloring time, and no greening phenomenon after coloring. The fruit stone is small, with 5–7 locules. The fruit stalk is thick and short, with an average of 2.1 cm. Due to its features, the fruit is basically not easy to fall off under natural conditions. The flesh is crisp and juicy, the taste is pure and sweet without astringency, and the character is stable and hereditary. The fruit stone is small and the edible rate is over 95%, with an average soluble solids content of 13.8%. The fruit has good storage properties, with a shelf life of 21 d at room temperature. It shows enhanced color and freshness during storage, with no significant browning of the skin or flesh after mechanical damage. The fruit mature in mid-to-late August in the Anning District.

2.6　Baozhu Pear

This local variety from Yunnan is cultivated in Chenggong and Jinning. The average single fruit weight is 258 g, with a vertical diameter of 10.5 cm and a horizontal diameter of 10 cm, and is nearly circular. The skin is green and thick, with a large fruit stone containing 5 locules. The flesh is white, coarse, crisp, and juicy, with a sweet taste and no fragrance, and the soluble solids content of 12.3%, with fair average quality. The tree has strong vigor, an open posture, and strong budding and branching abilities. It begins yielding 6–7 years after planting, with occasional biennial bearing. The one-year-old branches are gray-brown, while the elliptical leaves are 12.2 cm long and 7.2 cm wide, tapering at the tips and wedge-shaped at the base. The flower buds have a light pink margin. There are 5–7 flowers per inflorescence, with an average of 6 flowers, and 24–33 stamens, with an average of 27 stamens. The corolla diameter is 4.1 cm. The fruits mature in late September in Chenggong area District.

2.7　Hongxue Pear

This variety is produced in Weishan County at an altitude of 1 800–2 300 m. It's large nearly round. The average single fruit weight is 280 g and has a maximum weight of 600 g. The peel is rough, greenish-yellow surface, with over half covered in a red blush. The long fruit stalk measures 4.6 cm, has a shallow calyx, and sheds calyx leaves, with a brown base. The flesh is yellowish-white, coarse, slightly crisp, slightly acidic, and juicy, with the soluble solids content of around 12.3%, with fair average quality. It is slightly astringent at first harvest, and becomes fragrant after storage. It begins to mature in early November and can be stored until February or March of the following year. The tree has strong growth, a round or conical large form, and a height of over 8 m. It has strong budding and branching abilities, with reddish-brown 1-year-old branches. The elliptical leaves are 10.9 cm long and 7.3 cm wide, tapering at the tips and wedge-shaped at the base. It mainly bears fruit on medium and short branches, and can also fruit from axillary flower buds. It has high yield, strong adaptability, and weaker disease resistance.

3 Cultivation and management techniques

3.1 Planting planning and orchard construction

3.1.1 Orchard construction

The suitable cultivation areas are mainly flat or gently sloping areas with an altitude of 1 800–2 200 m. The orchard should be positioned with a windbreak, facing the sun, and have convenient irrigation and drainage facilities. The optimal soil pH value is 5.5–6.5, with a slightly acidic nature. Sandy loam or loam is preferred, and light clayey soil can be cultivated after appropriate improvement with organic materials such as straw. During orchard establishment, based on soil texture and cultivation mode, the width and depth of the excavation should be 60–80 cm. It is necessary to dig planting ponds or trenches. After disinfection and sun-drying, it is recommended to apply 4–6 t of organic fertilizer per mu, mix it thoroughly with the bottom soil, cover it with the upper fertile soil to a height of about 20 cm from the ground, and complete the planting and covering of seedlings. After planting, it needs to be irrigated with a large amount of water to compact the soil. Watering should be done once every 5 d in the early stage of planting to ensure survival.

3.1.2 Pollination variety configuration

Pears are fruit trees with gametophytic self-incompatibility. Most varieties cannot set fruit through self-pollination, so it is necessary to select pollination varieties. This can be achieved through configuration of pollination trees or artificial pollination assistance. There are three types of pollination tree configurations: equal-row configuration, unequal-row configuration, and checkerboard configuration.

3.2 Shaping and pruning

3.2.1 Pear-shaped cylinder tree form

The cylindrical tree form refers to a single main trunk structure, with a tree height of 3.0–3.5 m. The central trunk is robust and erect, reaching 50–60 cm in height. There are no secondary lateral branches or distinct stratification. A total of 20–25 small branches sprout uniformly directly from the central trunk, each measuring within 1.2 m in length. These small branches directly bear flower buds. Upon entering the peak fruiting period, central trunk tipping is performed.

Key techniques for shaping involve bud-notching, branch support, and branch arrangement. Firstly, based on the climate and variety characteristics in southern China, an optimal spacing of (1.2–1.5) m × (3.0–3.5) m between rows is chosen. When the trunk grows to around 1.6 m in height, bud-notching can start, 7–10 d before or after bud sprouting. The trimming distance is from 30 cm below the terminal bud to 60 cm above the root neck in the middle section of the trunk. Trimming is done about 0.5–1.0 cm above the bud, covering half of the trunk's width, and penetrating into the wood. Branch support aims to adjust branch angles, moderate growth vigor, promote flower bud differentiation, and accelerate flowering and fruiting. Branch support occurs between May and June before lignification, with optimal new shoot lengths around 20 cm. Angle support is facilitated by using toothpicks or small bamboo pieces, requiring small branch basal angles at approximately 45° and waist angles around 60°. During the peak fruiting period, it is advisable for the tree to have around 20–25 small branches. These branches should efficiently utilize space, forming an approximate spiral distribution along the main trunk. Adjacent branches on the same side should have a spacing of over 20 cm, avoiding overlap, forked branches, and a thickness not exceeding one-third of the main trunk's diameter at the point of attachment.

Pruning can be performed after the deciduous period and before bud emergence. Pruning a single main trunk has three main considerations: first, ensuring the rational distribution of fruit-bearing branch groups on the main trunk without mutual obstruction, maintaining a uniform spiral arrangement; second, retaining only one terminal shoot, removing others except for the weakest; and third, ensuring the extension of fruit-bearing branches along a single axis.

3.2.2 Pear-shaped happy tree form

The main trunk height ranges 60–80 cm, lacking a central trunk. From 60 cm upwards, evenly spaced 3–4 main branches extend outwards at a 45° angle from the ground. Approximately 25 cm apart on both sides of the main branches, primary lateral branches cross-distribute, carrying medium and small-sized fruit-bearing branch groups or fruit-bearing branches uniformly. The entire foliage has a thickness of 200–250 cm. Planting spacing is set at 3 m × 4 m, with around 50 plants per mu.

This tree form develops quickly, bears fruit early, and facilitates easy replanting. However, the cost of shaping and pruning is relatively high. The pruning focuses on letting growth occur naturally, only selectively retracting excessively robust or weak branches to adjust the growth vigor between long and short branches. For vigorously growing long branches, bamboo poles can be used to bind and open the branch angles, with main branch angles at 60° and secondary support branch angles at 70°. The basal axis is approximately 30 cm long, with generally two long branches growing from each axis. In total, the tree develops around 10–12 long branches, including those directly growing from the central trunk without basal axes. The top two branch groups should extend vertically between rows, while the lower branch groups form a 70° opening angle. The top two branch groups are bent downward to form a 90° angle.

3.3 Water and fertilizer management

Employ soil testing an formulated fertilization, configuring the types and amounts of fertilizers based on deficiencies in soil nutrient elements. Chemical fertilizers are generally used as supplementary fertilizers, with advantages such as high nutrient content, rapid effectiveness, and ease of use. However, they suffer from shortcomings like short-lasting efficacy, high costs, and environmental pollution. Organic fertilizers, typically applied as base fertilizers, exhibit characteristics of slow release, prolonged efficacy, diverse types, and soil improvement benefits. Hence, it is essential to coordinate the application of chemical and organic fertilizers according to the nutrient status of the trees. Base fertilization is usually applied early after harvesting in autumn, facilitating prompt nutrient absorption and storage by the root system. The fertilizer quantity should ideally match the fruit yield, with a possibility of incorporating small amounts of nitrogen and phosphorus fertilizers. Supplementary fertilization is typically determined based on the tree's nutrient diagnosis in conjunction with phenological phases, including pre-flowering, fruit enlargement, and flower bud differentiation stages. Fertilizers can be applied through surface broadcasting, foliar feeding, or utilizing integrated water and fertilizer technology. Supplementary fertilization is usually synchronized with irrigation. The optimal soil moisture content for pear tree growth is when the soil holds 60%–80% of its field water-holding capacity in the field. It is crucial to manage key water demand periods for pear trees, such as the bud initiation period, young fruit formation period, fruit enlargement period, flower bud differentiation period, post-harvest period, and the pre-soil freezing period. Skillful water management during these stages plays a pivotal role in achieving high-quality and abundant yields.

3.4 Prevention and control of major diseases and pests

The main principles of prevention and control of major diseases and pests in orchards involve a proactive approach with a combination of prevention and treatment, enhanced management of nutrients and water, and emphasis on organic fertilizers with balanced application, which can ensure adequate nutritional supply to trees, and boosts tree resistance. Additionally, maintaining reasonable fruit loads, controlling yields, promoting orchard ventilation and light penetration, increasing biodiversity, and introducing natural enemies contribute to effective pest and disease management. Specific measures include applying latex to mechanical wounds for isolation, reducing paths for pathogen invasion, and using Baume 3−5°Bé of lime sulfur compound or crystal lime sulfur compound 20−30 times liquid during dormancy to effectively prevent diseases such as pear scab, red spot disease, and brown spot disease. A combination of physical, chemical and biological control is used to control the occurrence and spread of pests and diseases in an integrated manner. In the major pear-producing areas of Yunnan, seven common pests and diseases have been identified as follows.

(1) Pear scab: Pear scab primarily affects various green tender tissues of pear trees, causing harm from post-flowering to fruit maturation. It forms elongated oval, spindle, or elongated stripe-shaped lesions, developing a black mold layer. The diseased tissue withers, gradually becoming concave and ulcerated, leading to leaf shedding and, later, cracked fruits. It mainly occurs from May to July, with higher rainfall exacerbating the severity. Before pear tree budding and flowering, spraying 25% tebuconazole EC (1 : 5 000), 25% difenoconazole EW (1 : 6 000), or 12.5% diniconazole WP [1 : (2 000−3 000)] helps eliminate the overwintering conidia produced by diseased parts.

(2) Pear black spot disease: This disease mainly affects leaves and new shoots, forming nearly circular or polygonal lesions. Surrounded by a yellow halo, the center appears gray-white to gray-brown, resulting in numerous diseased and prematurely dropping fruits. Before pear trees sprout in spring, spraying lime sulfur on branches helps eliminate overwintering pathogens. During mid-to-late June and rainy season from July to August, spraying 3% polyoxin WP (1 : 300) or 35% fluopyram · tebuconazole SC (1 : 3 000) is recommended to be applied 2−3 times continuously.

(3) Pear anthracnose: Pear anthracnose mainly affects fruit, leading to pear rot. Lesions appear as water-soaked spots, causing soft rot and continuous expansion from the pulp to the core, which results in a cone-shaped rotting. Disease occurrence and prevalence closely correlate with rainfall. Early onset happens during cloudy and rainy periods from April to May, with more severe outbreaks during continuous rains from June to July. Options for control include spraying 50% carbendazim WP [1 : (600−800)], 75% chlorothalonil WP [1 : (500−800)], 450 g/L prochloraz EW [1 : (800−1 500)], or 250 g/L azoxystrobin SC [1 : (800−1 500)]. It is advisable to spray once every 7−10 d, continuously for 2−3 times.

(4) Pear rust: It usually occurs in April and naturally diminishes without encountering rainy seasons. It primarily affects leaves, and in severe cases, can harm fruits. Affected areas turn orange-yellow, and a typical symptom is the growth of reproductive and rust spore structures on the lesions, resembling whiskers. You can use 25% flutriafol SC (1 : 3 000), 25%

triadimefon WP (1 : 2 000), or 430 g/L tebuconazole SC (1 : 3 000), applying continuously for 2–3 times.

(5) *Grapholita molesta* (Oriental Fruit Moth): It mainly harms fruits, with more severe damage to enlarged fruits. Initially, it harms the shallow layers of the fruit flesh, gradually tunneling towards the core and excreting waste inside, forming a "red bean paste" appearance. It is recommended to use sex attractants to lure and kill male adults or release parasitic wasps. The releasing frequency can be once every 3–5 d and consecutively for 3–4 times, and the amount of releasing can be 30 000–50 000 per mu each time. Chemical control options include spraying 10% efficient cyhalothrin EC [1 : (1 000–1 500)], 1.8% avermectin EC [1 : (1 000–1 500)], or 8 000 IU/μL *Bacillus thuringiensis* (Bt) fungicide [1 : (800–1 000)]. It is kindly advised to apply once every 7–10 d, continuously for 2–3 times.

(6) *Psylla chinensis* Yang et Li (Pear Psylla): You can use 24% abamectin · chlorpyrifos SC (1 : 4 000), 20% clothianidin SC (1 : 2 000), 20% fenvalerate EC (1 : 3 000), or 5% avermectin microemulsion (1 : 5 000), spraying continuously for 2–3 times.

(7) Fruit Flies: Chemical control options include using 2.5% efficient cyfluthrin EC (1 : 1 000), 2% avermectin EC (1 : 4 000), 75% cyromazine WP (1 : 4 500), 5% cypermethrin EC (1 : 1 000), and 2.5% deltamethrin EC (1 : 1 000).

(8) Scarabaeoidea (Scarab Beetles): Chemical control options involve using 5% chlorantraniliprole SC [1 : (2 000–2 500)], 15% indoxacarb EC [1 : (1 000–1 500)], and 5% methylamino abamectin benzoate microemulsion [1 : (2 000–3 000)]. It is recommended to apply once pesticides every 7–10 d, continuously for 2–3 times.

3.5 Harvesting and post-harvest treatment

3.5.1 Determination of harvest period

The harvest period is closely related to the yield, quality, and storage resistance of the fruit. Harvesting too early not only affects yield but also results in low sugar content, bland flavor, poor color, and inferior quality. Prematurely harvested fruits are prone to dehydration, wrinkling, and browning in fruit core during storage. Harvesting too late leads to rapid aging of pear fruits, increased sensitivity to carbon dioxide, and a higher probability of physiological disorders such as core browning and skin darkening. The harvest time is determined based on a combination of factors such as the target market distance, storage expectations, and the developmental stage of the fruits. The following indicators are utilized in production.

(1) Background color of fruit peel: As the fruit approaches maturity, chlorophyll in the peel gradually breaks down or transforms, revealing the background color. The peel color

changes from green to light green or green-yellow, with a slightly waxy appearance. When the fruit surface becomes glossy, it indicates that the fruit is ready for harvest.

(2) Hardness and soluble solids: Fruit hardness and soluble solids (usually referring to sugar content) are two major factors affecting taste. As the fruit matures, hardness decreases, and soluble solids increase. Recommended levels for hardness and soluble solids vary among different varieties: early-maturing varieties should have at least 9%, mid-maturing varieties 11% or above, and late-maturing varieties 12% of soluble solids. For short-term ventilated storage of duck pears, the average hardness at harvest should be above 3.81 kg/cm^2, and for refrigerated fruits, it should be above 4.67 kg/cm^2.

(3) Seed color: As the fruit matures, seeds gradually change from white to brown, floral or completely black from the tip.

(4) Stalk shedding: In the later stages of growth, the base of the stalk forms a separation layer, making it easy to pick.

3.5.2 Harvesting methods

Harvesting is done in stages, with 2–3 rounds of picking based on fruit maturity and practical considerations. Fruits on the outer periphery and those exposed to sunlight are harvested first, followed by those in the inner cavity and shaded areas. Pears have high water content, thin skin, and delicate flesh, so it is recommended to wear gloves during harvest, handle the fruits gently, and cut off excessively long or thick stems to minimize mechanical damage during picking and transportation.

3.5.3 Post-harvest treatment

After harvesting, removing surface dirt from the fruit and calyx cavity is needed. If possible, it is suggested to clean the fruits in a wash tank and dry them. Mechanical or manual grading can be performed. Before packaging, considering the target market is needed. And it is recommended to pre-cool the fruits as quickly as possible to eliminate field heat, store them in refrigerated warehouses, and prepare them for sale. Refrigeration temperatures are set at 1–4 °C, with humidity maintained at around 90% based on the target market distance and market time. Regular ventilation is essential. During transportation, it is advisable to utilize a cold chain to preserve freshness.

References

CAO Y F, LIU F Z, HU H J, et al., 2006. Standardization of Pear Germplasm Resources

Description and Data Standards [M]. Beijing: China Agriculture Press: 20−35.

HE Y Y, CHEN S Y, LI Y P, et al., 2020. Current status and countermeasure analysis of pear industry development in Yunnan [J]. China Fruits & Vegetable, 40(12): 63−66.

LI X G, ZHANG S L, 2020. Pear Tree Chronicle of China [M]. Beijing: China Agriculture Press: 205−307.

SU J, CHEN X, LI L, et al., 2016. New variety of red sand pear 'Caiyunhong' [J]. Journal of Horticulture, 43(S2): 2687−2688.

SHU Q, QIU M H, 2005. Current status of Yunnan pear industry and industrial development [J]. Agricultural Science and Technology (1): 25−27.

ZHANG S L, 2013. Pear Science [M]. Beijing: China Agriculture Press: 417−419.

ZHANG S L, TAO S T, ZHOU Y H, 2010. Current Status of Pear Production, Processing, Trade, and Basic Trends in Industrial Development [M]. Beijing: China Agriculture Press: 1−7.

ZHANG S L, XIE Z H, 2019. Current status, trends, problems, and suggestions for countermeasures in China's pear industry [J]. Journal of Fruit Science, 36(8): 1067−1072.

ZHANG Y X, 2011. General Introduction to Fruit Tree Cultivation [M]. 4th edition. Beijing: China Agriculture Press: 58−99.

❖ Key Techniques for Peach Cultivation

1 Overview of the development of the peach industry

The peach (*Prunus persica* L.) is a perennial, medium-sized deciduous tree in the Rosaceae family, subgenus *Amygdalus* of the genus *Prunus*. With a long history of cultivation in China, peaches have been a symbol of auspiciousness and longevity since ancient times. The fruit has brightly colored skin, diverse shapes, and a juicy or crisp texture. Peaches are rich in nutrients, containing sugars, proteins, fats, organic acids, calcium, phosphorus, iron, vitamin C, and vitamin B. In addition to being consumed fresh, peaches can be processed into jam, preserves, dried fruit, juice, wine, and canned goods. Furthermore, according to traditional Chinese medicine works such as the *Compendium of Materia Medica*, various parts of the peach tree, including the roots, leaves, flowers, fruits, and seeds, have medicinal applications. Currently, the peach gum secreted from the trunk has also gradually become a favored nutritional health product among consumers.

Peaches are native to western China, with Yunnan being one of the origin regions. Peach trees favor light and have strong adaptability, making them suitable for cultivation throughout the province of Yunnan. At present, peach tree production in Yunnan is primarily distributed in mountainous and semi-mountainous areas. As of the end of 2021, the planting area of peaches in Yunnan Province reached 834 600 mu, with a total output of 842 000 t, ranking seventh in the country for planting area. The cultivation of peach trees has become an important means of adjusting the agricultural industry structure in Yunnan's mountainous and semi-mountainous areas, as well as increasing farmers' incomes. In particular, the early-ripening peaches that enter the market from April to May, and the late-ripening peaches that mature from November to December, arrive during the off-season for peach production in other provinces and foreign peach-producing regions, providing a distinct market competitive advantage. The peach industry in Yunnan Province enjoys spatial and temporal advantages, with significant market prospects.

2 Main cultivars and characteristics

Yunnan has a variety of peach cultivation, with a predominance of hairy peaches and nectarines, and fewer cultivation of flat peaches. In terms of ripening periods, early-maturing and mid-early-maturing peaches ripen from late April to early July, and late-maturing peaches ripen from late September to early December, occupying a larger share of the market. The main cultivated peach varieties currently include the following.

Chunxue: Marketed in Yunnan under names such as Zitao、Hongtao and Hongxuetao, this new early-maturing peach variety was introduced from the United States by the Fruit Tree Research Institute of Shandong Province in 1998. The tree has an open canopy and vigorous growth, producing large and round fruits with an average single fruit weight of 180 g and up to 280 g. The fruit has a purple color when not bagged and a light red color when bagged. The white flesh is juicy, crisp, and has a soluble solids content of over 12%, sweet flavor, and strong aroma. And it has a sticky kernel and is resistant to storage and transport. It ripens in mid May to early June in Kunming and early May in the Luxi, Shilin, and Southern Yunnan regions, making it suitable for cultivation in areas with minimal winter and spring frosts and good sunlight resources.

Xiacui: This mid-maturing peach variety, bred by the Horticultural Institute of Jiangsu Academy of Agricultural Sciences, has a strong and semi-open tree growth. It begins to blossom from late February and comes to full flower in mid March. Its pollen is fertile, and the flowering period is not regular. It has high fruit setting and yield. It is relatively unaffected by aphids and leaf curl disease. There are some deformed flowers, but most of them can be self-thinning, and the rate of deformed fruits after fruiting is less than 10%. Individual differences are obvious. The oval fruits have over 90% coloration, with white, flavorful, and crisp flesh containing over 12% of soluble solids. With an average single fruit weight of over 150 g, it begins ripening in mid-to-late June in Central Yunnan, with a long hanging period and a harvest period of 30–45 d, which is resistant to store and transport. It has good adaptability and resistance to pests and diseases and a stable high yield which achieves 1 000 kg above when saplings are in fall fruit period, making it suitable for cultivation in central Yunnan Province, where the rainy season starts relatively late.

Jizao No. 518: This early-maturing nectarine variety is harvested and enters the market in late April to early May in the Shilin area of Kunming. The nearly round-shaped fruits, with an average single fruit weight of 90–120 g and a maximum single fruit weight of 180 g, have a vivid red color. Its flesh is white, juicy and sweet, and has a crunchy texture with good resistance to storage. With a high fruit set rate and good yield, it has a hanging period of around 20 d after ripening.

Zhongyou No. 5: An early-ripening peach variety was developed by the Zhengzhou Fruit Research Institute of the Chinese Academy of Agricultural Sciences. The large, highly productive fruit has strong adaptability, with white flesh that is firm, sweet with a moderate fragrance. The fruit development period is about 72 d, with an average single fruit weight of 166 g and a maximum single fruit weight of 270 g. The skin is green-white, with most of the fruit surface or all showing a rose-red color, making it very attractive and of excellent quality. Its fruit stone is clinging to the flesh. It matures from mid to late May in the Shilin area of Kunming.

Yingzui Honey Peach: Also known as Yingzui Peach, Kaiyuan Yingzui Peach, or Kaiyuan Honey Peach, it is a mid-ripening peach variety. It is a nationally designated geographical indication product. The vigorous tree has an open canopy. The fruit is round with a protruding top like an eagle's beak, and the two halves of the fruit are symmetrical. When it ripens, the fruit has a greenish skin turning purplish-red on the sunny side, and the flesh is green or whitish-green. The fruit is juicy, with a soluble solids content of over 16%, and has good resistance to storage. It's suitable for cultivation in regions with abundant winter-spring light and heat resources, dry climate, and a late onset of the rainy season.

Zhonghua Longevity Peach: It is a late-ripening peach variety. Initial poor performance in Lijiang led to a significant increase in quality and yield by grafting onto local stone walnut rootstocks, with extensive promotion by planting enterprises, forming a brand. The tree is vigorous and yields an average single fruit weight of 320 g, with a maximum single fruit weight of over 1 000 g. The nearly round fruit has a protruding top, deep sutures, milky-white flesh, a clinging stone, and a soluble solids content of 16%. It is suitable for cultivation in regions with low winter temperatures and sufficient chilling requirements for peach trees.

Zhonghua Winter Peach No. 2: It is a particularly late-ripening variety, introduced by the Donghong Green Industry Co., Ltd. in Kaiyuan, and cultivated across the entire province. The strong-growing tree is upright with many flower buds and a high rate of self-pollination, requiring significant thinning of the fruit. The fruit is nearly round, with a protruding top and obvious suture, and the two halves of the fruit are symmetrical. The average single fruit weight is 220 g and the maximum single fruit weight is 435 g. The flesh is whitish with a reddish hue near the fruit stone, crisp, free-stone, and has a soluble solids content of over 16%. The fruit has good storage resistance, early and abundant yields. It is suitable for cultivation in regions with low rainfall in November, good sunlight conditions, and light or no frost.

Fodu Winter Peach No. 1: It is a particularly late-ripening variety. A bud mutation of Zhonghua Winter Peach No. 2 is jointly developed by the Horticultural Research Institute of Yunnan Academy of Agricultural Sciences and Binchuan Fodu Winter Peach Cooperative. The tree has a wide open-canopy and strong growth vigor, and high budding and branching capabilities. The fruit, which is mainly borne on medium and short fruiting branches, has strong self-pollinating ability, a high fruit set rate, and good yield performance. Currently, the main cultivation area is in Binchuan County, Dali Prefecture and its neighbouring areas, where it buds from late February to early March, flowers in the mid March lasting for 10−16 d, matures in the mid-December with fruit development period of 260 d and falls leaves in the early January with an annual growth period of 321 d. The fruit is large, with an average single fruit weight of 220 g and a maximum single fruit weight of 589 g, nearly round with a flat top. When bagged, the fruit skin has a white base color, while the sunny side becomes bright red with over 50% coloration. The fruit flesh is white, non-melting, and has a bright red color near the fruit stone. The fruit has good marketability, a soluble solids content of 14.39%, acrisp texture, and good quality. It has good storage resistance and is suitable for cultivation in regions with minimal rainfall, warmth, good light conditions and no or light frosts from November to December. It is not suitable for regions with a short frost-free period and low autumn-winter temperatures.

3 Cultivation and management techniques

3.1 Planting planning and orchard construction

Peaches are light-dependent plants, and Yunnan, located on a low latitude plateau, has abundant sunlight, making it suitable for peach orchards to be established in areas with no obstruction to sunlight. Based on the geographical and climatic conditions of Yunnan Province, peach orchards should be established in regions with an altitude of 1 600−2 400 m, convenient transportation, and sufficient water supply for irrigation in the winter and spring seasons, such as terraced areas, semi-mountainous areas, and mountainous areas. The proposed orchard areas should have loose soil. These areas, such as valleys, low-lying areas, swamps, and other areas prone to water accumulation, severe winter and spring frosts, frequent late spring cold spells, rainy days during fruit ripening, and hail-prone areas are not suitable for high-quality peach production and are not recommended for orchard establishment.

Peach orchards should be equipped with roads that allow for agricultural operations and transportation vehicles for harvesting. It is advisable to integrate the irrigation and drainage system with road planning. For flatland orchard construction, a north-south orientation is preferred, while in hilly areas, the size of the orchard plots should be determined based on the terrain, and the orientation should follow the contour lines. For plots with a slope greater than 15°, terraces with a width of 2.5−10 m should be formed, slightly inclined inwards, with drainage ditches on the inner side.

The type of planted species should be chosen based on the target market, the environmental conditions of the cultivation site, and other factors. Peach trees in dormancy require an accumulated number of hours below or equal to 7.2 °C for normal flowering and fruiting. In Yunnan province and its southern regions, where the winter temperatures are relatively high, it is advisable to plant varieties with low chilling requirements. It is recommended to obtain seedlings from reputable nurseries to ensure quality and purity.

3.2 Pruning and shaping

For large-scale productive peach orchards (50 mu and above), it is recommended to adopt the wide-row and narrow-canopy cultivation model to facilitate agricultural operations, promote environmental ventilation and light transmission, reduce pest and disease occurrence, and improve fruit quality. For small-scale household peach orchards, a cultivation model that emphasizes convenience of operation and ventilation and light transmission is preferred. In Yunnan, the open-vase and "Y" shape are two suitable tree shapes for peach cultivation. The choice of tree shape and planting density should be based on planting and cultivation requirements, site conditions, water and fertilizer conditions, and management levels. For the open-vase shape, the planting distance is 3–4 m, with row spacing of 4–5 m. For the "Y" shape, the individual distance is 2 m, with row spacing of 4–5 m. The main stem height of the peach tree should be kept at around 60 cm, with the tree height not exceeding 3 m and the fruit-bearing part kept below 2.5 m.

Modern cultivation of peach trees advocates the use of the long pruning technique as the primary simplified pruning technique, which involves not conducting short pruning, not cultivating large fruit-bearing branch groups, and having the tree body composed of the main trunk and fruit-bearing branches with small fruit-bearing branches. The pruning intensity, locations, and methods should be selected based on the geographical and climatic characteristics of the peach orchard, as well as the management model. Summer pruning, performed 2–4 times based on the frequency of rainfall during the growing season in the orchard's region, focuses on thinning out vertical branches and overly dense branches to maintain the tree's ventilation and light. Winter pruning is carried out after the peach tree has shed its leaves and entered dormancy, with a focus on thinning out vertical branches, overly dense branches, diseased and pest-infested branches, and making necessary adjustments to the tree shape.

3.3 Water and fertilizer management

Peach trees, compared to some other fruit trees, exhibit strong drought resistance.

However, to achieve satisfactory economic yields, it is essential not to lack water during cultivation, especially during key periods such as flowering, fruit setting, and fruit enlargement. Adequate soil moisture is required during these critical stages. From bud sprouting to pre-flowering, it is necessary to irrigate once with sufficient water that the water amount ideally penetrates the soil to a depth of approximately 60 cm. During the flowering period, appropriate air humidity is also needed. Too low humidity can lead to pollen drying, poor pollination, and fertilization issues. During the hard stone formation period, the irrigation amount should moisten the soil to a depth of 50 cm. The regions with rainfall during this period can adjust irrigation accordingly. During the fruit enlargement period, it is necessary to pay attention to uniform irrigation, especially in the case of nectarine orchards, to avoid drastic fluctuations in water content, which may lead to fruit cracking. In regions with frequent rainfall and high air humidity during the fruit ripening period, special attention should be given to disease occurrence. Irrigation practices should be tailored based on local economic conditions, water sources, irrigation facilities, and topography. Various methods, such as basin irrigation, furrow irrigation, sprinkler irrigation, and drip irrigation can be employed. If conditions permit, it is suggested to prioritize drip irrigation to enhance water use efficiency.

Peach trees require a balanced application of nitrogen (N), phosphorus (P), and potassium (K) fertilizers, along with an appropriate increase in organic fertilizer application. The fertilizer requirements vary based on soil fertility, early-mid-late maturing varieties, and yield levels. The fertilizer demand for early-maturing varieties is 20%–30% less than that for late-maturing varieties. The coordinated application of calcium (Ca), magnesium (Mg), boron (B), and zinc (Zn) should be emphasized. The ideal yield for peach orchards in Yunnan is 1 500–2 000 kg per mu. It is recommended to apply 3–4 m^3 of organic fertilizer, 12–16 kg of nitrogen (N), 7–9 kg of phosphorus (P_2O_5), and 17–20 kg of potassium (K_2O) per mu annually. Base fertilizer is best applied in the fall, with organic fertilizer as the main component. The key periods for topdressing are around the fruit enlargement period, using high-potassium compound fertilizer primarily. The base fertilizer including all organic fertilizer, 30%–40% of nitrogen, 100% of phosphorus, and 50% of potassium should be applied by using trenching after peach harvest in autumn. The remaining 60%–70% of nitrogen and 50% of potassium should be applied in spring during bud sprouting, hard stone formation, and fruit enlargement periods (1–2 times for early-maturing varieties and 2–3 times for late-maturing varieties). Foliar spraying with an integrated water and fertilizer system or drip irrigation can be used for topdressing. In arid regions, covering individual or entire rows of basin plates with plastic film and employing hole storage of fertilizer and water can be employed for water and fertilizer management. Regardless of the method used, irrigation is necessary after fertilization to facilitate the transport and absorption of nutrients.

3.4 Prevention and control of major diseases and pests

In modern peach orchard, pest and disease management should follow the principle of "prevention as the main approach, comprehensive control". The main focus is on predicting and forecasting major pests and diseases, using physical control, agricultural measures, and biological control as the primary methods. High-efficiency, low-toxicity, and low-residue pesticides are selected to control pests within economically acceptable levels. The selected pesticides should comply with the guidelines and should be rotated and alternated to avoid the development of pesticide resistance.

3.4.1 Scab disease

Based on agricultural control, emphasis is placed on chemical control. The main objective is to control fruit damage, and bagging the fruits whenever possible is encouraged along with pre-bagging chemical control. Since the incubation period of the disease is more than 20 d, timely chemical control is essential.

In terms of variety selection, it is advisable to avoid planting varieties that mature during the rainy season. If choose such varieties, it is suggested to implement bagging to prevent fungal invasion. The orchard should maintain good ventilation and lighting, avoid waterlogging, and promptly remove diseased fruits and branches. Infected plant materials should be buried or burned to reduce field sources of infection.

Chemical control should be carried out from the fruit enlargement period until 20 d before maturity. Pesticides like prochloraz, difenoconazole, mancozeb, or thiophanate-methyl can be used alternately 2–3 times with an interval of about 10 d. If bagging is required, two rounds of chemical treatment should be applied 20 d in advance.

3.4.2 Brown rot disease

Similar to scab disease, a combination of agricultural measures and chemical control is employed for prevention and control. The main objective is to control fruit damage, and pre-bagging chemical control is emphasized. When using pesticides for control, high-efficiency, low-toxicity, and low-residue pesticides are selected to ensure fruit safety.

The orchard should maintain good ventilation and lighting, avoid waterlogging, and promptly remove diseased fruits and branches. Infected plant materials should be buried or burned to reduce field sources of infection.

The control period starts after the flowers have fallen. Pesticides such as prochloraz-manganese chloride complex, oxazole·fluoroamide, and oxazole·metiram can be used alternately 2–3 times with an interval of about 10 d. In the later stages of fruit maturation, the use of the pesticides mentioned above can be continued based on rainfall conditions.

3.4.3 Leaf curl disease

In addition to removing diseased leaves, chemical control should be emphasized, with timely spraying in early spring being a key factor.

Diseased leaves should be promptly removed and buried or burned to reduce field sources of infection.

Additional foliar fertilizers can be applied to promote leaf growth and help the trees recover.

Spraying should be performed 1–2 times after leaf bud emergence but before flower bud opening, with an interval of 7–10 d. Pesticides such as lime sulfur, prochloraz, difenoconazole, carbendazim, or thiophanate-methyl can be used.

3.4.4 Bacterial perforation disease

Based on agricultural preventive measures, emphasis should be placed on effective chemical control.

Diseased leaves, branches, and fruits should be eliminated from the field. They should be buried or burned to reduce the source of bacteria.

Appropriate planting densities and tree shapes should be selected to enhance ventilation and light transmission within the trees. Timely drainage to reduce field humidity is essential. Rational fertilization practices should be employed, emphasizing organic and compound fertilizers while avoiding excessive use of nitrogen-based fertilizers.

Spraying should commence after leaf expansion, repeating the process 3–4 times. Available medications include Chunlei quinolinic copper, Zhongsheng mycins, quinolinic copper, and thiediazole copper, with an approximate 10 d interval between applications.

3.4.5 Peach aphids

During the emergence of leaf buds, prolonged dry spells can lead to severe infestations of peach aphids. It is important to strengthen the management of soil, fertilizer, and water in peach orchards, and implement the growth of natural grass or sow grass within the orchard to enhance biodiversity, thereby safeguarding aphid predators like syrphid flies, asaphes vulgaris walker, and ladybugs. Spraying insecticides should be avoided when the ratio of beneficial insects to harmful ones reaches 1 : 30 or higher. For effective results, it is advisable to apply treatments during the budding of peach leaf buds and the peak hatching period of overwintering eggs. During the budding period, combining treatments for various pests and diseases, 3–5°Bé of sulfur lime mixture should be sprayed. During the initial hatching period of overwintering eggs, biological agents such as matrine and azadirachtin should be applied. Upon discovering aphid infestations, along with spring pruning, branches and leaves infested with aphids should be removed and disposed of properly. It is recommended to use low-toxicity pesticides like imidacloprid, acetamiprid, and emamectin benzoate. If resistance develops to certain agents like imidacloprid, alternatives such as imidacloprid, acetamiprid, thiamethoxam, pymetrozine, spirotetramat, ethyl acetate, and highly effective cyhalothrin can be selected.

3.4.6 Mulberry mealybug

Fertilization and water management should be enhanced to fortify tree vigor. It is proposed to combine winter pruning to remove branches severely infested by scale insects and burn them. In addition, a stiff brush can be used to remove scale insects from branches, primarily focusing on branches aged 2–3 years.

Grass could be sown within the orchard to increase biodiversity, protecting the natural enemies of mulberry mealybugs. Indiscriminate pesticide application during the adult stage of scale insects should be avoided to prevent harm to their natural predators.

Chemical control during the bud stage can employ sulfur lime mixtures to combat dormant stage mulberry mealybugs. During the emergence and peak infestation of nymphs, pesticides such as highly effective cyhalothrin can be used for control.

3.4.7 Peach leafhopper

The peach leafhopper overwinters in evergreen plants near peach orchards. After winter, it is essential to thoroughly clean up fallen leaves and debris in the orchard and burn them. Before the adult leafhoppers emerge, it is suggested to promptly remove any curled bark. Combining winter and spring pest control measures, stone lime sulfur mixture or other insecticides can be sprayed on surrounding evergreen plants.

It is needed to enhance water and fertilizer management to strengthen tree vigor. Pruning appropriately in winter and summer can prevent the orchard rows and tree canopy from becoming too dense.

Grass in the orchard can attract natural enemies of the peach leafhopper, such as small black spiders, ladybugs, seven-spot ladybirds, turtle-shaped ladybirds, large lacewings, and spiders. If natural enemies are scarce, releasing them strategically can be considered.

Chemical control is recommended during three key periods: when overwintered adults migrate, during the peak hatching period of the first generation nymphs, and at the peak nymph occurrence period. It is preferable to spray insecticides on peach trees and grassy areas on overcast or late afternoon sunny days. Recommended insecticides include imidacloprid, fenvalerate, avermectin, bifenthrin, isoprocarb, and methyl. During high-temperature seasons, insecticides with strong internal absorption like acetamiprid or thiamethoxam can be selected.

3.4.8 *Pseudococcus comstocki* Kuwana

Control measures should start with winter orchard cleaning. After harvest, it is advised to promptly remove fallen leaves and weeds, especially diseased fruits, leaves, and branches, which should be buried or burned. Manually scraping off remaining scale insects on tree trunks can be employed. During winter pruning, thinning out dense branches can ensure good orchard ventilation and light penetration. After leaves fall, it is suggested to mix with chlorothalonil and fenvalerate when whitewashing tree trunks. In winter, spraying the orchard with lime sulfur 1-2 times is preferred.

Applying efficient cyhalothrin (mixed with a small amount of laundry powder) before bagging the fruits is recommended, and reapplying if the interval is long or in case of rain.

3.5 Harvesting and post-harvest treatment

Peaches should be harvested at the right time, considering their variety characteristics, purpose, market distance, and transportation methods. It is suggested to choose a cooler time of the day for harvesting.

It is better to wear gloves during harvesting, handle the fruits gently, and stack them in picking containers to prevent skin damage. Harvest with moderate force, avoiding breaking the fruit stems or damaging the main trunk and branches. Harvested fruits should be stored in a well-ventilated and sanitary place, protected from sun exposure, rain, and animal damage.

Based on their characteristics, the fruits should be graded, inspected as required, and then proceed with packaging. It is better to choose packaging materials that minimize the risk of mechanical damage during transportation. The label of the outer box should include clear information such as fruit grade, weight, specifications, quantity, and origin.

Depending on the type of peach and market demand, short-term storage is possible, but long-term storage is not recommended. Rain-season ripe peaches should be consumed locally. Early, mid, and late-season peaches should be harvested on time, pre-cooled appropriately, and then transported for long distances.

References

CAO S Y, XIE S X, LU X P, et al., 2018. Local Variety Atlas of Chinese Peaches [M]. Beijing: China Forestry Publishing House.

CHEN Y X, TANG M W, LIU X W, et al., 2020. Standardized production management techniques for high-quality Lijiang snow peaches [J]. South China Agriculture, 14(27): 24–25, 29.

MA Z S, JIA Y Y, WANG Y H, et al., 2014. Peach Industry Technology in the Central and Southern Regions of Hebei [M]. Beijing: China Agriculture Press.

National Peach Industry Technology System, 2016. Research on Sustainable Development Strategy of China's Modern Agricultural Industry: Peach Volume [M]. Beijing: China Agriculture Press.

WANG L R, 2012. Genetic Resources and Varietal Atlas of Chinese Peaches [M]. Beijing: China Agriculture Press.

WANG L R, ZHU G R, 2005. Description Specifications and Data Standards for Peach Germplasm Resources [M]. Beijing: China Agriculture Press.

YU F, ZHANG Y, LU L, et al., 2017. New late-maturing peach variety 'Fodu winter peach No. 1'[J]. Journal of Horticulture, 44(S2): 2627–2628.

YUAN B, XIONG M G, YIN M, 2017. Large-Scale Production and Operational Management of Peaches [M]. Beijing: China Agricultural Science and Technology Press.

❖ Key Techniques for Strawberry Cultivation

1 Overview of industrial development

Strawberries (*Fragaria* × *ananassa* Duch.) have been widely cultivated small berries globally. They boast vibrant colors, a soft, juicy texture, and a delightful sweet-tart taste, making them one of the top ten fruits worldwide. They are esteemed as the "first fruit of early spring" and the "queen of fruits," holding significant economic, nutritional, and cultural value. Strawberries have claimed the second spot on the World Health Organization's list of top ten foods. Through continuous breeding by breeders, there are now more than 2 000 cultivated varieties of strawberries, enabling their widespread cultivation in the vast majority of countries worldwide due to their robust adaptability. China is the world's leading producer and consumer of strawberries. According to statistics, in 2018, China cultivated strawberries on 2.56 million mu, yielding 5 million t. China, characterized by its extensive geographical expanse, adopts a dual approach to strawberry cultivation—employing both open-field and protected cultivation methods. In the northern regions, protected cultivation is predominantly facilitated through forcing culture in greenhouses and semi-forcing culture in plastic greenhouses. Meanwhile, in the southern regions, cultivation is primarily carried out through forcing and semi-forcing culture in plastic greenhouses. Meanwhile, in areas below 1 700 m altitude in Yunnan province, open-field cultivation takes precedence.

2 Requirements for environmental conditions

The environmental conditions that influence the growth and development of strawberries are primarily related to soil, temperature, water, and light.

(1) Soil. Strawberries have a strong adaptability to various soil types, and they can be grown in most soils. However, for optimal cultivation, it is advisable to choose fertile and well-drained slightly acidic soil. This choice contributes to the production of higher-quality strawberry fruits. Strawberries may experience poor growth in marshy areas, saline-alkali soils, or heavy clay soils. The root system of strawberries has a limited tolerance to high fertilizer concentrations in the soil, and the lower the electrical conductivity in the soil, the

better the development of the strawberry roots.

(2) Temperature. Strawberries demonstrate a relatively rapid adaptability to temperature changes. They exhibit a certain degree of cold resistance, being able to withstand low temperatures of −15− −10 ℃ during dormancy. Additionally, they have a degree of heat tolerance, enduring high temperatures of 40−45 ℃ during the summer. The stolons of strawberry typically appear a lot at temperatures between 20 ℃ and 25 ℃, with slower and limited stolon development when the temperature falls below 15 ℃ or exceeds 28 ℃. The differentiation of strawberry flower buds generally takes place at temperatures between 10 ℃ and 17 ℃. Extreme temperatures, whether excessively high or low, prove detrimental to the differentiation of flower buds. Elevated temperatures result in a shortened fruit ripening period, yielding smaller and more acidic fruits. Conversely, lower temperatures lead to an extended ripening period, producing larger and sweeter fruits, albeit with a delayed maturation that will impact overall yield.

(3) Moisture. The majority of strawberry roots are mainly distributed within the top 20 cm of soil. Strawberries are neither tolerant to waterlogging nor resistant to drought, so it is crucial to pay attention to timely irrigation and drainage during production. Different growth stages have varying moisture requirements. Generally, during the flower bud differentiation stage, 60% of field moisture capacity is deemed suitable, approximately 70% during flowering, and around 80% during the fruit swelling period. Insufficient irrigation can inhibit growth, resulting in decreased yield. Excessive irrigation not only diminishes fruit hardness and flavor but also impacts root growth, potentially leading to diseases. Soil moisture significantly influences strawberry production, and air humidity is equally vital. Typically, an air humidity of around 60% is suitable, and during the flowering period, it should not exceed 80%. Excessive humidity can affect pollen dispersal and fertilization, leading to deformed fruits and promoting the occurrence of gray mold. Low humidity can impede plant growth and affect fruit appearance.

(4) Light. Strawberries thrive in conditions of both light and shade. The intensity, duration, and spectral composition of light all influence the growth and development of strawberries. Abundant light results in robust plants with large, dark green leaves, high-quality fruits, and a high yield. Insufficient light leads to weaker plant growth, characterized by slender stems, small and sparse flowers, delayed fruit ripening, small and poor-quality fruits, and a low yield. The light compensation point for strawberry leaves generally ranges from 5 000 lx to 10 000 lx, and the light saturation point typically falls between 20 000 lx and 30 000 lx. It is crucial to ensure an adequate amount of light during cultivation; However, prolonged exposure to excessively intense light may scorch the leaves. The duration of light

exposure affects plant growth, flower bud differentiation, dormancy, and so on. Generally, low temperatures and short sunshine conditions favor flower bud differentiation, while higher temperatures and longer sunshine conditions promote the development of stolons.

3 Main cultivars and characteristics

According to the different responses to photoperiod and fruiting period, strawberry cultivars are generally classified into three types: short-day, long-day, and mid-day. Currently, the main type of strawberry cultivated in China is the one-season type, also known as winter strawberry or short-day strawberry. The harvest period is mainly concentrated from December to May of the following year. Summer strawberries, also known as everbearing or four-season strawberries, are primarily mid-day varieties. They are not sensitive to photoperiod and can continuously flower and fruit in temperatures ranging from 4 °C to 29 °C. In cool regions, they can achieve fresh fruit production in summer and autumn, with the fruit harvest period from May to November.

Based on the variety's origin and characteristics, strawberries can be categorized into European and American varieties, as well as Asian varieties. European and American strawberry varieties exhibit significantly larger sizes compared to their Asian counterparts. The berries from European and American varieties are known for their more acidic taste, making them primarily suitable for deep processing. Conversely, Asian varieties are recognized for their sweeter flavor, making them well-suited for fresh consumption.

At present, in China, there are several extensively cultivated varieties of strawberries, specifically as follows.

3.1 Winter strawberry (Short-day) varieties

Zhang Ji: This is a Japanese variety. The fruits are elongated and cone-shaped, with a vibrant red color, presenting a well-proportioned and tidy fruit shape. The average single fruit weight is 18 g, with a soluble solids content ranging from 9% to 14%. The fruit attains exceptional quality when fully ripe. Due to the soft and juicy nature of this variety, it has limited resistance to storage and transport, so it is advisable to harvest when the fruit is 70% ripe for long-distance transportation. This variety exhibits strong resistance to wilt verticillium and gray mold, while its resistance to powdery mildew and anthracnose is relatively weak. Therefore, careful attention is needed during cultivation for disease prevention and control.

Hong Yan: A Japanese variety known for its robust plant growth and upright plant type. The fruits are cone-shaped, featuring a bright red color on both the surface and interior, exhibiting a glossy appearance, and a red fruit core. The fruit has a well-proportioned and orderly shape with consistent coloring and minimal occurrence of malformed fruits. Average single fruit weight is 21 g, offering a delightful balance of sweetness and tartness, accompanied by a strong fragrance, resulting in exceptional quality. The fruit maintains a moderate hardness and demonstrates good resilience during storage and transport. The variety has short time intervals between each cluster of flowers, ensuring robust continuous yields and excellent productivity. While it is less tolerant to heat, humidity, and drought, and faces challenges in seedling cultivation, it boasts strong cold resistance, performing well in continuous fruiting under low winter temperatures.

Xiang Ye: The plants exhibit a tall and relatively upright plant type, characterized by vigorous growth and elongated inflorescence. With shallow dormancy, it easily transitions into the flowering phase, exhibiting abundant blossoms and strong continuous fruiting capabilities, resulting in early maturity and high productivity. The fruits are conical in shape, with an average single fruit weight of around 25 g, and some fruits reach a maximum single fruit weight of over 100 g. However, the larger fruits may show signs of hollowness. The fruit skin is a vibrant orange-red, while the flesh is a creamy yellow, offering a crisp and tender texture with a pronounced and fragrant honey-like aroma. With the sugar content ranging between 10% and 12%, the taste is exceptional, though quality noticeably diminishes at higher temperatures due to a shorter growth period. The fruit's firmness ensures resilience during storage and transportation. The plants exhibit strong resistance, particularly against anthracnose and powdery mildew. However, they tend to have a relatively sparse number of stolons, resulting in a lower seedling coefficient.

Christmas Red: A Korean variety with an upright plant type. The surface of the fruit is smooth, highly glossy, and exhibits a bright red color. The average single fruit weight in the first and second inflorescences is 35.8 g, with the maximum single fruit weight reaching 64.5 g. The coloring of the fruit surface under the calyx is moderate, with backscrolling sepals, green in color. The seeds slightly protrude on the fruit surface, displaying a yellow-green color combination with moderate density. The flesh of the fruit is orange-red, with a white pith heart, and without any hollow areas. The flesh is delicate, with a velvety texture, and has a sweet taste. It contains soluble solids ranging from 10% to 13.1%. It possesses moderate resistance to storage and transportation.

Sweet Charlie: This American cultivar features nearly circular, thick leaves with a prominently serrated and blunt-toothed leaf margin. The plant is robust with a well-developed root system. This strawberry variety demonstrates remarkable resistance to gray mold and powdery mildew, and it displays strong resilience against other diseases and pests, with few diseases and pests occurring. With broad adaptability, it experiences a relatively short dormancy period of around 45 h. The fruit is cone-shaped, uniform in size, and characterized by minimal occurrence of malformed fruits. The surface displays a deep red, glossy appearance, while the seeds are yellow. The flesh, pink in color, emanates a robust fragrance and contains 11.9% of soluble solids. It exhibits significant hardness and is resilient during storage and transportation.

Tao Xun: This Japanese variety is the world's first cultivated strawberry with white fruits. The heart-shaped fruit has a slightly pale surface with a hint of pink, and its flesh is white. The fruit is known for its soft and juicy texture, delivering a distinct flavor accompanied by a rich fragrance reminiscent of honey peach. The plant grows upright, and the back of its leaves is densely covered with fine hairs. They exhibit robust growth characteristics. Notably, this variety is the sole decaploid material among current strawberry cultivars. In terms of production, the drawbacks include a lengthy cycle for flower bud differentiation and a delayed fruit maturation period. Furthermore, there is a notable high deformity rate observed in the first and second-order fruits of the first inflorescences.

Miao Xiang No. 7: The variety was developed by Shandong Agricultural University. The fruits are conical in shape, with an average single fruit weight of 35.5 g, a 25.9% increase compared to the control variety, Hong Yan. The fruit surface is vibrant red and glossy, and the flesh is fine-textured, vividly red, with a strong fragrance and high hardness. Under conditions of forced culture, it yields an average of 3 427 kg per mu.

Snow White: The variety was bred by Beijing Academy of Agriculture and Forestry Sciences, featuring a small plant type and moderate-to-weak growth vigor. The fruits are of medium size, with a maximum single fruit weight of 48 g. They are cone-shaped or wedge-shaped, with a white color and a glossy surface. The seeds are red and lie flat on the fruit surface. The flesh and core of the fruit are white, with small hollow spaces within the fruit. It contains soluble solids ranging from 9% to 11%. It possesses a unique flavor and strong resistance to powdery mildew.

Yue Xin: It is a cultivar developed by Zhejiang Academy of Agricultural Sciences. The plant exhibits robust growth, standing upright and displaying resilience to low temperatures and weak light. It possesses a strong capability for the emergence of stolons, has shallow dormancy, and matures early, making it well-suited for forcing culture. The fruit is of medium size, taking on a short conical or heart-shaped appearance. The fruit surface is smooth, with a light red color and uniform coloring, containing soluble solids ranging from 12% to 14.5%. It offers an exceptionally pleasing flavor, with a well-balanced sweet and sour taste, an enticing fragrance, and a thick peel that imparts durability during storage and transportation.

Furthermore, it exhibits strong resistance to anthracnose, gray mold and powdery mildew.

Ning Yu: This is a cultivar bred by Jiangsu Academy of Agricultural Sciences. This variety features a semi-upright plant structure with robust growth and a strong capability for stolon development. The inflorescence is of moderate length, resulting in a high fruit-setting rate. The cone-shaped red fruits are highly glossy, with a delightful sweet taste and a soluble solids content of 10.7%. The average single fruit weight in the first and second inflorescences is 24.5 g. The fruits are uniform, showcasing robust continuous flowering and fruiting, with early yield occupancy rate ranging from 40.5% to 57.9%. It exhibits strong resistance to both heat and cold, has shallow winter dormancy, resists dwarfing, and is less prone to excessive growth in high spring temperatures. Additionally, it demonstrates moderate resistance to powdery mildew and anthracnose.

Fen Yu: It is a cultivar developed by Hangzhou Academy of Agricultural Sciences. This variety features conical-shaped fruit with a pink surface and solid, white flesh, occasionally with a small or no hollow core. The average single fruit weight in the first-order of the first inflorescence is approximately 28.0 g, while in the second inflorescence, it is around 17.1 g. The soluble solids content ranges from 13.1% to 18.0%. The fruit exhibits moderate hardness and excellent storage and transportation resistance. This early-maturing variety demonstrates robust continuous flowering and fruiting capabilities.

3.2 Summer strawberry (mid-day) varieties

Monterey: It is a high-yielding cultivar bred by the University of California. Under conditions of forcing culture, it exhibits early fruiting with vigorous plant growth. The average single fruit weight is 33 g, reaching a maximum single fruit weight of 60 g. The fruit boasts exceptional quality, and demonstrates robust disease resistance.

Saint Andrews: This is a cultivar bred by the University of California, USA. The fruit is conical, average single fruit weight is 32.3 g and maximum single fruit weight is 125.6 g. Its surface is red and glossy. With high yield, firmness, and excellent storage tolerance, this variety is well-suited for transportation. This variety not only withstands high temperatures, but also thrives in them. It shows considerable resistance to powdery mildew and wilt verticillium.

Portola: This variety, cultivated by the University of California, USA and developed in 2008, falls under the long-day type. It exhibits a remarkable capacity for continuous flowering and fruiting, resulting in a high yield. The fruit is characterized by high uniformity, with single fruit weight concentrated in the range of 10−25 g.

4 Cultivation and management techniques

4.1 Planting

For planting, it is recommended to use plug seedlings, which are planted in a double-row pattern with a staggered and interlaced arrangement, maintaining a plant spacing of 18–25 cm. During planting, ensure that the young seedlings have their backs arched outward. This helps the inflorescence grow outward, facilitating insect pollination and harvesting. If using bare-root seedlings, it is advisable to plant them on cloudy days or during cooler temperatures. After transplantation, it is advisable to spray water onto the plant leaves within the first week to prevent dehydration and wilting. Regardless of whether plug seedings or bare-root seedlings are chosen, it is essential to water thoroughly after planting to ensure sufficient soil moisture.

4.2 Pruning for shaping

(1) Pruning of older leaves. Once the plants have successfully survived after transplantation and have sprouted more than two new leaves, it is crucial to promptly carry out the first round of leaf pruning. The process of removing older leaves is pivotal throughout the entire cultivation period, with specific requirements varying at different stages. Prior to the onset of the first inflorescence, when temperatures are still elevated, and plant growth is vigorous, maintaining four leaves per plant is essential to stimulate the emergence of the first inflorescence and the differentiation of the second. Subsequent to the emergence of the first inflorescence, to ensure plants have sufficient nutritional resources to support the substantial energy consumption associated with later flowering and fruiting, it is advisable to retain 6–7 leaves per plant. As temperatures rise noticeably post the Spring Festival, accelerating plant growth, it is recommended to keep 5–6 leaves per plant.

(2) Management of lateral buds. After the planted seedlings successfully have taken root, lateral buds will continuously sprout throughout the entire winter and spring production period, especially in white fruit varieties like Taoxun and Yutu, where the quantity of lateral buds is relatively higher. An excessive number of lateral buds can lead to the dispersion of plant nutrients, resulting in more flowers but smaller fruits and a reduction in commercial value. In cultivation, it is advisable to remove all lateral buds except for one (the growing point). If an increase in early yields is desired, at most, one lateral bud can be retained, allowing for two shoots (growing points).

(3) Management of stolons and inflorescences. stolons are organs for nutritional growth, and their sprouting during fruiting is an unnecessary drain on nutrients, so they should be promptly removed. Managing inflorescences is also crucial in the production process, as many varieties require flower thinning to concentrate plant nutrients and increase the average weight of fruits. After thinning a single inflorescence, it is advisable to retain a maximum of 7 fruits (one primary fruit, two secondary fruits, and four tertiary fruits). In case of weak plant growth, ensure subsequent inflorescences' productivity by keeping 3–5 fruits per single inflorescence. Additionally, after harvesting fruits, promptly remove aging inflorescences to support plant growth.

4.3 Water and fertilizer management

The management of water and fertilizers should be tailored based on factors like the growth stage of plants, climatic conditions, and production objectives. After the successful planting of young seedlings, it is important to control water and fertilizers appropriately to prevent excessive plant growth that might delay flower bud differentiation. After confirming the completion of flower bud differentiation, it is crucial to promptly enhance water and fertilizer levels to stimulate nutritional growth, facilitating the subsequent flowering and fruiting of the plants. Seasonally, during winter with lower temperatures and reduced plant transpiration, irrigation should be reduced, while in summer, it should be increased. From the perspective of production practice, the need for watering plants is not entirely determined by soil moisture alone. A crucial indicator to determine whether watering is needed is to observe if the edges of plant leaves exhibit "guttation" in the morning. If water droplets are present, indicating bleeding exudation or guttation, it suggests sufficient water and strong root absorption. Conversely, the absence of water droplets suggests water deficiency or poor water absorption, requiring timely watering. In terms of plant growth and development, strawberry fruit production encompasses processes like nutritional growth, flower bud differentiation, and flowering. During the cultivation of strawberries, there is a high demand for fertilizer. It is essential to maintain a balanced ratio of nitrogen, phosphorus, and potassium with a suitable proportion of $1 : (0.25-0.4) : (1.3-1.8)$ (by mass fraction). Additionally, ensure an ample supply of moderate elements such as calcium and magnesium, where calcium fertilizer should constitute more than 50%–80% of the nitrogen fertilizer, and magnesium fertilizer should make up 15%–20%. Micronutrients like boron and iron must not be deficient.

4.4 Prevention and control of diseases and pests

4.4.1 Prevention and control of diseases

4.4.1.1 Powdery mildew

Powdery mildew is the foremost disease in the process of strawberry cultivation, characterized as a fungus disease. The spores of the fungus primarily spread through the air and overwinter in the soil with infected plants and leaves. It spreads most rapidly within temperatures of 15-25 °C, while high temperatures and dry environments impede its growth. Powdery mildew primarily damages the leaves but can also infest stems, flowers, flower stalks, and fruits. Initially, leaves exhibit varying-sized dark spots, followed by the development of a powdery substance resembling thin frost on the undersides. As the disease progresses, reddish-brown lesions form, causing leaf curling and yellowing. Young fruits can be affected, appearing covered in a white powdery substance, leading to halted development and loss of commercial value. In severe cases, the entire plant may perish. Preventive measures primarily encompass agricultural, physical, and chemical control methods.

(1) Agricultural measures of control. By Heat Suffocation, the initial source of powdery mildew infection within the enclosure can be eliminated. It is recommended to choose varieties that are resistant to the disease. In addition to selecting clean seedlings, techniques such as raised-frame cultivation, substrate mulching, and the installation of side and top windows along with ventilation fans are employed in a coordinated manner, to control the microenvironment of the greenhouse regarding temperature and humidity. These measures aim to suppress the occurrence and spread of powdery mildew. It is advisable to implement drip irrigation under mulch, regulating water quantities, and appropriately adjusting the nitrogen fertilizer formula. Timely removal of old and diseased leaves enhances the ventilation around the plants.

(2) Physical measures of control. It is recommended to use sulfur fumigation. 10-15 pieces of 40 W sulfur fumigators can be installed per mu, with fumigation conducted for 1-2 h each night.

(3) Chemical measures of control. Different types of agents are employed at various stages to prevent and control powdery mildew. ①Prior to flowering and fruiting, broad-spectrum protective fungicides like mancozeb, carbendazim, benomyl, thiophanate, thiophanate-methyl are chosen, offering effective protection against strawberry powdery mildew and other diseases. ②During the flowering to fruit swelling period (or diseased period), systemic, fast-acting, and long-lasting fungicides are selected, including triazole fungicides such as triadimefon, myclobutanil, tetraconazole, triflumizole, tebuconazole,

sulfur, carbendazim·sulfur, pyraclostrobin, isopyrazam+azoxystrobin, etc., for effective disease prevention and control. ③ During the fruit harvesting period, it is recommended to use fast-acting fungicides with excellent therapeutic and eradication properties, such as fluopyram+trifloxystrobin (luna sensation), bupirimate, boscalid+azoxystrobin (switch), and fluxapyroxad+pyraclostrobin (kendat), to effectively control the spread and expansion of diseases. When the disease is relatively mild during the fruit harvesting period, it is advisable to use biological fungicides such as fructus cnidii, glucosan, kresoxim-methyl, and biointerferons (pyrimidine nucleoside class antibiotics). These options are safe for strawberry flowers and fruit harvesting, with no adverse effects on growth. For strawberries cultivated with drip irrigation facilities, during the cultivation process, 200 mL of 25% pyraclostrobin SC can be added per mu to the irrigation solution and applied through drip irrigation. Utilizing pyrazofurin's excellent systemicity, the fungicide can effectively be transported to various parts of the plant to control diseases. However, it is crucial to use it sparingly during fruit harvesting to avoid any impact on fruit ripening.

Precautions: Ensure an ample supply of medication, thoroughly penetrate the interior of plant leaves. Note that triazole fungicides exhibit inhibitory effects on plant growth; refrain from consecutive use in winter. During summer, consider employing them to curb excessive plant growth in seedlings.

4.4.1.2 Gray mold

Gray mold, caused by the fungus *Botrytis cinerea*, is a fungal disease. The optimal temperature and relative humidity ranges for the disease's development are 18−23 °C and over 80%, respectively. Conditions such as frequent cold spells, water accumulation on greenhouse culch films, continuous overcast and rainy weather, dense planting, and inadequate ventilation and sunlight can intensify the disease's onset and propagation. This disease primarily damages leaves, flowers, fruit stems, and fruits. It usually begins its infection and growth from wounds or areas of the plant that have died. Initially, it targets and damages withered branches and older leaves, eventually endangering the fruits. Once the flowers or fruits become infected, they quickly decay, leading to rapid and extensive spread throughout the plant. This can significantly reduce yields, with potential reductions exceeding 40% in severe instances.

Preventive measures are as follows.

(1) Promptly and carefully remove diseased leaves, flowers, and fruits, placing them in plastic bags for proper disposal outside the greenhouse. This helps prevent the spread of pathogens to other plants.

(2) Regulate the temperature and humidity within the greenhouse. During the strawberry's flowering to fruit swelling period, the daytime temperature should be maintained

at or above 25 °C, and the nighttime temperature should be kept at or above 12 °C. Simultaneously, efforts should be made to extend ventilation time to maintain the relative humidity inside the greenhouse at 60%–70%.

(3) Minimize the application of nitrogen fertilizer.

(4) Chemical Control. Recommendations suggest employing alternating sprays of boscalid, pyrimethanil, iprodione, flusilazole, polyoxin, procymidone and others. Apply the spray every 7 days, repeating the process three times. In the event of rainy days, utilize procymidone smoke agent for fumigation.

4.4.1.3 Anthracnose

Anthracnose is the primary disease affecting strawberries, transmitted through soil or seeds. It is a fungal disease that, once established, can result in plant death and complete crop failure. The pathogen overwinters in infected tissues or residual diseased material on the ground. Planting strawberries consecutively in the same area allows the pathogen to accumulate, contributing to the increasing severity of the disease. The spread of the disease through infected seedlings is also a major contributing factor to its occurrence.

Preventive measures are as follows.

(1) Select clean seedlings, and before purchasing each batch of seedlings, samples should be taken. Horizontally and vertically cut the short stems to examine if there are any brown disease spots on the cut surfaces.

(2) Disinfect the planting area. before transplanting, use disinfectants like tiazon and calcium cyanamide to sanitize the planting zone. Strawberries should not be cultivated in the infected area for the subsequent 3 years.

(3) After planting, if diseased plants are detected, they should be promptly removed. The suggestion is to use spray treatments with pyraclostrobin, prochloraz, azole ether·flutolanil, difenoconazole, azoxystrobin, dicyano-pyraclostrobin, bromothalonil, and other substances for spray control.

4.4.1.4 Bacterial angular leaf spot

Bacterial angular leaf spot disease is one of the main diseases in strawberry cultivation in Yunnan province in recent years. It primarily impacts the plant's leaves, causing irregular water-soaked spots on the lower leaf surface when infected. As the disease spots enlarge, they assume an angular shape constrained by the small leaf veins, hence it is also referred to as angular spot disease or angular leaf spot disease. When held against the light, these spots appear translucent, and in severe cases, bacterial ooze may be present. Another manifestation of bacterial angular leaf spot disease is more subtle, primarily affecting the vascular bundles. In the early stages, symptoms are minimal, making detection challenging. By the time symptoms such as yellowing and wilting of the leaves manifest, it is typically in the advanced stage. When these wilted seedlings are pulled out, they may break easily with gentle force. Examination of the root surface reveals an intact epidermis but a hollowed medulla. The vascular bundles turn reddish-brown, accompanied by the flow of a fluid resembling pus, albeit without a distinct foul odor. Some plants may also display angular leaf spots on their leaves.

Preventive measures are as follows.

(1) Before transplanting strawberries, clear the fields and surrounding areas of weeds, and dispose of them by burning or wet-composting in a centralized manner. Deeply plow the soil to eliminate crop residues, facilitate the decomposition of diseased and residual organisms, and thereby reduce the sources of diseases and pests. Implement timely pest control measures to minimize plant wounds and decrease the transmission pathways of pathogens.

(2) Practice sensible crop rotation, with a preference for alternating between wet and dry seasons. It is advisable to choose fields with convenient irrigation and drainage systems and establish well-functioning drainage ditches, in order to lower the groundwater level and then ensure no water accumulation after rainfall. Promptly cleaning the drainage system after heavy rains is essential to prevent the retention of stagnant moisture and reduce field humidity. This constitutes a critical measure for disease prevention, as many diseases are provoked by high humidity. Caution in this regard is paramount.

(3) In the initial stages of the disease, cuaminosulfate, copper abietate (effective against both bacteria and fungi), bacillus subtilis, chloroisobromine cyanuric acid, zhongshengmycin, thiodiazole copper, copper quinolate, zinc thiozole can be applied through foliar spraying and root irrigation. It is suggested to apply every 5 d, continuously for 2−3 times. In many cases, leaf spot disease is a mix of fungal and bacterial types. Therefore, a combination of difenoconazole+zhongshengmycin, thiophanate-methyl+zhongshengmycin, kasugamycin can be used for prevention and control.

4.4.2 Prevention and control of pests

4.4.2.1 Red spiders

Red spiders, with their small size, are challenging to identify. They have a preference for extracting juice from young, undeveloped leaves or the underside of leaves, often requiring careful scrutiny or the aid of a magnifying glass for detection. Red spiders have piercing-sucking mouthparts. Initially, their impact on the leaf surface of affected plants manifests as clustered small yellow spots. As their damage intensifies, plants may experience dwarfing and premature aging, with leaves turning reddish-brown and ultimately drying out. In severe cases, one might observe white web-like structures on the leaves of affected plants. Considering the safety of edible fruits, different strategies for prevention and control are implemented before and after flowering.

(1) Before flowering and fruiting, chemical control methods are primarily employed. Using a combination of chemicals like bifenazate+hexythiazox, cyflumetofen (or ethofenazole) +hexythiazox, abamectin+bifenazate, etc. is recommended for prevention. Ensure that the spray is evenly applied to the underside of the leaves with a concentration of 90% or higher. It is advisable to include additives like organosilicon. The spraying volume of the pesticide solution for 6 000 seedlings per mu should be no less than 60 kg. Prior to chemical control, it is advisable to remove the old leaves from the plants. Place them in sealed bags and take them outside the greenhouse. This practice can, to a certain extent, reduce the density of insect populations. Removing old leaves enhances the permeability of the plants, thereby improving the effectiveness of pesticide spraying.

(2) After flowering and fruiting, emphasis is placed on biological control, primarily using natural enemies. Predacious mites (*Neoseiulus californicus*) are released monthly for pest control at a rate of 5 bottles per mu (with each bottle containing 20 000 individuals). Select one large, flat leaf per plant and sprinkle the mites mixed with bran onto the leaf surface. Following the introduction of natural enemies, efforts should be made to minimize the use of insecticides in the greenhouse. During sunny and dry seasons, spraying to increase humidity inside the greenhouse can, to some extent, suppress the reproduction and damage caused by red spiders.

4.4.2.2 Aphids and thrips

Aphids primarily extract sap from the young leaves, flowers, heart leaves, and the undersides of strawberry plants. Affected leaves curl and deform, impeding the growth of strawberries. Thrips mainly harm the stamens, petals, and young fruits of strawberries, resulting in poor pollination, stiff or deformed fruits, adversely affecting both yield and quality.

During the peak of pest infestation, fungicides like spinetoram+pyriproxyfen, thiamethoxam, acetamiprid, imidacloprid, pymetrozine, dinotefuran, spirotetramat, flonicamid can be used for spray control. For the harvesting period, it is advisable to utilize biocontrol agents such as matrine, spinetoram and spinosad for control.

❖ Key Techniques for Blueberry Cultivation

1 Overview of industrial development

Blueberry, scientific name *vaccinium* spp., also known as *vaccinium* uliginosum (blueberry), belongs to the ericaceae family *vaccinium* genus. Blueberries are rich in nutrients, containing various vitamins and micro elements, as well as special components such as pectin, superoxide dismutases, and cyanine that are rare in other fruits. They have health benefits such as relieving eye fatigue, improving vision, and delaying brain aging. Blueberries have a bright color and a unique flavor, and can be consumed fresh or processed into various foods and health products, which are well-loved by consumers. It has been recognized as one of the five health foods for humans by the Food and Agriculture Organization of the United Nations, and has gained widespread attention in the domestic and international markets, occupying an important position in the industry.

In the past 20 years, with the improvement of people's living standards and the deepening of agricultural industrial structure adjustments, the blueberry industry has developed rapidly. In 2020, the global blueberry cultivation area reached 126 144 hm^2, with a total output of 850 886 t. In recent years, the global blueberry cultivation areas and acreage have continued to increase.

With its unique nutritional value, high economic benefits, diverse product forms, and the popular health concept, blueberries have become a fast-growing new fruit tree industry with enormous development potential. Currently, the blueberry industry has become the world's second-largest berry industry after strawberries, and the global blueberry industry development is showing a trend of diversified production areas.

2 Main cultivars and characteristics

2.1 Highbush blueberry

It includes three types: Northern Highbush, Southern Highbush, and Half-high Highbush. Generally, the trees are 1.5–3.0 m tall and grow well in sandy soil with rich organic matter, adequate water content, and a pH value below 5.5. This variety produces large, high-quality berries with a good taste that are suitable for fresh consumption or processing.

2.1.1 Northern Highbush blueberry

This variety prefers a cool climate and has strong cold resistance, with some varieties able to withstand temperatures as low as −30 °C. It has lower cold requirements than Southern Highbush blueberry and is suitable for cultivation in coastal humid areas and cold regions. The main excellent varieties and characteristics are as follows.

2.1.1.1 Bluecrop

Bred in 1952 by crossbreeding in New Jersey, USA, this is a mid-mature variety with large, sky-blue fruits, thick bloom, hard flesh, slight tartness small and dry tip mark, with light aromatic flavour. It tastes slightly acidic when not fully ripe, good flavour, and good storability. It appears slightly cracked fruit and dropping fruit during harvesting. It has strong growth and an open crown. the branches are softer in young trees. This variety has good frost resistance, drought resistance, and soil adaptability, with high and stable yields, making it an excellent variety for fresh consumption.

2.1.1.2 Yunlan

Selected and bred by the Horticulture Research Institute of Yunnan Academy of Agricultural Sciences and the Institute of Modern Agricultural Technology, Dalian University in 2015, this is a mid-mature variety with large, light blue fruits, and abundant bloom. It has a more round shape, and a few spheres, hard flesh, small and dry fruit scars, sweet taste, excellent flavor, and good storage resistance. It has strong vigor, is open-branched, has good fruit productivity, strong cold and heat resistance, and is an excellent variety for fresh consumption.

2.1.2 Southern Highbush blueberry

This variety prefers wet and warm climate conditions with lower cold requirements and is suitable for cultivation in warm regions. The main excellent varieties and characteristics are as follows.

2.1.2.1 Flordablue

Bred by the University of Florida, USA in 1976, this is a mid-to-late-mature variety with large fruits, aroma, small and dry fruit scars, medium flesh hardness and acidity. It has moderate growth, open canopy, and good productivity with a cold requirement of 150−300 h.

2.1.2.2 Sharpblue

Bred by the University of Florida, USA in 1976, this is a mid-mature variety with large fruits, medium blue color, aroma, small and wet fruit scars, and juicy flesh suitable for making fresh fruit juice, which is not resistant to transport. It has moderate growth and an open canopy with a cold requirement of 150–300 h. this variety has a strong adaptability of soil and good productivity with fruit and tree characteristics similar to Flordablue.

2.1.3 Half-high Highbush blueberry

It is a variety type obtained by crossbreeding highbush blueberries and lowbush blueberries. Generally, the trees are 0.5–1.0 m tall, with fruits larger than Lowbush blueberry but smaller than Highbush blueberry. It has strong cold resistance, able to withstand temperatures as low as −35 °C, and is suitable for cultivation in cold regions. The main excellent varieties and characteristics are as follows.

2.1.3.1 Northland

Selected and bred by the Agricultural Experiment Station of the University of Michigan, USA in 1968, it is a mid-early-mature variety with large, blue, round fruits, thick bloom, firm and juicy flesh, medium acidity, and good flavor. It has medium and dry tip marks, and a concentrated ripening period. It has strong tree growth and a moderately open canopy, and the height of the mature tree can reach about 1.2 m. The soil is widely adaptable and the tree has high productivity, and strong cold resistance, making it an excellent variety for cold region cultivation.

2.1.3.2 Northcountry

Selected and bred by the University of Minnesota, USA in 1986, it is a mid-early-mature variety with large, bright blue, sweet and sour fruits, owning excellent flavor. It has moderate tree vigor, about 1.0 m tall, 1.0 m wide canopy, strong cold resistance, able to withstand −37 °C. Early fruit has good productivity, with a yield of 1.0–2.5 kg per plant. It has small, dark green leaves that turn red in autumn, and an elegant tree shape. It is suitable as an ornamental variety for cultivation in high-altitude cold areas.

2.2 Rabbit-eye blueberry

This variety group has strong tree vigor, with trees generally over 3.0 m tall, a long lifespan, resistance to wet and hot conditions, and not strict soil requirements. They are tolerant to drought but not very cold-tolerant, susceptible to freeze damage at temperatures as low as −27 °C. They generally require 18–35 d of cold temperatures below 7.2 °C. It is suitable for cultivation in warmer climate areas with higher humidity. When cultivated in colder areas, consideration needs to be given to potential frost damage during the flowering period and winter freeze damage. The main excellent varieties and characteristics are as follows.

2.2.1 Baldwin

Developed in Georgia, USA in 1985, it is a late-maturing variety. The fruit is medium-large size and its peel is dark blue, with little bloom, firm flesh, medium acidity, excellent flavor, and small, dry fruit scars. It has strong, upright tree vigor with a large canopy. It has good productivity, requires 18−25 d of cold temperatures, and strong disease resistance. The maturation period of the fruits can last 6−7 weeks, suitable for cultivation in gardens and sightseeing parks.

2.2.2 Woodard

Developed in Georgia, USA in 1960, it is an early-maturing variety. The fruit is medium-large size, bright blue color and oval shape, with thick bloom and large, dry fruit scars. The flavor is excellent when fully ripe, but slightly sour before full ripeness, and the flesh is soft, not suitable for long-distance fresh fruit sales. The tree has weak vigor in its early stage, with an open canopy. It vigorously grows after reaching maturity. It has low cold temperature requirements, flowers rapidly with the rise of spring temperature, and is susceptible to frost damage. Pruning is primarily focused on weak pruning to ensure fruiting.

2.3 Lowbush blueberry

This variety group is cultivated from wild species or their derivatives from the original habitat. The main characteristic is the small tree size, generally 0.3−0.5 m tall. They have strong drought resistance and very strong cold resistance, able to withstand temperatures as low as −40 °C, and can grow normally in barren hills, rocky exposed hills, and flatlands. They have simple cultivation and management requirements and are extremely suitable for large-scale commercial cultivation in high-altitude mountainous areas. Due to the small size of the fruits, they are mainly used as processing materials. When commercially cultivated on a large scale, consideration should be given to the development of processing capabilities in conjunction with the fruit. The main excellent varieties and characteristics are as follows.

2.3.1 Blomidon

Developed by the Kentville Research Centre of the Canadian Department of Agriculture through hybridization of wild lowbush blueberry varieties, it is a medium-maturing variety. The fruit is round, light blue color with thick bloom and a faint fragrance, and excellent flavor. It has strong tree vigor, high productivity, with an average yield of 0.83 kg per plant at 5 years old, with a maximum of 1.59 kg. It is extremely cold-resistant, making it be selected as the preferred variety for development in high-altitude mountainous areas.

2.3.2 Chignecto

Developed in Canada, it is a medium-maturing variety. The fruit is almost round, blue color with thick bloom. The leaves are narrow and long. The tree has strong tree growth, easy propagation, relatively high productivity, and strong cold resistance.

2.3.3 Fundy

Developed in Canada, it is a medium-maturing variety. The fruit are slightly larger than Blomidon, with light blue color, bloom, and high productivity.

3 Cultivation and management techniques

3.1 Planting planning and orchard construction

3.1.1 Site selection

The base should be situated away from major traffic routes, with a distance of at least 60 m from highways, and within a 3 km radius free from direct and indirect pollution sources such as industrial and mining enterprises. It is proposed to opt for well-drained flat lands or gently sloping sunny areas with a slope less than 15°. The soil should be slightly acidic with a pH value range of 4.0–5.5. The site must have deep, fertile soil with good aeration, and an organic matter content of at least 3%. Irrigation facilities should also be available.

3.1.2 Soil improvement

One year before planting, it should perform deep plowing to a depth of 20–25 cm. And it is advisable to remove large rocks, tree roots, and other debris, and level the land. For wetlands, meadows, or marshy areas, it is suggested to clear the forest and set up drainage channels. For terrace planting, it is better to create a raised bed with a height of 25–30 cm and a width of 1 m, preferably done in conjunction with land preparation a year before planting. Blueberries have strict requirements for soil acidity. When the soil pH value is greater than 5.5, measures such as applying sulfur powder or citric acid should be taken to lower the pH value. For every unit decrease in pH value, it needs to apply 65 kg of sulfur powder per mu. A specific measure is to evenly distribute the sulfur powder to the soil and to perform deep plowing. Conversely, if the soil pH value is below 4.0, it is recommended to adjust it by applying 500 kg of lime powder per mu for every unit increase in pH value. A specific measure is to evenly distribute the lime powder to the soil and to perform deep plowing. When the organic matter content is less than 3%, it can mix organic materials such as peat, decomposed bark, and powdered decomposed agricultural straw to improve soil structure and increase organic matter. The mixture of peat, pine needles, sawdust, and other acidic substrates can also enhance soil acidity. It is suggested to mix them in a 1 : 1 ratio with the existing soil and fill the planting holes.

3.1.3 Variety selection and configuration

For open-field cultivation, it is advised to choose varieties of Northern Highbush, Half-high Highbush, or Lowbush blueberries that have large fruit, high quality, high yield, and strong adaptability. Northern Highbush and Half-high Highbush blueberries require pollination trees, while lowbush blueberries can be planted with a single variety. It is recommended to choose pollination varieties with a flowering period consistent with the main cultivated variety and a large pollen quantity. The ratio of main cultivated varieties to pollination varieties should be 1 : 1 or 2 : 1, and they can be planted in alternate rows or alternating plants.

3.1.4 Planting time and density

3.1.4.1 Spring planting

It should be carried out from the thawing of the soil to just before the seedlings sprout, typically from late March to early April.

3.1.4.2 Autumn planting

It should be done after the leaves have fallen and before the soil freezes. The row spacing for Northern Highbush blueberries is generally 1.5 m × 3.0 m, for Half-high Highbush blueberries, it is about 1 m × 2.0 m, and for lowbush blueberries, it is about 0.6 m × 1.5 m. Planning for dense planting is possible, with thinning after the plants have grown densely.

3.1.5 Seedling selection and planting

It is suggested to choose 2–3-year-old seedlings with well-developed and intact root systems, a main stem diameter greater than 0.6 cm, plant height greater than 30 cm, 3–5 branches, and no diseases or injuries. For seedlings transported over long distances, it is better to soak the roots in water for 12 h before planting, and prune damaged or dead branches. It is necessary to dig a planting hole with dimensions of 0.3 m × 0.3 m × 0.4 m. It should mix garden soil and organic materials in a 1 : 1 ratio, fill the hole evenly, and water to compact the soil. And then it can dig a small hole with a size of 20 cm × 20 cm in the compacted planting hole. It can plant the seedling into the small hole, burying 3/4 of the hole with soil, lightly tamp the soil, create a water-retaining basin, immediately irrigate with about 0.5 kg of water, and cover the soil once the water has completely infiltrated, ensuring the original soil mark on the seedling is level with the ground. Finally, it is suggested to use furrow irrigation, watering thoroughly in one go, and cover the soil after complete water infiltration, with a soil covering thickness of about 3 cm.

3.1.6 Soil mulching for cold protection

In cold regions, soil mulching for cold protection should be carried out before the ground freezes. When the average temperature is around 5 °C, and nighttime frost begins, it can start mulching. When the temperature drops to −3 °C, it can initiate cold protection mulching, bend the branches gently and bury them with a thickness of 10−15 cm, completely covering the upper part of the shrub in the soil.

3.2 Shaping and pruning

3.2.1 Pruning periods

Pruning is divided into dormancy pruning and growth pruning. Dormancy pruning occurs from late autumn after leaf fall, to early spring before budding. Growth pruning takes place during spring and summer, with an emphasis on dormancy pruning.

3.2.2 Pruning methods

3.2.2.1 Highbush and Half-high Highbush blueberry pruning

(1) Sapling stage. After planting, to encourage rapid crown expansion, all branches are allowed to grow naturally. During the first winter pruning, 5−6 vigorous primary branches should be retained, cutting them to 40−50 cm to cultivate main branches. Additionally, it is suggested to remove densely growing, thin, and flower bud-bearing branches.

(2) Initial fruit stage. In the second winter pruning, specific methods include trimming the main branches to 50−60 cm, thinning out densely growing and weak branches. Thinning of flower buds on fruiting branches is performed, keeping 4−5 vigorous branches, 2−3 moderate branches, and removing flower buds from weak branches. In the third summer pruning, it is advisable to remove excess primary branches, pinch the tips of vigorous and elongated branches to stimulate branching, and increase the number of fruiting branches. During winter pruning, measures such as removing excessively long, crossing, overlapping, densely growing, and weak branches should be taken. Thinning of flower buds on fruiting branches is performed.

(3) Peak fruit stage. Renewing and rejuvenating the main trunk and fruiting branch structure are conducted. It is necessary to employ techniques such as retracting, slowing down, shortening on fruiting branches, and following standards for retaining flower buds. Clearing away primary branches, removing crossing, overlapping, densely growing, and weak branches, and pinching the tips of vigorously growing branches should be performed.

(4) Decline stage. It is advised to retract again, remove aging main branches, and, if necessary, perform rejuvenation pruning. Selecting and retaining primary branches to cultivate new main branches, renewing and rejuvenating the fruiting branch structure are needed.

3.2.2.2 Lowbush blueberry pruning

Lowbush blueberry pruning involves cutting the above-ground part of the plant from the base using a flat cut. This is typically done once every 2 years, and the removed portions can be used to cover the ground in the orchard.

3.3 Water and fertilizer management

3.3.1 Water management

For young blueberry orchards, it should maintain optimal soil moisture conditions, with soil relative moisture levels ideally between 60%–70%. When soil relative moisture falls below 60%, irrigation should be applied appropriately. In mature orchards, it is recommended to reduce water supply during fruit development and before maturity, and to recover proper water supply after harvest, restoring relative moisture levels to 60%–70%. From mid autumn to late autumn, the water supply should be reduced to facilitate plant dormancy. It can consider filling frost irrigation once before winter. In addition, pine needles, sawdust and broken straw can be considered to cover the soil to keep the soil moisture.

Irrigation methods can include furrow irrigation and bed irrigation, drip irrigation or sprinkler irrigation are usually used for larger cultivation. Orchards should have drainage systems to ensure water retention and drainage.

3.3.2 Fertilization management

It should follow the principle of emphasizing organic fertilizers with supplementary use of inorganic fertilizers. Soil physical and chemical analysis should be conducted every 3–5 years before and after planting to determine fertilization quantities based on soil testing formulas. It is better to avoid using fertilizers containing chlorine, calcium, and nitrates. It is recommended to apply fertilizer twice a year, before and after flowering and after fruit harvest. Before and after flowering, fast-acting fertilizers are mainly used, such as ammonium sulfate and specialized compound fertilizers for fruit trees. The nitrogen (N), phosphorus (P_2O_5), potassium (K_2O) ratio in compound fertilizer is usually 1 : 1 : 1, with 20–30 kg applied per mu, supplemented with a small amount of well-rotted organic fertilizer. After fruit harvest, it is suggested to use nutrient-rich, well-rotted farmyard manure or commercial organic fertilizer as the main source, applying 2 000–2 500 kg per mu, supplemented with a small amount of specialized compound fertilizer and ammonium sulfate.

For sandy soil orchards, broadcast spreading is suitable. For loam and clay soil orchards, it is advisable to apply fertilizers in strips or holes. When nutrient deficiencies are observed in fruit trees, foliar spraying of deficient element fertilizers should be conducted. Orchards with good conditions can combine drip irrigation with fertilization.

3.4 Prevention and control for major diseases and pests

Diseases and pests infestation significantly affect the leaves, stems, roots, flowers and fruits of blueberries. This results in stunted growth, decreased yield, reduced commercial value of the fruit. And in severe cases, this causes complete loss of market value, leading to substantial losses in production. The primary disease affecting blueberries include twig blight, canker disease, root cancer, and gray mold. Common pests that pose a threat to blueberries encompass various moth species such as Lepidoptera, including Noctuidae (owlet moths), Arctiidae, and Pyralidae, as well as beetles from the Coleoptera order, leaf beetles, bugs from the Hemiptera order, and flies from the Diptera order.

The control and prevention of pests and diseases in blueberries should adhere to the principle of "prevention first and comprehensive control". When considering comprehensive pest and disease control measures, specific actions can be categorized into five major classes.

3.4.1 Plant quarantine

Blueberry plant quarantine encompasses two primary aspects. Firstly, external quarantine aims to prevent the introduction of hazardous organisms like diseases, pests, or weeds along with blueberry plants and products (e. g., seedlings) from abroad or their export from domestic sources. Secondly, internal quarantine involves measures dictated by laws when hazardous organisms like diseases, pests, or weed infestations have been introduced from abroad or are localized in specific domestic regions. In such cases, it becomes necessary to restrict and contain them within certain boundaries to prevent their spread to unaffected areas. Vigorous and effective measures are then implemented to eradicate or eliminate them thoroughly.

3.4.2 Agricultural prevention and control

Agricultural prevention and control entail adjusting cultivation practices or artificially modifying the growth conditions of blueberries. This includes enhancing cultivation management, applying organic fertilizers, scientifically pruning, and implementing intercropping to bolster the trees' vigor and improve their resistance to diseases. These adjustments aim to alter the growth status or environmental conditions of blueberries to minimize or eradicate pest and disease infestations and regulate their population. Agricultural prevention and control measures applied to blueberries primarily involve refining and utilizing tillage systems, enhancing field management, employing facility cultivation, and utilizing disease-resistant varieties. Please refer to Table 2−1 for a comprehensive overview of major diseases and pests of blueberries and their control methods .

Table 2-1 Major diseases and pests of blueberries and their control methods

Diseases and pests	Control period	Agricultural or physical control	Chemical control
Fungi, bacteria, overwintering pests, and insect eggs	Late January to Early February	① Orchard cleaning ② Remove diseased and pest-infested branches, overwintering pests, insect pupae, clear fallen leaves and weeds, and destroy or bury them	Use lime sulfur mixture. Spray 29% lime sulfur AS (1 : 50), 50% carbendazim WP [1 : (600–800)], or 80% mancozeb WP (1 : 600) on tree trunks
Soil-dwelling pests	Early March	① Artificially kill adult insects ② Free-range chickens in the orchard to catch pests	Spray 40% phoxim EC (1 : 1 000) on the ground; broadcast 15% chlorpyrifos GR 3 000 g per mu
Mummy (Dead fruit disease)	① Before winter ② Early spring ③ Before flowering	① Orchard cleaning ② Spring shallow plowing and urea application	Spray potassium dihydrogen phosphate + trace elements (1 : 800), 450 g/L prochloraz EW [1 : (800–1 500)], 25% pyraclostrobin SC (1 : 1 500)
Anthracnose	① Winter pruning ② Tender shoot period ③ After flower falling	Prune and remove diseased and pest-infested branches during winter pruning	Spray 50% carbendazim WP [1 : (600–800)], 75% chlorothalonil WP [1 : (500–800)], 450 g/L prochloraz EW [1 : (800–1 500)], 250 g/L azoxystrobin SC [1 : (800–1 500)]. Spray once every 7–10 d, continuously for 2–3 times
Powdery mildew	June to July	① Orchard cleaning ② Reduce orchard humidity	Spray 25% azoxystrobin SC [1 : (1 000–1 500)], 25% difenoconazole (1 : 1 500), 25% pyraclostrobin (1 : 1 500). Spray once every 7–10 d, continuously for 2–3 times
Fruit fly	June to August	Use sugar-vinegar solution and yellow boards for trapping	Spray 0.1% avermectin bait (1 : 200) for spot spraying, 60 g/L spinetoram SC [1 : (1 500–2 500)]
Cutworm	Early April	① Blacklight trapping ② Use small leaf roller nematodes for control	Mix 2 000–3 000 g per mu of 5% avermectin·chlorpyrifos GR for soil application, broadcast 3 000 g per mu of 15% chlorpyrifos GR

Continued Table

Diseases and pests	Control period	Agricultural or physical control	Chemical control
Aphids	March to April	① Use natural enemies such as ladybugs, parasitic wasps for control ② Hang yellow sticky traps, insect lamps	Spray 70% imidacloprid WDG (1 : 5 000), 70% acetamiprid WDG (1 : 4 000), 21% thiamethoxam SC [1 : (4 000–5 000)]
Fall webworm	May to October	① Manual removal of netting ② Tie grass on tree trunks to collect larvae ③ Blacklight trapping	Spray 10% effective cyhalothrin EC [1 : (1 000–1 500)], 1.8% avermectin EC [1 : (1 000–1 500)], 8 000 IU/μL *Bacillus thuringiensis* bacterial agent [1 : (800–1 000)]. Spray once every 7–10 d, continuously for 2–3 times
Bird damage	May to July	Use bird nets or bird repellents	—

Note: The information provided is a detailed translation of the Chinese text. The effectiveness and appropriateness of the control methods may vary based on local conditions and regulations.

3.4.3 Biological control

Various biological control methods are employed in blueberry cultivation, as outlined below.

3.4.3.1 Use of natural enemies to control pests

During the emergence period of parasitic wasps, it's recommended to refrain from spraying any chemical pesticides. Instead, it is suggested to protect and utilize parasitic wasps to control gall wasps, and artificially release parasitic wasps to control measuring worms. It can also provide shelter or artificially introduce beneficial insects like ladybugs to control soft scales and root mealybugs.

3.4.3.2 Use of microorganisms for pest control

Measures include adopting techniques such as soaking seedlings in K84 bacterial suspension or root burning before planting or after disease occurrence to prevent the development and impact of root cancer. In addition, employing trichoderma to control gray mold, using *Bacillus thuringiensis* against noctuid pests affecting leaves, flower organs, and fruits, and utilizing beauveria to control soil-dwelling pests.

3.4.3.3 Use of insect hormones for pest control

For example, it can implement sex pheromones to attract and kill pests.

3.4.3.4 Use of other beneficial organisms

It can control pests by raising chickens and ducks.

3.4.4 Physical and mechanical control

Commonly used physical and mechanical control methods in blueberry cultivation include the following.

3.4.4.1 Trapping and killing

It is suggested to utilize manual and mechanical methods to trap and kill pests based on their occurrence patterns and habits, such as blueberry fruit moths, weevils, and flower beetles.

3.4.4.2 Luring and killing

Based on the tendencies and characteristics of pests, it is advisable to employ suitable methods to lure and eliminate them. Measures include using sugar-vinegar solution to attract and kill fruit flies, employing black light or frequency oscillation lamps with high-voltage grids to lure and kill moths, controlling aphids by using yellow sticky traps, and managing thrips with blue sticky traps.

3.4.4.3 Utilization of temperature and humidity

Exploiting the adaptability of pests and diseases to temperature and humidity, the population of pests and diseases can be reduced by controlling these factors.

3.4.4.4 Application of new technologies

It is recommended to adopt innovative technologies such as radiation to eliminate pests and diseases.

3.4.5 Pesticide control

Ensuring food safety in blueberry production involves the judicious use of chemical pesticides in an uncontaminated environment (soil, water, air). It is essential to strictly adhere to regulations governing the use and prohibition of chemical pesticides. Application should occur when necessary and at the optimal timing, with the careful selection of pesticides and the use of correct application methods and techniques. While mastering dosage techniques, attention should be given to controlling the frequency of use and observing safety intervals to prevent human and livestock poisoning. Safe use of chemical pesticides is crucial, ensuring that the presence of foreign chemical substances in blueberry products complies with the standards of the consumer countries.

3.5　Harvesting and post-harvest treatment

3.5.1　Harvesting period

The berries are considered ripe when the skin turns black or purple-black. Blueberries do not ripen uniformly, so harvesting needs to be done in multiple stages. For Highbush and Half-high Highbush blueberries, harvesting needs to be done once during the peak fruiting period, which lasts 2–3 d. Additionally, it is suggested to harvest once every 4–6 d during the early and late fruiting periods. Lowbush blueberries have a more consistent ripening period, and the ripe berries do not fall off. Harvesting can be done after all the berries have fully ripened.

Generally, berries intended for fresh consumption and short-distance transportation with good storage conditions are harvested before reaching full ripeness. Berries intended for processing are harvested after reaching full ripeness. Sunny mornings are generally the best time for harvesting, and in regions with high summer temperatures, it is advisable to choose cooler temperatures in the morning, noon, or evening for harvesting.

3.5.2　Harvesting methods

For Highbush and Half-high Highbush blueberries, tools used for harvesting should be cleaned, disinfected, and air-dried before harvesting. Pickers should wear rubber gloves and handle the berries gently, placing diseased or deformed berries separately. The harvesting sequence involves picking the outer crown first, followed by the inner crown, and the upper layer before the lower layer. For Lowbush blueberries, a comb-like artificial harvester is used for harvesting. In large-scale orchards, handheld electric harvesters are used for Highbush and Half-high Highbush blueberries, while large comb-like harvesters with shaking devices are used for Lowbush blueberries.

3.5.3 Precooling and storage

Blueberries are prone to spoilage at room temperature. To maintain fruit quality and extend the shelf life, fresh blueberries should undergo low-temperature storage. After harvesting, it is suggested to precool the berries at 10−12 °C for 10−12 h, bringing the fruit temperature below 10 °C. After precooling, it is better to store the blueberries in a cold room at around 1−3 °C or freeze them for storage at temperatures below −18 °C.

3.5.4 Packaging and transportation

To prevent damage to the quality of blueberries during transportation, shallow, breathable containers such as baskets, cardboard boxes, or fruit trays are commonly used. Small packaging with multilayer stacking helps avoid compression and vibration. For fresh consumption, berries are often packed in foam boxes or cardboard boxes with ventilation holes, with each box containing up to 1.5 kg of fruit. Blueberries intended for processing are packed in larger breathable containers and transported directly to processing plants.

References

JIA Y X, LI L Q, MU X D, et al., 2021. Technical regulation of blueberry cultivation [J]. Journal of Fruit Resources, 2(5): 49−51.

SU J M, SHA Y F, DUAN X N, et al., 2007. Introduction to major excellent blueberry varieties [J]. Shanxi Fruits (1): 17−18.

WAN H, WANG L R, 2016. Common diseases and pests of blueberries and their control techniques [M]. Kunming: Yunnan Science and Technology Press.

WANG L R, TAO B, KONG L M, et al., 2013. Blueberry production techniques in Kunming region [J]. Practical Forestry Technology (5): 50−51.

WU L, 2016. Thirty-five years of research and industry development of blueberry in China [J]. Journal of Jilin Agricultural University, 38(1): 1−11.

❖ Key Techniques for Plum Cultivation

1 Overview of industrial development

Plums, a small deciduous tree, are classified in the genus *Prunus* L. and family Rosaceae. Chinese began cultivating plums around 3 000 years ago, making plums the oldest cultivars in Chinese history. Nearly 450 000 mu of plums were planted in Yunnan in 2021. The total production was 300 000 t. Plums are planted in almost all of Yunnan, though their management methods are rough. That is because the planting techniques are accessible, and there will be good production only with fewer inputs. In that regard, plums are significant to residents in economic undeveloped region, remote ethnic minority areas for increasing the farmers' incomes, economic growth, and improving life conditions. In terms of cultivating modes, open culture is adopted by Plum Cooperatives in large producers, such as Suijiang County and Yanshan County. Apart from that, an open culture is scarcely applied in the rest of the counties in Yunnan province.

2 Clement weather conditions

Plum trees are intensely adaptable to the environment. Hence, plum trees are planted in 1 100 m or higher areas, except in the southern parts of Yunnan, whose altitudes are low, and both temperatures and humidity are high. Plains, slopes, and mountains are suitable for plum growing.

2.1 Temperature

Plums vary in temperature due to different categories and varieties. The best temperature ranges from 20 ℃ to 30 ℃. The optimal temperature for flowering is 12−16 ℃, during which flowers are prone to frozen disasters. The low temperature in different development stages is also different, such as −5.5− −1.1 ℃ in the bud stage, and −2.2− −0.5 ℃ in the flowering and young fruit stages.

2.2 Water

Plum trees are shallow-rooted, with medium drought resistance and humidity preference. Watering varies from stage to stage of plant growth. Water is greatly needed in the emergence of new shoots and expanding of fruits. Sufficient moisture is crucial for the roots of developing plants: extreme dryness and too much water cause the drop of both flowers and fruits. Little water is needed in the flower bud differentiation stage and dormant stage. Plum trees grow naturally and usually when relative humidity remains at 60%–80%. When the absolute moisture content (AMC) remains at 10%–15%, the overground parts cede to grow. When the AMC is less than 7%, the root system stops growing. Therefore, irrigation facilities should be widely applied in dry areas for plums growing. Drainage should be prioritized in rainy-season areas where plums are grown in clay soil. It is proper to plant plum trees in areas with a low underground water level and free from waterlogging. Humidity plays another vital factor in plum growth.

If the air is too dry, water loses to evaporation. Branches will be wilt once the moisture content is reduced to 50% or above of normal content. Pollination will be affected when it is dry during the flowering period, with little air humidity. If it is dry in winter, plant death will occur, especially in shoots in autumn.

2.3 Sunlight

Plum trees need a location in full sun. Plums are well-coloured if grown in ventilated orchids with solid branches, full buds, and sweet fruits.

2.4 Wind

Plum trees are susceptible to winds. A strong wind will badly affect these trees. Hot and dry wind may dehydrate stigma, leading to pollination failure. If wind speeds exceed 10 m/s, branches may be torn down, and fruits may be torn off the trees. Winter wind is typically dry, causing frozen plum trees. In that regard, plum orchards should be positioned in shelters with fast-growing forests planted in advance in typhoon-prone areas.

2.5 Soil

Plum trees are not strict with soils. Chinese varieties are strongly adapted to soils. Chernozem, red earth, and yellow earth are ideal for growing plum trees. In addition, well-drained, fertile, moist, and low sandy loam soil is optimal. The common plum trees are strongly adapted to a pH value of 4.7–7.4.

2.6 Slopes

Moisture and heat conditions vary on different slopes due to different solar intensities and times of sunlight, which further lead to soil's physical and chemical natures. Leeward and sunny slopes are ideal for plum trees growing. It is essential to position plum trees on the south face of slopes. For one thing, facing the south means higher temperature and strong light. For another, a spot on the south shelters from dry and cold wind from the northwest in wintertime. The best slope direction for receiving light is 5° from south to east, but 5° from south to west in the alpine region is suitable.

3 Main cultivars and characteristics

There are over 100 plum varieties in Yunnan, among which are long-time cultivated local varieties. The main varieties are introduced as follows.

3.1 Early Golden

An early maturing variety for fresh food. It is oval-shaped with blushed skin when it is fully ripened and orange flesh. The average single fruit weight is 45.5 g. The soluble solids content is 12.1%, titratable acidity content is 0.94%, and vitamin C content is 2.8 mg/100 g. Fruit cracking and fruit splitting are hardly seen before harvest. As its kernel is small, the edible rate is 94.4%.

3.2 Red Beauty

An early maturing variety for fresh food. It is round shaped with red skin. The average single fruit weight is 50 g. The flesh is red, juicy and thick. The plum is medium in fibre. Red Beauty tastes sweet and slightly sour without a tart flavour. The soluble solids content is 12.6%, titratable acidity content is 0.95%, and vitamin C content is 5.0 mg/100 g. In terms of quality, Red Beauty stands out with a small kernel.

3.3 Hong Feng

A late maturing fruit for fresh food. The average single fruit weight of this variety is around 100 g. Hong Feng is round-shaped with red skin. The flesh is yellow, juicy and thick. The plum is medium in fibre. Hong Feng tastes sweet and slightly sour without tart flavour but solid and aromatic. The soluble solids content is 13.3%. In terms of quality, it is with a small kernel. The edible rate is 98.5%. It excels in production.

3.4 Golden-red plums

A late maturing variety for fresh eating. The average single fruit weight of this variety is 40 g. Golden-red plums are oval-shaped with both skin and flesh yellow and firm meat. Intensely aromatic, the plum tastes sweet, slightly sour, and not tart. The soluble solids content is 12.8%, titratable acidity content is 1.07%, and vitamin C content is 2.86 mg/100 g. In terms of quality, it is a freestone type. The comprehensive traits are merit.

3.5 Guofeng No. 2

A large and late maturing variety for fresh eating. It is a round-shaped variety with an average single fruit weight of 115 g. The meat is yellow, and the skin tends to be red. As the meat is crisp, juicy, and aromatic, it enjoys a high edible rate. It is a semi-clingstone variety. The soluble solids content is 13.2%, soluble sugar content is 7.4%, titratable acidity content is 0.8%, and vitamin C content is 3.1 mg/100 g. It has a maturation date of 117 d and is bountifully harvested at the end of August. The variety is superb and excels in storage. Other traits include long storage, large size, productivity, and adaptability.

3.6 Guofeng No. 7

A large and late maturing variety. It is flat and rounded with an average single fruit weight of 100 g. Flat and even, and a semi-clingstone variety, the fruit skin tends to be red, and the flesh is crisp and firm with a strong aroma. The soluble solids content is 18.0%, which will increase to 21% when plums are planted with facilitations, the titratable acidity content is 1.6%, and the vitamin C content is 6.0 mg/100 g. It has a maturation date of 115 d or more and is typically harvested at the end of August. The variety is aromatic, has extended storage, is firmly cold-resistant, early fruiting, and has a high yield. It is suitable either for planting in greenhouses or open cultivating.

3.7 Guofeng No. 17

A large and late maturing variety for fresh food. It is rounded and with an average single fruit weight of 118.5 g. The skin tends to be red, the flesh is yellow, crisp and firm, high edible rate, juicy, aroma and a semi-clingstone variety. The soluble solids content is 14.8%, titratable acidity content is 1.1%, and vitamin C content is 5.1 mg/100 g. It has a maturation date of 120 d or more, and is normally harvested at the beginning of September. The variety is of good quality, long storage, large in size, good-looking and strong cold-resistance.

3.8 The Suijiang Half Red Plum

The plum tree's vigor is medium and slightly firm, and the profile tends to be half open, with the crown in a canopy shape and a distinctive feature of diverting and branching. It usually comes out in early-mid March and ripens in early-mid July. The variety is flat-rounded or tends to be circled, with an average single fruit weight of 30.7 g. The flesh is crisp and firm, tastes juicy, and soft until it fully matures. In addition, it tastes aromatic and sweet. It is of superb quality and semi-clingstone. The soluble solids content is 11.43%. It keeps longest for about 14 d when stored indoor at room temperature in Kunming City, with long storage.

4 Site planning and design

4.1 Site location

When selecting a site for plum growing, choose a location with deep soil that is loamy, fertile, well-drained, and a low underground water level. Those places include plains, hills, mountainous areas, and beaches. When positioning in a mountainous area, one should prioritize selecting the south slopes in full sun, which is inward of the wind. And then plough the land later. Plant a late flowering variety in a lower-lying area to prevent late frost damage.

4.2 Orchard designing and layout

Spotting and mapping are the first steps for orchard building. Once information on soil, topography, weather, and hydrology is collected, orchard areas, windbreakers, roads, irrigation facilities, and surroundings are certified as well.

4.2.1 Site arrangement and distribution

The orchard can be divided into different plots referring to practical locations. The plainer the orchard is, the bigger the plots will be, or vice versa. An orchard in a low-lying area is situated best in the south-north direction for complete and average sun. Plots in mountainous orchards should be horizontally arranged with bounds parallel to the contour for water and soil reservation.

4.2.2 Lanes and routines

An orchard is linked through a main routine, small lanes and branch paths. As the name indicates, the main routine should be centred in 4–6 m width for transporting fertilizers and fruits. The zig-zag road is circling and upward to the top of the orchard. Also, windbreaks and canals should be arranged alongside roads to maximize the use of space. Usually, roads better account for 3%–5% of the total area of orchards.

4.2.3 Irrigation system

It is canals, branch canals, and conveyance ditches that make up the irrigation system of orchards. Canals should be set at the top of the orchard for better control of the whole. Branch canals should be set up in the bounds of plots, in which water may be induced to the pitch of trees via conveyance ditches. Facilities equipped with spraying irrigation and drip irrigation may be widely introduced for water saving and loss of watering. Reservoirs or water tanks may be built in orchards on mountains and hills. Orchards with a higher underground water level should drain the water via ditches, sheltering from the drainage in the long run.

4.3 Plough and refine soil

If plum orchards are positioned in plains, plant holes may be fixed in terms of the number of plants and spacing, of which the organic fertilizers are set. If plum orchards are positioned in sandy and loamy places, the first step is to refine the soil. Mix the soil (one portion of clay and 2–3 portions of loams) with organic fertilizer. After fully merging, one fills the mixture of soil into planting pitches. Expanding the pitches, filling soils, and fertilizing may effectively change the physical conditions of the soil in the loamy orchards. If plums are planted on mountains, soil can be improved by upgrading terrace fields and fish scale pits.

5 Planting

5.1 Methods and spacing

Rectangular plots: These are easy to manage due to wide spaces for ventilation and sunlight through plants.

Square plots: Evenly spaced spaces of plants are suitable for sunlight and management.

Contour farming: This is suitable for planting in mountainous orchards and tilling sloped land along the lines of consistent elevation.

In good soil fields with standard management requirements, apply 3 m × 4 m or 3 m × 5 m for plant spacing and rows and 2 m × 4 m in mountains or beaches where the soil is barren.

5.2 Pollinators

Some varieties have low self-fertility; therefore, it is necessary to plant pollinizer trees alongside the main cultivars. Pollinizers and main varieties bloom periods must overlap or be similar, and their pollen must be sufficient biennially for plum varieties. The plum trees and pollinators ratio should be 2 : 1 or 3 : 1, sometimes 8 : 1.

5.3 Planting methods

5.3.1 Digging a planting pitch

Generally, dig a hole of 80 cm in depth and 100 cm in diameter. When digging a hole, put the topsoil on one side and the core soil on the other. Before trees are settled into pitch, it is necessary to sterilize the pitch by using lime or directly drying it under the sun. At the same time, mix the organic fertilizer with the soil, refill the topsoil and subsoil, and seal it by watering. Mark the planting position at the centre of the pitch.

5.3.2 Planting

When planting in spring, dig a small hole at the planting point where to set plum seedlings in the middle. By doing so, roots can fully stretch. Earth-up is the second step. Fill the soil the same as the growth trace in the nursery. Pull seedlings slightly up for a complete stretching of roots. Step firmly into the soil for a complete integration of roots and soil when filling. Set up tree trays around the trunk, and water the sufficient root water. Sprinkle a layer of fine soil on the tree trays when water is thoroughly infiltrated. Straighten the seedlings afterwards.

5.4 Post-planting management

5.4.1 Heading trunks

Stems should be fixed in a timely manner after planting, with 50–60 cm reserved in height. Apart from that, 20 cm should be reserved for repair. There will be 70–80 cm high reservations. It is required that the buds on the repairing belt are complete and the rest of the incomplete branches and buds are cut off in time.

5.4.2 Earth mounting for cold protection

In severely cold areas in winter, in order to prevent freezing injury in winter and spring, crescent-shaped mounds can be piled up in the northwest 50 cm high from seedlings before winter to prevent cold and then removed after seedlings sprout in spring.

5.4.3 Irrigation

Plum trees planted in autumn should be filled with winter irrigation water before winter, and the soil should be loosened in time after water infiltration. Trees should be irrigated in time before sprouting in spring to facilitate bud germination.

5.4.4 Checking survival rate and timely filling the seedlings

When plum trees planted in autumn sprout in spring, check the survival rate in time and replant seedlings of the same age in time when dead seedlings are found.

5.4.5 Diseases and pests control

When seedlings germinate in early spring, they are vulnerable to scarabs and aphids. Therefore, manual capture or chemical control is required once those pests are found after timely observation.

6 Pruning

6.1 Shapes

6.1.1 The open-centered shape

There are three scaffold branches, each with a distance of 10–15 cm. Evenly distributing to 120°, those branches open to around 45°. Remove twigs to 1–2 of them left. Ultimately, prune trees until they become vase-shaped with no scaffolds in the centre to 30–50 cm.

6.1.2 "Y" shape

The shape is adapted for plum trees, which are densely planted but in a wide row. The space and row of trees are usually at 1.5 m × 4 m. The trunk is typically about 40 cm tall without the central scaffold, but 2 large scaffolds oblique between rows and at 45°, which make trees look like the letter "Y". The branchlets are straight or curved, and their bases are on the outer, oblique side or outward facing of the trunk base; 1–2 subbranches can be kept, while subbranches are free to grow on the middle and upper parts. The upper trunks will be prime and compact. Once trees are in a row, their height is better to remain at approximately 3 m. Canopies, no thicker than 2.5 m, are stretching longer towards the row but no more than 3 m. After planting and pruning for 4–5 years, trees are formed. "Y" shaped trees allow better circulation and light penetration for good cropping and management.

6.1.3 Slender spindle shape

The shape is suitable for planting high-density trees. Trunks are 50–60 cm high with 10–12 scaffolds at 70°–90°. Canopies' diameters are around 3 m. Scaffolds are almost horizontal and stretch around. There is no distinctive division among scaffolds with a 10–15 cm distance in between. The vertical distance between the main branches on the same side is not less than 50–60 cm, the lower main branches are 1–2 m long, the upper main branches are gradually shortened, and the shape is spindle-shaped. Prune until small and medium-sized fruit-bearing shoots are left.

Prune and manage plum trees in line with varieties and planting locations. Generally speaking, open heart shape and small canopy shape are for varieties with weak trunks but strong branching. The cup shape or "Y" shape is for upright, and strong vigor varies. Slender spindle shapes are for strong truck and weak branching varieties.

6.2 Pruning

6.2.1 Pruning young plum trees

Pruning is performed for reshaping. To slightly cut off extension leaders in each layer and form the shape and collar by nurturing scaffold branches and side branches. Plum mainly bears from short fruit spurs and cluster fruit spurs, so it is advisable to prune long branches lightly, ease the growth potential, promote the germination of short branches, and then cut them short according to the number of flower buds and the need for bearing, to achieve the purpose of early bearing and early high yield. The competitive branches and over-dense branches are thinned.

6.2.2 Pruning in fruit yielding period

Pruning's purpose is to restrict the size, encourage the robust branches, and be conducive to the tree's long-term yielding. Removing dead, diseased, and damaged to minimise redundancy comes first while cutting short comes second. Appropriately cut or trim dense, upright, overlapping, and crossing limbs. Cut-off water shorts barely grow from the base: removal of all but the best and most vigorous branches outside and upper of collar. Shorten extension leaders for wide opening of the canopy and maintaining the canopy. All this work is good for maximising fruit production due to the extension of robust branches. Pruning also means retracting perennial branches for bountiful harvesting.

7 Fertilizing and watering management

7.1 Top dressing

7.1.1 Top dressing before budding or flowering

Topdressing before budding and flowering can increase the chances of fertility, reduce falling flowers and fruits, and promote the vigorous growth of new shoots. The method of fertilization is to dig three ditches with a length of 60 cm, a width of 20 cm and a depth of 40 cm at the outer edge of the canopy. The fertilization amount of each tree (at the first fruiting stage) is 0.4–0.7 kg, with nitrogen, phosphorus and potassium as the main fertilizers, and the ratio of 1 : 2 : 1.

7.1.2 Topdressing in early fruit set

It is mainly to promote the growth of young fruits, shade from fruit dropping, and increase leaf growth due to the increase in the area of photosynthesis. Quick-acting nitrogen fertilizer is the main fertilizer, and some potassium dihydrogen phosphate compound fertilizer is appropriately added, with 0.5 kg per plant. It is also possible to apply topdressing outside the root and spray 0.4%–0.5% urea solution to make the leaves green, the branches grow rapidly, and the fruits develop rapidly.

7.1.3 Post-picking

During this time, it is better to apply organic fertilizers with phosphorus fertilizer and potash fertilizer for productive yield.

7.2 Base fertilizers

7.2.1 Time and varieties

Apply base fertilizers in autumn. Organic fertilizers are mainly applied in this period. Human or animal manure, bean cake, fried flour cake, stalk straws, weeds, and leaves are sources of base fertilizers, which can be applied once they are fully fermented and decomposed.

7.2.2 Application amount

When applied, determine the fertilizing ratio in line with tree ages, vigor, fruiting, soil fertility, and previous fertilizing ratio. For first-year planted young plum trees, apply 50 kg or more of base fertilizers. Once they become fruiting, the fertilizing ratio is equal to the fruit yield, sometimes two times.

7.2.3 Methods

Circular trench application method: Dig an annular trench at the outer edge of the canopy. The trench is 30–40 cm wide and 40–60 cm deep. Apply fertilizer into the trench and cover the soil. This method is simple to operate, and the fertility is concentrated and effective.

Trench fertilizing method: Dig 2 trenches with a width of 30–40 cm and a depth of 50–60 cm on both sides of the canopy. Apply fertilizer to cover the soil, change the direction of trenching in the next year, and apply fertilizer alternately. This method is suitable for large trees with large areas and full fruit periods.

Whole orchards fertilizing method: Fertilizer is scattered throughout the orchards and then ploughed into the soil. This method is suitable for dense planting orchards and ploughing deep. Otherwise, it is easy to cause the root system to move up.

7.3 Irrigation

7.3.1 Irrigating before flowering

The arid and dry weather in the early spring in Yunnan hinders the budding, flowering and fruiting of plum trees. Therefore, irrigate before flowering, which provides rich moisture and nutrients for full buds. Those steps set fruit buds for better fermentation and high yield. In this period, topdressing is implemented as well.

7.3.2 Irrigation in early fruit set

Roots, trunks, leaves, flowers, and fruits form in early fruit sets, and so are other plant tissues and organs, which are greatly against vegetative growth and reproductive growth. Such demands will be more apparent when it is dry and hot. For instance, moisture loss is evident due to evapotranspiration of overground and profound root growth. If irrigation fails to supply or soak the soil, fruit buds and fruits will stop growing and sometimes drop off.

7.3.3 Irrigation ahead of seasonally frozen ground

Irrigate with deep ploughing and fertilization ahead of the winter-like weather. It may begin from the late autumn to the time when the ground is frozen. This should be done as early enough instead of being late. The application of organic fertilizers is good for refining soil, and roots absorb nutrients, which delays the time of roots becoming dormant. Be aware of roots when ploughing deeply. Try to do the least harm to roots and not let roots expose longer. Irrigate in time after deep ploughing.

8 Prevention and control of diseases and pests

8.1 Prevention and control of diseases and pests in spring

8.1.1 Clear orchards in early springs

In order to greatly reduce the number of fungus and pests through winter, it is necessary to clear the plum orchards inside and outside the orchards in early springs, including scraping bark, removing infected fruits, picking up fall-out fruits, which are burnt completely. At the same time, overwintering hosts outside the orchards, such as corn stalks, should be thoroughly removed and burnt together.

8.1.2 Controls ahead of budding

The first is to spray one time of 5°Bé of lime sulfur mixture crystal, or 300-500 times of 45% lime sulfur mixture crystal [1 : (300−500)], or 1 : 1 : 100 bordeaux mixture, for removing pathogens and pests, such as brown rot, plum bacterial shot-hole, anthracnose and plum bag fruit disease overwintering on branches and stems.

The second is to spray 95% Flubendiamide EC [1 : (100−150)] for killing of overwintering bug eggs without harming of natural enemies. In line with early occurrence of overwintering aphids and scale insects, spray 2.5% deltamethrin EC [1 : (2 500−3 000)], or 30% of esfenvalerate·malathion EC (1 : 2 500), or 10% of imidacloprid WP [1 : (1 200−1 500)].

Thirdly, control and prevent pests of spider mites by spraying 20% clofentezine SC [1 : (2 000−2 500)] in budding period.

Fourthly, to prevent and treat gummosis. One can scrape off infectious parts without hurting healthy skins. After that, apply 0.5% of berberine SL or 5 times diluted liquid on diseased plants once in every 3 d, continuously for 2 times. If it is more serious, spray berberine SL [1 : (30−60)] on trunks, side trunks and braches. Also, before germination, scrape off the infected tissues at the gum-flowing part with bamboo chips (as bamboo chips are not easy to hurt healthy tissues), until the tender skin and xylem are found. Coat with 45% lime sulfur mixture crystal (1 : 30) and 21% peracetic acid AS [1 : (3−5)] for futher protection. Alternative chemical controls include: 30% tebuconazole·suspoemulsion SE (1 : 1 100), 50%

tebuconazole (1 : 800), or 50% iprodione WP (1 : 1 000), or 50% procymidone WP (1 : 1 200), which are all functioning well.

8.1.3 Prevention before flowering

Gummosis, perforation, leaf shrinkage and fruit sac are main disease to be controlled and prevented ahead of flowering. Smear 0.5% of berberine liquid on infected parts by gummosis. Spray before flowers come out. Apply one time of 5°Bé of lime sulfur mixture cryystal, or 1 : 1 : 100 bordeaux mixture, or 30% of basic copper sulfate SC, or 70% of mancozeb WP (1 : 500).

8.1.4 Prevention after flowers fall-off

Prevention of plum weevils and white peach scale: 90% trichlorfon technical (1 : 1 200) or 10% a bamectin·pyridaben (1 : 1 000), or 22.4% spirotetramat SC (1 : 2 000)+30% thiamethoxam SC (1 : 2 000).

Aphid: 70% imidacloprid WG [1 : (2 000–3 000)], or 10% acetamiprid EC [1 : (2 000–2 500)], and 25% pymetrozine WP [1 : (2 000–2 500)], after flowers falling out or buds emerge.

Plum red spot disease: Cracking of ascomyces follows the end of flowering with spores scratching around by wind and rain. Therefore, spray multiplier formula bordeaux mixture (copper sulfate : quicklime : water = 1 : 2 : 200) for eliminating large amount of ascospores. Alternative formulas include: 10% difenoconazole WP (1 : 1 200), or 80% mancozeb WP [1 : (800–1 000)], or 70% thiophanate-methyl WP [1 : (1 000–1 200)], or 50% carbendazim WP [1 : (800–1 000)].

8.2 Prevention and control of diseases and pests in summer

8.2.1 Prevention ahead of early fruit set

Plum bacterial shot-hole and red spot disease: Symptoms are mild in a dry May. If it rains much, timely control is necessary. Spray 70% mancozeb WP (1 : 500) every 10 d for once each.

Plum fruit moth: At the beginning of May, before the emergence of overwintering larvae, apply 0.25–0.5 L of 50% phoxim EC [1 : (300–500)] per mu on the ground which is under the crown for killing the adults and larvae.

Aphids: They are hugely booming in mid of May, greatly detrimental to plums. Spray 10% of imidacloprid EC [1 : (1 500–2 000)] before massive proliferation.

Spider mites: They are best prevented in May when adult pests die, before newly born female insects lay eggs. Spray 5% hexythiazox EC (1 : 1 500) first, and then spray 20% pyridaben WP (1 : 2 000), or 48% bifenazate SC [1 : (2 000–2 500)], or 40% etoxazole SC [1 : (2 000–3 000)].

Leaf miners: For orchards once inflicted, apply 20% fenpropathrin EC (1 : 2 000) mixing with 50% dichlorvos EC (1 : 1 000) or 25% chlorobenzuron WP (1 : 2 000) before eclosion of larvae, which is effective in killing adults, pupae and larvae. Alternatives include: 35% chlorantraniliprole WG (1 : 8 000) or 5% emamectin benzoate(D)(1 : 3 000) + 5% hexaflumuron EC (1 : 1 000).

8.2.2 Pest and disease control in seed hardening, fruit well and ripening period

Plum fruit moth and yellow peach moth: This period of time is at the right time of occurrence of larvae. So, spray 50% fenitrothion EC (1 : 1 500), or 2.5% deltamethrin EC [1 : (500–800)], or 35% chlorantraniliprole WG (1 : 8 000).

Spider mites and plums aphids: The same medication as the above mentioned.

Cracking or splitting disease: To spray palmityl alcohol. That is effective in preventing splitting/cracking, and also in pest control. Apply calcium and chelated calcium before fruit swells to help to reduce cracking in plums.

Apply 40% flusilazole EC (1 : 800) + 10% clofentezine WP (1 : 2 000) + calcium solution (1 : 300) once prior to harvest in ripening for preventing re-occurrence of pests and diseases.

8.3 Prevention and control of diseases and pests in autumn

8.3.1 Diseases and pests to be controlled

To control pests and diseases: spider mite, aphid, tortrix moth, plum bacterial shot-hole, plum tree red spot, brown rot, and anthracnosis.

8.3.2 Prevention and control methods

One is pre harvest spraying, using a mixture of fungicides and insecticides with Meilin calcium, for a total of 3 sprays. The first spray is at the end of June, the second at the beginning of July, and the last at the end of July. The second is post harvest medication, which mainly focuses on sterilization. Medication should be applied before packaging or storage to prolong the storage resistance of fruits. The method of soaking fruits with liquid medicines such as isoniazid and prochlorpene or spraying on the fruit surface can be used.

8.4 Pest and disease control in winter

Prevention should be prioritized in winter, and treatment is the second.

8.4.1 Clearing orchards

Diseased leaves, branches affected by diseases and insect pests, residual branches and weeds are all places where various diseases and insect pests overwinter. Therefore, it is necessary to thoroughly remove all fallen leaves, dead branches, diseased fruits and branches, and weeds in and around the park and burn them centrally.

8.4.2 Whitewashing trees

The main whitewash should be at the beginning of December. Scrape off the old skins of the main trunks and branches, and then paint the tree. Whitewashing is usually taken one week after the removal of old tree skins. Whitewashing formulas are 10 portions of quicklime, one portion of sulfur dust powder, 0.2 portions of salt, 40 portions of water and 0.5 portions of stickers. The sulfur dust powder can be alternated by calcium polysulfide residues.

8.4.3 Medication after deep tillage

Medication after deep tillage can kill overwintering pests deep in the soil, as pest eggs and pupae may be turned out and eaten by birds. Dust power is the main soil pesticide, and contact pesticide is the most effective.

9 Plum harvesting, package and storage

9.1 Harvesting

Harvesting at the proper time for the finest quality and commercial value.

9.1.1 Harvesting guidelines

The changing colour of skin is one of the indicators of pick-up. For red varieties, when half of a plum is turning red, it is the right time for long-distant transporting and processing; when red makes up four-fifths of the whole of a plum, it is a proper time for fresh food. For yellow varieties, when the greenish skin becomes yellowish-green, it means picking for processing and long-distance transporting; when yellowish-green turns yellow, it means harvesting for fresh food. Even fruits on the same tree differ in maturing; the harvest consists of several go-throughs, each of which picks up some fruits for a stable yield and supreme quality. Moreover, bountiful harvests may prolong supply time. Other indicators include firmness of fruit flesh or the soluble solids content, which can be tested and analyzed.

9.1.2 Harvesting requirements

To pick up separately: the earliest ripened as the first, the latest ripened as the last.

9.1.3 Harvesting time

Harvesting usually begins on cool and sunny days after dew is dry. It usually starts at 8:00—11:00 and 15:00—18:00. It is not recommended to harvest either on rainy, windy days or hot days.

9.2 Plums sorting and grading

There are two grades in terms of maturity and size. Plums free from damage caused by disease, insects, and growth cracks can be priced. Grade A plums consist of plums that are well-formed, mature, clean, spotless, bright, and without bruising. Over 90% of plums are large.

9.3 Package

Well-designed, beautiful and light packages are a promotion for plum sales. Wrapping paper should be soft, tender, smooth, clean, odourless and strong. Paper strips, shavings and sawdust are filled in the bottom of the package box as a buffering of the bumps. After filling, seal tightly and write the information tag, including origin, variety, gross weight, net weight, harvest date, inspector, and so forth.

The package should be manageable, with 5–10 kg at recommended.

9.4 Transportation

9.4.1 Cooling before transporting

Rapid cooling after harvest to the proper storage temperature is important for transportation. Cooling may be in fridges; if they are equipped with blasts, the plum temperature will cool down soon.

9.4.2 Light on-off loading

Light on-off loading is essential during transportation in case of bruising and damage.

9.4.3 Cold-chain transportation

Modern transportation tools such as refrigerated trucks, refrigerated trains and refrigerated ships can meet the requirements of temperature and humidity in the transportation process of plum fruit and reduce the losses in transportation.

References

CAI D R, 1992. Productivity Cultivation of Plum Trees [M]. Beijing: Jindun Publishing House.

CHEN J, 2006. Guidance of Plums Pruning [M]. Beijing: Jindun Publishing House.

LIU W S, 2004.Pollination-free and High Efficient Cultivation [M]. Beijing: Jindun Publishing House.

LIU W S, 2005. Introduction Guidance of Improved Varieties of Apricot Trees [M]. Beijing: Jindun Publishing House.

LÜ P H, 2003. New Techniques for Plums Annual Care and Management [M]. Yangling: Northwest A & F University Press.

LÜ P H, HE G L, 2013. Key Techniques of Annual Caring of Plums [M]. Beijing: Jindun Publishing House.

❖ Key Techniques for Kiwifruit Cultivation

1 Overview of industrial development

1.1 Nutritional value of kiwifruit

The kiwifruit belongs to the Actinidiaceae family *Actinidia* Lindl., commonly known as the kiwi, furry pear, furry kiwi, or hairy peach, is a perennial, deciduous, woody vine. The flesh of the kiwifruit is rich in dietary fiber, vitamins, organic acids, polysaccharides, proteins, amino acids, and other nutrients. The content of vitamin C in kiwifruit flesh is generally 50–400 mg/100 g, with the highest amount reaching 1 500 mg/100 g in the hairy kiwifruit, which is several times to tens of times higher than other fruits. Kiwifruit flesh also contains various essential trace elements for the human body, such as calcium, selenium, zinc, germanium, etc., and thus has biological activities such as lowering blood lipids, anti-oxidation of lipids, scavenging reactive oxygen free radicals, and inhibiting tumor cells. Due to its rich nutritional content, moderate sweet and sour taste, and delicious juicy flesh, kiwifruit is deeply loved by consumers.

1.2 Kiwifruit industry dynamics

According to the Food and Agriculture Organization of the United Nations (FAO), kiwifruit is produced in 23 countries across five continents, with Asia as the main producing region, accounting for more than 60% of the world's harvest area and output. By the end of 2019, the global kiwifruit harvest area totaled approximately 268 800 hm^2, with China at 182 600 hm^2, Italy at 25 100 hm^2, and New Zealand at 14 900 hm^2, representing 67.93%, 9.34%, and 5.54% of the world's total harvest area respectively. The global kiwifruit production was 4 348 000 t, with China producing 2 166 700 t, New Zealand 558 200 t, and Italy 524 500 t. The average global kiwifruit yield was 16.18 t/hm^2, with New Zealand's average yield at 37.41 t/hm^2, significantly surpassing other countries.

Early kiwifruit planting varieties were primarily the delicious Hayward and Bruno kiwifruit varieties. Currently, besides New Zealand, Italy, Greece, and Iran continue to cultivate predominantly delicious green-flesh varieties represented by Hayward. New Zealand's yellow-flesh variety covers 42% of the area, while the green-flesh variety covers

58%, with a slightly lower yield than the yellow-flesh variety. Aside from green and yellow-flesh varieties, Italy also has a small amount of red-flesh varieties. In China, kiwifruit varieties are diverse, with green-flesh varieties accounting for approximately 40% of the total planting area and yellow and red-flesh varieties each accounting for about 30%. Summarizing the planting situation in the world's major producing countries, green-flesh varieties remain the dominant cultivated type, accounting for around 55%; yellow-flesh varieties at about 25%, and red-flesh varieties at approximately 20%.

1.3 Suitable region for kiwifruit cultivation

Most kiwifruit varieties are suitable for planting in subtropical or warm-temperate humid and semi-humid climate areas, with elevations of 800−1 800 m. The most suitable region for kiwifruit growth has an altitude of 1 000−1 600 m, an average annual temperature of 11.3−16.9 °C, extreme high temperature of 42.6 °C, extreme low temperatures of −20.3 °C, effective accumulated temperature over 10 °C of 4 500−5 200 °C, frost-free period of 160−270 d, annual precipitation of over 800 mm with even distribution, and relative humidity of over 70%. Kiwifruit seedlings prefer shading and cool environments, while mature fruit-bearing trees require abundant sunlight, with a preferred sunshine duration of 1 300−2 600 h, and a preference for diffuse light, with an optimal light intensity of 40%−45%. Kiwifruit prefers deep, loose, well-drained sandy soils with high organic matter content and a pH value of 5.0−7.9. Kiwifruit requires gentle wind, so when establishing kiwifruit orchards, it is advisable to choose south-facing, southeast-facing, and southwest-facing slopes with a slope not exceeding 30°, while also considering windbreaks to prevent strong winds and ensure ventilation and light permeability.

2 Main cultivars and characteristics

China has a rich variety of kiwifruit, with the main varieties used in production include Chinese kiwifruit, delicious kiwifruit, furry kiwifruit, and baby kiwifruit (*Actinidia arguta*), also include pollinated male varieties.

2.1 Varieties of Chinese kiwifruit

2.1.1 Hong Yang

It is suitable for areas below 1 300 m altitude, with an average annual temperature of 13−16 °C and annual rainfall of 1 000−1 500 mm. This variety has a high fruit setting rate, and achieves early and high yield. The fruits are large, uniform, cylindrical or oblong, with an average single fruit weight of 60–90 g, green or brownish-green skin, yellow-green flesh, and bright red radiating stripes along the placenta. The juice is abundant, the flesh is delicate, pure sweet, and has a clear and refreshing aroma. It is excellent for fresh consumption and processing. The fruits ripen in late August, have good storage ability, with a ripening period of 10–15 d. It is not tolerant to high temperatures and high humidity in the summer which affects the red coloration of the flesh. Additionally, it has poor disease resistance and drug resistance.

2.1.2 Dong Hong

The fruits are elongated cylindrical, with an average single fruit weight of 70–75 g, brownish-green skin, golden-yellow flesh, and bright red color around the core. The flesh is tender and juicy, with a rich sweet flavor and a strong aroma. It has a high calcium content and better storage ability than Hong Yang. The fruits will not ripen until 30–40 d after harvest, can be eaten when slightly soft, and have an edible period of over 15 d. The fruits ripen in early to mid September, have a wide adaptability and stronger resistance to soft rot and ulcer disease compared to Hong Yang.

2.1.3 Jin Yan

The fruits are elongated cylindrical, large and uniform, with an average single fruit weight of 101–110 g and a maximum weight of 175 g. The skin is yellow-brown, with golden-yellow, tender and juicy flesh, and a sweet flavor. This variety has good early fruiting and high yield. The fruits ripen from late September to mid October. They have excellent storage ability, and can be stored at room temperature for 3 months or 120–160 d under low-temperature conditions.

2.1.4 Jin Tao

The fruits are elongated cylindrical, uniform in size, with an average single fruit weight of 90 g and a maximum single fruit weight of 160 g. The skin is yellow-brown, smooth and hairless when ripe, the flesh is initially yellow-green, turning to golden yellow, with a small and soft core. The flesh is crispy, juicy, and has a moderate sweet and sour taste, with excellent quality. The fruits have good storage ability, with a ripening period of up to 25 d, and the flavor improves after storage. This variety has good heat tolerance.

2.1.5 Hua You

The fruits are elliptical in shape, with an average single fruit weight of 80–110 g, brown or yellow-brown skin, and yellow or yellow-green flesh. The flesh is delicate, juicy, with a rich aroma and a sweet taste of high quality. It has good yields, with production reaching 2 t per mu at peak harvest. The fruits have a post ripening period of 15–20 d and a shelf life of about 30 d at room temperature and can be stored for around 5 months at 0 °C. This variety exhibits some resistance to ulcer disease.

2.1.6 Cui Yu

The fruits are oval in shape, with an average single fruit weight of 85–95 g and a maximum single fruit weight of 129 g. The skin is greenish-brown, and the flesh is green, dense, tender, and juicy, with a rich sweet flavor and abundant nutrients of top quality. It has good storage ability, with a shelf life of over 30 d at room temperature and a storage period of 4–6 months under cold conditions. This variety is early bearing, high yielding, and exhibits strong resistance to adversity, including high temperatures, drought, and wind.

2.1.7 Ganmi No. 3

The fruits are oval or elongated oval in shape, with an average single fruit weight of 81.8–107.3 g and a maximum single fruit weight of 163 g. The skin is brown or yellow-brown, and the flesh is yellow, fine, juicy, with a pleasantly sweet and slightly fragrant taste of above-average quality. The fruits have a shelf life of about 40 d at room temperature following a ripening period of approximately 15 d. It exhibits strong resistance to wind and high temperatures, making it a versatile late-maturing variety suitable for fresh consumption and processing use.

2.1.8 Wuzhi No. 3

The fruits are elliptical in shape, with an average single fruit weight of 80–90 g and a maximum single fruit weight of 156 g. The skin is thin and dark green, while the flesh is green, delicate, with abundant juice, a rich and clear fragrance, and good storage ability, with a shelf life of over 20 d at room temperature following harvest. It exhibits good heat tolerance, high and stable yields, and a strong resistance to disease and pests, as well as drought resistance, making it an excellent all-around variety.

2.2 Varieties of delicious kiwifruit

2.2.1 Shizong No. 1

The fruits are oval in shape, with an average single fruit weight of 81 g and a maximum single fruit weight of 150 g. The skin is brown, while the flesh is yellow-green, delicate, with abundant juice, a good balance of sweetness and sourness, and a mild fragrance. The fruit ripening period is 15–20 d after harvest. The fruits can be stored for 36 d at room temperature of 12 °C and under the conditions of relative humidity of 75%, and can be stored for 5–6 months under refrigeration at 1–3 °C. This variety exhibits high and stable yields, with a peak yield of over 3 000 kg per mu, and shows strong resistance to drought.

2.2.2 Hayward

The fruits are oval in shape, with an average single fruit weight of 80–110 g and a maximum single fruit weight of 165 g. The skin is greenish-brown, while the flesh is green, sour and sweet, can be eaten before fully ripening, with a slightly mild taste but intense fragrance. The fruits have a long shelf life and a long ripening period, which can be stored for around 30 d at room temperature after harvest.

2.2.3 Qin Mei

The fruits are oval in shape, with an average single fruit weight of 100 g and a maximum single fruit weight of 115 g. The skin is greenish-brown, while the flesh is light green, fine in texture, with abundant juice, a fragrant taste, and a good balance of sweetness and sourness and good quality. Storage life is moderate, lasting 15–20 d at room temperature. It has a high yield, with a peak yield of up to 3 000 kg per mu.

2.2.4 Xu Xiang

The fruits are cylindrical, with an average single fruit weight of 60–70 g and a maximum single fruit weight of 137 g. The skin is yellow-green, while the flesh is green, with abundant juice, delicate texture, and a variety of fruit fragrances such as strawberry, with a well-balanced sweetness and sourness. The fruit ripening period is 15–20 d and it has a shelf life of 15–25 d. The fruits can be stored for around 30 d at room temperature and can be stored for over 3 months at 0–2 °C.

2.2.5 Gui Chang

The fruit is cylindrical and elongated, with an average single fruit weight of 84.9 g and a maximum single fruit weight of 120 g. The skin is brown. The flesh is light green, fine and crisp, with plenty of juice, and has a moderate sweet and sour taste, fragrant and delicious. It is of high quality and suitable for both fresh consumption and processing. It has early fruiting, high yield, strong resistance to adversity, resistance to nutrient deficiency disease (chlorosis), strong resistance to pests and diseases, as well as resistance to low temperatures, drought, and fruit cracking.

2.2.6 Cui Xiang

The fruit is oval, with an average single fruit weight of 92 g and a maximum single fruit weight of 130 g. The skin is yellow-brown. The flesh is green, fine and juicy, sweet and sour, refreshing, fragrant, and of high quality. It has a ripening period of 12–15 d after harvesting, and can be stored for 100–120 d under refrigeration conditions. This variety has a good yield, strong early fruiting, generally bearing fruit in the 3rd year after planting, with a peak yield of 2–3 t per mu. It has broad adaptability and strong resistance to cold and ulcer disease.

2.2.7 Huamei No. 2

The fruit is long and conical, with an average single fruit weight of 112 g and a maximum single fruit weight of 205 g. The skin is yellow-brown. The flesh is yellow-green, fine and juicy, with a moderate sweet and sour taste and a rich aromatic flavor. The fruit is resistant to storage and can be stored for 30 d at room temperature without ripening. The variety has stable and high yield, and strong resistance to drought and disease.

2.3 Varieties of furry kiwifruit

Hua Te

The fruit is elongated and cylindrical, with an average single fruit weight of 82–94 g and a maximum single fruit weight of 132.2 g. The skin is brownish-green. The flesh is green, fine and slightly sour, with a vitamin C content of 628.37 mg/100 g, and superior quality. It has good storage resistance and can be stored for 3 months under room temperature conditions. This variety has strong adaptability, tolerance to poor soil and drought, high fruiting ability, and stable and high yield.

2.4 Varieties of baby kiwifruit

2.4.1 Kui Lü

The fruit is flattened oval, with an average single fruit weight of 18.1 g and a maximum single fruit weight of 32 g. The skin is smooth and green. The flesh is green, fine and juicy, with a balanced sweet and sour taste and a rich flavor, suitable for both fresh consumption and processing. This variety has stable and high yield, strong resistance to adversity, no frost damage in areas with an absolute low temperature of −38 °C, and is suitable for cultivation in areas with a frost-free period of more than 120 d and an annual accumulated temperature above 2 500 °C at a temperature of 10 °C or above.

2.4.2 Feng lü

The fruits are round, with an average single fruit weight of 8.5 g. The skin is smooth and green. And the flesh is delicate, juicy, and has a balanced sweet and sour taste, with good quality. It is a dual-purpose variety suitable for both fresh consumption and processing. This variety has stable and high yields, and is suitable for cultivation in regions with a frost-free period of over 120 d and an annual accumulated temperature above 2 500 °C at a temperature of 10 °C or above.

2.5 Pollinated male varieties

2.5.1 Moshan No. 4

This variety has compact plants, with each inflorescence usually having 5 flowers, up to a maximum of 8 flowers. The bud period is about 35 d, with a flowering period of 13−21 d. It has a large amount of pollen, with a 75% pollen germination rate. This variety has a long flowering period and can be used as a pollination tree for early, mid-season, and late-ripening Chinese kiwi and delicious kiwi varieties.

2.5.2 Matua

This variety has weaker tree vigor and can blossom in the second year after planting. It has an early flowering period, a large amount of flowers, a high pollen quantity, and a 64% pollen germination rate. It has a long flowering period of 15−20 d and can be used as a pollination variety for early to mid-flowering periods.

2.5.3 Tomli

This variety has a relatively late flowering period, a large amount of flowers, a 62% pollen germination rate. It has a concentrated flowering period of 5−10 d. It is primarily used as a pollination tree for late-flowering varieties.

3 Cultivation and management techniques

3.1 Orchard planning and construction

3.1.1 Orchard planning

The selection of the orchard site should take the environmental requirements for kiwifruit growth and the convenience of transportation into consideration. Based on the conditions of transportation, water sources, slope, and climate, the planning of the orchard should include road systems, drainage and irrigation systems, functional zoning, and pollination tree arrangements.

3.1.2 Roads, drainage and irrigation systems

The main roads are 5–7 m wide and can accommodate cars for transportation. Branch roads are 3–5 m wide, allowing passage of small vehicles and farming machinery, and also serve as boundaries for operational areas within the orchard. Operational paths are 1–1.5 m wide and primarily for pedestrian use.

The drainage system is generally integrated with the road system to form a comprehensive drainage network for the entire orchard. The drainage ditch along the main road serves as the main drainage channel for the orchard, while the ditches along the branch roads are for drainage in various operational areas.

3.1.3 Functional zoning

Structures within the orchard, including rest areas, tool rooms, fruit sorting and packaging areas, production material storage, caretaker's quarters, medicine storage, and septic tanks, should be built in areas that are convenient for transportation and overall orchard management and operations. The size of operational areas depends on the terrain, ideally taking a rectangular shape. For flat orchards, an area of 15–20 mu is suitable, while in mountainous orchards, the operational area can be slightly smaller. Each area should be equipped with a manure pit, composting area, and small-scale reservoir. Drip irrigation, a water-saving technique that can also be combined with fertilization, should be given priority, followed by furrow irrigation, sprinkler irrigation, and seepage irrigation.

3.1.4 Pollination tree arrangement

Kiwifruit is a dioecious plant and requires pollination trees to be arranged. The principle for selecting pollination trees is to choose male varieties that flower at the same time or slightly earlier 1–2 d than female varieties, with a large quantity of flowers, abundant pollen, a high pollen germination rate, and good compatibility. The ratio of male to female plants should be $(5-8):1$.

3.1.5 Cultivation mode selection

The choice of cultivation mode should be based on local weather conditions, topography, variety characteristics, etc.

Commonly used support structures include "T"-shaped trellises and pergolas. Both share a similar basic structure, composed of columns, crossbeams, and framework, with the main difference being that the "T"-shaped trellis forms a "T" support at the top of the column, usually leaving only two main branches forming a "T" shape for growth. Pergolas, on the other hand, allow for three or more main branches evenly distributed across the framework. Pergolas are often used in flat orchards, while "T"-shaped trellises are more common in mountainous orchards, with both structures sometimes used in combination.

3.2 Planting

The general practice is to perform these procedures after the leaf drop until the sap flows the next year. The dimensions for the planting pit are 0.6–0.8 m in length, width, and depth, and during filling, 20 kg of organic fertilizer and 1 kg of calcium superphosphate should be applied to each pit, with topsoil first followed by subsoil. Select robust, disease-free, and undamaged seedlings for planting according to the plan. Roots should not directly touch the fertilizer during planting, and the root collar should be slightly above ground level. After planting, create a tree plate with a diameter of 50–60 cm, higher around the edges and slightly lower in the middle, water the plant thoroughly, cover with a layer of fine soil, and then cover the tree plate with plastic film or straw to reduce water evaporation. After planting, when the seedlings reach a certain height, conduct shaping pruning. Cut the stem at the third or fourth full bud from the ground, and the first bud below the cut should be full. After the seedlings sprout and survive, timely check and replant as needed. Shade should be provided during the early seedling stage, and nitrogen and compound fertilizer should be applied in small amounts multiple times at 0.8%–1%. After the young seedlings produce new shoots, promptly set up bamboo poles or sticks beside the young trees and tie the main trunk to them to lead it to a trellis or support structure.

3.3 Structural pruning

3.3.1 Tree structure and shaping

The tree structure consists of the main trunk, main shoot, lateral shoots (fruit-bearing branches), and fruiting branches. After planting, select a vigorous, vertically growing shoot as the main shoot from the new shoots that have sprouted after dry pruning, and remove all other buds. When the shoot has reached a height of 20–30 cm from the support structure, cut off the apex to promote the growth of new shoots. Select 2–4 shoots with appropriate orientations as permanent main shoots for training. For trellis or pergola cultivation, select 3–4 evenly distributed new shoots to train as permanent main shoots. In the case of "T"-shaped trellis cultivation, select two main shoots respectively from the centre of the trellis surface along the rows of cross-reversed bondage in a "T" shape. In the following spring, select one lateral shoot every 30–50 cm along the main shoot as fruit-bearing branches, and remove the rest. Lateral shoots should be as vertical as possible to the main shoot. Choose a fruiting branch or group of fruiting branches with a distance of approximately 20 cm between each on the lateral shoot for fruiting.

3.3.2 Winter pruning

After leaf fall and before the second year's fruiting period, pruning is generally carried out. The main techniques include short pruning, thinning, and sparse pruning. Thin and weak branches, dead branches, diseased and insect-infested branches, overly dense large branches and vines, crossing branches, overlapping branches, competitive branches, and underdeveloped lower branches without value should all be removed. As a result, the distance between fruiting branches is maintained at 30–50 cm, evenly distributed on the trellis, forming a good fruiting system with 3–4 fruiting branches per m^2 of trellis and 10–15 buds left on each fruiting branch.

3.3.3 Summer pruning

Summer pruning can be conducted from bud break to leaf fall. The methods include bud rubbing, shoot pinching, thinning, flower thinning, fruit thinning, and tying of new shoots. It is carried out in four stages, with the first stage at bud break, the second stage one week after flowering, the third stage around one month after flowering along with thinning of fruit, and the fourth stage during the rapid growth period of the fruit along with thinning of overly dense or deformed fruit, and diseased or insect-infested fruit. After summer pruning, the leaf-to-fruit ratio per plant is maintained above (6–7) : 1, and the leaf area index is controlled at 3.5–4.0.

3.3.4 Pruning of male plants

Summer pruning for male plants should be conducted promptly after flower shedding, with a focus on thinning and lengthening, and the shaping of the tree structure according to the trellis system. Winter pruning mainly involves the removal of dead and slender branches, as well as the trimming of the top portions of developing branches, diseased and crowded branches, and unnecessary long shoots.

3.4 Water and fertilizer management

In the process of high-quality and efficient kiwifruit cultivation, it is important to adhere to the principle of prioritizing organic fertilizers with supplemental use of chemical fertilizers. The use of nitrate-based nitrogen fertilizer is prohibited, and the use of chlorine-containing fertilizers and compound fertilizers is restricted. During the seedling period, nitrogen fertilizer takes precedence, with each plant being supplemented with 5-10 kg of organic fertilizer and 20-50 g of compound fertilizer, to be applied in small amount and multiple times using a concentrated 0.8%-1% nitrogen fertilizer and compound fertilizer.

For mature trees, base fertilization should be conducted promptly after harvest, preferably in October to November, with an emphasis on organic fertilizer which should be applied after full fermentation. The amount of organic fertilizer applied should be based on fruit yield, with each plant receiving 20-60 kg of organic fertilizer. Timely supplemental fertilization during the growth period primarily involves the use of quick-acting fertilizers, with pre-flowering nitrogen fertilizer applied 10-15 d before bud break. In May, potassium-based fertilizer is applied to promote fruit development, in combination with nitrogen, phosphorus, and micro-nutrient fertilizers. About 25 d before fruit maturation, premium fruit fertilizer, primarily consisting of phosphorus and potassium, is applied. Fertilization is mainly carried out using trench fertilization, hole fertilization, and irrigation-based fertilization. After fertilization, timely irrigation should be carried out to improve fertilizer utilization. Based on the growth of the tree and the fruit, foliar fertilization, especially micro-nutrient fertilization, can also be conducted. For example, during the peak flowering period, spraying with a dilution of borax (1 : 350) can enhance fertilization capability. Starting from the fruit enlargement period, spraying with a dilution of potassium dihydrogen phosphate [1 : (300-350)] every 15 d can help increase sugar content and quality of the fruit.

Kiwifruit leaves are large and have high growth requirements, preferring water but avoiding water logging. During the growing season, soil moisture content should be monitored every 3-4 d, and irrigation should be carried out when the soil moisture content falls below 60% of field moisture capacity. Before bud break, pre-flowering, and post-flowering, irrigation should be conducted based on soil moisture conditions, with water control needed during flowering. The fruit enlargement period is critical for water requirements and may require 2-3 rounds of irrigation, with the possibility of inter-row misting to increase air humidity if conditions allow. Around 15 d before fruit maturation, irrigation should be ceased, and one final irrigation should be conducted before dormancy. The amount of water applied in each irrigation should bring soil moisture content to over 85% of field moisture capacity, with an infiltration depth of over 40 cm.

3.5 Prevention and control of major diseases and pests

The major diseases of kiwifruit include ulcer disease, blossom blight, fruit rot, and root rot, while the main pests include golden tortoise beetles, leaf roller moths, scale insects, and leaf hoppers. Integrated pest management practices such as agricultural control, physical control, biological control, and chemical control should be employed for daily management. Common diseases & pests and their control methods are detailed in Table 2-2 and Table 2-3, respectively.

Table 2-2 Major diseases and their control methods

Major diseases	Control period	Control method	Other control measures
Ulcer disease	Early spring	Remove diseased branches less than 2 years old; scrape off lesions on large branches; apply 30% copper hydroxide SE (1 : 500), or 20% thiazole zinc SE (1 : 200) to wounds	1. Clean the orchard thoroughly in winter, whitewash trunks, prune and remove diseased shoots, and burn them; 2. Choose well-drained planting sites and implement rain-sheltered cultivation; 3. Disinfect cutting tools with 75% ethanol
	Budburst to pre-flowering	Use 1.5% benziothiazolinone EW [1 : (600-800)], 77% copper hydroxide [1 : (600-800)], or 80% mancozeb WP (1 : 800) for fungicide treatment, spraying 2-3 times	
	May to August	Regularly inspect and remove bark curling; apply copper oxychloride or Bordeaux mixture to lesions	
	After fruit harvest to pre-defoliation	Apply 3% zhongshengmycin WP (1 : 600), spray every 10-15 d and last for 3-4 times	
	After defoliation	Apply EM bacteria solution or 15% Wuningmycin technical (1 : 800) to the ground	
	After winter pruning to budburst	Evenly spray 3-5°Bé of lime sulfur mixture, 1-2 times 1.5% benziothiazolinone EW [1 : (500-600)] or 80% amobam WP (1 : 800)	
Blossom blight	Budburst period	Spray the entire orchard with 3-5°Bé of lime sulfur mixture	1. Improve ventilation and light conditions for flower buds; 2. Strengthen orchard fertilization and irrigation management; 3. Remove diseased flower buds
	Flowering and leaf expansion period	Spray the whole tree with 65% Zineb WP (1 : 500), or 50% tuzet (thiram、ziram、urbacide) WP (1 : 800), or 0.3°Bé of lime sulfur mixture, every 10-15 d	

Continued Table

Major diseases	Control period	Control method	Other control measures
Fruit rot	Budburst period	Spray with 3–5°Bé of lime sulfur mixture	1. Thoroughly clean the orchard in winter, eliminating disease-carrying agents; 2. Improve orchard management, apply additional basal fertilization, and enhance tree vigor; 3. Improve ventilation and light conditions, and reduce shading
	Two weeks after blossom to fruit enlargement period	Spray with 80% thiophanate-methyl WP (1∶1 000), or 80% captafol WP (1∶1 000). Apply 2–3 times 0.2%–0.3% calcium for 2–3 times	
Root rot	March and late May	Drench the roots with 58% metalaxyl and mancozeb WP (1∶500)	1. Ensure proper drainage during the rainy season; plant at an appropriate depth, and ensure organic fertilizers are well-rotted; 2. Amend heavy clay soils with sand

Table 2-3 Major pests and their control methods

Pest name	Control period	Used pesticides	Other control measures
Golden Tortoise Beetle	1. When emerging from soil 2. May to July	Use 50% phoxim EC [1∶(800–1 000)] to kill the larvae in the soil 2–3 times; use 20% fenvalerate EC (1∶1 500), spray 2–3 times in 2–3 d before flowering and after flowering	1. Use black light or insecticidal lamp for attraction; 2. Knock the branches to dislodge and capture the pests; 3. Use bait of sugar and vinegar liquid with dipterex to attract and kill the pests; 4. Turn over the soil in the orchard in winter to dig out overwintering larvae
Leaf Hopper	Mid-late May to mid-late July	Use 70% imidacloprid WDG [1∶(4 000–5 000)] for nymph control; use 1.5% pyrethrin EW [1∶(600–1 000)] for adult control	Clear the orchard and scrape off egg masses and burn them
Fruit Fly	Larval stage	Use 0.1% avermectin lure point spray on plants	1. Immediately remove and dispose of infected fruits;
	Adult emergence period	Ground spray with 50% phoxim EC (1∶700); deep till the soil;	

Continued Table

Pest name	Control period	Used pesticides	Other control measures
Fruit Fly	Before adult egg laying	Spray 90% crystalline dipterex plus 3%–5% sugar, spray once every 5 d, 2–3 times in a row	2. Bag the fruits, use pheromone traps, etc.
Dienerella	Fruit swelling period	Spray 522.5 g/L cypermethrin chlorpyrifos EC [1 : (1 500–2 000)] or 2.5% highly effective cyhalothrin EC (1 : 2 000)	1. Eliminate overwintering sources of pests; 2. Control weeds in the orchard to reduce pest hiding places; 3. Properly thin the fruits to avoid excessive overlapping and crowding
Fruit-feeding moths	Mid to late June to mid to late October	Spray 1% celastrus angulatus EW [1 : (1 000–1 500)]	Use black light or bait of sugar and vinegar to attract and kill the pests
Piercing Moths	March to October	Spray 20% fenpropathrin EC [1 : (1 000–1 500)]	Use a small wire to puncture and kill the larvae when holes or insect feces are found
Scale Insects	February to November	Spray 25% buprofezin emulsion EC [1 : (1 000–1 200)] or 48% chlorpyrifos EC [1 : (800–1 000)] spray	1. Artificially eliminate overwintering female adult and larvae; 2. Use barrier methods to reduce tree borer activity; 3. Apply lime sulfur mixture to the scale insect affected branches; 4. Collect and destroy insect eggs in the soil around tree trunks in winter

3.6 Harvest

The harvesting period of kiwifruit is influenced by various factors. Currently, the most widely used method for determining the harvest period is based on the content of soluble solids. Generally, when the content of soluble solids reaches 6.5%–8.0%, it is considered ready for harvest. After harvesting, the fruits need to be graded, which can be done manually or mechanically. Manual grading can be done using fruit sorting boards, based on the different transverse diameters of the fruits, typically using a 5 mm difference in diameter to categorize the fruits. Mechanized weight grading is based on the weight of the selected products compared to a pre-set weight for grading.

References

CHEN D Y, HUANG J M, 2004. Pollution-Free and Efficient Cultivation of Kiwifruit [M]. Beijing: Jindun Publishing House.

HUANG H W, 2001. Efficient Cultivation of Kiwifruit [M]. Beijing: Jindun Publishing House.

HUANG H W, ZHONG C H, LI D W, et al., 2013. Actinidia Germplasm Resources in China [M]. Beijing: China Forestry Publishing House.

LANG B B, ZHU B, XIE M, et al., 2016. Variation and probability grading of the main quantitative characteristics of wild *Actinidia eriantha* germplasm resources [J]. Journal of Fruit Science, 33(1): 8−15.

LIU L Q, WANG D, 2016. Cultivation and Pest Control of Kiwifruit [M]. Beijing: China Agricultural Press.

ZHONG C H, HUANG H W, 2018. Forty Years of Kiwifruit Research and Industry in China [M]. Hefei: University of Science and Technology of China Press.

ZHONG C H, HUANG W J, LI D W, et al., 2021. Analysis of the global kiwifruit industry development and fresh fruit trade dynamics [J]. China Fruits (7): 101−108.

❖ Key Techniques for Ginseng Fruit Cultivation

1 Overview of industrial development

Ginseng fruit (*Solanum muricatum* Aiton) is a perennial herbaceous plant in the genus *Solanum* L. under the Solanaceae Juss. family, native to the Andes of South America. It was introduced into China in the 1980s and is now mainly distributed in Yunnan, Qinghai, Gansu, Guizhou, Sichuan and other provinces, making it an emerging characteristic fruit. The cultivation of ginseng fruit was introduced in Yunnan in the early 1990s. Yunnan's unique geographical location, sufficient light, and suitable temperature provide excellent climate and ecological conditions for the development of the ginseng fruit industry in Yunnan Province, enabling the annual supply of ginseng fruits produced in different regions of the province, and the yield has steadily ranked first in China for many years.

According to statistics, the cultivation area of ginseng fruit of China in 2022 is about 320 000 mu, including 270 000 mu in Yunnan, 20 000 mu in Gansu, and 30 000 mu in other regions. Yunnan's main seasonal cultivation area of ginseng fruit is mainly concentrated in Shilin County, Kunming City, with an area of 200 000 mu, while the sporadic cultivation is distributed in Qujing City, Yuxi City, Chuxiong Prefecture, Dali Prefecture, Baoshan City, with an area of about 20 000 mu. The scaled off-season cultivation began in 2017, mainly distributed in Honghe Prefecture, Wenshan Prefecture, Chuxiong Prefecture, Dehong Prefecture, Pu'er City, Yuxi City, Lincang City, etc., with an area of about 50 000 mu. The product is launched from December to May of the following year, during the slack season for fruits, with high prices, ideal economic benefits, and strong development momentum. It is expected that the area of off-season cultivation will reach 250 000 mu within 5 years, equal to the seasonal cultivation area.

Yunnan has 27 leading enterprises of ginseng fruit planting and circulation, 45 cooperatives of ginseng fruit specialized planting, including 38 in Shilin County. "Xijiekou ginseng fruit" has been successfully registered as a geographical indication certification trademark (China).

2 Main cultivars and characteristics

At present, the cultivars of ginseng fruit in Shilin include Yuanguo No. 1, Yuanguo No. 2 and Dazi. Yuanguo No. 2 is the main cultivar, which accounts for 95% of the cultivated area in the county. It is a fresh fruit variety with good taste and high nutritional value, which is popular among consumers. Dazi is a dual purpose variety for both fruit and vegetable use with high yield and strong disease resistance, but it only has a small share of the market.

2.1 Yuanguo No. 1

Botanical characteristics: It is a raceme with star-shaped calyx, purple with white lace in the middle. There are 5 yellow stamens surrounding the pistil and there is 1 green pistil, with its stigma higher than the stamens. The fruit is round or heart-shaped with a round or slightly pointed tip, young fruit is white while the ripe fruit is yellow with purple stripes or patterns, and the pulp is yellow, fruity and slightly fishy. The blade is lanceolate and dark green, with purple tender petiole. The stem is cylindrical, while tender stem colored by purple, gradually turning green. The root protrusion is prone to occur when humidity is high. The plant has poor heat resistance, and the pollen activity is low when the temperature exceeds 28 °C, which cannot complete self-pollination, resulting in low parthenocarpy rate, stiff fruits and a prolonged ripening period. The suitable temperature for fruiting period is 18−25 °C during the day, 12−18 °C at night, and the air relative humidity is 50%−75%.

Production performance: It can be harvested 110−120 d after transplanting and final planting.

Planting area: It is mainly distributed in Shilin County, Kunming City.

2.2 Yuanguo No. 2

Botanical characteristics: It is known as the Yuanguo in Shilin County, which is a fresh-eating type. The plant is an indeterminate growth type, with vigorous growth and strong germination ability of young shoots, which can be infinite and continuous germination after being removed. It is difficult to set fruit in the early stage, and the fruiting period is delayed. The fruit shapes round or oblate, with an average single fruit weight of 100 g, and a maximum single fruit weight up to 260 g. The young fruit is green and white, with purple stripes when it grows up. When it is fully ripe, the pericarp is orange, and the purple stripes turn to purplish black. The pulp is golden yellow, fragrant and refreshing, sweet and juicy, and the flavor is pure.

It has normal heat resistance, temperature above 30 °C is not conducive to bearing, parthenocarpy rate is low, easy to form stiff fruit, and only some can ripen normally. The large temperature difference between day and night is conducive to improving quality, unripe fruits have a sour taste, with ripe fruits being the main ones on the market. The pulp is golden yellow and fruity, soluble solids content of 10%–13%, with high quality.

Production performance: It can be harvested about 110 d after transplanting and final planting.

Planting area: The seasonal cultivation is mainly distributed in Shilin County of Kunming City, while sporadic cultivation is distributed in Qujing City, Yuxi City, Chuxiong Prefecture and Dali City. The scaled off-season cultivation is mainly distributed in Honghe Prefecture, Wenshan Prefecture, Chuxiong Prefecture, Dehong Prefecture, Pu'er City, Yuxi City, Lincang City, etc.

2.3 Dazi

Botanical characteristics: It is also called vegetable fruit, with a raceme, star-shaped calyx, mauve with white lace. There are 5 yellow stamens surrounding the pistil and 5 green sepals. The fruit shapes long and oval, with a pointed tip and a large size, with an average single fruit of 250 g, and a maximum single fruit weight of 800 g. The young fruit is white or with light green stripes, while the ripe fruit has purple stripes or patterns. After fully ripening, the surface is faint yellow with purple stripes or flake-like patterns. The pulp is golden yellow and fruity, smells fishy. The flowering and fruiting temperature is 18–30 °C, with a high fruiting rate. The plant grows vigorously and is an indeterminate growth type with strong branching ability. The blade is lanceolate and green. The stem is cylindrical and green, and the root protrusion is prone to occur when the humidity is high. This variety has strong heat resistance, fertilizer resistance and cold resistance, which is mainly used as vegetables, and the soluble solids content of ripe fruit is about 9%.

Production performance: It can be harvested 95–100 d after transplanting and final planting, with a high fruiting rate and strong disease resistance.

Planting area: It is mainly cultivated in protected areas of Gansu, because of a relatively low price in the Shilin County, the cultivation area is decreasing year by year.

3 Cultivation and management techniques

3.1 Environmental requirements for the production area

3.1.1 Natural environment

Ginseng fruit is easily affected by climate, environment and cultivation techniques. Data show that the annual average temperature of South American ginseng fruit origin is 15.2 °C, the average temperature of the hottest month is 19.6 °C, the extreme maximum temperature is 32.4 °C, the extreme minimum temperature is −0.1 °C, the annual precipitation is 1 242 mm, the annual evaporation capacity is 762 mm, and the annual sunshine duration is 2 094 h. Therefore, ginseng fruit needs a climate environment which is cool, dry but not droughty and has sufficient sunlight. Appropriate environmental conditions are an important prerequisite for obtaining high-quality ginseng fruit products. According to the requirements of environmental conditions for ginseng fruit, the planting plot should be well lit, with unobstructed water supply and drainage, an annual precipitation of about 1 000 mm, a daytime temperature of 20–28 °C, and a nighttime temperature of 15–20 °C. In order to prevent low temperature and late frost in spring planting, the nighttime temperature should be above 8 °C for more than continuously 10 d before starting to transplant. In winter frost-free areas, the Beginning of Autumn can be selected for planting.

3.1.2 Soil requirement

The soil should be convenient for drainage and irrigation, with good air permeability, high organic matter content, loose texture, and not easy to clump after watering. The root system of ginseng fruit is well developed, mainly distributed in the soil layer of 0–30 cm. Therefore, the cultivation of a fine root system is the key to improving the quality of ginseng fruit and loose soil is conducive to the sustainable growth of the root system. Yuanguo No. 1 and Yuanguo No. 2 have poor resistance to fertilizer, especially sensitive to nitrogenous fertilizer. Immoderate nitrogenous fertilizer is prone to excessive growth, less flowering, and difficult bearing fruit.

3.1.3 Irrigation water requirements

Irrigation water shall be implemented in accordance with the GB 5084—2021 standard, which means that the basic water quality shall meet the requirements in Table 2-4.

Table 2-4 Limits of basic control items for farmland irrigation water quality

Serial number	Items		Crop type: vegetables	
			Processing, cooking and peeling vegetables	Raw vegetables, melons and herbal fruits
1	pH		5.5–8.5	
2	Water temperature/°C	≤	35	
3	Suspended solid/ (mg/L)	≤	60	15
4	5-day biochemical oxygen demand (BOD_5) / (mg/L)	≤	40	15
5	Chemical oxygen demand (COD_{Cr}) / (mg/L)	≤	100	60
6	Anion surfactant/ (mg/L)	≤	5	
7	Chloride (in Cl^-) / (mg/L)	≤	350	
8	Sulfide (in S^{2-}) / (mg/L)	≤	1	
9	Total salt/ (mg/L)	≤	1 000 (non saline-alkali land), 2 000 (saline-alkali land)	
10	Total lead/ (mg/L)	≤	0.2	
11	Total cadmium/ (mg/L)	≤	0.01	
12	Chromium (Ⅵ) / (mg/L)	≤	0.1	
13	Total mercury/ (mg/L)	≤	0.001	
14	Total arsenic/ (mg/L)	≤	0.05	
15	Fecal coligroup count/ (MPN/L)	≤	20 000	10 000
16	Ascaris eggs count/ (/10 L)	≤	20	10

3.2 Cultivation technique

3.2.1 Variety selection

There are varieties including Yuanguo No. 1, Yuanguo No. 2 and Dazi, etc.

3.2.2 Seedling raising

3.2.2.1 Virus-free tissue culture seedling

The excellent individual plant that is cultivated in the same year and grows well is selected, and 2-3 cm young branches are cut as spare materials on a sunny afternoon. Removing leaves from the branches, the 1.5 cm stem tip is cut and put into a clean beaker of 250 mL, adding 1 drop of vegetable detergent and 150 mL tap water, stirring and soaking for 5 min, then rinsing with flowing tap water for 10 min, and transferring it to a super clean workbench for disinfection. After being treated with 70% ethanol for 30 s, the samples should be washed three times with sterile water, then placed into a sterilized 100 mL beaker and immersed in 0.1% mercury dichloride for 8 min. During the immersion, the bottle is continuously shaken to ensure complete disinfection of the outer branches. Finally, the samples are washed repeatedly with sterile water for five times.

The shoot tip growing point that is less than 0.2 mm is stripped from the sterilized material and inoculated on the induction culture medium of MS+KT0.1+IAA0.1+GA0.1+3% sucrose+5 g/L agar for induction of sterile seedlings. When the sterile seedlings are 5−6 cm high, a virus test is performed. After testing, the virus-free sterile seedlings are cut and transferred to the subculture medium, which can be MS culture medium without hormones, with 20−25 °C culture temperature and 2 000−3 000 lx light intensity, and the subculture is conducted every 25−30 d. The ginseng fruit seedlings are easy to root, and the MS culture medium without hormones can be used as the rooting culture medium. The 2 cm shoot tip is cut and inoculated into the rooting culture medium with 20 plants per bottle, rooting begins after 10 d, and the seedlings can be out of the bottle and hardened off after 15−18 d.

The seedling culture basin with 98 holes is adopted, and the cultivation medium is a formula matrix of peat and perlite with a ratio of 9 : 1. After bottle seedling cultivation, the seedlings should be watered thoroughly by spraying, covered with a small arched shed of 70% shading, and the temperature should be kept at 20−30 °C with relative humidity of 85%−90%. After a week, 0.1% foliar fertilizer should be applied every 3 d, the light is gradually increased after 15 d, and the seedlings can be transplanted into the field when they are 15−20 cm after 30−40 d.

3.2.2.2 Cuttage propagation

Ginseng fruit has many lateral branches, which is easy to cuttage survive. Cuttage propagation is the simplest way of reproduction, and is also the most important way of reproduction of ginseng fruit seedling raising at present.

Establishment of female parent plant garden: The female parent plant garden should be established in a well-isolated area far from the ginseng fruit planting area for cuttage and cutting. The female parent plant garden is cultivated in a protected area, using virus-free tissue culture seedlings as mother plants, with a cultivation density of about 8 plants per square meter, and about 5 000 plants per mu. After 70−90 d of cultivation, the shoots can be done cuttage in batches.

The 98-hole seedling culture basin is adopted, and the cuttage medium is a formula matrix of peat and perlite with a ratio of 9 : 1. The 8−12 cm strong lateral branch is cut as cuttage branch. Before cuttage, the cutting branch need to be trimmed, 1−2 leaves are left on the top of the branch, while the rest of the leaves are cut off, and the branch is cut into 6 cm shoot tip, with 1−2 internodes. The cuttage depth is 2−3 cm, and 1 stem segment is inserted into each hole. After the cuttage of each plate is done, it should be placed neatly into the seedbed and sprayed with water to make the matrix reach the maximum moisture capacity, covered with a small arched shed of 70% shading, and the temperature should be kept at 25−32 °C with relative humidity of 85%−90%. After 10 d, 0.1% foliar fertilizer should be applied every 3 d, the light is gradually increased after 15 d, and the seedlings could be sold as commercial use after 40 d when they are 20 cm and strong.

3.2.3 Cultivation

It can be implemented in accordance with the T/YGIIA 001—2023 standard.

3.2.3.1 Preparation before planting

It is recommended to rotate crops reasonably, avoiding continuous cropping with solanaceae crops. The land should be prepared to make the furrows flat and make the soil fine, and 1 000−1 500 kg of composted farmyard manure and 10−20 kg of balanced compound fertilizer should be applied per mu, using hole applying as base fertilizer and fully mixed with soil. The furrow depth should be 1.2 m, with 20−30 cm in height and 50 cm in width.

3.2.3.2 Planting density

One row should be planted in each furrow, a single plant should be planted with a plant spacing of 60 cm and a row spacing of 120 cm, 800−1 000 plants should be planted per mu.

3.2.3.3 Transplanting in spring (main season)

Transplanting and final planting time vary slightly due to different altitudes. Planting should be started after the end of the local frost period in principle. Early planting can prolong the fruiting period and obtain higher yields. In seasonal cultivation, final planting should be started in early April in low-altitude areas (about 1 700 m), and on April 15th in high-altitude areas (about 2 100 m). After final planting, the roots should be watered sufficiently and then covered with plastic mulching, covering black or silver plastic mulching. Plastic mulching can inhibit the growth of weeds, improve the soil temperature and retain moisture, which is beneficial to plant growth. However, plastic mulching with high quality and long service life should be selected, which can reduce the subsequent management investment and create a ideal environment for the healthy growth of plants. The mulching can be covered while planting, and then digging the seedlings after covering. With fine soil, the hole that seedlings dig out from the plastic mulching and around area are compacted to achieve thermal insulation and moisturizing effect.

3.2.3.4 Transplanting in autumn (off-season)

Off-season cultivation is carried out in warm areas without frost in winter using the way of field culture. The suitable areas are mainly distributed in Honghe Prefecture, Wenshan Prefecture, Yuxi City, Chuxiong Prefecture, Pu'er City, Lincang City and Dehong Prefecture with an altitude of 600−1 200 m. For large-scale cultivation, mountainous or semi-mountainous areas with irrigation conditions and convenient transportation should be selected. Final planting is carried out from August to October. The remaining information is the same as seasonal transplanting.

3.2.3.5 Seedling digging

After planting under plastic mulching, seedlings should be dug out quickly and timely, especially in sunny weather. If the seedling is not dug out in time, the temperature under the plastic mulching will be too high, causing dehydration and burn of seedlings. In the mild case, some function leaves will be burned, and the seedling need a long time for recovery, affecting the yield, while in the severe case, the seedling will die of dehydration during transplanting. Therefore, the seedling should be dug out while covering the plastic mulching, and the seedling digging and sealing should be conducted out at the same time, otherwise, the high temperature gas from the hole of digging seedling will also cause seedling burn. In addition, the seedling digging hole should not be too small, with a diameter of about 10 cm in general, so as to facilitate later fertilization and rainwater infiltration.

3.2.4 Cultivation of setting up support and inducing vine

3.2.4.1 Preparation before planting

It is the same with 3.2.3.1.

3.2.4.2 Planting density

One row should be planted in each furrow, a single plant should be planted with a plant spacing of 30 cm and a row spacing of 120 cm, and 2 000 plants should be planted per mu.

3.2.4.3 Transplanting and final planting in spring (main season)

It is the same with 3.2.3.3.

3.2.4.4 Transplanting and final planting in autumn (off-season)

It is the same with 3.2.3.4.

3.2.4.5 Setting up vine-inducing support

Ginseng fruit branches have strong germination ability and weak erectness, which causes a heavy burden after bearing fruits and difficulty in erect growth. Without pruning, hanging branches on support may easily cause the plant lodging and overshadowing, which affects ventilation and light transmission and leads to diseases and pests, so as to affect bud differentiation, flowering, and fruiting. The cultivation of setting up support and inducing vine can significantly improve the bearing rate and fruit quality.

The method is to use cement columns, wooden stakes or bamboo poles as vertical columns, and to build a support that is parallel to the ridge by iron wires, 1.5−1.8 m high, one ridge per support. When the seedlings are about 30 cm high, the vines should be lifted in time. 1−2 strong branches are selected from each plant, clamped by wire clips, using wire to twine the branches and pulling them onto the support, and all the other branches should be trimmed. The stability and support of the vine-inducing support built by this method are rather strong, but the cost of human and material resources is relatively high. Three bamboo poles can be directly built into a triangle as the vertical column of each cluster of ginseng fruit plants, or only a single bamboo pole as the vertical column, and the ginseng fruit branches can be directly glued to the vertical column with strong adhesive tape. Although the above methods are simple and easy to practice, they have poor stability and are easy to be blown down by the wind.

3.2.4.6 Vine-inducing

When the plant height reaches 30−35 cm (the plant begins to tilt), the vine can be induced, with nylon and cotton thread, etc. With the elongation of the vine, every 10−15 d, the hanging thread is wound around the branches in the shape of "S" to induce the vine on the support. The vine-inducing method of ginseng fruit is similar to the conventional tomato vine-inducing, and the hanging thread cannot be tightened too much immediately (the plant stem just can be basically straight), because with the growth of the plant, it is necessary to continuously wind the vine, and at the same time, the thread is too tight, which will cause damage to the plant skin, resulting in bacterial infection and disease. In addition, the knot should be a noose, so that when the thread is not long enough, the induced vine can be put down. The cultivation of vine-inducing is conducive to ventilation and light, reducing diseases and achieving higher yields.

3.2.5 Pruning

3.2.5.1 Pruning for general cultivation

The lateral branches of ginseng fruit have strong germination ability. After 60 d of cultivation, the plants grow to about 20 cm. According to the soil fertility, 4−6 strong bearing branches are generally reserved, following the principle of keeping more branches when the soil has excellent fertility, and fewer branches when the soil has low fertility. The bearing branches are evenly distributed, which is conducive to ventilation and light transmission. In the cultivation process, the old and diseased leaves at the lower part should be removed promptly according to the actual situation, so as to provide sufficient sunlight and well-ventilated conditions for the development of ginseng fruit.

3.2.5.2 Pruning for vine-inducing cultivation

Spring pruning: When the plant grows to about 20 cm, the main branch and trunk of ginseng fruit shall retain 2–3 strong bearing branches. Others are the same as conventional cultivation.

Thinning fruit: Thinning fruit should be conducted after the fruit setting. Large fruits with a nice shape should be reserved, and small, malformed, and diseased fruits should be thinned out. Generally 2–3 fruits are left for each inflorescence.

3.2.6 Weeding

Weed should be removed timely by manual or mechanical methods, with minimal or no use of chemical herbicides to avoid soil accumulation of residues that damage the ecology and soil structure. Weed in areas not covered by plastic mulching and ridges or furrows should be removed or pulled out by hand promptly. In addition to manual removal of weeds on the soil surface, herbicides with low toxicity, few residues and short residual period can be used for weeding of road surfaces and edge weeds.

3.3 Water management

During the early stage of final planting, from mid April to the end of June before the rainy season, watering should be done in time according to the soil moisture to ensure the normal growth of the plants. After entering the rainy season, when there is a lot of rain, drainage should be conducted well to prevent water stagnation. In October, when the rainy season ends, water should be added in a timely manner according to the soil moisture condition.

3.4 Fertilization management

The use of fertilizer is implemented in accordance with the *Green food—Fertilizer application guideline* (NY/T 394—2023). Fertilization follows the principle of "the combination of organic and inorganic fertilizer, the coordination of nitrogen, phosphorus and potassium application in different periods, appropriate topdressing potassium fertilizer, and timely supplement of medium and trace elements", which advocates the integrated technology of water and fertilizer. After each fruit harvest, timely supplement of fertilizer and water is recommended, and high potassium water-soluble fertilizer containing amino acids is suitable. The fertilizers used in ginseng fruit production should have no adverse effects on the environment, be conducive to protecting the ecological environment, maintaining or improving soil fertility and soil biological activity, and have no adverse consequences on the nutrition, taste, quality and plant resistance of ginseng fruit. On the basis of ensuring the effective supply of nutrition for ginseng fruit, the amount of chemical fertilizer should be reduced. The selection of fertilizer types should be mainly organic fertilizer, supplemented by chemical fertilizer.

The times, types and the amount of topdressing during the whole growth and development period of ginseng fruit are determined by the soil fertility, the application of base fertilizer and the growth of the plant, and about 4 times of topdressings should be conducted during the whole growth and development period. The first topdressing is carried out 1 month after the survival of the ginseng fruit seedlings, combined with watering topdressing once, applying 10–15 kg compound fertilizer including nitrogen, phosphorus and potassium (15 : 15 : 15) per mu, which is diluted with water before applying. In the flowering and bearing period, 10 kg high potassium compound fertilizer or 8 kg potassium sulphate per mu should be applied, generally once every 1–2 times of fruit harvest. In the fruit setting period, leaf fertilizer can also be sprayed appropriately, mainly using potassium dihydrogen phosphate and medium or trace elements fertilizer.

3.5 Integrated prevention and control of diseases and pests

3.5.1 Types of diseases and pests

The main diseases in the cultivation process of ginseng fruit include blight, sooty blotch, blackspot, etc., while the main pests include aphid, red spider, whitefly, thrip, etc.

3.5.2 Principles of prevention and control

In accordance with the principle of "prevention first, combining comprehensive prevention and control", based on agricultural prevention and control, ecological regulation, physical prevention and control, and biological prevention and control should be comprehensively adopted, to create unfavorable environmental conditions for the occurrence of diseases and pests, and reduce the losses caused by various diseases and pests.

3.5.3 Agricultural prevention and control

Planting in low-lying land and sticky heavy soil should be avoided. Before final planting, the soil should be ploughed over and sun-dried at high temperature for 15–20 d. Resistant varieties and virus-free high-quality seedlings should be chosen to reduce the initial infection source of diseases and pests. Reasonable plant density should be adopted, while pruning, shaping, setting up support, the old, diseased and pest-infested leaves should be removed promptly, improving ventilation and light conditions and reducing field humidity. Fertilizer and water management should be strengthened, which is to apply more organic fertilizer, and reasonably combine the fertilizers including nitrogen, phosphorus, and potassium. Especially in the middle and late stages of plant growth, nitrogen fertilizer should be applied less to prevent plant succulent growth. It is necessary to promptly remove diseased plants in the field, concentrate on burning them to reduce the amount of pathogenic bacteria in the field, and control the spread of diseases.

3.5.4 Biological prevention and control

During the entire growth and development period of ginseng fruit, the release time of is when aphids occur in patches in the field, and 1 000−1 200 asaphes vulgaris walker are released per mu.

3.5.5 Physical prevention and control

Utilizing the taxis of insects, yellow and blue sticky card traps, sex hormones, and insecticidal lamps are used in the garden to lure and kill pests. A total of 30−35 yellow and blue sticky card traps are placed per mu, with a suitable ratio of 4 : 1 for yellow and blue.

3.5.6 Chemical prevention and control

The use of pesticides should be implemented in accordance with the standard of GB/T 8321 series.

When the prevention and control by agricultural, biological and physical fail to achieve effect, chemical prevention and control should be adopted. The chemical prevention and control measures for the major diseases and pests of ginseng fruit are shown in Table 2−5.

Table 2−5 Chemical prevention and control measures for major diseases and pests of ginseng fruit

Types		Prevention and control measures
Diseases	Blight	It is advisable to evenly employ 80% mancozeb WP (1 : 500), 64% oxadixyl WP (1 : 400), or 58% metalaxyl·mancozeb (1 : 500), 25% azoxystrobin SC [1 : (2 000−3 000)], and fluopicolide·propamocarb SC [1 : (900−1 000)], once every 7−10 d, continuously for 2−3 times, depending on the situation
	Viral disease	The current effective strategy for the prevention and control of viral disease is "viral disease agents+nutritional foliar fertilizer+conventional pesticide". For infected plants at the seedling stage, it is advisable to evenly employ 20% copper succinate·moroxydine hydrochloride (1 : 1 000), 30% Dufulin + 5% oligosaccharins + foliar zinc fertilizer [1 : (600−800)], 60% moroxydine·hydrochloride + lentinan [1 : (800−1 000)], once every week, continuously for 3 to 4 times

Continued Table

Types		Prevention and control measures
Diseases	Sooty blotch	It is caused by the honeydew secreted by piercing-sucking pests such as aphid, whitefly, and scale insect etc. Therefore, it is essential to control pest and prevent disease. It is advisable to employ 10% imidacloprid WP, or 25% thiamethoxam, 30% acetamiprid for pest control and prevention. In the stage of sooty blotch occurrence, it is advisable to employ 43% methionyl·pentazolol SC [1 : (1 000–1 500)], or 26% pyrimethanil·diethofencarb SC [1 : (500–800)], 40% Captafol WP (1 : 500), 40% carbendazim SC (1 : 600), 50% benomyl WP (1 : 1 500), 50% carbendazim·diethofencarb WP (1 : 1 500), and 65% metalaxyl·hymexazol WP [1 : (1 500–2 000)]. It is best to mix the fungicide with the above mentioned insecticides when evenly applying them, once every 10–15 d, continuously for 2–3 times
	Blackspot	In the early stage of the disease, it is advisable to alternately employ 75% chlorothalonil WP [1 : (500–600)] or 50% iprodione WP (1 : 1 000), 25% pyraclostrobin SC (1 : 1 500), 60% pyraclostrobin·metiram WDG [1 : (1 000–1 500)], 50% captan WP (1 : 400) or 70% mancozeb WP (1 : 400) for evenly spraying, once every 7–10 d, continuously for 2–3 times
	Gray mold	It is advisable to alternately employ 50% iprodione WP (1 : 1 000), 50% procymidone WP (1 : 1 500), 40% pyrimethanil SC (1 : 1 200), 50% boscalid WDG, 50% iprodione thiram WP (1 : 800), 70% thiophanate·methyl WP (1 : 700) for evenly spraying, once every 7–10 d, continuously for 2–3 times
Pests	Aphid	It is advisable to employ 10% imidacloprid WP [1 : (1 000–1 500)], or 20% acetamiprid SL (1 : 3 000), and 50% pymetrozine WP [1 : (2 500–5 000)], according to the number of insect-population and control effect, continuously for 1–2 times
	Red spider	In the early stage of disease, it is advisable to alternately employ 34% spirodiclofen SC (1 : 4 000) + 43% bifenazate SC [1 : (1 800–2 500)]. Spirodiclofen can kill eggs, nymph mites and imago mites, which can be sprayed evenly without being absorbed by the plant itself, with the 30 d effective period, and it can be used again at an interval of 7 d. When the disease is severe, the eggs, nymph mites and imago mites can be completely eliminated by continuously spraying 2–3 times. Use 20% cyflumetofen EC (1 : 2 000)+5% beta-cypermethrin·emamectin benzoate EC (1 : 1 500), or 20% avermectin·bifenazate SC [1 : (2 000–2 500)] can be used alternately at an interval of 7 d, then applying according to the control effect. Cyflumetofen can kill the eggs, nymph mites and imago mites, with a nice effect by spraying evenly
	Thrip	It is advisable to employ 25% thiamethoxam WDG (1 : 800) simultaneously by mixing with avermectin ME. The prevention and control can also be conducted by using 25% alexin WDG [1 : (2 500–3 000)], 20% fenpropathrin EC [1 : (3 000–5 000)] or 2.5% deltamethrin EC [1 : (2 500–3 000)]

Continued Table

Types		Prevention and control measures
Pests	Whitefly	It is advisable to employ 25% buprofezin WP (1 : 1 500) together with 2.5% bifenthrin EC [1 : (2 000–3 000)] or 25% thiamethoxam WDG [1 : (3 000–5 000)] by adding 10% pyriproxyfen EW (1 : 1 000) or 10% lufenuron SC [1 : (2 500–3 000)]. At the same time of spraying, 22% dichlorvos fumigant can be combined for fumigation treatment
	Yellow tea mite	It is advisable to employ 40% bifenazate·etoxazole SC (1 : 8 000) and 20% avermectin·bifenazate SC [1 : (3 000–4 000)] or 5% hexythiazox EC [1 : (1 500–2 000)] for prevention and control

3.6 Harvest

With promptly harvesting, the harvest of ginseng fruit generally starts from 110–120 d after final planting, and enters the harvest peak period around 140 d after final planting. The harvest time of off-season cultivation is generally from January to May of the next year. Due to different fruits in each inflorescence having different fruit setting times, the nutritional distribution is unbalanced in the growth process, resulting in different ripe stages. It is the appropriate time to harvest when the fruit surface presents obvious purple stripes and the pericarp and pulp become faint yellow. Harvesting needs to be done according to the need, fruits that need long-term storage or long-distance trafficking can be harvested in advance for short-term storage, while those for short-distance trafficking and local sales can be harvested when they are fully ripe. Wearing cotton gloves when harvesting, fruits should be gently picked up, cut out with scissors and graded according to size, putting packing net on each fruit and packing. The growth and development period of ginseng fruit is continuous flowering and bearing, so it should be ripened and harvested in stages.

References

CHEN X D, 2023. Planting technology of a-class green food ginseng fruit [J]. Bulletin of Agricultural Science and Technology (4): 223–225.

GAO F E, 2019. Discussion on ginseng fruit cultivation technology in solar greenhouse [J]. Agricultural Development & Equipments (1): 171.

HUANG X L, WANG L F, CHEN H X, et al., 2022. Key techniques for improving fruit quality of ginseng fruit by setting up support and inducing vines [J]. Journal of Changjiang Vegetables (24): 59–61.

YAN C Z, 2019. Cultivation techniques of ginseng fruit in greenhouse in Ledu District [J]. Qinghai Agro-Technology Extension (4): 14–15.

Yunnan Province Yunnan Geographical Indication Industry Association, 2023. T/YGIIA 003−2023 Technique Regulations for the Vine Cultivation of Xijiekou Pepino Melon[S]. Yunnan Geographical Indication Industry Association.

ZHANG L F, 2022. Efficient and high-quality cultivation technology of ginseng in open field [J]. China Fruits (6): 84−87, 82.

ZHANG L F, HUANG X L, CHEN X D, et al., 2021. Current situation, existing problems and development countermeasures of ginseng fruit industry in Shilin County [J]. Bulletin of Agricultural Science and Technology (6): 47−50.

❖ Key Techniques for Banana Cultivation

1 Overview of industrial development

Banana (*Musa nana* Lour.) is important economic and food crops in tropical and subtropical regions, and is the world's largest fresh fruit in terms of production, trade volume, and trade value. The core banana-growing regions are distributed in Asia, the Americas, and Africa. In 2020, the global banana cultivation area was 5.203 5 million hm^2, with a production of 119.833 7 million t. The main producing countries are India, China, Indonesia, Brazil, and Ecuador. According to the data from the National Bureau of Statistics, in 2019, China's total banana cultivation area was 330 300 hm^2, with a total production of 11.655 7 million t, ranking second only to India, making China a major producer of bananas. At the same time, China is also a major importer of bananas. In 2019, China imported 1.94 million t of bananas, ranking second in the world after USA. From 2002 to 2019, China's cultivation of bananas showed a trend of "westward" and "southward". At present, the main banana-producing areas in China are mainly distributed in 8 provinces (autonomous region、municipality) including Fujian, Guangdong, Guangxi, Hainan, Chongqing, Sichuan, Guizhou, and Yunnan. Yunnan, one of the main banana-producing areas in China, is an important part of Yunnan's modern agriculture with plateau characteristic. From 2002 to 2013, the development of the banana industry in Yunnan Province entered the "fast lane". From 2014 to 2020, it experienced certain fluctuations due to factors such as banana wilt disease (Panama disease of banana) and cold waves. As of 2020, the banana cultivation area in Yunnan Province was 82 400 hm^2, accounting for 25% of the total planting area in the country, with a total production of 1.976 4 million t, accounting for 18% of the national total, ranking 2-3 in China. Its cultivation area and production value both rank first among various fruit types in Yunnan Province. The banana industry has become an important means for ethnic minority farmers in border areas to increase income and become prosperous. It plays a key role in consolidating the achievements of poverty alleviation and the implementation of the rural revitalization strategy, and has important political, social, and economic significance.

2 Planting conditions

2.1 Temperature

Bananas are tropical fruit trees, and their growth requires high temperature. Low temperatures are the primary limiting factor for high yields. The young fruit can be damaged below 13 °C in the bud sprouting stage and young fruit stage, and the leaves can be damaged at the temperature of 1−2 °C, especially susceptible to frost damage. The optimal growing temperature is 27 °C, and the maximum temperature should not exceed 38 °C.

2.2 Water

Bananas are large herbaceous plants with high water content, large leaves, and high transpiration rates, thus requiring adequate rainfall to meet their water needs. Typically, annual precipitation should range from 1 000 mm to 2 500 mm.

2.3 Soil

The root system of bananas consists of fleshy fibrous roots which grow from the corm. They are shallowly distributed, and as bananas have a short growth period, they need to accumulate a large amount of biomass in a short period. Consequently, they require soils with high organic content and good drainage to maintain optimal soil moisture and oxygen supply. This is to prevent the occurrence of root suffocation and diseases caused by excessive waterlogging. The soil should be neutral to slightly acidic, with a pH value typically ranging from 5.5 to 7.0. The soil depth should be at least 60 cm, and the water table should be below 1 m.

2.4 Planting areas

The suitable areas for banana growth are dry-hot river valleys and humid tropical areas with an annual average temperature of ≥10 °C, and an annual accumulated temperature of over 6 500 °C at an elevation of 90−600 m.

In the river valleys of the six major water systems in Yunnan (Irrawaddy River, Nujiang River, Lancang River, Yuanjiang River, Jinsha River, and Nanpan River), there is an area of approximately 81 000 km^2 which is generally suitable for banana production. Currently, bananas are planted in 60 counties across 14 prefectures (cities) in Yunnan, with the planting areas and yields in Honghe Prefecture, Xishuangbanna Prefecture, Wenshan Prefecture, and Pu'er City accounting for over 86% and 87% of the total area and production in the province, making them the main banana-producing areas in Yunnan.

3 Main cultivars and characteristics

The cultivation area of two varieties, Brazilian banana and Williams banana, accounts for over 80% of the total cultivation area in Yunnan province.

3.1 Brazilian banana

The original variety of the Brazilian banana, known as Nanicao, was introduced from Queensland, Australia, by the Zhanjiang Agricultural Biotechnology Research Center of Guangdong Province in 1989. It is a variant of the Dwarf Cavendish banana. The plant (Figure 2-1) is tall, with a pseudostem height of 240-330 cm and a base diameter of 81.7 cm and a middle diameter of 54.5 cm. The leaves are dense, open, sword-shaped, 210-212 cm long, and 95-102 cm wide. The inflorescence, usually located at the top of the plant, is composed of closely packed flowers. The fruit bunch (Figure 2-2) is 85.5 cm long, with relatively sparse hands of fruit. The fingers are 20-23 cm long, with an average yield of 18-34 kg per plant. The fruit contains 18%-21% of soluble solids and has a strong fragrance. Due to its high yield, good fruit quality, strong adaptability, cold and drought resistance, it is extensively cultivated in Yunnan Province, with strict requirements for fertilization and water management.

Figure 2-1 Plants of Brazilian banana   Figure 2-2 Fruit bunch of Brazilian banana

3.2 Williams banana

According to reports, Williams is a variety that was developed by selecting a mutant plant from the first introduction of banana plants from Taiwan of China by Peiji in the early 19th century. It was later introduced to Australia and became a leading variety. In the 1980s, it was further introduced to Yunnan Province. Williams belongs to the type of medium to tall stems, with tall and robust plants (Figure 2-3). The pseudostem is 250-300 cm tall, but the stem is relatively slender, with a circumference of 80-90 cm. The leaves are large and fan-shaped, green and broad, and standing upright. The fruit bunch is long, with relatively sparse hands of fruit. The number of fruit is not large, but they are closely arranged, with a straight shape. The total sugar content of the fruit is 18%-21%, and it has a strong fragrance. The yield is 20-30 kg per plant and can reach up to 43 kg. The fingers are 20-24 cm long, yellow in color, and the fruit flesh is soft and sweet. The bunch is neat and aesthetically pleasing, and the variety has strong adaptability.

Figure 2-3 Plants of Williams banana

3.3 Guijiao No. 6

Guijiao No. 6 was introduced from Australia in 1987 as a banana variety strain through tissue culture and selection of variant strains domestically. It belongs to the Cavendish banana type with a genotype of *AAA*. The growth period for spring, summer, autumn, and winter planting is 300-420 d, 360-400 d, 360-400 d, and 330-390 d, respectively. The first generation tissue culture plants have a pseudostem height of 220-260 cm, with a circumference of 75−95 cm at the base and 48−65 cm in the middle. The stem index ranges from 3.7 to 4.3. Each bunch has 7−14 hands of fruit, with 16−32 fingers per hand, and each bunch of fruit 20−30 cm long, with a yield of 2 400−3 500 kg per mu. The bunch is neat and aesthetically pleasing, with stable high production and high-quality characteristics, and strong adaptability. It has moderate wind resistance, is not frost-tolerant, and is susceptible to infection by *Fusarium oxysporum* f. sp. cubense race 4, which causes banana wilt disease. The plants are also susceptible to banana mosaic disease and banana bunchy top disease.

3.4 Jinfen No. 1

Jinfen No. 1 is the first cultivated banana variety independently bred by the Guangxi Academy of Agricultural Sciences and the Guangxi Xingwang Tissue Culture Seedling Company. The plants are tall with shiny pseudostems, scattered brown spots, and light green color. They grow to a height of 460−500 cm, with a basal girth of 94−101 cm and a middle girth of 61−67 cm, with a stem index of 7.6−8.2. The unripe fruit skin is light green with some uneven wax powder on the fingers, and its flesh is white. Each bunch contains 10−15 hands, and each hand has 18−20 fruits arranged in two rows. The yield is 20−30 kg per plant, with a maximum of 45 kg, and the yield is 2 000−3 000 kg per mu. The ripe fruit is golden yellow skin with sweet, soft, and smooth flesh. It is resistant to cold and wind, tolerates light frost but not heavy frosts and snow. It is susceptible to Fusarium and wilt disease but resistant to black Sigatoka, bunchy top disease and banana mosaic disease, and is attractive to the banana aphid and banana weevils. It is relatively drought-resistant but intolerant to waterlogging. The thin peel makes it unsuitable for long-distance transportation and storage.

4 Cultivation and management techniques

4.1 Orchard planning and construction

4.1.1 Site selection

The areas for bananas planting features no frost throughout the year, good air circulation, open terrain, facing north to south with good wind protection, and a slope of less than 15°, which are equipped with windbreaks and convenient traffic. The large-scale banana plantations of over 1 000 mu should be divided into different planting areas based on the topography, geology, and characteristics of banana varieties, including mother plant gardens and seedling bases.

4.1.2 Road system

Main and branch roads are designed within the plantation. The main road, used for large machinery, should have a base width of 3−4 m, which is adjustable according to the height of banana stems and the type of machinery to prevent damage to the plants. Branch roads should accommodate small machinery or be adjusted to them, with a width ensuring no damage to the plants.

4.1.3 Irrigation and drainage systems

Yunnan has distinct dry and wet seasons, with dry winters and springs and rainy summers and autumns. It is necessary to equip with water management facilities in banana plantations. For every 3−5 mu of land, a water reservoir with a capacity of 10 m^3 and a non-leaking manure pit should be constructed. For terraced plantations, it is suggested to dig a flood interception ditch with 70 cm wide and 50 cm deep on the top of the land, and drainage channels with a width of 60 cm and a depth of 40 cm can be established along the slope. For flat plantations, vertical ditches and branch ditches perpendicular to each other should be built along the roadsides. In suitable areas, drip irrigation or micro-sprinkler systems can be installed. For larger areas, main and sub-mixing fertilizer pits should be constructed, with the main pit located near the main irrigation pump for convenient water-fertilizer mixing, and the sub-mixing pit near the management post for fertilizer mixing.

4.1.4 Cableway

For sloping terrains, banana harvesting arched cable-ways and irrigation pipe networks can be installed, utilizing the cable-way's backbone pipes as the main irrigation pipes to save costs. The cable-way can be set up as a single or double track transportation line for uphill and downhill transport.

4.1.5 Land preparation

In Yunnan, most banana plantations are located on slopes with a gradient of ⩾ 5°, and the terrain is often complex with a height difference of more than 200 m. Therefore, before planting, terraces on slopes with a gradient of <25° need to be constructed by cutting into the hillsides, with a terrace width of 1−1.6 m, sloping inward at 5°, and raised 15−20 cm outward, with a small ridge of 20−25 cm along the edge, and planting trenches or pits 60−70 cm wide and 50−60 cm deep along the terrace. The spacing between plants should be adjusted based on the terrace width, with a planting density of 100−130 plants per mu being appropriate.

For banana plantations in flat, dry areas, they adopt a cultivation method of trenches plus subsurface drip irrigation under plastic film. The planting trenches are 1–2 m wide and 40–50 cm deep, creating ridges.

4.1.6 Planting

In Yunnan, banana planting usually takes place around the spring or autumn, with spring planting being more favorable. Spring planting of bananas can take place from early February to mid April, bud sprouting from September to the following spring, and harvesting from February to June. Autumn planting occurs from August to October, with the second year's harvest from August to December, allowing for market placement around the Mid-Autumn Festival or National Day. During planting, the soil should not be too dry or too wet. It is advisable to choose a cloudy day or after 16:00 for planting, while taking care not to damage the root system. Sufficient water should be given to the seedlings after planting. After the seedlings have grown 1–2 new leaves, spray 0.1%–0.3% water-soluble organic fertilizer such as potassium dihydrogen phosphate, amino acid, and seaweed acid.

After planting, regular checks for seedling survival should be carried out, with dead seedlings removed promptly and replaced. Check for pest infestations and apply preventive measures when detecting occurrences of banana weevil or aphids. In the initial stage of seedling growth, manual weeding should be done regularly, and ground coverage can be used to inhibit weed growth.

Shoot removal and bud retention: To ensure the vigor of mother plants, only two shoots can be retained for each plant. Other sprouted shoots should be completely removed from the ground and the growing point in the middle should be dug out to prevent further growth. From April to August, shoots should be removed once every two weeks, with frequency adjusted based on the growth of the sprouts in other seasons.

4.2 Water and fertilizer management

Banana is a fast-growing, high-yielding herbaceous plant that requires proper fertilization even when planted in fertile soil to ensure fruit yield and quality.

4.2.1 Water management

Throughout the growing season, it needs to keep the soil in the orchard moist. The soil relative humidity content should be 65%–75% from the seedling stage to the budding stage and 50%–60% during the fruiting period. The water demand increases sharply during the budding stage, and watering should be increased as needed. In the middle and late stages of fruiting, the soil moisture should be properly controlled, and watering can be stopped 7–10 d before harvesting.

4.2.2 Fertilization management

Key mineral elements needed for banana growth and development include potassium, nitrogen, phosphorus, calcium, magnesium, sulfur, and chlorine and micro nutrients such as manganese, iron, zinc, boron, copper, and molybdenum. Potassium is crucial for banana yield and quality, so there is a high demand for potassium fertilizer, with a ratio of $N : P_2O_5 : K_2O$ being 1 ∶ (0.2–0.5) ∶ (2–4). In addition, bananas have a high demand for nitrogen, calcium, and magnesium, with an absorption ratio of 1 ∶ 0.69 ∶ 0.2, and also for calcium and magnesium fertilizers.

4.2.2.1 Base fertilizer

The fertilizer is mainly used for improving the soil physical and chemical structure and ventilation, thus laying a good foundation for the entire growth period of bananas. Base fertilizer mainly consists of organic fertilizer, such as well-fermented pig, cow, or chicken manure. And it is supplemented with compound fertilizer, superphosphate. Each plant applies 5 kg of farmyard manure, 0.1–0.5 kg of calcium-magnesium phosphate fertilizer, and 0.2–0.5 kg of compound fertilizer. Alternatively, 1–2 kg of bio-organic fertilizer and 0.3 kg of compound fertilizer can be used. When applying base fertilizer, the fertilizer should be thoroughly mixed with the soil and filled back into the planting trench or planting hole.

4.2.2.2 Fertilization during the flower-bud differentiation period

When the sucker has 20 leaves and the tissue-cultured plant has 28 leaves, which means the pseudo stem is 1.5–2.0 m in height, each plant applies 5–7 kg of farmyard manure + 0.2 kg of compound fertilizer + 0.2–0.5 kg of potassium fertilizer. Use the topdressing method for additional fertilization, applying 0.1 kg of compound fertilizer and 0.05 kg of potassium fertilizer per plant every 20 d. Before the appearance of the flower bud, apply lime 4–5 times, approximately 15 kg per mu each time. After the appearance of the flower bud, apply 0.1 kg of compound fertilizer ($N : P : K=15 : 15 : 15$) per plant every 20 d and then water after application.

4.2.2.3 Fertilization during the fruiting period

As the root system of the banana plant ages during the fruiting period, warm water-soluble organic fertilizer should be sprayed on the leaves, and nitrogen fertilizer should be avoided. The focus should be on supplementing potassium and trace element fertilizers.

4.3 Prevention and control of major diseases and pests

4.3.1 Banana wilt disease

Banana wilt disease, caused by the tropical race 4 and race 1 strains of the fungus *Fusarium oxysporum* f. sp. cubense, is a devastating fungal disease also known as Panama disease or Fusarium wilt of banana. It is the most serious disease affecting the global banana industry. Currently, it cannot be effectively controlled by a single technique, and requires integrated control measures.

Symptoms: The pathogen survives in the soil and infects the banana roots, causing root rot with a long latent period that can last for months to over a year. Infected leaves turn yellow and the discoloration gradually spreads throughout the entire plant. Due to the xylem tissue being damaged by invasion of the pathogen, the vascular tissue in the pseudostem and leaf sheaths show continuous browning. As the disease progresses, the pseudostem may crack or rot, resulting in the death of the entire plant.

Disease pattern: Banana wilt disease has a long latent period, often lasting for months to over a year. It can occur throughout the year, but tends to be more severe from October to February of the following year. After initial infection, the pathogen spreads upward through the vascular tissue from the roots. It continues to multiply and spread within the stem, causing tissue decay and loss of root function. The disease gradually spreads throughout the entire plant, leading to overall wilting. The pathogen produces numerous macroconidia, microconidia, and chlamydospores. The chlamydospores are capable of surviving in the soil for several years to a decade. The pathogen can spread to healthy plants through infected soil, farming tools, water sources, and pests.

Control methods: Prevention is the key, and integrated disease management is essential. ① It is crucial for preventing banana wilt disease to select banana varieties with strong disease resistance. Before planting, it is necessary to test the quantity of *Fusarium oxysporum* f. sp. cubense spores in soil, and if the spore count exceeds 10^6 per gram of soil, current varieties will likely lose their disease resistance and should not be planted with bananas again. Depending on market conditions, suitable resistant varieties including Nantian Huang, Baodao Jiao, Guijiao No. 2, Guijiao No. 9, Zhongjiao No. 8 could be planted. ② It implements quarantine management in plantations and post-harvest treatment facilities. It sets up disinfection pools (containing 0.3% potassium permanganate) at the entrances and exits of banana plantations, as well as disinfectant-covered walkways. ③ Prohibits adopting seedlings from infected areas. It is advised to use seedlings of soilless and sterilized substrates or the seedlings of first-level histoculture bag to reduce the incidence of disease. ④ For plantations already affected by the disease, soil should be disinfected by using heat or chemical agents before planting in order to reduce the presence of the pathogen. ⑤ In areas where conditions permit, it is better to avoid continuous banana cropping in the same plot and practice rotational fallowing. For infected plants, it should be promptly isolated, removed and destroyed to prevent the further spread of the disease.

4.3.2 Banana Cordana leaf spot

Banana Cordana leaf spot is caused by the fungus *Cordana musae* (Zimm.) Hohn. Initially, the lower leaves began to appear dark brown or grayish-brown spots, semi-circular, oval, or irregular along the leaf margins with varying sizes and peripheral infiltration. As the disease progresses, the spots enlarge into elongated oval shapes with rounded ends, concentric rings, with a grayish-brown to gray center and a distinct bright yellow halo. Under high humidity, grayish-brown powdery spore-bearing structures of the pathogen can be found on the lower leaf surface, known as conidiophores and conidia. The fungus can be transmitted by wind, rain, and farming tools. The favorable conditions for the pathogen's reproduction and spread are high temperature and humidity. Banana plants during the rainy season from June to September, are particularly vulnerable during the growth and bud pumping stages. For its treatment, it is commonly used control measures include triazoles (such as propiconazole, difenoconazole, and tebuconazole), azoxystrobin, pyraclostrobin, carbendazim, chlorothalonil and mancozeb, as well as the use of specific beneficial microorganisms such as antagonistic bacteria (e.g., *Penicillium simulans*) or parasitic fungi (e.g., *Pseudomonas syringae*) to suppress the growth and reproduction of the pathogen.

4.3.3 Banana anthracnose

Banana anthracnose is a post-harvest disease caused by the fungus *Colletotrichum musae*. Early symptoms include long oval-shaped spots on the leaves, which gradually become covered with small black dots. When the green fruit is affected, long oval black-brown spots appear on the fruit skin, and many small black dots are distributed on the spots, but the phenomenon is not common. The affected fruit usually is the ripe fruit, with the spots being black-brown, pike shaped, and longitudinal in the centre, and the spots of some varieties being smaller and brown. The disease is mainly spread through wind and rain. Conditions of high humidity and temperature favor the growth and transmission of the pathogen.

To control banana anthracnose, preventive measures can be taken by applying fungicides such as carbendazim, thiophanate-methyl, prochloraz, and azoxystrobin 2–3 times, with 5–10 d interval before bagging the fruits. The spray should focus on wetting the fruits and nearby leaves without causing dripping.

4.3.4 Banana freckle disease

Banana freckle disease is caused by the fungus *Macrophoma musae* Cooke, which belongs to the deuteromycotina. It primarily affects the leaves and fruit. On the leaves, the disease starts with the appearance of numerous rough black spots along the leaf veins and eventually spreads to cover the entire leaf, causing it to turn yellow and wither. Older leaves are more susceptible to the disease than newer leaves, causing leaves to wither earlier. On the fruits, when the disease affects the fruit spike, the disease spot develops from the fruit axis to the fruit stalk to the inner curvature of the fruit, so the inner row of fruit is more serious than the outer row of fruit. Small black spots initially appear on the fruit skin, which later develop into larger lesions covering the entire fruit. In mature fruits, the lesions become surrounded by brown circular spots, with the central tissue becoming rotten and sunken, and small black spots appear on top.

High humidity and temperature favor the growth and transmission of the pathogen. The disease is more severe during the rainy season from June to September and after the plants have produced their flower buds. Commonly used fungicides for disease control include triazoles (such as propiconazole, difenoconazole, and tebuconazole), azoxystrobin, picoxystrobin, flusilazole, pyraclostrobin, carbendazim, chlorothalonil and mancozeb.

5 Harvest and post-harvest treatment

5.1 Harvest

5.1.1 Harvesting timely

The maturity of the fruit can be determined by changes in the angular edges, with the optimal harvest time being when the maturity is between 70% and 75% (Table 2-6). Alternatively, the time of fruit harvest can also be determined based on the time of inflorescence emergence; for plants with inflorescence emergence in spring and summer, the fruit should be harvested 80-90 d after the emergence, while for plants with inflorescence emergence in late autumn and winter, the fruit should be harvested 120-150 d after the emergence. During harvesting, it is best to avoid harvesting during the hottest part of the day and preferably harvest in the morning on cloudy or sunny days.

Table 2-6 Ripeness evaluation method

Ripeness		Characteristic of the fruits
Summer and autumn	70%–<75%	Approaching fullness, very distinct edges
	75%–<80%	Basically full, distinct edges
	80%–<90%	Round and full but still visible edges
Winter and spring	75%–<80%	Basically full, distinct edges
	80%–<90%	Round and full but still visible edges
	Above 90%	Round and full, basically no edges

5.1.2 Cutting the banana

Several people can work together for fruit harvesting. One person cuts down the pseudostem below the middle, while another person or two people support the bunch and cut off the fruit stalk 25–30 cm from the top of the fruit comb, then place the bunch on a dedicated rack and transport it to the banana orchard cableway, rail transport line, or roadside.

5.1.3 Field transportation

Orchards with the necessary conditions generally use cableway transport, where laborers transport the bananas to the side of the cableway and securely tie them to the axle with a rope, then hang the bunches on the gantry hook. The bunches are separated by bars, with 5–10 bunches forming a group, and are uniformly transported to the fruit processing area via the cableway. Some orchards also use trailers for transportation, with a layer of sponge at the bottom of the wagon to protect the fruit, and the bunches are transported to the fruit processing area in a single layer.

5.1.4 Debunching

After the bananas are transported to the post-harvest commercial processing site, the bunches are placed on a debunching rack to remove the remaining floral parts. A special knife is then used to cut the bananas from the fruit stalk and place them in a cleaning pool.

5.1.5 Washing and trimming

Submerge the banana bunches in a 0.1%–0.2% alum solution for deodorizing, sterilizing, and stopping sap flow or in a low-pressure water washing pool to remove dirt and sap from the surface of the fruit. During the washing process, trim the cuts to ensure smooth edges, and remove any defective fruit such as rotten, split, or misshapen fruit.

5.2 Grading and weighing

According to customer requirements, classify and weigh the bananas based on the standards outlined in *Bananas* (GB 9827—1988), *Banana Grading Specifications* (NY/T 3193—2018), and the Yunnan standard *Geographical Indication Product—Hekou Bananas* (DB53/T 912—2019).

5.3 Preservation treatment

For long-distance transportation, it is necessary for commodities to conduct a preservation treatment. The fruit bunches should be Hang-sprayed or immersed in a preservation solution for 10–30 s. It is strictly prohibited to abuse preservatives. It is better to choose specialized preservatives and bactericides such as prochloraz, Iprodione, or thiabendazole, which are safe, efficient, low toxicity, and have low residues. After being placed in the preservation pool, the solution should cover the surface of the fruit. The preservative solution should be prepared on-site and changed every 48 h. To reduce the occurrence of anthrax and agrobacterium tumefactions during transportation and storage, the fruit bunches need to be air-dried after preservation treatment.

5.4 Packaging

After air-drying, separate the banana bunches with absorbent paper or pearl cotton. Apply labels according to customer requirements, package the fruit bunches in plastic bags (Previously placed in cardboard boxes), vacuum-seal the bags, check for leaks, cover with a paper lid, seal the box, affix the brand, and then batch-send them to controlled atmosphere rooms or cold storage for pre-storage. The outer packaging and packaging materials should comply with national and industry standards, and the product packaging label should include the product name, net content, shelf life, product implementation standard number, packaging date, place of origin, and name of the producer.

5.5 Long-distance transportation

In Yunnan, bananas are mainly transported by trucks. The truck compartments and surroundings should be padded with soft materials to prevent damage. In summer, measures should be taken to prevent high temperatures, while in winter, precautions should be taken to prevent freezing. If conditions allow for modified atmosphere transportation, the temperature should be regulated to 13–14 °C with a relative humidity of 90%–95%.

References

LIN X Y, JIANG W X, XU S R, 2022. Spatial-temporal change of banana planting area and its influencing factors in China [J]. Acta Agriculturae Jiangxi, 34(7): 218−223.

ZOU D M, FAN Q, 2022. Present situation of globle banana production and trade and prospect for banana industry [J]. Guangdong Agricultural Sciences, 49(7): 131−140.

ZOU Y, LIN G M, MOU H F, et al., 2011. Breeding of new banana variety "Jinfen No.1." [J]. South China Fruits, 40(1): 47−48.

❖ Key Techniques for Mango Cultivation

1 Overview of industrial development

Since 1986, China has initiated large-scale cultivation of mangoes. Within a span of 30 years, mango (*Mangifera indica* L) has swiftly emerged as the fourth-largest tropical fruit, trailing only behind lychee, longan, and banana. Mango cultivation is predominantly concentrated in provinces (autonomus region) such as Guangxi, Yunnan, Hainan, Sichuan, Taiwan, Guizhou, Guangdong, and Fujian, etc. In 2021, the nationwide mango cultivation area reached 5.619 million mu, yielding 3.958 million t, marking a year-on-year increase of 9.1% and 19.5%, respectively.

1.1 Overview of Yunnan

1.1.1 Natural distribution

Yunnan boasts ecological and natural advantages for producing high-quality mangoes, ranking first in the nation in terms of mango cultivation area. Additionally, it holds the record for the longest supply period of mangoes in the country, with a supply season spanning from May to November, totaling seven months. In 2021, the total mango cultivation area in the province reached 1.721 million mu, yielding 1.329 million t, marking respective year-on-year increases of 13% and 38%.

Yunnan is the province with the broadest natural distribution and the most diverse varieties of mangoes in China. China has six species of plants belonging to the Mangifera genus and Yunnan is home to five of them, namely *Mangifera indica* L., *Mangifera simensis* Warby. ex Craib, *Mangifera sylvatica* Roxb., *Mangifera longipes* Griff., *Mangifera persiciforma* C. Y. Wu et T. L. Ming. Among these, the most widely distributed is *Mangifera indica* L, naturally distributed in 46 counties (cities) across 9 prefectures and cities in Yunnan.

1.1.2 Planting situation

The cultivation of mango in Yunnan is widespread. Mango is grown from Wenshan Prefecture in the east to Dehong Prefecture in the southwest, and from Qiaojia County in the northeast of Yunnan to Xishuangbanna Prefecture in the far south. The cultivation area spans coordinates of 97°51′–105°38′ E and 21°29′–26°53′ N, covering a vast tropical and subtropical region with elevations ranging from 76 to 1700 m. Yunnan Province comprises 16 prefecture-level cities and 129 counties/districts. Currently, mangosteen cultivation is found in 15 prefecture-level cities and 91 counties/districts. The current main cultivated varieties of mango include Tainong No. 1, Gui Fei, Seintalone, Dasai Yellow Flesh, PaLa Hin Tha, Yexiang, Nam Doc Mai No. 4, Jin Huang, Kiett, Sensation, and Hong Xiangya.

2 Main cultivars and characteristics

2.1 PaLa Hin Tha

2.1.1 Variety origin

Myanmar.

2.1.2 Variety traits

Intermediate maturity.

2.1.3 Variety characteristics

The tree exhibits medium vigor, with a round crown. The average single fruit weight is 260.2 g. The fruit is roughly ivory-shaped, with mature fruit having yellow skin and moderate fruit powder. The fruit ripens from late June to late July, with a growth period of 90–100 d, a harvesting period of 30 d, and a shelf life of 15 d. It has a sweet and fragrant taste, with good edible quality.

2.1.4 Distribution area and suitable range

Mainly distributed in the Longyang District and Yuanjiang County, it is most suitable for cultivation in areas with an altitude of 800–1 200 m.

2.1.5 Advantages and disadvantages

Advantages: ①High and stable yield; ②attractive appearance; ③resistant to storage and transportation, long shelf life; ④drought-resistant.

Disadvantages: This variety is prone to thrips infestation.

2.2 Jin Huang

2.2.1 Variety origin

Jin Huang, cultivated by Mr. Huang Jinhuang in Taiwan, China, is a hybrid progeny resulting from the crossbreeding of Nang klangwan and Kiett.

2.2.2 Variety traits

Intermediate maturity.

2.2.3 Variety characteristics

The growth pattern of this variety is upright, and it exhibits strong and vigorous development. The fruits are large, resembling ivory in shape, with an average single fruit weight of 650–1 500 g and a maximum single fruit weight of 2 400 g. The fruit shoulders are relatively small, exhibiting a reddish blush on the sun-exposed side. The fruit features a rounded and blunt tip, with a thick and lengthy stem, and smooth skin. It has sparse dots, thin peel, and a slightly raised pit. The flesh is yellow, dense, tender, and juicy, with a delicate fiber. The taste is sweet, and it emanates a rich aroma. In Yunnan, flower bud differentiation begins in December, reaching peak flowering in late February, and the fruits mature in late July.

2.2.4 Distribution area and suitable range

This variety is cultivated throughout the province, although the overall cultivation area is not large. It is most suitable for cultivation in regions with altitudes between 800 m and 1 200 m.

2.2.5 Advantages and disadvantages

Advantages: ①Attractive appearance; ②sweet and juicy fruits; ③high edible rate; ④early and abundant fruiting; ⑤long shelf life; ⑥strong resistance to stem rot, powdery mildew, and anthracnose.

Disadvantages: ①Not frost-resistant; ②susceptible to water blister disease.

2.3 Tainong No. 1

2.3.1 Variety origin

Tainong No. 1, originating from Taiwan, China, was meticulously cultivated through the hybridization of the Hayden and Erwin varieties at the Fengshan Tropical Horticultural

Experimental Substation of the Taiwan Agricultural Research Institute.

2.3.2 Variety traits

Early maturity.

2.3.3 Variety characteristics

The crown of the tree is round and vigorous in growth. The male and female flowers are in a proportion of 46%, and the tree exhibits a tendency for secondary flowering. The fruit is broadly oval, with an average single fruit weight of 250−300 g. The fruit shoulder is slightly sloping, the fruit belly is raised, the fruit back is curved, the fruit socket is shallow, and the beak is large and blunt. The fruit skin has a green base with a reddish tint, turning yellow when fully ripe, and the upper part of the fruit shoulder shows a red color. The flesh is deep yellow to orange-yellow, with a tender and smooth texture, minimal fibers, and an intensely sweet and fragrant taste, making it of exceptional quality. The harvesting period for the fruit is in the early to mid July. In Baoshan, Yunnan, bud differentiation occurs in mid January, and trees with top grafting for replacing variety can produce 5−6 shoots in a year. After fruit harvesting and pruning, the tree can also produce 1−2 shoots in the autumn. The flowering period is from late February to mid March, with harvesting taking place in early July, requiring approximately 90−110 d from flower withering to fruit ripening.

2.3.4 Distribution area and suitable range

The cultivation is widespread across the province, primarily concentrated in Yuanjiang County and Honghe County. The most suitable regions for cultivation lie within the altitude range of 800−1 200 m.

2.3.5 Advantages and disadvantages.

Advantages: ①Dwarf plant size, easy to management; ②attractive appearance and good quality; ③high yield, stable production, resistant to storage and transportation, and resistant to anthracnose.

Disadvantages: Weak resistance to powdery mildew.

2.4 Seintalone

2.4.1 Variety origin

Myanmar.

2.4.2 Variety traits

Early maturity.

2.4.3 Variety characteristics

The tree has an expansive and umbrella-shaped canopy, with a smooth, gray-brown main trunk. The fruit is flattened and spherical, with mature fruit skin turning yellow and a moderate amount of fruit powder. The flesh is golden yellow. The fruit ripens from mid June to mid July, with a harvesting period of 30 d. The fruit has a storage tolerance of 8 d, featuring a sweet and fragrant taste, and excellent edibility.

2.4.4 Distribution area and suitable range

It is primarily distributed in Longyang District, Yongde County, Yuanjiang County, and Honghe County. The most suitable cultivation areas are below an altitude of 1 000 m.

2.4.5 Advantages and Disadvantages

Advantages: ①Attractive appearance; ②strong fragrance, excellent quality; ③stable production, durable for storage and transportation, resistant to anthracnose.

Disadvantages: Weak resistance to bacterial black rot disease.

2.5 Nam Doc Mai No. 4

2.5.1 Variety origin

Thailand.

2.5.2 Variety traits

Intermediate maturity.

2.5.3 Variety characteristics

This is a superior variety obtained through the selection of seeding trees in Thailand, exhibiting strong tree vigor and a rounded canopy. The branches are robust, and growth is rapid. The flowering period extends from mid-late November to early February of the following year, showcasing a unique characteristic as flowering in different seasons. The inflorescence takes on a tower-shaped to conical form. The fruit has a growth period of 100–120 d, with ripening occurring from early July to early August. Average single fruit weight is 300–400 g. When the fruit is mature, the skin is orange-yellow with a thin wax layer, offering a sweet and juicy taste.

2.5.4 Distribution area and suitable range

It is scatteredly distributed in Longyang District, Yongde County, and Yuanjiang County, with the most suitable cultivation areas within the altitude range of 800–1 200 m.

2.5.5 Advantages and disadvantages

Advantages: ①High and stable yield; ②attractive appearance; ③fine-textured, juicy and sweet-tasting flesh.

Disadvantages: Susceptible to powdery mildew and bacterial black rot disease.

2.6 Gui Fei

2.6.1 Variety origin

Gui Fei, cultivated by Mr. Zhang Mingxian in Taiwan, China, is a hybrid progeny resulting from the crossbreeding of Erwin and Nang klangwan.

2.6.2 Variety traits

Early maturity.

2.6.3 Variety characteristics

Gui Fei, also known as Hong Jinlong, exhibits robust growth, forming a canopy with a rounded crown. The fruits mature from April to May, typically with an average single fruit weight of 400–800 g. The fruit has an elongated-oval shape, with a fruit shape index (length : width) of approximately 1.74. The fruit has a spindle-shaped appearance without shoulders. Its ventral side protrudes while the dorsal side remains flat, and the fruit's beak is pointed. The thin skin becomes deep red when fully ripe, presenting a smooth and vibrant surface. The flesh, a luscious golden yellow, possesses a fine texture, abundant juiciness, and minimal fibrous content. The flavor is a delightful combination of sweetness and freshness, offering a pure and enjoyable taste. The fruit's growth cycle spans 100–110 d, with flowering occurring from March to April. Typically, the fruits reach maturity from early June to mid July, with peak ripening around late June.

2.6.4 Distribution area and suitable range

It is mainly distributed in Yuanjiang County and Honghe County, and is most suitable for cultivation in areas below an altitude of 800 m.

2.6.5 Advantages and disadvantages

Advantages: ①Attractive fruit shape and color; ②rich fragrance with a sweet and juicy taste; ③strong resistance to cold; ④resistant to scab disease.

Disadvantages: ①Thin skin, limited storage resistance; ②prone to seedless fruit development in low-temperature flowering periods.

2.7 Dasai Yellow Flesh

2.7.1 Variety origin

This variety was selectively bred from the wild mango resources in the Nujiang River Basin. Under standard cultivation practices, it matures from seed sowing to fruiting in just three years. The variety is named for its high yield, stable production, and golden fruit flesh.

2.7.2 Variety characteristics

The tree exhibits moderate vigor with a round-topped crown and a rough, gray-brown main trunk. Average single fruit weight of 135−265 g and the fruit has a kidney-shaped appearance. At maturity, the fruit acquires a golden-yellow peel with a moderate bloom, and the peel thickness measures 0.131 cm. The flesh is also golden-yellow, housing an oval seed with an edible rate of 80.56%. Flowering occurs from early February to mid March, while the fruit ripens from mid May to early July. The harvesting window spans 30 d and the fruit can be stored for up to 4 d. Notable attributes include a fragrant taste, excellent flavor, and an overall high quality. It is worth mentioning its early production and maturation, coupled with a distinctive alternate bearing phenomenon. However, it is not well-suited for storage and transportation. The tender leaves and flowers are susceptible to powdery mildew and anthracnose, and the fruit is prone to infestation by fruit flies.

2.7.3 Distribution area and suitable range

Mainly distributed in Jinggu and Longyang districts, it is most suitable for cultivation in areas with altitudes range of 800−1 200 m.

2.7.4 Advantages and disadvantages

Advantages: ①Early production and maturation characteristics; ②abundant and elongated fibers in the fruit flesh; ③sweet and sour taste, ideal for processing; ④intense fragrance with complex notes reminiscent of fruit, orange, spice, and pine.

Disadvantages: Thin peel, limited storage resistance. Susceptibility to fruit fly infestation and vulnerability to anthracnose.

3 Cultivation and management techniques

3.1 Planting planning and orchard construction

3.1.1 Orchard site selection

Choosing a sunny slope with deep soil is recommended. The groundwater level should be low, ensuring good drainage. The soil should consist of slightly acidic to neutral loam and sandy loam. During the flower bud differentiation period, a moderately dry climate with sufficient water supply and convenient irrigation and drainage is preferred. The establishment of a large-scale mango orchard is most crucial in suitable eco-climate district with favorable production and marketing conditions. The guiding principles should prioritize cost-effectiveness, minimal labor input, quick returns, and high profitability.

3.1.2 Orchard construction

The planning of the mango orchard includes the design of orchard zones, terraces, protective forests, drainage systems, and roads. The orchard can be divided into zones based on climate and topographical characteristics, with protective forests generally serving as boundaries between zones. Due to variations in terrain, it is advisable to limit each zone to 15–30 mu. On flat land, the cruciform method can be used for measurement and planning, and a rectangular design with intersecting parallels at a ratio of 2 : 1 to 4 : 1 is preferable. On sloping land, the long side should follow the contour lines, and a baseline measurement method can be used for contour planning to facilitate cultivation and prevent erosion. In gentle terrain, it is preferable to have a longer dimension from east to west and a shorter dimension from north to south (In regions prone to strong winds or frequent typhoons, the windward side should be shorter). In general, orchard planning should be adapted to local conditions, taking into account factors such as topography, terrain, soil conditions, and so on.

3.1.3 Orchard reclamation

During the establishment of the orchard, ensure the thorough removal of weeds and trees, followed by comprehensive plowing and harrowing. Plow the land deeply to a depth of 40 cm through 2 times of cultivation. This process aims to create a fine and loose soil structure, after which the land is sun-dried before planting.

3.2 Pruning for shaping

3.2.1 Post-harvest pruning

Thinning out excessive, weak, drooping, aging, and crossing branches, as well as shortening the tips of declining branches and excessively long nutrient branches.

3.2.2 Autumn pruning

Thinning out internal, crossing, weak, and overly dense branches, and removing ineffective branches from the upper-middle part of the tree crown.

3.3 Bagging

3.3.1 Pre-bagging preparation

Fruit thinning: Before bagging, thin out diseased fruits, insect-infested fruits, and deformed fruits before bagging, ensuring that 3–5 fruits remain on each flower.

Disease and pest prevention: Prior to bagging, take preventive measures against bacterial black rot disease and thrips.

3.3.2 Material and specifications

Bag materials: Foam net bags, nylon net bags.

Bag specifications: Foam net bags are 12 cm × 7 cm or 14 cm × 7 cm, and nylon net bags are 45 cm × 30 cm.

3.3.3 Bagging time

Bagging should be carried out within a week after the second physiological fruit drop, preferably on sunny days.

3.3.4 Method

Completely wrap individual fruits with foam net bags, and then encase whole ear fruits with nylon net bags, securing the bag opening tightly.

3.4 Fertilizer management

3.4.1 Management of young trees

3.4.1.1 Fertilization

It is recommended to apply compound fertilizer 2–3 times a year, with each plant receiving 0.8–1.5 kg of compound fertilizer (15 : 15 : 15). Starting from the second year after planting, additional organic fertilizer can be applied at a rate of 5–10 kg per plant annually. Another method is to apply fertilizer through trenching (sulcus or strip circularis), followed by covering the soil after fertilization.

3.4.1.2 Tree shaping

It is recommended to cultivate a round-shaped tree crown while simultaneously pruning excessively dense branches and thinning out weak branches. Shoots can be pruned 3–4 times a year. Through practices such as removal, shoot erasing, cutting back, branch bending, shaping and other pruning techniques, an early formation of a productive tree crown can be achieved.

3.4.1.3 Field management

Establishing an eco-friendly orchard involves maintaining weed height below 30 cm and removing dead branches and fallen leaves.

3.4.2 Management for fruiting tree

The flower bud differentiation period (November to December each year): Apply a nitrogen-potassium compound fertilizer to each plant, ranging from 1 to 2 kg. The recommended ratio for the compound fertilizer is 1 : 1 (urea : potassium oxide).

The fruit swelling period (March to April): Apply a compound fertilizer to each plant, ranging from 0.3 to 0.5 kg. The recommended ratio for the compound fertilizer is 1 : 1.5 (urea : potassium oxide).

Post-harvest fertilization: After fruit harvesting, apply 20–30 kg of organic fertilizer to each plant. Additionally, apply 0.8–1.0 kg of urea, 0.5–1.0 kg of potassium oxide, and 0.5–1.0 kg of calcium superphosphate per plant.

3.5 Prevention and control of major diseases and pests

3.5.1 Major diseases

3.5.1.1 Mango anthracnose (*Colletotrichum gloeosporioides* Penz.)

Mango anthracnose belongs to subphylum Deuteromycotina, phylum Colletotrichum.

【Damage situation】This disease is the most common fungal disease affecting mango cultivation in countries and regions around the world. It primarily affects the leaves, branches, inflorescences, and fruits of the mango. Symptoms include leaf spots, stem spots, flower infections, rough skin of fruit, stain spots, and decay.

【Occurrence patterns】The disease primarily overwinters on infected branches, leaves, diseased fruits, and fallen mango debris. Under conditions of high humidity, the fungus generates a substantial number of conidia disseminated through wind and rain. It infiltrates through the host's wounds, lenticels and stomata, directly penetrating the epidermal layer on the tender leaves of the mango. Mango Anthracnose exhibits robust re-infection capabilities and can endure for 2-3 years on host debris. The incidence and spread of mango anthracnose are intricately linked to meteorological conditions, tree vitality, field management, resistance of varieties, phenological periods, inoculum levels in the field, and ventilation and light conditions.

【Preventive methods】Use carbendazim, mancozeb, pyraclostrobin, prochloraz, azole ether· flutolanil, difenoconazole, azoxystrobin. Refrain from using EC during the young fruit stage. Application timing: During the flowering stage, spray once a week with protective agents like chlorothalonil, carbendazim, mancozeb. In the fruiting stage, spray every 2 weeks, alternating therapeutic agents such as pyraclostrobin, azoxystrobin, difenoconazole.

3.5.1.2　Mango bacterial black spot (*Xanthomonas campestrs* pv. *mangiferae indicae*)

【Damage situation】This disease is prevalent in provinces like Guangdong, Yunnan, Guangxi, Fujian in China, and has been reported in countries such as South Africa, India, Pakistan, and Brazil. It mainly results in early leaf shedding and the formation of black spots on the fruit surface, with potential harm during storage and transport.

【Occurrence patterns】The pathogen primarily survives winter on infected branches or residual tissues. In the following year, under favorable temperature and humidity conditions, diseased parts release pus, which spreads through rainwater or insects, entering the host through wounds or natural openings. Hot and rainy weather, particularly after storms, is conducive to the occurrence and spread of this disease. Mango orchards located on windy mountaintops are susceptible to this disease.

【Preventive methods】Application timing: Spray every 10-15 d during the spring shoot period, repeating 3-5 times, especially before and after storms. Suitable agents include bordeaux mixture, kasumin+bordeaux, thiosen copper, *Bacillus subtilis*, copper hydroxide, kasumin·copper quinolate, etc.

3.5.1.3　Mango powdery mildew (*Oidiun mangiferae* Berthet)

【Damage situation】This disease has occurred in various domestic regions including Guangdong, Hainan, Yunnan, and Guangxi. Similar occurrences have also been reported in international locations such as India, Pakistan, Brazil, South Africa, Australia, and Sri Lanka. The main targets of this disease are inflorescences, young fruits, tender leaves, and young branches, leading to significant flower and fruit drop.

【Occurrence patterns】The infestation of powdery mildew is more likely to happen in conditions of low temperature and high humidity, especially during prolonged periods of overcast and rainy weather in winter and spring. It primarily affects tender leaves, flowers, and young fruits.

【Preventive methods】Utilize 70% thiophanate methyl WP [1 : (300–500)], 20% triadimefon EC [1 : (1 000–2 000)], 50% sulphur colloidal SC [1 : (200–400)], 70% thiophanate methyl WP [1 : (600–800)], 10% difenoconazole WDG [1 : (800–1 500)], 12.5% diniconazole WP (1 : 2 000), and 25% pyraclostrobin EC (1 : 1 500). Caution should be exercised when using ECs during the young fruit stage to avoid scorching the fruit surface.

3.5.2 Major pests

3.5.2.1 Thrips

Thrips, a collective term for insects in the order Thysanoptera, are a common threat to mango fruits, with almost 20 different species causing damage. Among them, around 14 species are prevalent, including Anaphothrips theiperdus Karny (*Scirtothrips dorsalis* Hood).

【Damage situation】Period of infestation: The mango thrips can appear year-round, showing a distinct peak period. This peak incidence is closely tied to the phenological stages of mango fruits, causing significant harm during the flowering, young fruit, and various tender shoot periods. Damage areas: Thrips use their rasping-sucking mouthparts to penetrate mango tissues, extracting sap. Adults and larvae congregate, causing damage to the tender shoots, young leaves, flowering period, and young fruits of mango. Symptoms of damage: Adults and nymphs of thrips harm the young leaves of mango trees, causing numerous sooty black spots, darkening leaf tips, and curling leaf edges. Ultimately, the leaves shed entirely, leading to the complete withering of the plant.

【Preventive measures】From the flowering stage until the second physiological fruit drop, and each time new shoots grow to 3 cm before the leaves turn green, it is advised to use substances like spinetoram, Imidacloprid, Acetamiprid, Spirotetramat or Cypermethrin for spraying. Maintain a spraying interval of 7–10 d, applying them 2–3 times consecutively. Ensure rotation of these substances to prevent the development of resistance.

3.5.2.2 *Deporaus marginatus* Pascoe

【Damage situation】Period of infestation: The emergence period of adults coincides with the shoot growth period of the mango tree, mainly causing damage to the summer and autumn shoots. In Yunnan, the first peak damage period typically occurs from mid to late May to early to mid July, and the second peak period occurs from mid to late August to early-mid October. The diapause of mature larvae is classified as a short-day diapause type. In Yunnan, during early November and late March of the following year, under a short-day photoperiod, mature larvae enter a diapause state in the soil for overwintering. As daylight lengthens in

April of the following year, mature larvae pupate and undergo metamorphosis. Damage areas: Primarily targets tender leaves. Adults feed on the upper epidermis and flesh of young leaves, resulting in nearly circular feeding spots. The female insect lays its eggs on young leaves, biting off the leaf near the base with a neat incision resembling a cut made by a knife; the egg-laden portion falls to the ground, causing bare shoots. Symptoms of damage: Adult feeding on the upper epidermis of young leaves leads to leaf curling and withering, leaving the lower epidermis untouched. After egg-laying on young leaves, females bite horizontally near the base, leaving a distinct scissor-like cut at the leaf base.

【Preventive measures】It is advisable to employ antibiotics like abamectin, emamectin benzoate, lambda-cyhalothrin, thiamethoxam as control agents. When applying these pesticides, ensure a rotation of their use. Additionally, exercise caution to protect natural enemies such as hemipteran bugs and spiders.

3.5.2.3 *Erosomyia mangiferae* Felt

【Damage Situation】Period of infestation: Annually during the shoot elongation phase, it adversely affects young leaves. Damage areas: The larvae of Mayetiola sp. bite through the epidermis of tender leaves, entering to feed on the leaf tissue. Damage areas initially exhibit white spots that later turn into brown patches, accompanied by perforations and ruptures, causing the leaves to curl. In severe cases, the leaves wither and fall off, resulting in the withering of the shoots. Symptoms of damage: Larvae feed on tender leaves, causing brown spots, perforations, and leaf curling. In severe cases, leaves wither and fall off, leading to poor growth of tree crowns. The damage peaks with a 100% harm rate to plants and new shoots during the infestation period.

【Preventive measures】Recommended pesticides: abamectin, imidacloprid, fenvalerate, lambda-cyhalothrin, etc. Apply pesticides 2–3 times during each shoot elongation phase, ensuring uniform application with a 7–10 d interval between treatments.

3.5.2.4 *Chlumetia transversa* Walker

【Damage situation】Period of infestation: The severity of damage throughout the year is closely linked to temperature and the condition of plant shoots. The infestation is more pronounced when the average temperature exceeds 20 °C. There are typically four peak infestation periods, occurring from mid April to mid May, late May to early June, early August to early September, and early to mid November. Damage areas: Larvae feed on tender shoots and flower spikes, causing affected shoots or flower spikes to wither or weaken. Symptoms of damage: Widely observed in mango-producing regions, larvae feed on into tender shoots and flower spikes, resulting in withering or weakened growth of affected shoots or flower spikes.

【Preventive measures】During the egg and early larval stages, chemical treatments are applied. Spraying should commence when inflorescences and new shoots reach a length of 3 cm, with intervals of 7–8 d between applications, repeated 2–3 times. Commonly used chemicals include avermectin, fenvalerate, decamethrin, emamectin benzoate, chlorantraniliprole.

3.6 Harvesting

3.6.1 Harvesting time

The optimal time for harvesting is on a sunny day without dew.

3.6.2 Harvesting criteria

Harvest when the color at the top of the fruit changes from green to yellow.

3.6.3 Harvesting method

When cutting the fruit, leave a stem of 0.8–1.5 cm. Handle with care to avoid damage, and store in a cool, dry place.

References

DONG Y C, LIU X, 2006. Crops and Their Wild Relatives in China: Volume on Fruit Trees [M]. Beijing: China Agricultural Press.

NI Z G, 2016. New Cultivation Techniques of Mango [M]. Kunming: Yunnan Science and Technology Press.

NI Z G, CHEN Y F, XIE D H, et al., 2013. Research and development plan of Yunnan mango industry [J]. Chinese Journal of Agricultural Resources and Regional Planning, 34(3): 89–94.

NI Z G, ZHANG L H, LUO X P, et al., 2008. Investigation and analysis of the wild mango (*Mangiferca indica* L.) germplasm resources in Nujiang low and hot river valley [J]. Southwest China Journal of Agricultural Sciences, 21(2): 436–439.

South China Agricultural University, 1989. Special Discourse of Fruit Tree Cultivation (Southern Edition) [M]. 2nd edition. Beijing: China Agricultural Press.

ZHANG C X, et al., 2020. The Current situation of mango industry in Yunnan [J]. China Fruits (6): 112–117.

❖ Key Techniques for Citrus Cultivation

1 Overview of industrial development

Citrus, a woody plant belonging to the Rutaceae family and Citrus genus, is a tropical and subtropical evergreen fruit tree thriving in warm and humid climates. China, with a cultivation history spanning over 4 000 years, is a significant origin of citrus, boasting rich resources and a diverse range of superior varieties. The economically cultivated areas of citrus in China are mainly distributed in the latitude of 16°–37°N and the altitude of 700–1 000 m. Citrus grows optimally at temperatures between 23° and 29 °C, requiring a total sunlight exposure of 1 300–1 400 h per year. While adaptable to various soils, citrus flourishes in loose, fertile, well-drained, slightly acidic to neutral soils. Citrus fruits are not only sweet and juicy, but also rich in nutrients. Compared to other fruits, they hold significant economic value. Moreover, the flowers, fruits, and peels of citrus have high medicinal value. More than 30 kinds of human health-care substances have been isolated from citrus fruits, mainly including flavonoids, monoterpenes, coumarins, carotenoids, propanols, acridones, glycerol glycolipids, etc.

Yunnan, with a rich history of citrus cultivation, is a renowned producer of very early and late-maturing specialty citrus and one of the origins of citrus. The citrus industry in Yunnan boasts the broadest cultivation area in the province, with an annual yield second only to bananas in the fruit industry. In 2020, the citrus cultivation area in Yunnan reached 1.464 million mu, accounting for approximately 14.67% of the total fruit orchard area in the province. Citrus production amounted to 1.358 5 million t, contributing to 14.13% of the total fruit production in the province. Over the past years, Yunnan's citrus industry has witnessed rapid growth, with the cultivation area expanding from 514 500 mu in 2010 to 1.464 million mu, marking an increase of approximately 184.55%. Similarly, production has surged from 456 200 t to 1.358 5 million t, indicating an increase of about 197.79%. Yunnan boasts diverse climate types and abundant water and soil resources, enabling year-round citrus production and supply, providing a clear advantage in market timing compared to other domestic production areas. In addition, Yunnan's excellent ecological environment is conducive to the green organic production of the citrus industry, which has laid an excellent environmental foundation for the quality of Yunnan citrus. Yunnan Province boasts diverse climate types and abundant water and soil resources, facilitating year-round citrus production and supply. This grants a distinct advantage in time-to-market compared to other domestic production areas.

Yunnan's citrus planting has formed advantageous production areas dominated by dry-hot valleys and humid-hot valleys, such as Xinping County, Huaning County, Yuanjiang County in Yuxi City, Jianshui County, Mile City, Hekou County, Honghe County in Honghe Prefecture; Binchuan County, Heqing County in Dali Prefecture; Shizong County in Qujing City, Yongshan County in Zhaotong City; Jinghong City, Menghai County, Mengla County in Xishuangbanna Prefecture; Yongsheng County in Lijiang City, Jiangcheng County, Zhenyuan County in Pu'er City; Guangnan County in Wenshan Prefecture; Ruili City and Mang City in Dehong Prefecture.

2 Main cultivars and characteristics

2.1 Orah mandarin

Orah mandarin is a citrus variety developed in Israel, resulting from the crossbreeding of the Temple (tangor) and the Dancy (tangerine). It falls into the category of hybrid citrus. This particular variety is a late-maturing hybrid citrus known for its robust growth. Initially, the tree crown takes on a naturally rounded shape, which gradually expands after the fruiting stage. The fruits are of medium size, with a slightly flattened, round shape, and an average single fruit weight of 130 g. The peel of the fruit is smooth, colored in orange or reddish-orange, easily peelable, with fine and closely spaced oil sacs. The fruit surface might be slightly convex or flat, with few depressions. The orange-red flesh is tender, non-fibrous, juicy, and sweet, with a soluble solids content ranging from 13.3% to 14.5%. Each fruit typically contains 9–20 seeds and the seeds are elongated and single-seeded. Orah mandarin is usually available in the market from January to April. In some regions, like Yuanjiang County, it can be available for sale as early as late November or early December. In other areas like Binchuan County, the harvesting period may extend to May or June.

2.2 Wenzhou honey tangerines

2.2.1 Miyamoto

Miyamoto is a cultivar developed from a mutant branch on the early-maturing plant of Miyagawa. The trees of this variety have a short crown, short internodes on the branch tips, and thorns. The leaves are elliptical or ovoid-elliptical in shape. The flowers are solitary and white. The fruit is flat-round, with an orange-yellow color, smooth peel, flat or slightly raised oil sacs, easy peeling, and an average single fruit weight of 120 g. The flesh is orange-red, fine and tender, with abundant juice, easy to chew and swallow, and a soluble solids content of 8.0%−9.0%. The fruits are generally harvested and brought to market in mid-late September, with harvesting in Huaning County typically occurring from late July to early September.

2.2.2 Taepo No. 5

Taepo No. 5 is an especially early maturing Wenzhou honey tangerine bred from the branches of the branches of the early-maturing Wenzhou honey tangerine in Yamazaki. Taepo No. 5 is a small evergreen shrub with a low crown, short internodes on branches, and thorns. Its leaves are evergreen unifoliate compound leaves, elliptical in shape, with a dark green, smooth surface. The flowers are solitary and white, and the fruit is parthenocarpic, producing wide-skinned citrus suitable for fresh consumption. It demonstrates good early fruiting, a short growth period, and rapid development. The fruit has a flattened-round shape and turns orange-red when fully ripe, with a smooth surface. It has a rich flavor, and the pulp is easy to chew and swallow. The average single fruit weight is 115 g, with the sugar content of 8.5% and the acid content of 0.57 g/100 mL. The fruit generally ripens in mid September, and in Huaning County, harvesting usually takes place from late July to early September.

2.2.3 Xingjin

Xingjin is a consistently green, small evergreen tree characterized by evenly distributed branches, vigorous growth, and thorns. Its leaves are diamond-shaped and deep green, while the flowers, relatively large in size, bloom singularly in white with five petals. The fruit takes on a tall, flattened round shape, with an average single fruit weight of 180 g. The flesh of the fruit is delicate and juicy, with the soluble solids content ranging between 9.0% and 10.0%, along with the acidity of 0.7−0.8 g/100 mL. Typically, the fruit ripens in the early mid October, with harvesting commencing in late September in Huaning County for market availability.

2.3 Bingtang Orange

Bingtang Orange, originally from Hongjiang City, Hunan Province, is renowned as one of China's distinctive local sweet orange varieties. Its fruits are characterized by sweetness, crispness, juiciness, easy peeling, few or no seeds, high soluble solids content, and excellent flavor. The Bingtang Orange tree is robust, with an open canopy and sturdy branches. The leaves are elliptical and relatively large. The fruit is nearly round, with an orange-red color and smooth peel. Average single fruit weight is 150–170 g, with a soluble solids content of 14.5%, sugar content exceeding 12 g/100 mL, and acid content of 0.6 g/100 mL. The flavor is characterized by richness, sweetness, and a fragrant aroma, with minimal seeds present. Typically, the oranges reach maturity in mid to late November, with certain regions such as Yuanjiang entering the market before the National Day holiday.

2.4 Sanhong honey pomelo

Sanhong honey pomelo is produced in Pinghe County, Fujian Province. Its outer peel, under appropriate leaf coverage, exhibits a rare light pink coloration. The sponge layer beneath the peel is pink, while the flesh appears rose-red, hence the name "Sanhong" (Three Reds). It is a new variety resulting from the mutation of Guanxi Honey Pomelo to red-fleshed pomelo, belonging to the series of Guanxi Honey Pomelo varieties. The latest variety of red-skinned and red-fleshed honey pomelo shows early ripening, high yield, and exceptional quality. The fruit is large with thin skin, seedless pulp, abundant and tender juice, mildly sweet with a hint of sourness. Its shape is generally an inverted oval, with an average single fruit weight of 1 680 g. The peel is red-green, the sac peel is pink, the juice sacs are red, with soluble solids content of 11.55%, and total acid content of 0.74%. It is generally available in the market from September onwards.

3 Cultivation and management techniques

3.1 Planting planning and orchard construction

3.1.1 Planting planning

Citrus production hinges on the diversity of its varieties. When selecting varieties for cultivation, it is crucial to consider a comprehensive range of factors, including local climate and water conservancy. Proactive cultivation is recommended for new varieties recognized for their high quality and abundant yield. Before undertaking large-scale planting, it is imperative to assess the adaptability of the chosen varieties to local conditions. Conducting small-scale experimental plantings is advisable to observe the quality and yield of the crops. Based on these observations, decisions regarding expanding cultivation or implementing large-scale planting can be made.

3.1.2 Orchard construction

3.1.2.1 Orchard site selection

For the best orchard site, it is recommended to choose a mountainside with a slope less than 25° that receives ample sunlight. The soil should be either neutral or slightly acidic. It is essential that the soil is deep and fertile, with a high content of organic matter. Moreover, the soil must exhibit good permeability to ensure efficient drainage within the orchard. Accessibility to transportation is crucial, and the region should maintain a well-preserved ecological balance, creating an ideal environment for cultivating citrus fruits.

3.1.2.2 Soil improvement

For orchards with a slope less than 15°, a gradual slope cultivation method is employed. When the slope ranges between 15° and 25°, terrace construction is implemented to maintain a width of 2.5–3 m. Soil improvement involves deep plowing, with a depth exceeding 0.8 m, burying organic materials such as straw and medicinal residues, constituting at least 60% of the mixture. For each mu, apply a layer of crushed dry straw and medicinal residues with a thickness of at least 20 cm for burial. Alternatively, a dedicated organic fertilizer for soil improvement can be used, comprising 5 000 kg of organic fertilizer, 2 000 kg of dry livestock and poultry manure, 500 kg of oil cake, and an additional 1 000 kg of bio-bacterial fertilizer, or directly spray bio-fertilizer. The mixture is spread evenly and compacted, and the main and back furrows are established along the terraced surface in a standardized and leveled manner.

3.1.2.3 Water conservancy construction

It is advisable to procure intelligent, fully automated water-saving irrigation equipment with robust water-saving capabilities. Drip irrigation involves the use of plastic pipes to deliver water through holes or drippers with a diameter of approximately 10 mm, directing it to the roots of crops for localized irrigation. This method currently stands out as the most effective means of water-saving irrigation in arid and water-scarce regions. It adapts well to different terrains and soils, allowing precise control over fertilizers and water. With a water and fertilizer utilization rate exceeding 90%, it brings about significant advantages in terms of water and fertilizer conservation, reduced labor, and time efficiency.

3.1.2.4 Road construction

In the construction of roads, a main road is built at intervals of 200 m, with a width of 3 m. Additionally, at intervals of 100 m, an operational road with a width of 2 m is constructed. In areas where the terrain is unsuitable for road construction within orchards or garden sections, a single-track mountain transport vehicle is installed. The track machine room is constructed adjacent to the main road, facilitating the transportation of production materials and fruits. Internal roads within the garden are interconnected, and the development of the garden area is aimed at becoming a scenic spot. This facilitates tourists' sightseeing, allows them to experience fruit-picking, and also helps reduce production costs.

3.1.3 Planting seedlings

The planting of seedlings is generally done from late February to early March before the spring shoots sprout, or from late October to early December. The planting density varies based on the characteristics of the variety, terrain, soil, rootstock, and cultivation management. For early-maturing Wenzhou honey tangerines on flat land, the planting density is around 40–50 plants per mu, and in mountainous areas, it is 70–80 plants per mu. For mid-maturing Wenzhou honey tangerines on flat land, the planting density is similar, ranging from 40–50 plants per mu, and in mountainous areas, it is 50–70 plants per mu. Sweet oranges are planted at a density of 20–40 plants per mu on flat land and 40–60 plants per mu in mountainous areas. Ponkans are planted at a density of 50–60 plants per mu on flat land and 70–80 plants per mu in mountainous areas. Pomelos are planted at a density of 20–40 plants per mu on flat land and 40–60 plants per mu in mountainous areas. Before the actual planting process, it is advisable to apply planting fertilizer to each hole. This involves using 3 kg of cake fertilizer and 1 kg of phosphate fertilizer per hole. These should be thoroughly mixed with the soil, applied to the bottom of the hole, covered with 20 cm of fine soil, and then the seedlings are placed in the hole. The root system should be spread out, and after filling with a small amount of fine soil, the seedlings are gently lifted upward to ensure a close and secure connection between the soil and roots. The soil should be filled, pressed firmly, leveled with the root collar, and each plant should be watered with 20 kg of water. Finally, the plants should be covered with soil, leaving the graft union exposed.

3.1.4 Planting management

Watering and moisturizing: After planting, the seedlings usually begin to thrive after approximately two weeks. During this time, when the soil is relatively dry, it is advisable to water every 1–2 d to keep the soil moist and promote the growth of new roots. Before the seedlings take root, only watering is necessary, and fertilization should be avoided. After successful survival, implement a regular application of light fertilizer to stimulate the growth of both roots and branches.

Soil covering: It is advisable to cover the ground around the base of the tree with straw, stalks, green manure, or mulch film to preserve the soil's looseness and moisture, ensuring the safe wintering of young seedlings.

Staking for wind protection: Newly planted seedlings, with their root systems not yet firmly established, are prone to being swayed by strong winds. Therefore, it is recommended to securely insert a bamboo stake deeply next to the seedling and fasten it with plastic rope.

Pinching: For young trees exhibiting vigorous growth, it is necessary to timely pinch the long branches and remove shoots that sprout in inappropriate positions or on the main trunk.

Fertilizer management, prevention and control of diseases and pests: After successful survival, it is essential to conduct frequent inspections and apply liquid fertilizer diligently but sparingly. Upon detecting issues such as red spiders, leaf miners, aphids, anthracnose canker disease and others, prompt prevention and treatment are advisable according to the specific situation.

3.2 Pruning for shaping

3.2.1 Tree form

The natural open-center model typically comprises three main branches without a central trunk. The trunk extends outwards without exposing its core, and the shaping process spans three years. The first year: Perform a leader branch cutting at a height of approximately 50 cm above the ground, select and arrange the main branches, remove the central bud, wipe off excess buds, and eliminate shoots. The second year: Cut back the extension branches of the main branch, select and match the secondary main branches, pinch off the growing tips, wipe out buds, and thin out flower buds. The third year: Cut back the extension branches of the main branch, select and match the secondary main branches, pinch off the growing tips, wipe out buds, and thin out flowers.

The natural round shape is the most natural form of tree shaping. The leader branch cutting is set at a height of 40–50 cm, with 2–3 robust large branches naturally branching out from the main trunk. These large branches are spaced 10–15 cm apart, developing in different directions. In the second or third year, an additional 1–2 branches are retained, ensuring no overlap between upper and lower branches. Each main branch has an approximate angle of 45° at its base, developing diagonally in four directions, resulting in a total of 3–5 main branches. Depending on available space, 1–3 secondary branches are then retained. On each main branch, large, medium, and small branches are further arranged. The tree can be shaped after three years.

3.2.2 Pruning of young trees

Young trees exhibit vigorous growth, with multiple shoot flushes in a year and rapid crown expansion. When pruning, a light touch is advised, starting by defining the trunk based on the tree's structural form. It is advisable to select and cultivate main branches in the shaping zone, retaining a certain number of supporting branches to increase leaf count for nutrient supply, enlarge the crown, and promote early production.

After planting, in the first two years, you can trim the shoots 3–4 times annually, with a focus on summer and autumn. Prior to pruning, eliminate sporadic shoots 1–2 times. Once overall shoot growth reaches about half, cease removing shoots and conduct uniform pruning. It is advisable to conduct one thinning before the new shoots turn green, removing stronger upper branches and retaining moderate ones in the upper part, and removing weaker branches while keeping stronger ones in the middle and lower parts. After the new shoots leaf out and turn green, excessively long shoots can be pruned by removing the tips, leaving around 10 leaves to promote branch maturity. Shorten the tips of main, sub-main, and lateral branches by 1/3–1/2 before each pruning, ensuring that 2–3 robust buds below the cut are in the middle section of the shoot. As the tree grows, the canopy continues to expand, and depending on the situation, secondary branches on the lower trunk and main branches are pruned in stages. If the elongated branches on the trunk and main branches of young trees disrupt the tree's form, they should be promptly trimmed. If the main trunk is positioned too low or if the arrangement of the main branches is not ideal, and there's a notable hollow space within, it is recommended to substitute it with an elongated branch. Subsequently, timely removal of the terminal bud and encouraging the emergence of lateral branches are necessary to promote the shaping of the tree's structure. During the juvenile stage, flowering and fruiting in young trees result in nutrient depletion, impeding the crown's expansion. It is essential to promptly remove flower buds to ensure optimal growth. If flower buds are not entirely removed, it is advisable to timely eliminate young fruits after their formation to minimize nutrient consumption.

3.2.3 Pruning of initial fruiting trees

During the initial fruiting stage, it is essential to continue expanding the tree canopy for shaping purposes. Pruning should be done with the premise of maintaining further expansion of the canopy, aiming for early production and high yield. Initial fruiting trees exhibit vigorous growth, and the conflict between nutritional and reproductive growth is pronounced. To address this, strategies such as thinning spring shoots, controlling summer shoots, releasing autumn shoots, and wiping winter shoots are employed to harmonize these conflicting aspects. Emphasis is placed on preserving flowers and fruits, balancing tree vigor, and striving for early fruit setting and high yields. For young citrus trees, the bearing base branches are predominantly autumn shoots. Excessive fruiting on these shoots can result in lower quality.

About 10–15 d before releasing autumn shoots, it is advisable to remove the large terminal fruits from the upper and outer parts of the canopy, or cut them along with the calyx, to promote shoot emergence and cultivate high-quality autumn shoots, laying a solid foundation for fruitful results in the coming year. Varieties with a low fruit-setting rate often experience significant flower and fruit drop. While pruning is typically done in winter for branches with flower and fruit drop, to minimize nutrient consumption, it is recommended to conduct short pruning before the release of autumn shoots in summer, encouraging the emergence of autumn shoots and thereby forming high-quality bearing base branches. In a year with good autumn shoot growth, the following year tends to have abundant flowers, leading to a prolific fruiting season. Therefore, during winter pruning of the current year, it is essential to heavily cut back around one-third of the autumn shoots, removing weak shoots to reduce the flowering quantity for the following year and prevent excessive fruiting. If the terminal branches are summer shoots, they should also be cut back to control. Unused branches, including dead, diseased, insect-infested, densely grown, thin, crossed, overly elongated, and hanging-to-the-ground branches, should be handled in stages. Branches that significantly affect the tree's growth, such as diseased or excessively long ones, should be pruned during the growing season, while the rest can be removed during winter pruning.

3.2.4 Pruning of mature fruit-bearing trees

For mature fruit-bearing trees entering the high-yield period, the key focus of shaping pruning is to nurture bearing base branches, renew fruit-bearing branch groups, balance the ratio of nutrient branches to fruit-bearing branches, prevent alternating years of large and small yields, and achieve both high and stable yields while extending the period of abundant production. During the abundant production phase, the upper part of the crown grows vigorously, leading to potential density and shading issues. When pruning, large branches should be selectively thinned to increase internal light exposure, while paying attention to cultivating a concave-convex tree shape, with fewer large branches and more small branches, ensuring that the upper canopy is not too sparse and the lower canopy is not too dense, allowing light to penetrate from top to bottom and enabling three-dimensional fruiting. Each year, approximately one-third of the weak branches should be heavily short pruned to renew the branch structure and promote the growth of spring shoots. A small number of summer shoots should be retained, promoting shoot growth by removing apical buds and rubbing off lateral buds to increase the number of bearing base branches. The fruit-bearing branch group needs to be rotated and renewed annually to maintain tree vigor and extend the years of abundant harvest. Once the tree enters the abundant harvest period, spring shoots replace autumn shoots as the main bearing base branches. The quality of the spring shoots directly affects the following year's yield. Spring shoots can be promoted through branch renewal. Additionally, summer and autumn shoots that do not require elongation can be pruned short to the base of the spring shoots, encouraging the development of high-quality spring shoots. After the fading of flowers until early to mid July, the vigorous growth of early autumn shoots can be promoted by selectively cutting back the branches with fallen flowers and fruits, encouraging the formation of high-quality bearing base branches for fruiting. During the high-yield season, citrus trees tend to grow tall and dense, making it easy for the canopy to become closed. It is important to promptly manage and make good use of the inner branch space, as neglecting this can significantly impact the yield and quality of the inner-branch fruits.

3.3 Water and fertilizer management

Water and fertilizer management is a crucial task in citrus cultivation. Effective management ensures that citrus plants receive sufficient nutrients throughout the entire growth process. Adequate nutrition helps prevent issues such as fruit shrinkage, thereby ensuring the quality of citrus cultivation.

During the dry season, special attention should be given to the water supply for citrus trees, and artificial irrigation can be employed during this period. In practice, the use of drip irrigation or micro-sprinkler irrigation techniques can effectively prevent water loss during the citrus cultivation process. The application of drip irrigation technology can significantly reduce the waste of water resources, promoting the rational use of resources. It is preferable to adopt a strip irrigation method to avoid saturating the entire soil where citrus is planted. Citrus cultivation has specific water requirements, and excessive water can lead to rotting, adversely affecting the normal growth, ultimately impacting the yield and quality of citrus fruits.

Fertilization should be conducted based on various factors, including citrus variety, rootstock, soil type, climate, fertilizer type, tree vigor, and planting density, with an emphasis on a rational and economical approach. The fertilization of citrus fruit trees mainly includes: flowering fertilization, fruit-setting fertilization, fruit-maturing fertilization, and post-harvest fertilization. During the flowering stage, it is best to utilize quick-acting fertilizers as the primary choice, alongside the application of organic fertilizers. For fruit-strengthening fertilization, the primary emphasis is on nitrogen fertilizers, supplemented by the application of phosphorus and potassium fertilizers. For fruit-strengthening fertilization, the main focus is on quick-acting fertilizers, complemented by the application of organic fertilizers. When it comes to post-harvest fertilization, the emphasis is on using organic fertilizers, supplemented by the application of quick-acting fertilizers.

3.4 Prevention and control of major diseases and pests

3.4.1 Major diseases

3.4.1.1 Yellow shoot

Agricultural control: To prevent the entry of infected seedlings and scions into disease-free areas, it is necessary to implement quarantine measures for seedlings. Timely removal of diseased trees is crucial, especially during each of the spring, summer, and autumn shoot periods, thorough inspections on a plant-by-plant basis are essential. Once diseased trees and suspicious ones are identified, they should be immediately removed and disposed of in a centralized manner through incineration. Before removing diseased trees, apply an insecticide for citrus psyllids.

Chemical control: In the process of citrus cultivation management, it is crucial to promptly prevent and control citrus psyllids. During the spring and autumn seasons, when sporadic pest infestations are detected in the fields, it is recommended to use pesticides such as imidacloprid, thiamethoxam, acetamiprid, buprofezin, dinotefuran, flonicamid, thiamethoxam+pyriproxyfen, and others. Timely spraying should be conducted for effective prevention and control. Through enhanced cultivation management and the reinforcement of tree vigor, the resistance to diseases can be heightened.

3.4.1.2 Canker disease

Agricultural control: To prevent the entry of infected seedlings and scions into disease-free areas, it is necessary to implement quarantine measures for seedlings. In winter, it is necessary to clean the orchard by pruning diseased branches and removing fallen leaves and fruits, which should be centralized and incinerated to reduce overwintering sources of disease. Additionally, it is important to enhance water and fertilizer management to promote the neat emergence of new shoots. Precautionary measures should be taken against pests such as leaf miners. Planting windbreak forests can also help reduce wind damage.

Chemical control: It is recommended to use a 10 billion spores/g *Bacillus subtilis* WP [1 : (700−800)], 30% thiazole zinc SC [1 : (500−750)], 46% copper hydroxide WG [1 : (1 000−2 000)], 30% thiosen copper SC [1 : (750−1 000)], 77% copper calcium sulphate WP [1 : (400−600)], 47% kasumin·bordeaux WP [1 : (380−470)], or other copper-based agents and their compound agents for spray control. To achieve better results, it is advisable to rotate the use of different agents.

3.4.1.3 Anthracnose

Agricultural control: Through enhanced cultivation management and the reinforcement of tree vigor, the resistance to diseases can be heightened. Adequate water and fertilizer management, along with measures to prevent pests and sunscald, should be implemented while avoiding mechanical damage. Pruning of diseased and overgrown branches is advisable. Additionally, clear away fallen leaves and burn them in a concentrated manner on the ground.

Chemical Control: It is advisable to employ 30% zopiclone·tebuconazole SC [1 : (1 500−2 000)], 40% zopiclone·prochloraz EW [1 : (3 000−4 000)], 60% pyraclostrobin. metiram WG [1 : (1 000−1 500)], 15% tebuconazole · trifloxystrobinat SC [1 : (2 000−2 500)], 10% difenoconazole WG [1 : (2 000−2 500)], 70% iprovalicarb WP [1 : (600−700)], and 75% mancozeb WG [1 : (370−470)], and other compound formulations for effective spray control.

3.4.1.4 Sunscald

Agricultural control: Choose varieties with lower susceptibility to sunscald; advocate for grass management within the orchard to regulate the microclimate; properly dense planting, or encouraging the growth of summer shoots in young fruit trees at the conclusion of physiological fruit drop, serves to shield the fruit and alleviate the severity of sunscald. To prevent and control, attach paper, yellow tape, bags, or apply lime paste to the fruits on the outer periphery of the canopy; maintain soil moisture and improve the microclimate within the orchard; prompt the delayed growth of summer shoots to use them as a natural cover for the fruit.

3.4.2 Major pests

3.4.2.1 Woodlice

Chemical control: It is recommended to employ 21% thiamethoxam SC [1 : (3 360−4 200)], 5% acetamiprid microemulsion ME [1 : (4 000−5 000)], 10% biphenthrin EC [1 : (1 666−3 333)], 2.5% cyhalothrin EW [1 : (3 000−4 000)] and their compound formulations such as spinetoram+Avi-pyridyl ether (1 : 2 000), and 30% pyriproxyfen·chlorfenapyr SC [1 : (2 000−2 500)]. Alternatively, the use of 2% chlorfluclothianidin GR + 10% pyriproxyfen SC [1 : (1 000−2 000)], 15% tolfenpyrad SC (1 : 1 500)+ 5% emamectin benzoate SC (1 : 3 000) for spray control is suggested, and the rotation of different pesticides is recommended.

3.4.2.2 Red spider

Agricultural control: Proper pruning of orchards can ensure ventilation and light penetration; additionally, Sod-culture provides a habitat for predatory mites (natural enemies of red spider mites).

Chemical control: During winter orchard cleaning, it is recommended to use 99% mineral oil EC [1 : (150–200)] or 24.5% abamectin·mineral oil EC [1 : (1 000–2 000)] for spraying. When the average number of insect mouths per 100 leaves ranges from 1 to 2, comprehensive control measures should be implemented. Options for control include using 43% bifenazate SC [1 : (1 900–2 400)], 34% Spirodiclofen SC [1 : (2 000–3 000)], 110 g/L etoxazole SC [1 : (2 000–2 500)], 5% fenproximdined SC [1 : (1 000–1 500)], 22% abamectin·spirodiclofen SC [1 : (4 000–6 000)], along with other compound agents for spray control. It is advisable to rotate the use of different agents.

3.4.2.3 Leaf miner

Agricultural control: In late autumn and early summer, branches with larvae or pupae should be pruned slightly. Water and fertilizer management should be strengthened, while sporadic early summer branches are thoroughly pruned, then followed by concentrated monitoring to disrupt the food chain of pests.

Physical control: Use black light or frequency trembler grid lamps to lure and kill pests.

Chemical control: Utilize 10% imidacloprid EC [1 : (2 000–2 500)], 40% chlorfluthiamethoxam WG (1 : 2 000), 20% tebufenozide SC [1 : (1 500–2 000)], or 10% chlorfenapyr SC [1 : (800–1 200)], in addition to their compound agents such as 1% emamectin benzoate EC [1 : (1 500–2 000)], 20% fenpropathrin EC [1 : (1 500–2 000)], 1.8% avermectin EC [1 : (1 500–2 000)], or 40 billion spores/g *Beauveria bassiana* WP [1 : (1 600–2 000)] or 16 000 IU/mg *Bacillus thuringiensis* WP [1 : (200–300)] for spray control. It is important to rotate the use of these agents.

3.4.2.4 Small fruit fly

Agricultural control: Gather fallen and wormy fruits in orchards and dispose of them by burying them deeply in a centralized manner. When clearing orchards in winter, the soil is turned over to destroy theirs overwintering environment, thus effectively reducing the insect source.

Physical control: Set up small fruit fly attractant adhesive boards for trapping; alternatively, employ methyl eugenol bait in bottles to lure and eliminate adult flies.

Chemical control: Apply 80–100 g of 1% thiamethoxam bait per mu, 70–120 g of 1% imidacloprid bait per mu, or 180–270 mL of 1% avermectin concentrate bait per mu for effective pest control.

3.5 Timely harvesting

3.5.1 Maturity indicators

The optimal harvest time for citrus fruits depends on their intended use, whether for fresh sale, processing, or storage. Fresh Sales: The harvesting process follows the principle of selecting yellow fruits instead of green ones and harvesting in batches. The harvested fruits should exhibit a ripeness of over 90%, continuing until they reach full maturity according to the inherent indicators of the specific variety. The maturity indicators for harvesting are as follows: Navel oranges ≥11% (total soluble solids, the same below), Bingtang oranges ≥16%, Citrus sinensis osbeck ≥8.5%, Early-maturing Tangerines ≥8.5%, Mid-maturing Tangerines ≥12%, Ponkans ≥10%, and Pomelos ≥12%.

3.5.2 Harvesting techniques

It is advisable to choose a clear day for harvesting, preferably after the surface moisture on the fruits has dried. Utilize a specialized citrus picking stool to harvest in a sequence from top to bottom and from outer to inner areas. Harvesting employs a double-pruning method. In the initial cut, the fruit is snipped approximately 1 cm from the stalk, followed by a second cut along the base of the stalk, ensuring a smooth surface and preserving the intact sepals. Fruits, whether placed in the fruit-picking basket or transferred from the fruit-picking basket to the fruit box (or wicker basket), must be handled delicately. The fruit box (or wicker basket) should only be filled up to 90% capacity. Harvested fruits should be shielded from rain and sunlight, and it is not advisable to leave them exposed outdoors overnight. Damaged fruits and those affected by diseases or pests should be kept separately.

References

ANONYMOUS, 2003. Brief introduction of the Ri Nan No.1 variety [J]. Jiangxi Horticulture (4): 39.

DENG X F, 2009. High-standard cultivation of citrus orchards [J]. Anhui Forestry, 163(5): 34.

HUANG L, 2020. Effects of altitude and slope aspect on soil nutrients and fruit quality of citrus [D]. Changsha: Hunan Agricultural University.

LI S J, LONG G Y, YANG H, et al., 2010. Characteristics and cultivation techniques of Taepo No.5, the very early-ripening satsuma mandarin [J]. Hunan Agricultural Science (3): 101−102, 106.

LU H F, LU Q C, 2022. High-standard cultivation technology for citrus orchards [J]. Modern

Agricultural Research, 28(8): 122−124.

PENG S Y, 2016. The evaluation and analysis of the very early-ripening stsuma mandarin biological characteristic [D]. Changsha: Hunan Agricultural University.

SHEN P, SHEN C, 2018. Study on pruning techniques of citrus [J]. Seed Technology, 36(11): 76, 78.

SHU X Y, XIAO D X, 2014. Harvesting and storaging of citrus [J]. Hunan Agriculture (10): 32.

SONG W, WANG J, LI H X, et al., 2022. Preliminary exploration of pruning techniques of citrus in the hilly area of southern Sichuan basin [J]. Sichuan Agricultural Science and Technology (5): 25−28.

WU X M, 2012. Characteristics and cultivation methodology research of the very early-ripening satsuma mandarin, Taepo No.5 [J]. China Fruit News, 29(6): 62−63.

❖ Key Techniques for Grape Cultivation

1 Overview of industrial development

As one of the World's Four Major Fruits, the grape enjoys a long history of propagation, covering a wide range of growing. There are over eleven thousand (11 000) known grape varieties, which may be eaten as table fruits, dried, crushed into juice, and made into jam or vinegar following advanced management skills, various methods of cultivation, and a fully completed industrial chain. The versatile fruits, juicy and palatable along with their by-products, make grapes a unique position in the fruit industry. Until very recent years, the readjustment and transformation of the agro-industry has given grape planting a push, due to planting grapes bringing great earnings rapidly. By doing so, the grape industry has become one of the leading pillar industries in Yunnan province, whose unique geographic location, sufficient sunlight and suitable temperature provide good climate and ecological environment conditions for the growth of the grape industry. Grapes can be grown all year round even in different areas of Yunnan Province, making Yunnan province as a major grape producer in China. It is acknowledged that the industry plays an essential part to consolidate the achievements in poverty alleviation and achieve rural revitalization in Yunnan.

1.1 Production

Since 2010, there has been a stable growth of grape planting areas. It is till the year of 2019 that Yunnan Province ranked 5th in China in grape planting areas, as the 4th in production volume. Grapes planting in Yunnan Province. In 2020, the grape planting area in Yunnan Province accounts for 5.28% of the total fruit planting area as the 7th in the province, following citrus, mango, apple, banana, pear and peach. The production volume constitutes 10.21% of the total volume in Yunnan Province, ranking 5th after banana, citrus, pear and apple, sourced from Green Food Development Center, Yunnan.

1.2 Production distribution

112 counties (cities/districts) are grape producers out of 16 prefectures (municipals). Over 100 thousand mu of grapes are planted in Honghe Prefecture and Dali Prefecture. About 10 thousand to 100 thousand mu of grapes are planted in Chuxiong Prefecture, Qujing City,

Kunming City, Diqing Prefecture, Wenshan Prefecture, Zhaotong City, Lijiang City and Yuxi City. The rest 6 prefectures or municipals plant less than 10 thousand mu of grapes on average. In terms of production volume, Dali Prefecture, Honghe Prefecture and Chuxiong Prefecture produce more than 100 thousand tons of grapes, Qujing City, Kunming City, Zhaotong City, Yuxi City, Lijiang City and Wenshan Prefecture produce from 10 thousand to 100 thousand t, other prefectures (municipals) less than 10 thousand t.

Binchuan is the largest county growing grapes in Yunnan Province, followed by Mile City. There are only two counties (municipals, districts) as well with which planting areas accede 100 thousand mu. Fresh grapes (raw) are cultivated in Binchuan County, and wine-making grapes in Mile City. There are 50 000−100 000 mu of grapes being cultivated in Jianshui County, while 10 thousand to 50 thousand mu in Yuanmou County, Luliang County, Mengzi City, Deqin County, Qiubei County and Yongsheng County.

1.3 Participants

There are 20 leading provincial enterprises in grape growing and processing and 37 specialized farmers' cooperatives in Yunnan Province. Above twenty local grape brands, such as "Yunnan Secrete" "Guo Xianfeng (The Fruit Pioneer)" and "Lou Tie Yuan", released the market, of which "Yunnan Secrete" and "Guo Xianfeng (The Fruit Pioneer)" have been entitled with "Top 10 Yunnan Brands" in the consecutive ten years. Mile Grape has been registered as an "Agro-product Geographic Indication", and "Binchuan Hongti Putao" has been registered as a "China Geographic Indication Certification Mark".

2 Main cultivars and characteristics

To date, Yunnan is a cultivation home to more than 50 varieties. The planting area of table grapes accounts for more than 85%, mainly distributed in Honghe, Dali, Chuxiong, Qujing and other prefectures (cities). Mainly cultivated table grapes include Red Globe, Summer Black Grape, Crimson Seedless Grape, Centennial Seedless, Kyoho, etc. Due to the fluctuating table grapes markets, the planting areas of traditional main varieties, such as Red Globe and Summer Black, have decreased in recent years. After grafting in the top has been applied in Jianshui, Yuanmou, and Binchuan counties, some supreme varieties have grown. The planting area of wine grapes accounts for about 15%, where Deqin County, Weixi County of Diqing Prefecture, Mile City of Honghe Prefecture and Qiubei County of Wenshan Prefecture are significant producers. The wine grape varieties cultivated in those areas include Cabernet Sauvignon, Red Rose, Vidal, French Wilde, and so on, with Niagara Grape developed as well, used as both table grapes and for wine.

2.1 Red Globe

General botanical characteristics: It is also known as the American Red Grape, the Late-ripening Red, and the big Red Globe, belonging to Eurasian species. In early spring, the plant's growing tips are purplish red, and the young blades are light purplish red. At the same time, the upper surfaces of the blades are smooth. Epicuticular hairs may be found on the underside of the leaves. The adult leaves are medium-sized, heart-shaped, and 5-lobed with lobate in the upper and the parted in the bottom. There is no villus on both sides of the front and back of the adult leaves. The leaf edge is purely blunt serrated, and the petiole is longer in light red. The petiole is arched and bisexual. Self-flowering fruit setting rate is high, and the ears are purplish red, long cone-shaped, the size of fruit grains is on average and even, and in round or oval. The fruit powdery is medium thick, the peel is red or deep red, the fruit stalk is long, which is closely combined with the fruit and is not easy to crack; The fruit brush is soft, extremely firm, extremely tensile and non-threshing. The berries' peels are thick, and the flesh is both hard and crisp, which can be cut into thin slices without juice. It tastes sweet and refreshing and has good quality, with more than 17% of soluble solids content.

Production performance: A kind of table grapes. The variety is supplied to the market throughout the year. The tree's vigor is medium, with a strong fruiting capability. Grape ears and fruits are extremely large, conical and well-filled. The grapes tolerate alkaline soil but are quite delicate with disease. The grapes show salt-alkali tolerance but is prone to pathogenic infections, including powdery mildew, downy mildew, gray mold and sour rot. The fruit endures long transit and storage. The fruit stalk is slender, and the combination of fruit grains and fruit stalks is firm. It is of a strong pressure resistance, and tension resistance is not easy to thresh.

Planted in: Yunnan Province with the largest areas in Binchuan County, Luliang County and Yuanmou County.

2.2 Summer Black

General botanical characteristics: A European and American hybrid. Summer Black is an early-ripen. Growing tips are in green and yellow. Few epicuticular hairs may be found on those tips. The young leaves are in light green with a soft purple halo. The surface of the blades is shiny, and the back of the leaves contains dense woolly epidermal hairs. The adult leaves are large, nearly round, slightly concave in the middle and convex at the edge. Most are 4-lobed, parted. Blunt serrations are on the leaf edges. Petiolar sinus is deformed. New shoots grow upright and turn into canes in red and brown (reddish-brown) within a year. Varieties come out bisexual flowers. Bunches are conical, occasionally with two shoulders. The fruit grains are nearly round, in purple-black or blue-black, closely attached. Coated in fruit powder, berries have thick peel, so does hard flesh. They taste sweet and refreshing, with strawberry-like flavour. When the fruit is fully ripened, the soluble solids content retains 17%–24%. The variety is feature in low fruit acid content as well as no seeds.

Production performance: A kind of table grapes. The variety is supplied to the market throughout the year. This kind tolerates a variety of soil types in prime growth. Flower buds come out quickly, making the grape crops good yields. The resistance to powdery mildew and grey mold is weaker than that of Kyoho. No fruit cracking, no threshing, storage and transportation resistance.

Planted in: Most of Yunnan, with the largest areas in Jianshui County, Mengzi City, Yiliang County and Yuanmou County.

2.3　Niagara Grape

General botanical characteristics: Perennial vines are dark brown, while newly grown vines are light brown or light green and haired. Tendrils are typically green, while a few are purple-brown. The upper end of the growing tips is bronzed and shiny, and the back of the blades are covered with white epidermal hairs, which gradually turn grey as the leaves grow. The leaves are thick, heart-shaped, and dark green. There are five lobes either parted or divided. Sharp serrations are on the leaf edges. The main veins and lateral veins are protruding on the back of the leaf. Bunches are conical and barely have shoulders. Clusters are compact. The ears weigh about 200 g, and the maximum ears weigh 300 g. The fruit grains are round and even, with a single berry weighing 4–5 g in green. When the berries are fully ripened, they tend to be yellowish-green. Thickly coated in fruit powder, Niagara Grape is crystal clear and lustrous. Grape pulp is grass-green with a tender texture and tastes sweet and aromatic. The soluble solids content retains 14%.

Production performance: Both for wine making and table grapes. It is on the market from May to September. High, deep and well rooted, the variety is of a strong resistance, barren resistance and wide adaptability. It is featured by disease resistance, easy cultivation, good and stable production, and good quality.

Planted in: In Yunnan, Mile City has the largest cultivating area, with Qiubei County coming as the second.

2.4 Cabernet Sauvignon

General botanical characteristics: A Eurasian variety. It is a late ripened, originated in France. Growing tips are green, with purple stripes, and without epidermal hairs. Newly developed canes are brown, whereas internodes are short and thick. Young leaves are orange-red, whose surfaces are smooth, and on whose backside grow dense epidermal hairs. Adult leaves are small and heart-shaped. There are five lobes on which sparse epidermal hairs are. Blunt serrations are on dome-shaped leaf edges, curling upward. Petiolar sinuses are closed into circles. Cabernet sauvignon comes out with bisexual flowers. The grape ear, conical, is small, with an average ear weight of 165 g. The clusters, round, purple-black and grass flavour, are of medium density, with an average grain weight of 1.9 g. The soluble solids content retains 16.3%–17.4%, the acid content 0.456%, and the sugar content high.

Production Performance: A kind of wine grapes. It is commonly ripened and is on the market from July to October. Medium in tree's vigor, it has a strong fruiting capability, adaptability, disease resistance, and freeze resistance. Well-draining soil and fertilizer provide a stable harvest. In Diqing Prefecture, it will be ripened in late August and early September.

Planted in: Mile City, Deqin County and other counties in Yunnan Province.

2.5 Crimson Seedless Grape

General botanical characteristics: Also known as Crimson, and C102−26. It is a Eurasian variety and a late-season grape. Growing tips are reddish-green, shinny, and without epidermal hairs. Tender shoots are purple-red and their serrations are green. Adult leaves, 5-lobe, are medium in size and green. Serrations are medium protruding of two sides. The petioles are long. Petiolar sinuses are closed into oval or round shape. Crimson Seedless Grape comes out with bisexual flowers. Bunches are conical with an average weight of 500−700 g. Berries are cylindrical-oval well fill to compact or slightly compact with an average weight of 5−6 g. Crimson Seedless Grape, coated in white powdery, are bright red, with a thick, though skin, and firm crisp flesh The berries are crisp and firm, oval and seedless with a thick, tough skin, and firm crisp flesh with a sweet flavour. The flesh is yellowish-green, and semi crystal which is tightly attached to the skin. Berries are aromatic and seedless with a longer brush which barely falls. The soluble solids content retains 18%−22% and the acid content 0.6%.

Production performance: A kind of table grape. Harvest and marketing time is from May to December in Yunnan Province. Plants thrive because of strong germination and branching ability. It is slightly strong in disease resistance, though, not immured to white rot, downy mildew and powdery mildew. Neither berries fall before harvest nor post-harvest;

hence, it has excellent storage capability.

Planted in: Binchuan County, Mile City, Anning City and other cities in Yunnan Province.

2.6 Centennial Seedless

General botanical characteristics: A Eurasian variety. It is an early ripening variety. Growing tips are green with scarce hair. Tender shoots tend to be light with scarce hair. There are five lobes, divided and circled, of adult leaves in heart size. Serrations are large and sharp. Hairless the adult leaves are. The petiolar sinus opens into an arch. Long oval and chicken-heart-like grains on the same bunches ripen simultaneously and are well-filled to compact. Unripen berries taste slightly bitter and much sour. When fully ripened, berries turn yellowish-green and taste sweet. The soluble solids content retains 17.0%–20.6%. Berries are seedless, as the name identified.

Production performance: A kind of table grape. The two seasons of harvesting are from March to July and October to December. Cultivars grow well with a high fruit-setting rate. It excels in increased production and drought resistance. Resistance to powdery mildew, downy mildew, and grape spot anthracnose is medium. Overripened grains are easy to fall. The storage capability is on average.

Planted in: Yuanmou County, Binchuan County, Shilin County and other parts of Yunnan Province.

3 Cultivation and management techniques

3.1 Planting planning and orchard construction

The orchard should be constructed in a clean site in which the ecological and environmental conditions are good, and the pH value remains from 8.5 to 8.2. The tillage layer should be above 80 cm and groundwater level below 2 m. Good texture, well-draining, and fertility should be prioritized so that the organic content above 1.5%. The facilities such as tractor roads, ditches, pools, guard rooms, storage rooms and packaging rooms are planned and implemented based on the scale of the site. Barbed wire is preferred for fence building. When planting, leave 2.3–3 m of spaces between the rows and 0.8–1.2 m between the grapes.

3.2 Shaping and trimming

To set up a "Y"-shaped column deep into 50 cm, which is 2.6–3 m high with a frame height of 2 m. If the frame is 1.2 m tall, one uses a 40 cm rung double iron wire. If the frame is 1.5 m high, one uses an 80 cm rung double iron wire. If the frame is 1.7−1.8 m high, one uses a solid bamboo or steel pipe column to connect two ends of the column. An iron wire is hung from under the beam at 1.1 m on both sides of the column as the third stringing line. The height of the fixed stem is 1.2−1.4 m, and the width of the "V"-shaped upper mouth is 2.0−2.4 m. The column must be 2.5 m high, 0.5 m deep into the earth and erected 2.0 m high. Solid bamboo or steel pipe is used as the beam, and the beam is connected from 1.1 m of the column to 1.1−1.2 m of the upper side of the beam with solid bamboo or steel pipe, and the two sides are symmetrically bound to form a "Y"-shaped frame surface. The "Y"-shaped frame consists of three belts, with ventilation belts below 1.2 m, fruit hanging belts from 1.2 to 1.5 m and production belts from 1.5 to 1.8 m. Summer Black Grape grows quickly and has a low flower bud node. "Yi"-shaped or "H"-shaped shaping is recommended. Red Globe and Crimson Seedless Grape grow moderately and have high flower bud nodes. The "Two fans" shape is recommended to replace the fruiting branch groups to prevent the slow growth of growing tips and hinder growth for the following year.

They are pruned in winter time. "Yi"-shaped pruning is recommended in winter. Prune the canes both longer and shorter so as to reserve buds three times higher than the bunches. When pruning in winter, prune 5−7 buds away from the main canes. Cut the buds every 2−3 buds. The total buds are estimated to be reserved 5 000−7 000 per mu, with 2 000−2 500 fruits born. "Two fans" are pruning for shaping as well. When pruning in winter, prune everything off with 5−6 buds being left per vine. The total buds are estimated to be reserved for 2 000−2 500 per mu, and four buds for being shortened.

3.3 Water and fertilizer management

Drip irrigation under plastic film is preferred. Water thoroughly to promote sprouting, properly control water before flowering, keep the soil moist in the young fruit period, alternate dry and wet in the colour change period, and pay attention to drainage.

In autumn, we recommend applying 3−5 t of barnyard manure or 1 000−1 500 kg of refined organic fertilizer with 80−100 kg of formula fertilizer which is controlled released, 600 g of exogenous boron and 500 g of exogenous zinc. Before flowering, 3 kg of urea and 3−5 kg of high calcium and magnesium water-soluble fertilizer are applied. After flowering, 5 kg of calcium and magnesium water-soluble fertilizer is applied. 80−120 kg high potassium controlled release formula fertilizer is applied in the young fruit stage (hardcore stage). During the colour change period, 3−5 kg of high potassium water-soluble fertilizer is applied once every 7 d, 3 times in total. In other periods, water-soluble fertilizer should be adjusted in time according to seedling conditions. Foliar supplement: Spray boron, zinc, iron and calcium fertilizer, magnesium fertilizer, potassium dihydrogen phosphate, amino acid and other foliar fertilizers four times after fruit picking, three-leaf stage, before flowering and after flowering, two times before flowering and after flowering, and several times from young fruit stage to mature leaves. Look at leaf and fruit fertilization: To achieve tree phase balance, we must observe the growth of branches, leaves and fruits. When the internodes are too long, and the leaves are large before flowering, they can be adjusted by controlling fertilizer and water. On the contrary, when the growth is weak, the amount of fertilizer and water should be slightly strengthened to reach the basic tree phase. That is, the internode length is 8–10 cm, the petiole is 8–10 cm, and the leaf opening is 12–15 cm. During the colour change period (secondary expansion), the shoots basically stopped.

3.4 Disease and pest control

Downy mildew, powdery mildew, grey mould and aphids are diseases and pests commonly seen on grapes. Agricultural prevention, physical prevention, and bio-control are the main treatments. Chemical control is used in a practical way. The diseases and their treatment are listed in Tabele 2–7.

3.5 Harvesting and post-harvest management

When grape maturity criteria (colour, flavour, aroma, and taste) fit in, harvesting is coming. The coloured grapes of clusters of coloured varieties should be more than 80%, and the soluble solids should meet the requirements of grape-grade standards. Harvesting should begin in the morning when the dew is wiped off. While it is not recommended to harvest when the temperature is too high. Also, the harvesting is arranged according to maturity. To reserve 3–4 cm of stalk when one cuts. To be aware of long-time exposure to the sun and to place them gently. Some poorly coloured fruits, infected grapes, and mechanically damaged and bruised berries should be removed.

Pre-cooling is the first step for the grapes, which will be stored longer. The refrigerated room should be turned on 3 d ahead of storage and kept temperature to −1 °C. The best storage temperature is maintained between −1 °C and 0 °C. The room temperature should be stable with fluctuation less than 0.5 °C, and the accurate measurement is ±0.2 °C. Humidity: grapes should be restored at a humidity of 90%–95%, which the precise measurement is ±5%. The qualified grapes will be sorted and packed. In accordance with *Green Food—Guideline on Packaging* (NY/T 658−2015), those who use calcium plastic double corrugated boxes will be matched with provisions of *Single and double corrugated boxes for transport packages*

Table 2-7 Common diseases and pests of grapes and their prevention methods

Diseases and pests	Symptoms	Prevention period	Chemicals needed	Dose	Appling by	Interval days (d)
Downy mildew	On leaves, initial symptoms appear, as crystalized spots show on and later expend into chlorotic spots on the upper leaf surface. When being dampen, whitish lesions appear on the back of the leaf surface. Infected spots can cause distortion, drying, and premature drop. On shoots, tendrils, and spike-stalk, infected areas begin as oily translucent spots and later have the appearance of brown/black diffuse patches. When being dampen, whitish lesions appear on the surface of the leaf. On canes, infected areas can cause stagnation, distortion, and wilt. Infected flowers and young berries can become covered with the mist-like white fungus, may turn dark brown, shrivel, and split soon, and/or may not ripen properly. Berries infected in the mid of growing to pea-like size may turn reddish brown at the very beginning, drying and splitting at the end	Before the DM establishment, and the initial stage of flowering	80% Bordeaux Mixture WP	1 : (300–400)	Spraying	—
			25% azoxystrobin SC	1 : (1 000–2 000)		
			722 g/L propamocarb AS	1 : (600–800)	Spraying	7
			90% fosetyl-aluminium(D) WP, 10% cyazofamid SC	1 : (2 000–2 500)		
Powdery mildew	When leaves are infected at the very beginning, grey-white powdery patches appear, and later floury patches come out. If the disease is severe, the mildewed foliage may be stunted and distorted. When berries are infected, they may develop a netlike pattern of russet covering with powdery patches. Infected berries may stop growing, distorted, small and sour. If berries are infected in rainy seasons, they may crack open, and rotten. When canes, growing tips and pedicels are infected, mildew usually appears first as whitish powdery patches, which later expand into irregular brown patches. Infected canes, growing tips and pedicel become too crisp to grow	Initial stage of infection	29% lime sulfur AS	1 : (6–9)	Spraying	15
			25% triadimefon EC	1 : (1 000–1 500)		
			25% difenoconazole EC	1 : (2 000–2 500)		
			40% sulphur colloidal SC	1 : (400–500)		
			40% flusilazol EC	1 : 8 000		

Continued Table

Diseases and pests	Symptoms	Prevention period	Chemicals needed	Dose	Appling by	Interval days (d)
Gray mold	The infected pedicel and rachis turn watery as burned by boiled water and appear light brown in the very beginning, later dark-brown and soft. When it is dry, the infected pedicel and rachis shrivel, and drop. Under relative high humidity and moisture, infected pedicel and rachis become covered with the grey fungus. When berries of white cultivars are infected, they turn brown with lowering spots which later expand to the whole with gray-brown sporulating growth of the fungus on the surface of infected berries	Prior to and after flowering	50% iprodione WP	1:(500–800)	Spraying	14
		Initial stage of infection	50% cyprodinil WG	1:(700–1 000)	Spraying	7
			50% boscalid WG	1:(1 000–1 500)		
		Prior to or in the initial stage of infection	43% procymidone SC	1:(600–1 000)	Spraying	14
Aphids	Colonies on the downsides of shoots, and leaves. When it is serious, aphids cause the shrink and deformity of leaves, hindering the growing tips	Initial stage of infection	1.5% matrine SL	1:(3 000–4 000)	Spraying	10
			25% pymetrozine SC	1:(2 000–2 500)		
			70% imidacloprid WG	1:(2 000–3 000)		

(GB/T 6543—2008). For those who are using plastic boxes, those boxes will be matched with provisions in *Packing containers—Composite intermediate bulk container* (GB/T 19161—2016).

References

Ministry of Agriculture and Rural Affairs of the People's Republic of China. Agricultural Statistics of China (2011—2020) [M]. Beijing: China Agriculture Press.

National Bureau of Statistics of the People's Republic of China. China Statistical Yearbook (2011—2020) [M]. Beijing: China Statistics Press.

Office of Leading Group for Building World-class "Green Food Brand" in Yunnan Province, 2021. 2020 Development Report of Key Industries of "Green Food Brand" in Yunnan Province [R]. 2021.7.

❖ Key Techniques for Jackfruit Cultivation

1 Nutrition facts

Jackfruit is known for its aromatic, thick pulp and fresh, juicy flavour. Most fruits, such as bananas, mangoes, pineapples and papayas, offer little protein. Jackfruit servings are relatively high compared with those fruits. It is packed with essential nutrients, including starch, protein, fatty acids (FA), calcium, ferrum, and vitamin B_1, as well as beneficial elements, namely, calcium, magnesium, zinc, ferrum, sodium, manganese, as well as plant active ingredients, namely, carotene, flavonoids, volatile acid sterols and tannins. Every 100 g of raw jackfruit provides 16.0%–25.4% of carbohydrate, 1.2%–1.9% of protein, 0.1%–0.4% of fat, and 1.0%–1.5% of fibre.

Starch and fibre contained in jackfruit seeds are good sources of diet. The edible seeds are rich in functioning and active agents such as resistant starch, protein, saponins, alkaloids, organic acids, amino acids, trace elements and essential fatty acids for the human body. Every 100 g of seed provides 25.8%–38.4% of carbohydrates, 6.6%–7.04% of protein, 0.4%–0.43% of fat, and 1.0%–1.5% of fibre.

2 Growing environment

Jackfruit is a typical tropical fruit tree, requiring a warm and humid tropical climate. Cold-intolerance nature makes the planting temperature ≥21 °C, mean temperature of the coldest month ≥13 °C and absolute temperature ≥0 °C. Young trees are susceptible to cold stress. If it is at 0 °C and −1 °C, leaves and branches will be damaged and killed between −3 °C and −2 °C. Elder trees are slightly stronger in cold tolerance when lower to −3.89– −3.33 °C. Trees will die instantly when the temperature is below −6.67 °C. Jackfruit grows well at 27–31 °C all year round.

Flowering twigs emerge from the trunk and large branches. As male and female flowers are on the same trees, the tree is known as "monoecious". Fruits are oval-sized and bear due to the pollination of female inflorescence. Flowers bloom every year; some are many times a year, usually from February to April. Within 120–150 d of growth, jackfruits mature from

June to August. It can be harvested 3-5 years after planting, and full fruiting period in 5-6 years. About 30-100 fruits hang on the trunks, weighing 10-20 kg. The smallest fruits weigh 1-2 kg, while the biggest weigh over 40 kg. The yield volume achieves to 4 500 kg per mu.

Jackfruits are best planted in full sun with abundant rainfall (annual rainfall above 1 200 mm), which is evenly distributed in seasons. Soil requirements include deep, organic soil and well-draining soil, pH value level is 6.0-7.5. They grow well in lower hills or plains, which are under 600 m in altitude.

3 Overview of Industrial development

India was the earliest Jackfruit producer. Later introduced into other tropical and subtropical areas worldwide, jackfruit was produced in China, India, Indonesia, the Philippines, Thailand, Bangladesh, and Brazil. According to a rough calculation, the total area for jackfruit planting is about 3.9 million mu, with a total volume of 3.62 million t. China has a 1 400-year history of Jackfruit planting. In the early years, jackfruit has mainly been planted in courtyards or as street trees. Due to extensive management, no large-scale planting has been formed. Trees are too high, significant difference in plant yield, varying in fruit setting and low in fruit quality. The increasing demand for supreme, rare and special fruits is in line with the booming tourism industry and improving livelihood, and so is the popularity of jackfruit. Jackfruit is deeply loved by consumers because of its aromatic, nutrient-rich, and sweet meat. Sales of fresh fruit and processing products are rising. At the same time, supportive policies for growing small and special tropical fruits are expanding as well. The industrial chains have been continuously integrated, followed by the completion of planting, processing and transportation, embarking on a golden time in the jackfruit industry. There is 20% of annual growth in planting areas. Large and commercial planting is in these superior producers, such as Yunnan, Hainan, Guangdong, and Guangxi. Grossly calculating, the total area for jackfruit planting is over 500 000 mu, which yields more than 600 000 t of jackfruits and creates revenues exceeding 9 billion by the end of 2021. Jackfruits are cultivated in tropical and subtropical areas of Hainan, Guangdong, Guangxi, Fujian, Yunnan and Sichuan. Until the end of 2021, more than 100 000 mu of jackfruits were cultivated in Yunnan. It is one of the prioritized fruits in hot and dry areas, increasing the income of local peasants and being one of the pillar industries of rural revitalization.

4 The main cultivars and characteristics

4.1 Domestic varieties and characteristics

Malaysian No. 1 (Qiongyin No. 1): It is long and oval. Fruit average equatorial diameter is about 47.63 cm, and transverse diameter 25.06 cm. Fruit shape index (SI) is 1.9, average single fruit weight is 20–30 kg. Berries are golden yellow. Its soluble solids content is about 16.0%.

Qiongyin No. 8: It is oval. Fruit average equatorial and transverse diameters are 39.75 cm and 24.22 cm. Fruit SI is 1.65, average single fruit weight is 13.70 kg. Berries are in between golden yellow and orange. Its soluble solids content is 25.53% and edible flesh percentage is 38.6%.

Xiangmi No. 17: It is oval. Fruit's equatorial diameter is 35.5 cm on average, and transverse diameter 23.7 cm. Fruit SI is 1.5, average single fruit weight is 4.5–18.5 kg. Berries are orange-red. Its soluble solids content is 23.5%–27.8%, and edible flesh percentage is 45.5%.

Haida No. 2: Berries are oblong with an average single fruit weight of 7.35 kg. Pulp is yellow and crisp, with strong and full flavor. Its soluble solids content is 21.5%, and edible flesh percentage is 58.0%.

Red Jackfruit: It is long and oval, with an average single fruit weight of 9.5 kg. The pulp is reddish-orange. Its soluble solids content is 18.87%, and edible flesh percentage is 78%.

Changyou Jackfruit: It is oval shape, with an average single fruit weight of 4–6 kg. Pectin is hardly seen when it ripen. Pulp is golden yellow, thick, crisp, and aromatic. Its soluble solids content is 26.88%–28.13%, and edible flesh percentage is 75%.

Autumn Red (Qiuhong): It is oblong, with an average single fruit weight of more than 25 kg. After its maturity, hardly any pectin is seen. The meat is pink. On the outside, its skin is thin and has spines. Its soluble solids content is over 18%.

4.2 Main varieties and characteristics worldwide

J-30: A selected Malaysian variety. The flesh tends to be dark orange and yellow, frim and fragrant. It has a pleasantly sweet flavor. Edible flesh percentage is 38%.

J-31: A selected Malaysian variety. The flesh is deep yellow, firm and sweet. It has a strong and earthy flavor. Edible flesh percentage is 36%.

NS1: A selected Malaysian variety. Average single fruit weight is 4–5.5 kg. The fruit production is 90 kg per tree. The ripen time is between May and June. The flesh is dark orange and firm, with a rich and sweet flavor. The spines flatten on the skin. Edible flesh percentage is 34%. There are 63 seeds in total in each fruit, accounting to 5%. The flavor is strong and fragrant.

Chompa Gob: A selected Thailand variety. It was the best type in Thailand. The inside flesh comes in colors between deep yellow to orange and has crisp texture, and sweet flavor. Easy to chew due to less pectin. The edible flesh percentage is 30%.

Cheena: A selected Australian variety. It is a natural combination of Jackfruit and chempedak. The inside flesh is orange, fibrous, sweet, almost melting, and fragrant. The edible flesh percentage is 33%.

Lemon Gold: It was selected in Queensland, Australia. The inside flesh is lemon yellow and has trim and spiky skin outside. The flavor is sweet. The edible flesh percentage is 37%.

Honey Gold: It was selected in Queensland, Australia. The inside flesh comes in colors between deep yellow to orange and has a firm texture. Outside skin is spiky and trim. It has a strong aromatic flavor. The edible flesh percentage is 36%.

5 Cultivation and management techniques

5.1 Planting planning and orchards building

5.1.1 Orchard site selection

Orchards are in the ground where the annual temperature is above or equal to 21 °C, with the temperature of the coldest month no less than 15 °C, and the absolute lowest temperature is more than 5 °C. It is best to plant jackfruits in a sunny wind shelter whose slope is less than or equal to 45° and whose annual rainfall is above or equal to 1 200 mm. Trees prefer well-drained and deep soil enriched with organic nutrients. The underground water level is less than 1 m. The environmental conditions of the garden should be selected in the agricultural production area with an excellent ecological environment, far away from pollution sources and sustainable production capacity, in line with *Environmental Conditions of Pollution-free Agricultural Products Planting Areas* (NY/T 5010—2016).

5.1.2 Orchard planning

In planning garden plots, many factors such as garden plot scale, slope, topography, microclimate, topography and mechanization degree should be considered comprehensively.

Necessary roads (main roads, branch roads and field trails), irrigation and drainage, water storage and other facilities will be built to create shelterbelts. Shelterbelt should choose fast-growing wind-resistant tree species (*Acacia formosana*, etc.) that do not have the same main diseases and insect pests as jackfruit or native fast-growing wind-resistant tree species. The garden is divided into several communities by roads and shelterbelts, and the size of the communities is arranged according to the topography of the garden. The garden site is selected in the hillside zone, and the flood control ditch around the mountain should be excavated at the top of the slope. The standard is that the width of the ditch surface is 1−1.2 m, the depth of the ditch is 0.6−0.8 m, and the width of the bottom of the ditch is 0.6−1 m. When designing, the outlet should be controlled to connect with the drainage channel as much as possible so as to reduce soil erosion.

5.1.3 Orchards reclamation

Within four months before planting, the uncultivated garden land should be reclaimed and deeply ploughed, with a depth of 0.4−0.5 m, to improve the soil structure and make the soil mature. Before reclamation, the location of the shelterbelt should be planned. Once fixed, the land should be reserved. Moreover, the missing parts should be replanted and transplanted to remove other trees outside the shelterbelt. After deep ploughing, the garden can be levelled and built to conserve water and soil. Generally, plots with a slope of less than 5° are planted at the same height, and attention should be paid to building a ridge every 4−6 rows. Garden plots with a slope of 5°−45° should be reclaimed at the same height, and horizontal terraces with a width of 2−3 m or rows around mountains should be built, with the platform slightly inclined inward by 4°−5°, and planted in a single row. The spacing between terraced fields or rows around mountains is 5−6 m.

5.1.4 Digging the planting hole

Planting holes should be dug 60 days ahead of planting. The hole is 0.8 m×0.7 m×0.6 m deep. Separate the topsoil and subsoil when digging, which is easy for filling back the soil after sun and weathering. If the slope is less than 5°, planting holes are 6 m in plant spacing and 7 m between plant rows. If the slope is between 5° and 45°, planting holes are dug in a pyramid shape along two-thirds of the terraced fields or belts, with 18−20 plants per mu. Leave the hole open for 20−30 d under the sun, and fill the organic potting mix, namely, 20−30 kg of decomposed organic fertilizer, 1−2 kg of ternary compound fertilizer, and 3−5 kg of calcium magnesium phosphate fertilizer, which are evenly mixed with subsoil. Put 0.2−0.25 m deep soil into the base first, fill the mixtures then, and cover the topsoil last. Be mindful that refilling the soil should be 0.2–0.3 m higher than the ground for later planting.

5.1.5 Planting

Before planting, trim all leaves of seedlings which removal 2/3 of each leaf, with 1/3 left. Loose the planting hole, remove the nutrition bag, and gently place it in the middle of the planting hole. After that, start covering the soil. When covering soil, try to use fine soil, and gently compact the soil by hand while returning. Be careful not to squeeze the soil balls. The covering soil can be 2–3 cm from the bud interface. Take the seedlings as the center and make a water tray of about 50 cm. The water tray is covered with weeds or permeable grass-proof cloth, and water it in time. When planting, if it is exposed to direct sun, it needs shading protection, and it should be removed after the plant is restored.

5.2 pruning of jackfruit trees

5.2.1 Pruning of young jackfruit trees

When young trees are 1.5–2 m tall, cut stems to stimulate growing of side branches and subbranches. Reserve germinating full and growing branches evenly in the east, west, south and north direction, which are 1.2–1.5 m higher than the ground and 45°–60° with the trunk. The first layer of subbranches is that of 3–4 growth branches, hence removing extra buds. When the first layer is 0.8 m high, remove the top for stimulating subbranches. To trim branches to remove the canopy's inner branches to allow much light and air within the dwarfed canopy.

5.2.2 Pruning for adult trees

Two to three months before flowers bloom, remove the inner dead, diseased, damaged, infested, weak, overabounding, crossed and rubbing branches for air and light. After harvesting, cut off the fruiting branches and male flower branches remaining on the trunk and big branches, trim prolonged branches, and cut off crossed, hanging, dead, diseased, damaged, infested, weak, overabounding, crossed and rubbing branches which may hinder the growth, limiting the height to 4 m or below.

5.3 Fertilization and water management

5.3.1 Fertilizing principles

That aligns with fertilizing principles that apply frequently and less while applying more in prime time. Apply organic fertilizers as the major, along with proper inorganic fertilizer.

5.3.2 Fertilizer types

It is the manure fertilizers and chemical fertilizers that recommended to use referring to *Green food—Fertilizer application guideline* (NY/T 394—2023). Commonly used organic fertilizers: livestock manure, livestock manure urine, pond mud, cake fertilizer and green manure, etc. Livestock manure and cake fertilizer are generally retted into water and fertilizer; Livestock manure is generally fermented and retted with topsoil or pond mud to make dry fertilizer. Commonly used inorganic fertilizers: urea, compound fertilizer, potassium chloride and calcium magnesium phosphate fertilizer.

5.3.3 Methods of fertilizing and rations

5.3.3.1 Fertilising and its management for young trees

For the one-year-old tree, apply 50 g of urea or 100 g of balanced three-nutrient compound fertiliser to each plant every three months. Apply 10–20 kg of organic fertiliser and 0.5 kg of calcium magnesium phosphate fertiliser from October to December. For 2–3-year-old trees, apply 100 g of urea or 150 g of three-nutrient compound fertiliser to each plant every 3 months. Apply 15–25 kg of organic fertiliser and 1–1.5 kg of three-nutrient compound fertiliser from October to December. In the first year, apply at about 0.3–0.4 m away from the base of the trunk, and after the second year, apply at the periphery of the crown drip line.

5.3.3.2 Fertilising and its management for adult trees

Fertilising before flowering: Apply 2–3 months before the flower and bud differentiation period. Applying 15–25 kg of organic fertiliser to each plant and 1–1.5 kg of high potassium three-nutrient compound fertiliser is suggested.

Fertilising in promoting flower formation: Apply when flowers bud fleck. Use readily available fertiliser as the major fertiliser. Rations are 0.5 kg of urea and 0.5 kg of potassium chloride. Spray potassium dihydrogen phosphate (0.3%) and 2–3 times fulvic acid on the leaf surface. It is better to apply on cloudy days or after 16: 00 to late in the evening on sunny days.

Fertilising for fruit expanding: Apply after fruit formation and fixation. Use 1–1.5 kg of high potassium three-nutrient compound fertiliser and 1.5–2 kg of calcium magnesium phosphate fertiliser.

Post-harvest fertilisation: Apply immediately after harvest. Fertilising in this period restores and recovers the nutrition of trees and builds reserves back up to avoid early ageing. Manure is a major fertiliser, while chemicals are a minor one. Use 20–25 kg of organic fertilisers for each plant and over 2 kg of balanced three-nutrient compound fertiliser.

The fertilisers mentioned above are a general guide. Specific fertilising formulas should be applied considering the practical situation (varieties, ages, types of fertilizer, soil conditions and soil types.)

5.3.4 Watering management

Dry season: On time, water trees in flowering and fruit development to maintain moisture stress. Water in the morning or late evening sheltering from the midday heat.

Rainy season: The drainage ditch should be dredged. Fill the concave land to drain waterlogging.

5.4 Prevention and control of major diseases and pests

5.4.1 Major diseases and pests

Diseases are anthracnose, rhizopus rot, stem and fruit rot, and root rot. Pests are longicorns, *Diaphania caesalis* (Walker), scarabs, green weevil, and *Latoia sinica* Moore.

5.4.2 Prevention guidelines

In order to implement the plant protection policy-prevention in the first place and integrating prevention with control-agro-prevention, bio-prevention and physical prevention should be prioritized and comprehensively integrated for the betterment of the orchard ecological environment while enhancing cultivating management as the preliminary work. In terms of chemical prevention, it should be applied in a scientific way. That is, the application of pesticides should comply with the guidelines and provisions regulated in *General Guidelines for Pesticide Safe Use* (NY/T 1276 —2007).

5.4.3 Prevention measures

5.4.3.1 Agricultural prevention

To select the supreme varieties of strong adaptability and resistance. To enhance cultivating management for robust and vigorous trees which are vital for their disease resistance. Well pruning for a good ventilation and light through crowns. Clean orchards in winter. Gather dead, and infest leaves and burn those together to prevent infectious disease.

5.4.3.2 Physical prevention

Setting up solar insect trap lamps to trap and kill nocturnal pests. Arrange those within the range of every 20–30 mu.

Death feigning is a series of tactics of insects within antipredator. Considering of that, human capture by shaking trees is useful.

Set sticky pads outside of orchards if they are needed.

5.4.3.3 Biological prevention

To mimic a mini-environment for conducive to the reproduction of natural enemies by planting bee plants around orchards and in row. Breeding, releasing and assisting the migration of natural enemies of major insect pests, such as predatory ladybugs and predatory mites.

5.4.3.4 Chemical control

Apply chemicals in referencing to Table 2-8 or *Technical regulations for jackfruit cultivation* (NY/T 3008—2016).

Table 2-8 Chemical control of major diseases and insect pests

Prevention object	References and dosage
Anthracnose	In the initial stage of occurring of disease, apply 50% carbendazim WP [1 : (600–800)], 75% chlorothalonil WP [1 : (500–800)], 450 g/L prochloraz EW [1 : (800–1 500)], and spray 250 g/L azoxystrobin SC [1 : (800–1 500)]. Apply one time in every 7–10 d in consecutive 2–3 times
Rhizopus rot	Spray pesticides in flower blooming and early fruit set period to protect inflorescence and fruits in advance. Spray 77% copper hydroxide WP [1 : (600–800)], 80% liquid Bordeaux mixture WP [1 : (500–600)], 25% liquid prochloraz EC (1:1 000), 20% thiodiazole-copper SC [1 : (500–600)], and 47% copper oxychloride WP [1 : (500–700)]. Apply one time in every 7–10 d in consecutive 2–3 times
Stem and fruit rot	Spray pesticides in flower blooming and early fruit set period to protect inflorescence and fruits in advance. Spray 70% thiophanate-methyl WP (1 : 800), 50% carbendazim WP [1 : (500–600)], and 25% prochloraz EC (1 : 1 000). Apply one time in every 7–10 d in consecutive 2–3 times
Root rot	In the initial stage of occurring of disease, apply 70% thiophanate-methyl WP. The dosage is mixing 0.1–0.25 kg per plant with 25–50 kg of soil, spraying around roots. Pour root with 80% mancozeb WP (1:400) or 50% thiram WP (1:500). Apply one time in every 7–10 d in consecutive 2–3 times
Longicorns	When a new defecation hole is found in the trunk, the larvae borers in the branches can be killed with steel wire hooks. In holes of the borers may be sealed with cotton which is dipped in 40% phoxim (1:20). After that, seal the holes with mud, smoking and apply pesticide. Spray the trunk with 40% thiacloprid SC [1 : (3 000–4 000)]. For artificial killing, kill adults and scrape off eggs in bark crevices or under skin
Diaphania caesalis (Walker)	It is recommended to apply pesticide in budding, flowering and early fruit-set period, including 10% lambda-cyhalothrin EC [1 : (1 000–1 500)], 1.8% abamectin EC [1 : (1 000–1 500)], 8 000 IU/μL *Bacillus thuringiensis* (Bt) [1 : (800–1 000)]. Apply one time in every 7–10 d in consecutive 2–3 times
Scarabs	In the initial stage of occurring, spray 5% chlorantraniliprole SC [1 : (2 000–2 500)], 15% indoxacarb EC [1 : (1 000–1 500)]and 5% emamectin benzoate ME [1 : (2 000–3 000)]. Apply one time in every 7–10 d in consecutive 2–3 times
Green Weevil	In the initial stage of pests occurrence, spray 10% high efficient cyhalothrin EC [1 : (1 000–1 500)], and 1.8% abamectin EC [1 : (1 000–1 500)]. Apply one time in every 7–10 d in consecutive 2–3 times
Latoia sinica Moore.	In the initial stage of pests occurrence, spray 5% liquid chlorantraniliprole SC [1 : (2 000–2 500)], and 15% indoxacarb EC [1 : (1 000–1 500)]. Apply one time in every 7–10 d in consecutive 2–3 times

5.5 Harvesting and post-harvest treatment

5.5.1 When to harvest

Maturity signs of jackfruits: When the stalk tends to be yellow or a yellow leaf near the fruit stalk is off. Hollow sounds, like "Boop, Boop", can be heard when tapped, when the fruit is from light green to yellow-brown until it is quite brown—the sharp "spines" on the fruit turn blunt. Stabbed by a knife, a light fluid is out of fruit.

5.5.2 Harvesting

Pick and harvest jackfruits by different maturity levels, providing them to the markets due to their function and demands. Harvesting times are better on cloudy days or in the morning and the late afternoon on sunny days. It is better not to pick in the mid-day when the sun is strong and on rainy days. Harvest: The harvesting process should involve light mining, release, and transportation to avoid mechanical injury. Postharvest fruits are stored in a cool and dry place to avoid exposure to the sun. Store according to the specified conditions of *Jackfruit* (NY/T 489—2002).

References

HUANG X F, XIONG M Y, SU C Y, 2020. Characteristics and key points of cultivation techniques of an excellent jackfruit line of 'Qiu Hong' [J]. Southeast Horticulture, 8(3): 36−37.

SU L X, BAI T Y, Wu G, et al., 2019. Research staus and development trend of jackfruits cultivation [J]. Chinese Journal of Tropical Agriculture, 39(1):10−15,41.

TAN L H, 2017. Cultivation and Processing of Jackfruit, Breadfruit, and Champedak [M]. Beijing: China Agriculture Press.

WANG Z H, PAN D F, Deng Z Q, et al., 2012. Breeding and selecting of a new jackfruit cultivar 'Red Flesh Jackfruit' [J]. Journal of Fruit Science, 29(3): 518−519, 312.

YAN C B, HU F C, ZHAO Y, et al., 2023. Breeding report of a new excellent jackfruit variety Qiongyin No.8 [J]. Journal of Fruit Science, 40(3): 600−603.

YUAN H H, ZHONG H J, AO X Y, 2018. Determination and analysis of nutritional components in *Artocarpus heterophyllus* seed [J]. Food Research and Development, 39(24): 169−173.

ZHANG T, PAN Y G, 2013. Research progress on nutritional ingredients and pharmacological functions of jackfruits [J]. Guangdong Agricultural Sciences, 40(4): 88−90,103.

❖ Key Techniques for Passion Fruit Cultivation

1 Overview of industrial development

Passion fruit (*Passiflora caerulea* Linnaeus) is a perennial liane of the genus *Passiflora* L. in the Passifloraceae family. It is widely cultivated in tropical and south subtropical zones from the Tropic of Cancer to the Tropic of Capricorn, and it grows rapidly and has the advantage of being planted and put into production in the same year. In recent years, China's planting area of passion flower has been increasing. In 2018, the national planting area of passion flower reached 472 km^2 with an output of 635 000 t, mainly distributed in tropical and subtropical zones including Guangxi, Fujian, Guangdong, Hainan, Yunnan, Guizhou, etc. It is widely used in fresh food, ornamental, industrial ingredients, and medicinal fields.

When the fruit of the passion flower is ripe, it can emit rich aromas of various fruits such as strawberry, guava, banana, pineapple, lemon, etc, and has the reputation of "the king of juice". Fruit juice accounts for 30%–40% of fresh fruit weight with high nutritional value. The whole fruit is rich in various organic acids, dietary fiber, ascorbic acid, sugars, vitamins, minerals, and pectin, etc. The fruit juice contains nutrients and elements that the human body needs, such as sugar, fat, protein, vitamins, amino acids, calcium, potassium, iron, phosphorus, etc, so it has high edible value.

Due to the differences in the genetic background of the varieties, the requirements and reactions of the cultivation environment will also be different. ① Purple fruit type, it can adapt to the subtropical region between the north of the Tropic of Capricorn to the tropical climate zone and the subtropical region between the south of the Tropic of Cancer to the north of the tropical zone. It can also adapt to a certain cold weather in winter, tolerate 0 ℃ and light frost injury for a short time, and easily resume growth after cold days. This type likes the monsoon climate of alternating dry and wet, and can grow and bear fruit in areas with annual precipitation of 800–2 000 mm. If it is planted in tropical lowlands with perennial high temperature and rain, the vines will grow excessively, and the flowers and fruits will be less. ② Yellow fruit type, it is originally from the low altitude areas of the tropics, likes humidity and heat, can adapt to the environmental conditions of the tropical region with high temperature, rainy, high humidity. The plants grow, blossom and bear fruit multiple times in a year, can be harvested 2-3 times a year, but is more sensitive to low temperature frost, which they can not tolerate for a longer period of low temperature below 0 °C. In the case of -2 °C low temperature for a long time, it is easy to freeze to death, while in the case of high temperature above 35 °C, drought or untimely irrigation, the leaves will turn yellow, and even dry and fall off.

The growth and fruiting of passion flower require a temperature of 20-33 °C, the temperature of 8-15 °C will grow slowly, the fastest growth at 25-30 °C, the lowest overwintering temperature should be above 5 °C, in the event of low temperature and frost below 0 °C, it will be affected by freezing damage, and 25-30 °C is the most suitable temperature for flowering. The annual sunshine duration of 1 800-2 200 hours can meet the needs of its high quality and high yield. The annual precipitation should be 1 200-2 000 mm, an even distribution of rainfall is ideal, and no flooding because it is afraid of waterlogging. It is advisable to choose areas where there are fewer winds above force 8, to avoid strong winds blowing down the trellises and breaking the vines, flowers and fruits. The planted soil should be loose and fertile, with a pH value of 5.5-6.5 in loam or sandy loam.

2 Main cultivars and characteristics

2.1 Main cultivars

2.1.1 Purple passion fruit (*Passiflora edulis* Sims)

The fruit is dark purple after ripening, and it is adapted to the southern subtropical climate zone between the south of the Tropic of Cancer and the tropics, and the subtropical climate between the north of the Tropic of Capricorn and the south of the tropics, with slightly stronger cold endurance.

Purple passion fruit has strong growth potential, with pure green stems, tendrils, and leaves. The leaf is trifid, with finely serrated leaf margins. The stem is heart-shaped, and the length of the leaves is 10-18 cm. The flowers are slightly smallers, with a diameter of about 4.5 cm, and each new shoot can grow one flower. Each flower has 5 white petals and sepals, 2 rounds of linear corollas, with a dull purple base and white edges. It has 5 stamens and 1 large anther at the top. The ovary is located in the middle of the flower, and the style is divided into 3 stigmas. The fruit is round or oval in shape with a diameter of 4-5 cm and an average single fruit weight of 40-60 g. The juice has a strong aroma and high sweetness, making it suitable for fresh consumption. The average juice yield is around 30%. The fruit development period is 60-80 d, which is relatively cold resistant, but it is not resistant to extremely hot weather during the flowering stage, blooming at midnight and closing before the next day noon.

2.1.2 Yellow passion fruit (*Passiflora edulis* Sims f. *flavicarpa* Deg.)

After ripening, the fruit turns yellow and requires tropical ecological conditions.

The growth potential of yellow passion fruit is stronger than that of the purple type. The stems, leaves, and tendrils are the same characteristic red, pink, or purple color as the purple type, and the leaves are similar in type to the purple one, but larger. The flower is larger, with a diameter of about 6 cm, and the base of the filament is bright dark purple. The shape of the fruit is similar to that of a purple one, with a larger diameter of about 6 cm and a single fruit weight of 60−90 g. When ripe, the pericarp is dark yellow or bright yellow, and the star-shaped flecks on the outer pericarp of the fruit are more obvious. The juice content can reach 45%. The pulp has a higher acid content than purple type, and the seeds are dark brown. Bloom around noon and close from 21:00 to 22:00. Due to the staggered flowering stages of the yellow and purple fruit types, natural pollination of each other rarely occurs. Flowering can occur from spring to late autumn, with a short break in early summer, so mature fruit appears from early summer to winter. About 10 months after sowing, flowering occurs, and the fruit development period is about 70 d. Two seasons of fruit can be harvested each year.

Yellow passion fruit grows prosperously and blooms frequently with high yield and strong disease resistance. However, this type is not cold resistant, and it is easier to be frostbitten and frozen to death in case of severe low temperatures and frost. Its juice has high acidity, generally used as industrial raw materials to process juice, not suitable for fresh food. Most of the varieties and strains of yellow passion fruit are self-incompatible, so it is necessary to pay attention to the allocation of yellow fruit varieties with distant relationships during the establishment of the garden.

In addition, there are large-fruit passion fruit, banana passion fruit, sweet-fruit passion fruit, camphor leaf passion fruit, apple-shaped passion fruit, etc, only local cultivation.

2.2 Main varieties and strains

2.2.1 Tainung No. 1

Tainung No. 1 is the first generation of superior plant obtained by passion flower expert Mr. Lin Yingda in 1981 at Fengshan Tropical Horticultural Research Branch in Taiwan, China by crossing purple fruit as female parent and yellow fruit as male parent. The fruit is bright red and round, and the pericarp is not flecked, slightly smooth, with an average single fruit weight of 62.8 g. This variety has strong adaptability to the environment, with prosperous growth potential, self-compatibility, high fruit setting percentage, large fruit, high yield, high juice rate and excellent quality. The flowering stage in the northern tropical climate zone is very long, from mid-late March to late November, and a year can be harvested 2-3 times. The disadvantage is that they are not resistant to viral diseases.

2.2.2 Yellow-fruit selection 5-1-1

It has bright yellow and large fruit, with high yield. The fruit juice yield is high, with fresh color, strong aroma, high acid content and excellent quality, suitable for processing. It requires artificial pollination of different plants or varieties (strains) to achieve high yields. The fruit is round, the unripe fruit is green, the ripe fruit is bright yellow, the average single fruit weigh is 68.4 g and the natural fruit setting rate is 46.7%, which has excellent processing quality and fresh food quality, the juice rate is 39.3%, the juice contains soluble solids of 15.1%, each 100 mL contains 2.4 g acid, 30.32 mg vitamin C and 14.3 g reducing sugar.

2.2.3 Jilong No. 1

It has wide adaptability, strong stress resistance, relatively cold-resistant, light occurrence of diseases and pests, and early bearing and high yield. The fruit is purple or purplish red and egg-shaped or spherical, the growth is fast, planting and flowering in the same year, the self-pollination fruit setting percentage is more than 80%, no need for artificial pollination. Good fertility, average single fruit weight of 61.28 g, with excellent quality, sugar content of 15.4%-21%, rich in vitamin C with 49 mg per 100 g of pulp, and a strong aroma.

3 Cultivation and management techniques

3.1 Planting planning and orchard construction

3.1.1 Orchard site selection

According to the volume of fresh fruit sales in the planting area and the quantity and quality requirements of raw material demand in the processing line, the survey and forecast are made to determine the scale and target of construction, so the site selection should pay attention to the following points.

(1) The garden plot should be selected in an ecological suitable area for passion fruit to avoid severe natural disasters such as low temperature, high temperature and strong wind.

(2) It should have convenient transportation. There is no need to build a road to

connect the existing traffic. Only need to plan the road in the garden plot, to facilitate the transportation of production materials and products.

(3) Choosing low hilly land, sunny and well ventilated, deep soil layer with pH value of 5.5–6.5, soil organic matter content of 1%–1.5%, loose loam or sandy loam, it is appropriate to build terraces and drainage and irrigation system.

(4) Close to the water source can ensure that water can be diverted for irrigation during drought, and easy to drain when there is too much rain, so as to facilitate the construction of diverted irrigation canals and drip irrigation facilities.

(5) There is a distance of more than 2 km between the garden plot and cucurbitaceae and nightshade vegetable base, tobacco base and melon fruit base to prevent the transmission of virus diseases of the same pathogen.

(6) A seedling base should be prepared in advance to cultivate sufficient and robust seedlings for production.

3.1.2 Garden plot planning

(1) According to the topography, production and living habits and transportation network configuration, the location of office area, staff dormitory area, production materials warehouse, agricultural machinery warehouse, fresh fruit classification and fresh-keeping packaging workshop, and temporary turnover warehouse should be reasonably planned.

(2) The production area of the garden plot should be divided according to drainage and irrigation management.

(3) The main road, auxiliary road and pedestrian road network that meet the production and transportation needs of the area should be rationally distributed.

(4) A network of drainage and irrigation systems should be planned for convenient management.

(5) Set up a wind protection forest belt.

3.1.3 Hill slope orchard construction

(1) Hill slope orchards should be terraced according to contour lines. The width of the each terrace surface is not less than 2 m, and it is tilted about 3° towards the inside. In the construction of a terrace, the topsoil should be stacked in sections first, then the subsoil should be excavated, and the subsoil should be stacked to form the outer wall of the terrace. At the upper end of the outer wall, an anti-erosion embankment 20 cm higher than the surface and 25 cm wide is constructed. A bamboo joint groove with a width of 25 cm and a depth of 20 cm is built inside the terrace, which can store water to prevent drought when there is little rainfall,

while draining in time when there is much rainfall. The soil surface with a width of 1 m and a depth of 20 cm, 15 cm away from the embankment outside the terrace, should be dug up and loose.

(2) Final planting holes of 60 cm in length, width and depth should be dug according to the plant spacing, 10–20 kg of high-quality organic fertilizer should be applied to each hole, 1 kg of compound fertilizer with 15% nitrogen, 15% phosphorus and 15% potassium content, and 20 kg of broken green manure and grass should be fully mixed with topsoil for 1–2 months decomposition and set aside for use.

(3) According to the design, in addition to the terraces with particularly gentle slope and particularly wide surface, one row of trellises is generally set up, and only 2–3 rows of trellises are set up for especially wide surface.

(4) Passion flower trellis usually has three kinds: single-column single-line cylinder frame, single-column double-line "T"-shaped frame, double-column three-line in the shape of an A-frame, which are simple in structure, conducive to climbing expansion. The large light area of the leaves can make full use of luminous energy to grow rapidly, which is conducive to promoting blooming and strengthening fruits, convenient in management and operation, and low in construction cost.

3.2 Form pruning

Passion flower is a climbing vine with tendrils growing between the main vine and lateral vine node and near the axil of the leaves. Under natural conditions, it relies on tendrils to randomly twine around objects that can be twined in front of it, expand the occupied space, increase the light area of the leaf, and improve the amount of photosynthesis in order to obtain more organic matter. Reasonable form pruning can be guided by artificial binding and cutting, so that the plants can occupy the vacant space more quickly and make more reasonable use of the growth space.

3.2.1 Sapling form pruning

After the seedlings are planted and live, they should be inspected every 3–5 d to erase the lateral buds that sprout on the main vine. The weeds should be pulled out, and if there are seedlings lodging, small bamboo branches can be inserted to tie and right. When the main vine grows to 50–60 cm high, the sapling should be fixed to the climbing trellis with plastic rope or the tendrils of the sapling to lead it up and to the top of the trellis. When the main vine grows to 100 cm, the terminal buds should be removed, each main vine promotes 2–3 first-order lateral vines, and other lateral buds should be continuously removed. When the first-order lateral vines grow 20 cm higher than the top of the trellis, the first-order lateral vine should be divided into both sides of the trellis line or into 3 directions (such as "T"-shaped frame) and tightened. The terminal buds of the first-order lateral vines should also be removed, and each first-order lateral vine will promote 2–3 secondary lateral vines. When the secondary lateral vine grows to 50–60 cm, the terminal buds of the secondary lateral vine also need to be removed, and each secondary lateral vines promote 2–3 tertiary lateral vines.

The main and first-order lateral vines are the main vegetative shoots, and the secondary to fourth-order lateral vines are the main fruiting vines when bearing fruits. Therefore, cultivating the secondary to fourth grade lateral vines with robust growth, enrichment and sufficient quantity is the basic condition for achieving high yield in that year.

3.2.2 Bearing tree form pruning

When half of the first fruiting is harvested in the year, the vine which once bore fruit before should use cutting-back technique for pruning. The more robust fruiting vine should be selected, with 3–4 intact leaves in front of its branch base, and the fruited part should be cut off at 1 cm above the node eye to promote the extraction of 2–3 four-grade lateral vines, making the second fruit-bearing vine. As for the fruiting vine with weak growth potential, it is recommended to cut off only the diseased and pest-infected branches and the weak branches that grow too densely to ensure the fruit yield in autumn and winter.

3.2.3 Form pruning before overwintering

Before winter, it is necessary to carry out a cleaning pruning of passion flower, mainly to cut off the diseased and pest-infected branches and vines, cut back the branches and vines that seriously grow across the plant, and thin the weak branches and vines that are too dense, so that the whole plant can get sufficient light, and reduce the spacing competition and mutual interference between the plants. However, pruning should not be too heavy, which is prone to some cases of dying. If the original leading branch of the main vine is withered, the original leading branch should be replaced by the secondary shoot.

3.3 Water and fertilizer management

The orchards with high yield and high quality should attach importance to the application of organic fertilizer and implement comprehensive fertility measures combining organic fertilizer and inorganic fertilizer. Of the suitable amount of nitrogen, phosphorus and potassium is planned to be applied every year. The amount provided by organic fertilizer should account for about 40% of the whole year. From final planting, each plant should be applied high-quality organic fertilizer 20–40 kg per year, which can promote anthropogenic mellowing of soil, and the content of organic nutrients and inorganic nutrients of the soil are maintained at a high level.

The difference in environmental conditions in each planting area will directly lead to the different phonological phase of passion flower in each region, so fertilization can be guided according to the phonological phase of passion flower in different places. According to the cultivation performance in tropical and south tropical zone, in the same orchard of the same variety and the same plant at the same time under the same management, there are different phonological states, resulting in difficulty in determining the fertilization time. Therefore, the fertilization period of this phonology is determined according to the occurrence of about 60% branches and vines in the same phenological state (such as vegetative shoot, alabastrum stage, full bloom stage, small fruit rapidly increasing stage, etc.). In addition to meeting the needs of the main phonological phase, the combination of fertilization elements and their proportion should also be considered, and it is also necessary to pay attention to not causing damage or obstruction to the plant organs of other phonological phases at the same time (for example, applying a large amount of nitrogen fertilizer in order to extract spring vines will cause the excessive growth potential of the branches and vines, causing flower and fruit dropping). When increasing the proportion of nitrogen properly, the appropriate amount of potassium and phosphorus elements should be matched with other trace elements to fully meet the needs of all aspects in order to achieve the best results.

According to the phonological phase of purple passion fruit in the southern subtropical climate, the fertilization time of purple passion fruit based on the normal fruiting stage under the southern subtropical climate is referred to:

The 1st time, from mid January to late January, each plant is applied 150 g urea, 100 g compound fertilizer and 100 g potassium sulfate to promote spring buds;

The 2nd time, from early March to mid March, each plant is given 100 g compound fertilizer and 150 g potassium sulfate to strengthen buds and flowers;

The 3rd time, from mid April to late April, each plant is applied 50 g urea, 50 g compound fertilizer and 200 g potassium sulfate to promote fruit enlargement and high quality;

The 4th time, from mid August to late August, each plant is given 60 g urea, 50 g compound fertilizer and 100 g potassium sulfate to strengthen autumn buds;

The 5th time, from late November to mid December, each plant is applied 220 g calcium superphosphate, and organic fertilizer is also applied to improve the soil and strong trees to lay the foundation for next year's high yield.

According to the actual situation of annual phenology and variety characteristics, the specific fertilization period and fertilizer amount are adjusted. In orchards with drip irrigation, water-soluble fertilizers should be adjusted in the mixing tank in advance according to the safe concentration as far as possible, and then dribbled together with irrigation after passing the inspection, so as to improve the work efficiency and facilitate the timely, safe and efficient fertilizer absorption and utilization.

3.4 Prevention and control of major diseases and pests

The prevention and control of orchard diseases and pests adhere to the guiding principles of prevention first, integrated prevention and control, and unified prevention and control. Ecological health orchards should be built to reduce the risk of pests and diseases and reduce the economic losses caused by them. The major diseases and pests of passion flower are fusarium crown rot, virus disease, anthracnose, fruit fly, cletus punctiger, epicauta hirticornis, red imported fire ant, brown marmorated stink bug, etc.

3.4.1 Fusarium crown rot

Fusarium crown rot is caused by fusarium solani, which is a devastating disease of passion flower, mainly affecting the base of the plant stem 5–10 cm above the ground. At the beginning of the infection, dark brown lesions appear on the base of the stem of the plant. Then the pathological cortex is cracked, softens and rots, and easily separates from the xylem. When in high humidity, pink colonies often appear on the surface of the disease site. In the later stage of the disease, the upper leaves and vines turn yellow and wilt, and then the whole plant withers and dies.

Prevention and control methods: Management should be strengthened and drainage work should be done well. After the removal of diseased branches or plants, quick lime should be applied, and the main stem within 30 cm from the ground should be sprayed with 35 g/L metalaxyl-M·fludioxonil SC, 47% kasugamycin dicopper chloride trihydroxide WP, to kill the surface pathogenic bacteria, and the root should be irrigated.

3.4.2 Viral disease

Viral disease, also called mosaic disease, is common in planting areas, the incidence rate is usually 30%–40%, whose main symptoms are ring spot in leaves, crimped, mosaic, ring spot mosaic, tip necrosis and fruit lignification, etc. The passion fruit is propagated by using asexual cuttings. The pruning shears are not disinfected in time during the cutting process and the seedlings are infected with viruses, which are the main ways of the rapid spread of viral disease in China.

Prevention and control methods: The diseased plants should be removed, the intermediate insects that may spread the virus should be killed, the virus-free seedlings should be selected or the disease-resistant seedlings should be planted, reasonably dense planting, shortening the fruit-bearing cycle of a single plant, and updating the orchard once every 1–2 years. Oligosaccharides and lentinan can also be used to spray the plants to reduce the symptoms of disease.

3.4.3 Fruit fly

At present, there are mainly 4 kinds of common fruit flies causing serious harm: oriental fruit fly (*Bactrocera dorsalis*), melon fruit fly (*Bactrocera cucurbitae*), pumpkin fruit fly (*Bactrocera tau*), and striped fruit fly (*Bactrocera scutellata*). The oviposition under the pericarp and the larvae mining cause the fruits to hollow out, and resulting in a large number of fruit dropping in serious cases.

Prevention and control methods: A combination of agricultural comprehensive prevention and control, physical and chemical monitoring and control, and chemical control can be used to selectively use green control agents such as attractants, food attractants, or contact killing agents (such as pyrethroids, ethyl fungicides, and avermectin) and stomach toxic agents (such as thiacloprid and imidacloprid) for prevention and control.

3.4.4 Brown marmorated stink bug

After brown marmorated stink bug infecting the leaves and tips of passion fruit, the symptoms are not obvious. However, after the fruit is infected, the infected part becomes corkification, hardens, development ceases and sinks, causing bumps or deformity in serious cases, and loses economic value. The infected parts are leaves, buds, young shoots and fruits. It may also spread and carry passion fruit viral disease.

Prevention and control methods: Emphasis should be placed on controlling the overwintering imagoes and the nymphs of generation I beginning to copulate, which are more sensitive to chemicals at this time and are the best time for control. The medicament can be used for 1–2 times, with 4.5% beta-cypermethrin EC (1:1 000), 25% thiamethoxam WG (1:1 500) and 5% emamectin benzoate microemulsion ME (1:3 000).

3.5 Harvest and post-harvest treatment

3.5.1 Harvest

After the fruit ripens, the fruit and the stalk will generate an abscission layer and fall off automatically. Within 3 d from the time the fruit falls, the passion fruit has the best quality and flavor.

Selling fresh passion fruit for long-distance transportation, efforts should be made to minimize the damage to the fruit. If it is to provide raw materials for processing, in order to reduce labor and costs, it is necessary to inspect the orchard every 1−2 d during the fruit ripening period, pick up ripe fruits that automatically fall to the ground, wash the sediment with water in a timely manner, dry the moisture, weed out rotten fruits, and store them at lower temperatures and higher air humidity for later use after precooling and cooling down.

It is forbidden to pick unripe fruit and mix it with ripe fruit for marketing, processing and storage, because unripe fruit contains highly toxic cyanide, which is harmful to human health.

3.5.2 Post-harvest treatment

The ripe passion fruit is a typical type of respiratory climacteric, and ethylene can be released spontaneously in the fruit, so the ripe fruit is not resistant to storage.

The harvested passion fruits should be stored in a cool and ventilated place at room temperature and can only be stored for 3−5 d. If it is packed in plastic sheeting bags and stored at the temperature of 6.5−8 °C and the relative humidity of 80%−95%, it can be stored for 20−30 d. Therefore, controlling the temperature and humidity of the storage environment can extend the shelf life, and placing ethylene absorbent in the fresh-kept bag can extend its shelf life for at least 6 d.

References

CAI Z Y, DONG L, WANG H Q, et al., 2023. Pollen viability, stigma acceptability of passion fruit at different developmental stages and their effects on fruit setting [J]. Journal of Fruit Science, 40(5): 969−977.

GAN L S, LIAO Y L, CHEN X S, 2020. Color Illustrated Book of Passion Fruit Cultivation with High-quality and High-yield [M]. Guangzhou: Guangdong Science and Technology Press.

GUO Y F, LI X L, YANG D P, 2019. Analysis of volatile components of different passion fruit [J]. South China Fruits, 48(6): 59−63,71.

HUANG C S, WU C B, 1998. Breeding of a New Yellow Passion Fruit Strain 5−1−1 [J]. Guangdong Agricultural Sciences (5): 18−20.

HUANG X L, FENG J Q, MAO D N, et al., 2022. Research progress on preservation and processing of passion fruit [J]. Light Industry Science and Technology, 38(6): 1−4.

LI D P, CHUN D Z, LI J, et al., 2022. Effects of different nitrogen and potassium fertilizer dosage on growth, quality and yield of passion fruit [J]. Soil and Fertilizer Sciences in China (12): 123−132.

YI Z L, XU W Q, 2018. Introduction and cultivation status and prospect of passion fruit in hainan [J]. Chinese Journal of Tropical Agriculture, 38(7): 25−28.

YU D, XIONG B Q, YUAN J, et al., 2005. Germplasm resources of passion fruit and its application research status [J]. South China Fruits, 34 (1): 36−37.

ZHANG R, CHEN M, FENG H Y, et al., 2022. Research progress on diseases and pests occurrence rules and integrated control technology of passion fruit [J]. Modern Agricultural Science and Technology (22): 94−98.

ZHANG Z R, XIAO L Y, FENG H Y, et al., 2022. Investigation and identification of diseases and pests of passion fruit in Hainan and the damages [J]. Chinese Journal of Tropical Crops, 43(10): 2114−2121.

ZHOU H L, ZHENG Y Y, ZHENG J Z, 2015. Excellent varieties and cultivation techniques of passion fruit [J]. South China Fruits, 44(2): 121−124.

ZHOU Y J, TAN F, DENG J, 2010. Update review of passiflora [J]. China Journal of Chinese Materia Medica, 33(5): 1789−1792.

❖ Key Techniques for Watermelon Cultivation

Watermelon [*Citrullus lanatus* (Thunb.) Matsum. et Nakai] is an annual vining plant. Watermelon is widely cultivated in China, with many varieties and various forms of rind, pulp and seeds, among which Xinjiang, Lanzhou City of Gansu, Dezhou City of Shandong, and Dongtai City of Jiangsu are the most famous producers. As a kind of melon, vegetables and fruits, watermelon is renowned as "the king of midsummer". The refreshing, thirst-quenching, sweet and juicy taste makes watermelon a good fruit in midsummer. Watermelon not only contains no fat and cholesterol but also a lot of glucose, malic acid, fructose, protein amino acids, and lycopene. It is rich in vitamin C. The pulp tastes sweet and can cool down and beat the scorching summer heat. Seeds contain oil, which can be used as a snack food. The rind is used for medicine, which has the effects of cooling the body, increasing urination and lowering blood pressure. Hence, watermelon is a favourite summer fruit. It has become one of the important high-efficiency horticultural crops in China because of its refreshing thirst-quenching, sweet and juicy taste, strong adaptability in planting, short cultivation cycle, large market demand benefit and remarkable income increase effect.

1 Conditions and requirements

1.1 Temperature

Watermelons prefer full sun and hot temperatures but are not tolerant of cold temperatures. The appropriate temperature is between 18 °C and 32 °C. Watermelon seedlings stop growing when the temperature is below 10 °C and will be injured when it is around 5 °C. A proper temperature differential between day and night is needed in the growing period, and a suitable temperature difference between day and night is beneficial to the accumulation of sugar in fruits. The higher the sugar content of watermelon, the better the quality of watermelon, and the relatively low temperature for vegetative growth, so a higher temperature is needed for fruiting and fruit growth.

1.2 Water

Watermelons withstand droughts, not humidity. If humidity stress is high due to too much rain, watermelons are susceptible to diseases that lower productivity and quality. Watering is greatly needed in seedlings, stretching periods and fruit set periods. Given too much water, watermelon roots may rot.

1.3 Light

Watermelons require full sun. Ten to twelve hours of sun a day is proper and beneficial to gain sweetness.

1.4 Soil

Watermelons are not strict with the soil, and sandy loam with a deep, fertile soil layer and loose soil is best. Avoid continuous cropping and rotate with other non-Cucurbitaceae crops, among which paddy-upland rotation fields need to rotate for 3–5 years, and dry land needs to rotate for 5–7 years.

2 Watermelon Varieties

Selection of watermelon varieties should be considered natural conditions such as soil, weather, and water. Also, market demands are considered as well. Varieties with good productivity, quality, and environmental adaptability are other factors in determining the varieties to be planted.

In terms of stocks, pumpkins, gourds, and wild pumpkins can be selected, especially those with strong affinity, symbiosis and strong resistance. Pumpkin hybrids or wax gourd can be selected as root stocks in facilities cultivation or summer and autumn cultivation.

3 Cultivation and management techniques

3.1 Soil preparation

3.1.1 Tillage

Before soil preparation, removing the debris of previous stubble is necessary. Turn over 30 cm deep before winter, and dry the soil for about 15 d in autumn and winter. Finely crush the soil and level the land 15 d before planting.

3.1.2 Furrow plough

Plough one furrow, which is in the same direction. The ditch is 120 cm wide and 130 cm wide in between. The depth of the drainage ditch is 25 cm. Plough two furrows, which are in opposite directions. The ditch is 300 cm wide and 50 cm wide in between. The depth of the drainage ditch is 25 cm.

3.1.3 Base fertilizer

To apply 1 500–2 000 kg of decomposed manure or commercial organic fertilizer, 50 kg of calcium superphosphate, 30 kg of diammonium phosphate (N : P ratio is 18 : 46), 10 kg of potassium sulfate and 50 kg of decomposed cake fertilizer per mu. Plots that lack microelements should be supplemented with 1–2 kg of relevant element fertilizer.

3.1.4 Mulch

Film mulching cultivation plays a role in temperature, moisture and grass conservation. Lay drip lines first. Cover those ditches with films that are more than 0.01 mm in thickness and 120 cm in width. Tightly cover those films with compacted soil.

3.2 Seedlings nursery

3.2.1 Facilities requirements

3.2.1.1 Nursery houses

Plastic greenhouses or greenhouses are used for seedling nurseries, which are equipped with drainage and lighting. Heating and heat maintenance systems are equipped for seedling growth.

3.2.1.2 Seedling bed requirements

Generally, the length of seedbeds in industrial seedling raising is less than 4 000 cm. The width is divided into 3 series: 165 cm, 180 cm and 185 cm, and the height is 810 cm, which can be fine-tuned according to the ground conditions. The main structural materials of the seedbed are hot dip galvanized and aluminium alloy, and the materials of the bed surface are barbed wire, tidal plate and plastic net, which are adjusted according to local conditions. Seedbeds must be equipped with water supply and drainage facilities.

3.2.2 Direct sowing

3.2.2.1 Bases

Generally, 32-hole or 50-hole plug plates are selected, which are washed and dried repeatedly with clean tap water. Reusable plug plates are disinfected with 2% sodium hypochlorite aqueous solution after being cleaned. The substrate comprises non-polluting peat, vermiculite and perlite (peat : vermiculite: perlite = 3 : 1 : 1). 1.2 kg of balanced compound fertilizer of nitrogen, phosphorus and potassium and 25 g of 50% carbendazim WP are applied to 1 m^3 of substrate. The mixture is set for 2–3 d before applying. The substrates which are repeatedly used should be disinfected in advance.

3.2.2.2 Seeds treatment

Uncoated seeds require seed disinfection treatment. ① Sun exposure: dry those seeds on sunny days for four to six hours. ② Soak those seeds in 55 Celsius lukewarm water for 20 min. Repeatedly stir and clean them after the seeds cool down to room temperature. Wash out the sticky substances attached to the seeds. ③ Chemical Disinfection: Soak seeds in 0.1%–0.2% potassium permanganate, 0.1% carbendazim solution, 0.1% copper sulfate solution or 0.1% trisodium phosphate solution for 15 min, and then rinse them with clear water 4–5 times. Coated seeds do not need to be treated.

3.2.2.3 Seeds cracking

(1) Directly sow watermelon seeds without cracking them.

(2) The mesoembryo development of seedless watermelon seeds is incomplete, so the seed shell, seed coat and umbilicus are thick, leading to high temperatures for germination. Before seedlings are raised, it is necessary to crack the seed manually. Crack a small crack equivalent to one-third of the seed length from the umbilicus suture with pliers and other tools, and create a high-temperature environment to accelerate germination so as to improve the germination rate.

3.2.2.4 Sowing after accelerating sprouts

Place the disinfected seeds in an incubator between 33 °C and 35 °C for pre-germination. Be mindful to wrap those seeds in wet wipes. Inside the incubator, the relative humidity is maintained between 90% and 95%. When 75% of the seeds come out showing the white buds, sort them out. Eliminate these ungerminated seeds after 48 h of incubation.

3.2.2.5 Sowing

Sow and lie down seeds into one hole. After that, cover the propagator with a 1 cm deep substrate. Water well after planting. When seedlings come out, pick off the capsule on the cotyledons in case the cotyledons are yellow due to untimely extension.

3.2.2.6 Caring for seedlings

Temperature is maintained at 30 °C during germinating to boost the development of the propagators and the seedlings. After the first true leaves have emerged and in a period of seedling strength, reduce the temperature to prevent the propagators' overgrowing and improve their stress tolerance. After sprouts, keep the daytime temperature between 25 °C and 28 °C, and 15–20 °C at night. Also, air relative humidity remains 75%, and substrate relative humidity is between 50% and 60%. Add plant support when it is necessary. Apply 0.2% potassium dihydrogen phosphate solution 1−2 times on sunny afternoons. The application is 7−10 d apart.

3.2.3 Grafting

3.2.3.1 Sow rootstock and scion

Sow grafted seedlings 33−40 d prior to plant establishment. When adopting the hole insertion grafting method, sow the rootstock first and scion 5 d later. When using side grafting, sow the scion first and the rootstock after 5 d later. Both cell trays and grafting bowls can be used for grafting. The bowl is 10 cm tall with a diameter of 8 cm and a soil depth of 7−8 cm deep.

3.2.3.2 When to grafting

When both cotyledons and the first true leaf start to develop, the rootstock plant is ready to graft for hole grafting. For side grafting, when both cotyledons and the first true leaf start to develop, or cotyledons on the rootstock start to emerge, the rootstock plant is ready to graft.

3.2.3.3 Preparations before grafting

Grafting should be practised in a shaded shed greenhouse. Tools used for grafting are supposed to be prepared, including bamboo sticks, razor blades, and disinfectants, should be set in advance. Both manual workers engaging in grafting and tools should be sanitized with 75% ethanol solution or potassium permanganate solution in advance. Spray 50% chlorothalonil WP (1 ∶ 1 000) and 3% polymyxin WP (1 ∶ 1 000) as the mixed disinfected solution one day before the grafting. Water the seed beads well. The temperature for the grafting is better maintained between 25 °C and 30 °C.

3.2.3.4 Grafting methods

(1) Hole insertion grafting. Remove the growing point of a rootstock with a sharp probe. Make a slit with a 45° angle at the growing tip, from where insert a bamboo barbecue skewer. Cut down the scion one centimetre below the cotyledons. Cut the lower end of the scion into a 0.5 cm long "V"-shape. The scion is then inserted at once into the slit while the stick is removed. The cotyledon adopts a cross-shaped butt joint mode to ensure that the scion and the rootstock are entirely attached, and some of them should be left outside. Finally, they are fixed with grafting clips.

(2) Side grafting. Remove the growing point of a rootstock with a grafting knife. A slit is cut in the cotyledon of the rootstock 0.5−1 cm lower. An angel cut at almost 35° between the slit and the hypocotyl, reaching to 2/3 deep of the hypocotyl. Then, cut in the opposite direction between scion seedlings and rootstock incisions, about 1.5 cm below the cotyledon

at about 30°. The length of the incisions equals the rootstock, and the depth is 3/4 of the hypocotyl. Finally, the two tongue-shaped incisions of rootstock and scion are matched together and immediately fixed with grafting clips. Soon after grafting, the seedlings are planted in a nutrition bowl.

3.2.3.5 Post-grafting caring

Within three days after grafting, the seedbed should be covered and shaded, with the relative humidity of air above 90%, the shading rate of 70%–80%, and the temperature should be controlled at 25–28 °C during the day and 18–20 °C at night. After 3 d, sun-lit the seedlings both in the morning and evening. Proper ventilation is carried out. After 8–10 d of grafting, regular management is resumed when rootstock germination is removed in time during this period.

3.2.4 Nurturing robust seedlings

Five to seven days ahead of plant establishment, water properly. Maintain the substrate relative humidity at around 60% with the condition that the seedlings are not withering. Reduce temperature in the greenhouses by enhancing circulation and light. Three to four true leaves come out on the watermelon seedlings with short nodes and stems 3.5–4.5 mm wide. Leaves are dark green, with the root system in the hole wrapped around the substrate and forming a complete root lump. There are no mechanical injuries or pest decay of the seedlings. The scion for grafting is not overcrowded.

3.3 Plants establishment

3.3.1 Planting conditions

Planting meets the following temperature. The lowest temperature should be above 5 °C; At the same time, deep into 10 cm, the soil temperature should keep at 15 °C above.

3.3.2 Planting density

The planting density is determined according to variety characteristics, cultivation season and pruning method. In outdoor cultivation, the plant spacing of early-maturing varieties is about 40 cm. About 660 plants are planted per mu, while the plant spacing of middle-maturing and late-maturing varieties is about 45 cm, and about 530 plants are planted per mu. In greenhouse cultivation, the plant spacing of early and middle maturing varieties is about 30 cm, and 1 200–1 400 plants are planted per mu, while the plant spacing of late maturing varieties is about 45 cm, and 530 plants are planted per mu.

3.3.3 Planting methods

During planting, the nutrient clods carried by the stems, leaves and roots of seedlings should be ensured to be complete, and the planting depth should be level with the surface of nutrient clods and the border surface. The interface of grafted seedlings should be 1−2 cm above the ground. Pour water after planting. According to the needs, cover the plastic film and seal the surrounding area of the plastic film mouth for planting seedlings with fine soil.

3.4 Watering and fertilizing management

3.4.1 Water

After transplanting, water as much of the root fixation water as possible. Water once when the transplanted plants are established. As the vines begin to run, water once. Water once before flower formation, which is known as fruit setting water. Water once 3−4 d after fruits are set for swelling of fruits. Stop watering 7−10 d prior to harvest.

3.4.2 Topdressing

(1) At the initial stage of tendrils, nitrogen fertilizer is the main support, followed by potassium fertilizer. Apply 10−15 kg of urea with 4−5 kg of potassium sulfate per mu.

(2) Early fruit swell. Perform potassium and nitrogen fertilizer mainly. Apply 20−25 kg of urea and 10−15 kg of potassium sulfate per mu, spray 0.3%−0.5% potassium dihydrogen phosphate + 0.1% borax and 0.5% ferrous sulfate aqueous solution on the leaf surface.

(3) The Mid of fruit swell. 10−15 kg of ternary compound fertilizer (N : P : K= 15 : 15 : 15) is applied per mu. At the early and middle stage of fruit expansion, spray 0.3%−0.5% potassium dihydrogen phosphate + 0.1% borax and 0.5% ferrous sulfate aqueous solution on the leaf surface.

(4) After the first batch of watermelon is harvested. Apply 15 kg of ternary fertilizer compound (N : P : K=15 : 15 : 15) per mu, stimulating the restoration of the seedlings. When watermelons grow 3−5 cm in diameter, apply 20 kg of ternary compound of fertilizer (N : P : K= 15 : 15 : 15) per mu.

(5) The middle of the second batch swell. Fifteen kilograms of ternary compound fertilizer (N : P : K= 15 : 15 : 15). Spray 0.3%−0.5% potassium dihydrogen phosphate + 0.1% borax and 0.5% ferrous sulfate aqueous solution on the leaf surface.

3.4.3 Pruning

It is proper to prune in fruit-setting. There are three methods: single-vine, double-vine, and triple-vine pruning.

3.4.3.1 Single-vine pruning

Remove all side vines but leave one; as a matter of fact, only one melon remains.

3.4.3.2 Double-vine pruning

Remain the main vine. Choose the most robust side vine when side vines grow twenty centimetres. Even if 2 vines are reserved, watermelon only yields on the main vine. There is only one melon produced. Before the melon grows, all side vines should be cut off but the one. After fruit set, newly grown side vines can be reserved because the vigor of those new plants is too weak to overgrow. Both single and double vine pruning are suited to early maturing varieties.

3.4.3.3 Triple-vine pruning

Remain the main vine. Choose the most robust two side vines when side vines reach 20 cm. This method is adapted for large and late maturing sizes by low-density cultivation. Melons yield on the main vine, while two side vines are support nutrients.

3.4.4 Arranging and covering vines

When vines grow to 50–60 cm, arrange vines in the same direction. Pact every 4–5 vines with a stone on the mound. Lie stones 2–3 times to fix.

3.4.5 Clean the side branches

Before the watermelon fruit setting, the branches on the main tendrils and side tendrils should be knocked out in time to prevent an unbalanced nutrition supply, affecting the watermelon fruit setting and fruit expansion. However, the number of branches should be reduced after sitting melons.

3.4.6 Keep the melon

At 130–160 cm in length, the fruits with large and straight ovaries, straight and thick stalks and a second or third female flower are selected. When a single vine, double vine and three vines are used for pruning, one melon is left for each plant. When multi-vine pruning is adopted, 1–2 melons can be left on each plant. When the first crop of melons is harvested or near maturity, one more melon can be left, and the second can be left with side vines.

3.4.7 Pollination

Bees can be used for pollination for efficiency. In addition, manual pollination is adopted. Every morning between 7:00 and 9:00, when the flowers are fully open. Pick a healthy male flower and remove its petals to expose the inside of the bloom where the pollen is produced. Gently move the male flower into the fully open female bloom, which allows the pollen from the male anthers to transfer onto the female stigma.

3.4.8 Turn watermelon

Turning is for the even colour of the watermelon. In the fruit swelling period, rotate the melon in the same direction once every 2−3 d. Each time rotating is no less than 30°. Turn each melon 2−3 times. Rotating tines double when it rains.

4 Prevention and control of diseases and pests

4.1 Prevention and control of diseases

4.1.1 Fusarium wilt

4.1.1.1 Symptoms

Watermelon growers name the disease as "dead vines". Fusarium wilt of watermelons wilt during the daytime and rebound in the morning and at night. As the disease progresses, watermelons die within 4−5 d. Vertical fracture appears on the plant basal cortex, with gum out due to the diseased vascular tissues within. It is one of the common diseases in watermelon producers.

4.1.1.2 Prevention and control

In the initial stage of infection, fungicide application is to prevent. Once the disease is spreading, any of these fungicides can be used to water roots: 50% thiophanate-methyl WP (1∶600), 10% fludioxonil SC (1∶1 000), 1 billion CFU/g of sphaerotheca amyloliquefaciens WP, and 15% humeral AS (1∶300). To apply 300−500 mL of fungicide once every 7−10 d apart, continuously for 2−3 times in a row. Sanitize the fields that were heavily affected by the disease before planting.

4.1.2 Gummy stem blight (GSB)

4.1.2.1 Symptoms

The disease mainly affects stems and wines, and so do leaves and fruits. Black, wrinkled spots on which dark dots appear are signs of diseased leaves. Under humid or wet conditions, those spots rapidly expand to the entire leaves, darkening and collapsing. Grey-white and oval lesions on which dense black dots scatter appear around nodes when fruits and vines are affected. The black and wrinkled spots around stems and nodes are signs of severe infection.

When the fruits are infected, water-soaked spots at the beginning turn brown dead spots in the centre, showing star-shaped cracking, cork-shaped dry rot inside, and rots after being slightly blackened.

4.1.2.2 Prevention and control

In initial infection, spray 80% zineb WP [1 : (700−800)], 40% difenoconazole·pyraclostrobin SC (1 : 2 000), and 30% difenoconazole·kresoxim-methyl SC (1 : 1 200) once every 7 d, continuously for 3−4 times. Application of infections part with 50% thiophanate-methyl WP (1 : 500) and 40% formaldehyde (1 : 100) will be effective and well-treated.

4.1.3 Watermelon phytophthora blight

4.1.3.1 Symptoms

Seedlings, foliage, vines and fruits are highly susceptible. The diseased parts appear water-soaked and later wilt and die. The infection is emergent and aggressive when it is rainy, wet and waterlogged. The symptoms are minor or hardly noticeable when it is dry and it rains less.

4.1.3.2 Management

(1) Agricultural control. Strengthen drainage in the fields. Empty pumpkin patches wader off water splashing, which is a carrier for the disease.

(2) Chemical control. Spray 25% metalaxyl WP (1 : 1 000), 72% propamocarb hydrochloride AS (1 : 600), 60% pyraclostrobin·metiram WG (1 : 600).

4.1.4 Anthracnose

4.1.4.1 Symptoms

Pumpkins are susceptible to affecting under high temperatures and humid conditions. Spots and lesions turn up as wet patches that later darken to black on stems, leaves and fruits when the disease spreads. Meanwhile, the circular appears, and later the leaves wilt and die.

4.1.4.2 Management

In the initial stage of the infection, spray 80% mancozeb WP (1 : 500), 50% azoxystrobin WDG (1 : 2 000), 325 g/L difenoconazole·azoxystrobin SC (1 : 1 500) once every 7−10 d, 3 times in a row.

4.1.5 Powdery mildew

4.1.5.1 Symptoms

In the initial stage of infection, small white moulds appear, which later expand into white colonies of spores. In the latter stage of the disease, a wide powdery substance covers on which yellow and brown spores appear. As the white powder moves through, the upsides and backsides of the leaves are covered with white mould, and the leaves wilt, crisp and curl.

4.1.5.2 Management

In the initial stage of the infection, spray 10% difenoconazole WP (1 : 1 000), 70% thiophanate-methyl WP (1 : 800) and 40% difenoconazole·azoxystrobin SC (1 : 1 500). One application is 7 d apart, 2−3 times in a row.

4.1.6 Watermelon mosaic virus (WMV)

4.1.6.1 Symptoms

Also known as lobular disease and mosaic disease, the disease makes yellowish green mosaic appear on the leaves. The stems and leaves shrink and twist, the leaves are uneven, and the leaves of mosaic, melon, and vine tips become smaller and curled.

4.1.6.2 Management

The key to care and prevention is to control aphids and spider mites by applying pesticides, such as imidacloprid, acetamiprid, and thiamethoxam. In addition, strictly control any physical contact. The WMV is emergent for outdoor planted watermelon at the end of May to early June; Therefore, spray 1% lentinan AS (1 : 2 000), 1.5% dodecyl sodium sulphate WP and 20% moroxydine hydrochloride·copper acetate WP (1 : 500).

4.2 Prevention and control of pests

4.2.1 Aphids

4.2.1.1 Symptoms

Aphids cluster and hang out on the underside of foliage or tender buds, feeding on the plant's sap. They cause the infested leaves to curl, wither, and stunt when they are rampant. Sometimes, leaves wilt and die in seedlings. The sticky substances aphids produce cover the foliage, which is at high risk of sooty mould. All these symptoms hinder the normal carry-out of photosynthesis. Moreover, melon aphids transmit WMV, causing mosaic and discoloured leaves and deformation. Hence, aphids greatly decrease both the productivity and quality of watermelon.

4.2.1.2 Management

Place sticky traps, and colour film or mulch will deter aphids from visiting. Insect nets can be hung on the ventilation fans. When aphids occur, preventive measures include spraying 25% thiamethoxam WG (1 : 5 000), 3% acetamiprid ME (1 : 2 000), and 10% imidacloprid WP (1 : 1 000).

4.2.2 Thrips

4.2.2.1 Symptoms

As are leaves and vines, affected young leaves and shoots are twisted and hardened. Fruits barely develop. Even if fruits are set, they are covered with stippling and rusty stripes or hardened, deformed, and cracked. Nevertheless, thrips transmit diseases to which seedlings are highly susceptible.

4.2.2.2 Management

Place blue sticky traps to control. Application of pesticides is the same with aphids.

4.2.3 Whiteflies

4.2.3.1 Symptoms

Both whiteflies and their nymphs feed on sap on watermelon sap. Affected watermelon leaves are discoloured, yellowing and wilt, which causes the death of an entire plant.

4.2.3.2 Management

(1) Place yellow sticky traps to trap adult whiteflies in greenhouses.

(2) Before planting watermelon, spray the closed greenhouses with 80% chinomethionate EC (1 : 1 000) 3–5 d in a row. Alternatively, it can also be fumigated with 10% isoprocarb aerosol.

(3) Spray 25% thiamethoxam WG (1 : 6 000), 25% buprofezin EC (1 : 1 000), and 20% acetamiprid EC (1 : 2 500).

4.2.4 Spider Mites

4.2.4.1 Symptoms

Tiny yellow and brown dots are on the leaves at the beginning of the infection. Massive clustering, those pests spin thin bits of silk as they feed. Then, red and large spots appear on the backsides of the leaves. Infested leaves curl, yellow, and drop off. The entire leaves are decoloured and whitened.

4.2.4.2 Management

Timely and reasonable topdressing and irrigation can promote the growth of watermelon to increase its resistance. At the same time, the high humidity in the field is not conducive to the occurrence of red spiders. Avermectin, pyridaben and pargyne mites can be used for prevention and control.

4.2.5 Grubs

4.2.5.1 Symptoms

Larvae bite off the roots and stems of watermelon seedlings, causing the whole plant to die, resulting in a lack of seedlings and broken ridges. They eat roots and stems, which weakens the growth of watermelon seedlings and directly affects the yield and quality.

4.2.5.2 Management

(1) Light Trapping. Trap and kill adults in the emerging period, decreasing pest damage by lowering the numbers.

(2) Soil Management. Mix 2.5% trichlorphon GR with soil, and spread the mixture on the bottom of seedbeds or planting holes. The ratio of the 2.5% trichlorphon GR is 1.5–2 kg per mu.

(3) Chemical Control. Water roots with 90% trichlorfon crystalline WP (1 : 800) and 25% thiamethoxam WG (1 : 500). Alternatively, apply 35–45 kg per mu of 0.05% emamectin benzoate (D) GR.

5 Harvest

Watermelon should be harvested in batches according to the maturity characteristics of varieties and the marketing demands. Harvesting is carried out in batches in line with the Melon's maturity. The varieties which do have strict requirements for harvest maturity can be harvested earlier, while varieties with requirements for harvest maturity should reach the required maturity before harvest. 90% of ripe melons can be picked in the local market, and 75%–80% of ripe melons should be picked when transported to other places.

5.1 Harvest time

Generally, the first batch of watermelons cultivated in spring begins to be harvested in late April, or 40 d after flowering. After the second batch, watermelons are harvested 28–32 d after flowering. It takes 28–30 d for early-maturing varieties, 30–35 d for middle-maturing varieties and 35–45 d for late-maturing varieties. Harvest should be in the morning and evening to avoid picking melons on rainy days and to prevent the increase of cracked melons.

5.2 Signs of readiness

After watermelon matures, its rind is generally hard, smooth and shiny, and its fruit surface pattern is clear. The bottom and pedicle of the fruit shrink and sag inward, and the bristles at the fruit stalk are sparse but not noticeable. It is a ripe melon and should be harvested in time. In a pinch at the melons, when a hollow sound is heard, the melon should be harvested in time as well.

5.3 How to harvest

When harvesting, keep the handle and a section of melon vines, and prepare watermelons that have been stored for a long time. It is best to cut them together with a section of melon vines. After harvesting, it should be prevented from being exposed to the sun and rain and delivered for sale in time. If it cannot be transported temporarily, it should be placed in the field or in the shade of the roadside to ward off the heat in the field, and the melon should be handled gently, with some melons and vines or grass under the melons.

6 Post-harvest treatment

6.1 Pre-cooling

Placing melons in a cool, shady spot in the fields is the easiest way of precooling. Cool watermelons at night, as the temperature is relatively low to remove field heat. Put the melons in trucks or storage before the temperature rises in the early morning. Mechanical air-cooling can be used for precooling where conditions permit.

6.2 Grading

Watermelons are graded by quality, colour, size, weight, and freshness, and they are free from pests and disease decay.

Fancy: Similar varietal characteristics—typical size and colour, size consistent, clean and neatly arranged packages, within 4% of errors (numbers or weights).

No. 1: Similar varietal characteristics—typical size, colour, flavour. Some errors are allowed but do not affect the appearance and quality of storage.

6.3 Package

Watermelon can be packed or bulk with corresponding materials according to the size of the fruit type and commodity value. Generally, it is mostly in bulk, and a few are packed in baskets, boxes and plastic net bags.

6.4 Storage

If they are not sold fresh for the time being, watermelons can be stored in the storage warehouse, which should be cool and ventilated. Standard temperature storage mainly depends on ventilation to cool down. Generally, doors and windows are closed during the day and ventilated at night. The temperature is kept at 15−20 °C, and the relative humidity of air is 70%−80%. Varieties with hard peel, dense flesh and crisp flesh are more storable, while seedless watermelons are more storable. After the fruit is harvested, it should be treated with chemicals in time. If any diseased, rotten and injured fruits are found, they should be removed in time.

References

FENG C, LIU Y H, TIAN P F, et al., 2022. Green and efficient annual cultivation mode of early spring watermelon—summer loofah—autumn cauliflower [J]. China Vegetables (6):128−130.

MA J L, XU H, SUN X X, 2022. Different treatments affects continuous cropping soil and watermelon yield and quality [J]. China Cucurbits and Vegetables, 35(7):50−55.

MENG J L, WU S J, WANG X W, et al., 2020. Cultivation techniques of watermelon before rice cropping [J]. China Cucurbits and Vegetables, 33(1):79−81.

WANG C, YANG Y B, LIN Y, et al., 2021. Multi-cropping planting modes and key cultivation techniques of watermelon in facilities cultivation [J]. China Vegetables (11):117−121.

YANG L B, WANG H Y, ZHANG M Z, 2020. Cultivation techniques of selenium-rich watermelon in outdoor saline-alkali land [J]. China Cucurbits and Vegetable, 33(3):79−80.

Chapter Three

Vegetables

Key Techniques for Tomato Cultivation

Tomato (*Lycopersicon esculentum* Mill.) is an annual or perennial herb of the genus *Lycopersicon* in the Solanaceae family, nicknamed Xihongshi, Fanshi and Shizi in Chinese. Tomatoes are native to Peru, Ecuador, and Bolivia in South America, with the advantages of strong adaptability, easy to cultivate, high yield, rich in nutrition, and versatility. They are widely cultivated in both northern and southern China. In 2021, the cultivation area of tomatoes in China was 1 144 821 hm^2, accounting for 22.2% of the world's total cultivation area. The fruit of the tomato is rich in nutrition and has a special flavor, which can be eaten raw, cooked, processed into tomato sauce, tomato juice, or canned whole fruit.

1 Growth conditions

Tomatoes prefer warmth and light, are sensitive to frost and heat, tolerant to fertilizer, and semi-drought-tolerant. Factors such as temperature, light, moisture, and soil will affect the growth and development of tomatoes.

1.1 Temperature

Tomatoes have an adaptation range of 15–33 °C, and generally growes well at 20–25 °C. The response of tomatoes to temperature varies depending on their growth stage and development.

1.2 Light

Tomato is a crop of light-loving, hence insufficient light or continuous rainy weather often causes blossom and fruit dropping. Light intensity has a great impact on the growth and development of tomatoes.

1.3 Moisture

Tomatoes have a relatively developed root system and strong water absorption capacity, so the water requirements are semi-drought resistant. It is intolerant of both drought and flood, requiring excellent soil drainage, low groundwater level, and even water supply with various water requirements in different developmental stages.

1.4 Soil

Tomatoes have relatively low requirements for soil conditions, and prefers fertile loam soil with a thick layer, good drainage, and rich organic matter, with a pH value of between 6 and 7.

2 Types and varieties

2.1 Types

There are many types of tomato varieties, which can be roughly divided into the following types in horticulture: according to plant growth habits, they are divided into unlimited growth type and limited growth type; according to leaf type, they are divided into ordinary leaf type, potato leaf type and wrinkled leaf type; According to fruit size, they are divided into large fruit type (>150g), medium fruit type (100−149 g) and small fruit type (<100 g); According to color, they are divided into red fruit, pink fruit, yellow fruit, etc.

2.2 Cultivars

Tomato cultivars with high yield, heat resistance, and disease resistance that are suitable for seasonal cultivation, commercially suitable for local market sales or export, and resistant to storage and transportation should be selected for production. The early and medium-ripe varieties with low temperature and weak light resistance should be selected for spring and summer cultivation, while varieties with viral disease and heat resistance should be selected for summer and autumn cultivation. In recent years, tomato viral disease with yellowish curl leaf has occurred occasionally in China, thus it is recommended to select disease-resistant varieties to reduce losses.

3 Cultivation season and method

According to the local climate environment and market demand, the sowing time should be reasonably arranged. Parts of South China, Hainan, and parts of Yunnan have frost-free conditions all year round, allowing for open-field overwintering cultivation. But in other regions, open-field cultivation should be carried out during the frost-free period. Open-field cultivation and facility cultivation can be carried out, while annual cultivation can be realized in Yunnan.

4 Cultivation and management techniques

4.1 Sowing and seedling raising

Seedling raising is an important part of intensive vegetable production, while raising and applying age-appropriate seedlings is a vital guarantee for early maturity, high quality, and high yield of vegetables. At present, there are three commonly used seedling raising methods in production: plug-seedling, seedbed seedling, and nutritional bowl seedling. Among them, plug-seedling is conducive to factory seedling raising (convenient management, transportation, etc.). Except for a few remote areas, large-scale production areas of vegetables all adopt plug-seedling. The following is an introduction to the method of plug-seedling.

4.1.1 Pre-sowing preparation

In central Yunnan Province and similar climate areas, seedlings raising should be conducted in plastic greenhouses from January to February. At present, nursery trays with 50–72 holes are mostly used in tomato seedling raising. The sites for seedling raising should choose fields with excellent drainage and irrigation, and the field's previous crop should not have been planted with solanaceous vegetables (such as tomato, eggplant and pepper). Plug-seedling can not only reduce the occurrence of soil-borne diseases, but also ensure neat and robust emergence, which can be grown at once without the need for seedling separation or other operations. The ground for placing nursery trays should be leveled in advance to avoid uneven terrain that may affect the quality of seedling emergence. Where conditions permit, it is better to use seedling racks. The seedling substrate can be vegetable seedling substrate, tobacco seedling substrate, or self-formulated. For self-formulated, the decomposed pig manure can be mixed with field soil in a 1 : 1 ratio, and 100 g of compound fertilizer and 10 g of carbendazim can be added per cubic meter and mixed evenly. The field soil can be selected from the surface soil within 15 cm of the field that has not been planted with solanaceous vegetables within 1–2 years, or from red soil that has not been planted with vegetables. The day before sowing, the nursery trays should be filled with seedling substrate or soil, flattened with small planks, and then using a sprinkler to pour water on the bottom of the tray for later use. Two seedling trays are arranged neatly in groups, which is convenient for workers to sow and thin out seedlings.

In frost-free areas, seedlings are raised in August for overwintering cultivation. The temperature is high during the seedling stage of this crop of tomato, so measures such as covering with shade nets can be taken to cool down.

4.1.2 Seed soaking and germination

The sowing amount for tomato seedling raising is generally 10–15 g per mu. In order to achieve the goal of early emergence, complete emergence, and raising strong seedlings, seed treatment should be carried out before sowing, mainly soaking and disinfection (coated seeds do not need to be treated). The commonly used method is to soak the seeds in hot water. The specific method is to first put the seeds into a nylon mesh bag or gauze bag and immerse them in room temperature water for 15 min, and then soak them in hot water at 55–60 °C for 15 min, with a water volume of 10 times the seed volume. After 15 min, let it cool naturally and soak the seeds at room temperature for 4–6 h. The soaked seeds can also be soaked in 50% carbendazim WP (1 : 1 000) or 10% trisodium phosphate solution for 20 min to neutralize the virus carried by seeds. After fishing out seeds, they should be rinsed with clean water and wrapped in gauze or a damp cloth to germinate.

The processed seeds should be wrapped in a wet towel and placed in a bowl. To maintain humidity, the seed bag should be covered with a few layers of wet towels, then placed at a temperature of 25–30 °C for 3–5 d. As long as 50% of the seeds sprout, they can be sown. During the germination process, the seeds can be flipped once a day to keep them loose. If the seeds are found to be sticky, the towel should be rinsed with clean water immediately. Generally, they should be cleaned once a day, then excessive moisture should be removed after cleaning, in order to avoid redundant water absorption by the seeds, which may cause oxygen deficiency and poor germination. If there is no condition for germination, the seeds can be sown after soaking directly.

4.1.3 Sowing

The day before sowing, the nursery tray or bed should be watered thoroughly. Generally, one seed is sown per hole on the nursery tray. After sowing, it should be covered with 1 cm of fine soil which should not be too thin or thick. If the covering soil is too thin, the seeds will be easily exposed after watering or capped seedlings will emerge, while if the covering soil is too thick, it will affect the emergence of seedlings. If the seeds are exposed after watering, it is necessary to cover them with soil promptly. After sowing, the nursery tray should be covered with plastic mulching for moisture conservation. When seedlings emerge, the plastic mulching can be removed. In low temperature conditions, it is best to add a small arch shed inside the greenhouse for seedling raising, or build a small arch shed with bamboo pieces on the seedbed and cover it with a 0.08-0.1 mm thin film for heat preservation. In places with frost at night, straw mats should be covered to keep warm, which is conducive to seedling emergence.

4.1.4 Seedbed management

From sowing to emergence, it is required that the bed soil has sufficient moisture and good ventilation, and the seedbed temperature should be kept high to promote the germination of seeds as soon as possible. When the seedlings are unearthed and the cotyledons are open, the straw mat and thin film should be removed in time during the day to ventilate and reduce the seedbed temperature to prevent overgrowth, but the straw mat and thin film should still be covered at night for thermal insulation. If the seedlings overgrow, the shed should be opened in the morning and covered at night, and water should be controlled appropriately. In the process of seedling raising, it is necessary to topdress properly, and usually combine with watering using 0.1%-0.2% urea. Watering should be sprayed with a fine mesh watering can or sprayer, while avoiding pouring with a gourd ladle to cause excessive soil moisture and serious disease the seedling stage. The main disease in the seedling stage is damping-off. Once a diseased plant is found, it should be immediately removed, and carbendazim or original chlorothalonil powder should be sprinkled around the diseased plant to prevent the spread of the disease. Starting hardening off seedlings 10 days before final planting, strengthening ventilation, reducing watering and controlling growth can enable seedlings to adapt to outdoor environmental conditions as soon as possible in order to shorten the time for seedling recovery.

The watering of the seedbed is generally carried out from 9:00 to 11:00, once every 1-2 d. If there are symptoms of poor growth cased by lack of fertilizer, such as yellowing of leaves, small leaves, and thin stems during the seedling stage, the method of top dressing outside the roots can be adopted by spraying 0.3% potassium dihydrogen phosphate+0.2% urea solution on the leaves.

4.2 Site preparation and open furrows

Sandy loam or loam plots with loose, fertile, well-drained soil and no solanaceous crops planted for 2-3 years should be chosen, and rice and dry crop rotation is the best.

Timely plowing and sun drying the soil clod after harvesting the previous crop. 2 500-3 000 kg of decomposed farmyard manure, 50 kg of compound fertilizer, 40 kg of calcium superphosphate, and 15 kg of potash fertilizer should be applied per mu. The ground should be leveled to form a 1.2 m wide bed (including a ditch) with a depth of about 30 cm.

4.3 Timely final planting

Generally, tomato that is cultivated in the open-field in spring can be done final planting after the local spring frost ends. The density of tomato planting should be determined according to the characteristics of varieties, pruning methods, climate, cultivation methods and soil fertility level, etc. Early-ripening varieties always have a higher density than late-ripening varieties; Varieties with compact plant types have a higher density than varieties with large plant expansion; Plots with low soil fertility have a higher density than plots with high soil fertility.

Tomato seedlings can be done final planting when they have 5–7 true leaves. The general plant row spacing is (30–40) cm×60 cm, and two rows are planted in each bed. Using a small hoe to dig holes, the seedlings should be put into holes and covered with soil, then compacted around. The cultivation depth should be as deep as the cotyledon node. After final planting, the roots should be watered enough, which is conducive to seedling recovery. Final planting should be done on cloudy days, not on wet soil during rainy days, as it is difficult to grow new roots and slow to recover.

The standard for healthy tomato seedlings is that the cotyledons are intact, the seedlings are 15–20 cm high, and there are 5–7 true leaves which are thick, wide and dark green; The stems are stout and the internodes are short; The roots are developed and have many fibrous roots, which are white and tender, free of diseases and pests.

4.4 Field management

4.4.1 Moisture management

After final planting and recovering, water should be adjusted appropriately to control the growth of the aboveground plants, promote the root system to develop in depth, and regulate the balance between vegetative growth and reproductive growth, so as to accumulate more nutrients for the development of the fruit. After the first spike of fruits is set, watering amount should be increased to once every 5–7 d to keep the soil moist. Drainage should be done well in the rainy season.

4.4.2 Fertilization

Tomato is a continuous growth and fruiting vegetable with high yield and large fertilizer demand. On the basis of applying sufficient base fertilizer, early and frequent topdressing is advocated. According to different growth needs, timely and appropriate staged topdressing should be carried out. After final planting about 10 d later, raising seedling fertilizer should be applied, and an appropriate amount of urea can be applied 5–7 kg/mu. Starting from the first spike of fruit expansion, topdressing should be applied every 10–15 d, and 10 kg of urea and 15 kg of compound fertilizer should be applied alternately per mu. At the same time, in addition to appropriate soil topdressing during the fruiting period, foliar fertilization should also be combined to make up for the lack of root fertilizer absorption ability. A mixture of 0.2% urea and 0.3% potassium dihydrogen phosphate can be used for spraying once every 15 d. Foliar spraying should be carried out on sunny days, which can achieve the condition of 24 h without rain after spraying. When spraying, it is best to be conducted in the evening or morning when there is still dew, rather than at noon, so as not to accelerate the concentration speed of fertilizer solution due to high temperature and cause damage.

4.4.3 Intertillage, weeding and hilling

Generally before setting up supports, intertillage and weeding should be done in time and combined with hilling, which can ensure soil looseness and permeability, thus promoting the emergence of new roots and enhancing the absorption capacity of the root system. Starting from the beginning of final planting, intertillage should be done every 10–15 d. Intertillage should be done when the soil is semi-dry and wet. During the flowering and fruiting period, intertillage should be combined with topdressing and hilling. After setting up supports, the plants is shielded, mainly by clearing ditches and hilling, combined with weeding.

4.4.4 Pruning and set up supports

When the tomato plant is about 40 cm high, the ways of hanger and support-setting should be adopted for hanging or tying the vine. The bamboo supports are used for tying the vine, which is mostly in the shape of an A-frame, an 1.8–2.0 m high. It is necessary to tie the vine and prune in time. The pruning is generally adopted the ways of single-rod or double-rod (the main stem is only left in single-rod pruning, and all the lateral branches are completely removed. In double-rod pruning, in addition to the main stem, one lateral branch from the first inflorescence to the lower leaf axil is left, and the rest of the lateral branches are removed), and the lateral branches should be cut when they grow to 7.5–10 cm. Unlimited growth varieties are mostly adopted single-rod pruning. It is advisable to prune in the afternoon on a sunny day, because the wound is easy to heal. When the plant grows to the top of the support, 2–3 leaves are left with top removal, so as to concentrate nutrients and improve the fruit setting rate of the upper inflorescence. In the late growth period, old and diseased leaves at the base should be removed in time to facilitate ventilation and light transmission, reduce nutrient consumption and the occurrence of diseases and pests. The removed old, diseased and residual leaves should be cleaned out of the garden timely and buried deeply or burned.

4.4.5 Fruit and flower saving

The main reasons for tomato blossom dropping are insufficient light, too high temperature (night temperature above 25 °C or day temperature above 35 °C) or too low (below 15 °C). Malnutrition, water shortage, improper fertilization, excessive growth of stems and leaves, and pests and diseases can also cause blossom and fruit dropping. In addition to strengthening water and fertilizer management, pruning and leaf picking, flower and fruit saving can also be used to promote rapid fruit expansion. 10−15 mg/L of 2,4-D (2, 4-Dichlorophenoxyacetic acid) or 30 mg/L of 4-Chlorophenoxyacetic acid is used to soak flowers or apply on flower stalks every 3−4 d. Colored ink should be added into the prepared solution for marking, so as to avoid repeated marking and causing damage. 2,4-D (2, 4-Dichlorophenoxyacetic acid) will cause harm to tender leaves and growth points, so use with care.

4.4.6 Thinning of blossom and fruit

After fruit setting, thinning of blossom and fruit should be conducted promptly, that is, manual removal of redundant flowers, to ensure nutritional supply, develop into commercially valuable fruits, and avoid uneven fruit size, deformed fruits, and low commodity rate. Large tomato varieties are selected to leave 3−4 good-shaped fruits per flower spike, medium fruit varieties are selected to leave 4−5 fruits per flower spike, and small tomato varieties are selected to leave 8−15 fruits per flower spike.

5 Prevention and control of diseases and pests

Guided by the policy of prevention first and combining integrated prevention and control, the principle is centered by agricultural prevention and control, physical prevention and control, and biological prevention and control, supplemented by chemical prevention and control.

5.1 Agricultural prevention and control

The disease-resistant varieties should be selected to cultivate healthy seedlings, and crop rotation should be implemented; Deep furrow and high bedding cultivation should be adopted; Reasonable dense planting should be conducted; Pruning and supports setting should be done timely; Seriously diseased plants should be pulled out, cleaning the field; The soil should be deeply plowed to reduce the source of diseases and pests; Fertilization should be applied scientifically.

5.2 Physical prevention and control

Yellow and blue sticky card traps and frequency trembler grid lamps can be used to trap and kill imagoes; laying silver gray film or hanging strips of silver gray film in the field can be used to repel aphids; Insect-proof screens can be used in greenhouses to prevent pests from entering, and manually removing pest eggs, catching and killing pests are also considered as useful methods.

5.3 Biological prevention and control

The beneficial organisms should be protected. Biological agents from botanical source, microbial sources, and agricultural antibiotics and so on can be used to prevent and control the pests and diseases.

5.4 Chemical prevention and control

5.4.1 Prevention and control of major diseases

The main diseases of tomato include viral disease, late blight, early blight, bacterial wilt, gray mold, powdery mildew, etc.

5.4.1.1 Viral disease

(1) Symptoms. The main symptoms of the disease caused by the virus in the field are mosaic, fern-ash leaf, streak, clump, leaf roll, chlorosis and yellowing. The disease is more serious in the growth of autumn sown tomatoes.

(2) Prevention and control methods. Before or at early stage of the disease, 5% oligosaccharins AS (1 : 2 000), 1% lentinan AS (1 : 600), 20% moroxydine hydrochloride· copper acetate WP (1 : 1 000), and 5.9% xinjunan acetate moroxydine hydrochloride AS (1 : 300) and other agentia can be used for prevention and control, spraying once every 7–10 d, continuously for 2–3 times.

5.4.1.2 Late blight

(1) Symptoms. It mainly affects the leaves and fruits, as well as the stems and petioles. The lesions mostly start from the leaf tip or leaf edge, and are initially water-soaked chlorosis spots, which gradually expand, and can expand to most of the leaf even the whole leaf. In dry weather, the lesions are dry and brown, without white mold on the leaf back, brittle and easy to crack, and expand slowly. The stem cortex forms brown streaks of varying lengths, and a white mold layer is formed on the surface of the diseased part when the humidity is

high. The fruit is mostly infected during the green fruit period, and the affected parts can be the fruit stalk, sepal, and fruit. The initial stage of the disease is oil-soaked light brown spots, which mostly start from near the fruit stalk and gradually spread, causing the sepal to develop disease, and expanding around the fruit as cloudy irregular lesions, with no obvious boundaries at the edge of the lesions. The surface of the diseased fruit is rough, and the fruit is hard in texture. After expansion of disease, the lesions are dark brown, and a sparse white mold layer grows at the edge of the lesions when the humidity is high.

(2) Prevention and control methods. In the early stage of petiole and stem disease, 72% cymoxanil mancozeb WP (1 : 600), 68% metalaxyl-mmancozeb DG (1 : 600) or 75% chlorothalonil WP (1 : 500) can be used for prevention and control, spraying once every 5−7 d, continuowly for 2−3 times.

5.4.1.3 Early blight

(1) Symptoms. It mainly affects leaves, stems, and fruits. The leaves are dark brown or black at the initial stage, with small spots shaping circular to oval, which gradually expand into lesions with a diameter of 1−2 cm. The edges of the lesions are dark brown, and the center is grayish-brown, with obvious concentric rings. Some yellow haloes can be seen at the edges. When the humidity is high, the surface of the lesions is covered with a black mold layer. The disease often occurs in the lower leaves of the plant, and gradually spreads upward. When in severe condition, the lesions connect with each other to form irregular large lesions, and the lower leaves of the diseased plant will die and fall off. The lesions on the stem occur mostly on the branches of the stem. The lesions are grayish-brown, oval, slightly sunken, with concentric rings. When the disease is serious, the branches are broken. The lesions on the fruit occur mostly near the stalk and cracks, round or nearly round, dark brown, slightly sunken, with concentric rings. When severely infected, the diseased fruit often falls off early. Under wet conditions, black mold can grow on all affected parts.

(2) Prevention and control methods. For greenhouses with annual disease, after sealing and before final planting, 45% chlorothalonil or 11−13 g of procymidone fumigant for fumigation can be used per mu. For field culture, 72% cymoxanil mancozeb WP (1 : 600), 58% metalaxyl mancozeb WP (1 : 800), 10% difenoconazole WDG (1 : 1 000), 70% thiophanate-methyl WP (1 : 700), 50% iprodinone WP (1 : 1 000) can be used for spraying once every 7−10 d, and pay attention to the alternate use of pesticides.

5.4.1.4 Bacterial wilt

(1) Symptoms. It is a bacterial vascular bundle tissue disease. Initially, the symptoms appear on the top young leaves, especially wilt and sag at noon, and then return to normal after sunset. Then, the symptoms quickly spread to the whole plant causing wilt, which does

not recover and dies. When the diseased stem is cut open, the vascular bundle turns brown, and milky mucus exudes when squeezed by hand after crosscutting.

(2) Prevention and control methods. At present, there is no ideal agent for tomato bacterial wilt, and comprehensive measures should be taken in the prevention and treatment. When the early fruiting stage begins, field inspection should be strengthened. Once diseased plants are found, they should be pulled out promptly, collected and burned. with 77% cooper hydroxide SC (1 : 800) or 50% copper succinate WP (1 : 400) The roots should be irrigated once every 10 d, continuously for 3-4 times, with an interval of 7-10 d, and the roots should be fully irrigated (200-500 mL/plant).

5.4.1.5 Gray mold

(1) Symptoms. During the entire growth and development period, each part of the plant can be infected. Tomato gray mold mainly causes leaf and fruit rot, usually starting from the edge of the weaker cotyledon and true leaf, the leaf becomes soft and sagging, and a large number of gray molds are produced at the diseased part, and finally the diseased plant falls off. In severe cases, the young seedlings in the field rot in patches; In the adult stage, it can affect various parts above the ground. Tomato leaf infection mostly starts from the leaf tip and leaf edge, initially water-soaked, and then the color becomes faint, light brown, slight dark and light rings. The leaf lesions are mostly "V"-shaped, expanded into irregular or round rings, with obvious edges, the leaf surface produces gray mold, and sometimes the lesions break; The lesions are often not limited by the veins to continue to expand to the whole leaf, resulting in drying and death of the leaf. The stem infection initially appears as small water-soaked flecks, and then the lesions expand, and gray molds are produced on the lesions when the humidity is high, and serious cases lead to plant death. The pathogen mostly infects from the petals or stigmas, resulting in flower in rot, growing light grayish-brown molds, and causing flower dropping. Fruit damage can result in rotting or circular spots with white edges and green centers.

(2) Prevention and control methods. When dipping (or spraying) on tomato flowers, agents can be added for the prevention and treatment of gray mold, such as adding 50% iprodione WP, 50% procymidone WP, and 50% carbendazim WP diluent to the prepared 4-Chlorophenoxyacetic acid or 2,4-D (2,4-Dichlorophenoxyacetic acid) diluent, and then dipping (spraying) flowers. In the initial stage of disease, 50% procymidone WP (1 : 800), 40% pyrimethanil·boscalid SE (1 : 2 000), 65% metalaxyl-hymexazol WP (1 : 800), 25% procymidone·thiram WP (1 : 800) and other agents can be used for spraying for 2-3 times, with 7-10 d as the interval period.

5.4.1.6 Powdery mildew

(1) Symptoms. Tomato leaves, petioles, stems, and fruits can be infected with the disease, usually starting from the lower leaves and gradually developing upward. In the early stages of the disease, small chlorosis spots appear on the leaf surface, which expand into nearly round or irregular lesions with white powdery substances on the surface. At first, the white powder layer is relatively sparse, then gradually thickens and expands around. In severe cases, the entire leaf is covered with white powder, and the chlorosis leaf tissue can be seen after wiping away the white powder. Finally, the diseased leaves turn yellowish-brown and gradually wither and die. White powdery lesions can also occur in other parts of the plant when infected with the disease.

(2) Prevention and control method. With reasonable dense planting, the old leaves on the lower part of the plant should be removed in time. In the early stage of the disease, 40% flusilazole EC (1 : 6 000), 15% triadimefon WP (1 : 500), and 70% thiophanate-methyl WP (1 : 600) can be used for rotational spraying, continuously for 2–3 times, with an interval of 7–10 d.

5.4.2 Prevention and control of pests

The main pests are cotton bollworm, whitefly, aphid, thrip, etc.

5.4.2.1 Cotton bollworm

(1) Damage characteristics. The main symptom is that the larvae eat buds, flowers, fruits, also tender stems and leaves. After the bud is damaged, the bracteal leaves open and turn yellowish-green, and fall off 2–3 d later. The buds and young fruits are often eaten empty, causing rotting and falling off. Damage during the fruiting period can lead to fruit dropping, resulting in yield reduction.

(2) Prevention and control methods. For chemical control, during the peak period of larvae hatching, 2% emamectin benzoate EC (1 : 2 000), 32 000 IU/mg *Bacillus thuringiensis*(Bt) WP (1 : 300), 60 billion PIB/g HaNPV WDG for 3 g/mu, and 14% chlorantraniliprole·beta cypermethrin microcapsule SC [1 : (2 000–3 000)] can be used for spraying, once every 3–5 d, continuously for 2–3 times.

5.4.2.2 Whitefly

(1) Damage characteristics. Whitefly is an omnivorous pest with a high reproductive rate, and its population reaches its peak in autumn. Imagoes and nymphs of whitefly gather on the back of leaves and suck sap with piercing-sucking mouthparts, causing the leaves to fade and turn yellow, slowing the growth of plants, withering, and even dying. Whitefly can also secrete a large amount of honeydew, causing sooty blotch. It can also spread viral diseases.

(2) Prevention and control methods. In the initial stage, 25% imidacloprid WP (1 : 3 000), 70% acetamiprid WDG (1 : 5 000), 25 g/L bifenthrin EC (1 : 2 000), 1.8% avermectin EC (1 : 2 000), 25% buprofezin EC (1 : 2 500), and 20% fenpropathrin EC (1 : 2 000) can be used for spraying once every 3−5 d, continuously for 2−3 times.

5.4.2.3 Aphid

(1) Damage characteristics. Imagoes or nymphs of aphid gather on the back of leaves, tender leaves, tender stems, buds and near-ground leaves, sucking plant sap and nutrients with piercing-sucking mouthparts and secreting honeydew, often causing severe dehydration and malnutrition in plants. When young leaves are infected, they curl and wrinkle, with light cases losing green, having spots, and yellowing; When in severe cases, leaves are curled, deformed, withered, and even die. Aphids can also spread viral disease.

(2) Prevention and control methods. In the initial stage, 25% imidacloprid WP (1 : 3 000), 20% fenvalerate EC (1 : 2 000), 20% fenpropathrin EC (1 : 1 500), 28% avermectin·spirotetramat SC (1 : 3 000), and 14% chlorantraniliprole·beta cypermethrin microcapsule SC (1 : 4 000) can be used for spraying, once every 3−5 d, continuously for 2−3 times.

5.4.2.4 Thrip

(1) Damage characteristics. Imagoes or nymphs of thrip suck sap from the dorsal hair tuft of young leaves and flowers. The downy hairs and mesophyll on the back of the affected leaves turn grayish-brown, harden and age, and the buds are injured and shed. The affected plants grow slowly, and the internodes shorten. Thrips reproduce extremely quickly, with over 10−20 generations per year, generations overlapping and reproducing year-round. The peak period is from May to September, with autumn being the most severe. Imagoes are active and fly well, afraid of light, and nymphs fall into the topsoil and pupate.

(2) Prevention and control methods. 25% imidacloprid WP (1 : 3 000), 25% thiamethoxam SC (1 : 3 000), and 28% avermectin·spirotetramat SC (1 : 4 000) can be used for spraying, once every 3−5 d, continuously for 3 times. The spraying focus is on the growth point, the back of the tender leaves, the bud and other parts of the plant.

6 Timely harvesting

From flowering to fruit ripening, early-ripening varieties need 40−50 d, medium and late ripening varieties need 50−60 d. Timely harvesting and fruit storage for accelerating ripening should be conducted as needed.

Tomato harvesting time should be determined according to the tomato product sales market. Tomatoes have four ripe stages: green-ripening, color-changing, ripening, and full ripening. For long-distance transportation, fruits should be picked at the green-ripening stage (when the top and surface of the fruit turn white); For short-distance transportation, fruits can be picked at the color-changing stage (when 1/3 of the fruit turns red); For local sale or self-consumption, fruits should be picked at the ripening stage (when more than 1/3 of the fruit turns red). It is best not to twist the stalk during harvesting. Tomato scissors are used to gently cut along the root of the fruit stalk, without exposing above the fruit surface. Fruits should be picked and placed gently to avoid mechanical injury. After harvesting, the fruit can be graded for packaging, storage, or transportation and sales.

References

Institute of Vegetables and Flowers, Chinese Academy of Agricultural Sciences, 2010. Olericulture in China [M]. 2nd Edition. Beijing: China Agriculture Press.

LIN J R, LIU S Y, QIU M Y, et al., 2004. Colored Illustrated Book of Pollution-Free Production of Tomato and Eggplant [M]. Guangzhou: Guangdong Science and Technology Press.

National Agricultural Technology Extension and Service Center, National Major Vegetable Industry Technology System, 2017. Colored Drawing Spectrum of Efficient Cultivation and Disease and Pest Control for Tomato [M]. Beijing: China Agriculture Press.

❖ Key Techniques for Pepper Cultivation

Pepper (*Capsicum* spp.), also known as Fanjiao, Haijiao, and Lazi, is an annual or perennial plant of the *Capsicum* genus in the Solanaceae family, which is one of the most widely cultivated and consumed vegetables and spice crops around the world. Originating from the tropical and subtropical regions of Central and South America, it is one of the oldest crops cultivated by humans. As early as 7 000 BC, there were records of planting peppers in South America. It was introduced to Europe in the 16th century and introduced to China from Europe in the 17th century.

At present, there are 5 species of *Capsicum* genus domesticated and cultivated, including *C. annuum*, *C. chinense*, *C. baccatum*, *C. frutescens*, and *C. pubescens*. Among them, *C. annuum* and *C. frutescens* are widely cultivated worldwide.

According to the statistics of the National Staple Vegetable Industry Technology System, in recent years, China's annual planting area of pepper has remained stable at more than 32 million mu, accounting for 9.28% of the vegetable cultivation area, becoming the largest vegetable cultivated in China and an important vegetable economic crop in China. Not only is pepper used as a vegetable and condiment, but also as an important industrial raw material for industrial, medical, cosmetic, military, and navigation fields due to its rich bioactive substances such as capsaicin, capsorubin, vitamin C, and vitamin E.

1 Growth conditions

1.1 Temperature

The optimal temperature for the germination of pepper seeds is 25–30 °C, the growth and development temperature is 20–30 °C, and the fruit setting temperature is 20–25 °C; The fruit setting rate decreases when the temperature is below 15 °C or above 35 °C; The growth stops when the temperature is below 10 °C; The plant is vulnerable to cool damage when the temperature is below 5 °C; and the plant is vulnerable to freezing damage when the temperature is below 0 °C.

1.2 Light

Pepper is a thermophilic and photophilic crop, with a light compensation point of 1 500 lx and a light saturation point of 30 000 lx, but it is not sensitive to photoperiod. Under short sunshine conditions, pepper blooms and bears fruit quickly, so sowing in spring can promote early ripening.

1.3 Moisture

Pepper is a drought-tolerant crop among solanaceous vegetables, and should avoid waterlogging, otherwise it is easy to cause diseases and lead to large-scale dead plants; Pepper is sensitive to water demand during the seedling stage, and the water demand increases during the fruit setting stage. When soil moisture is 60%−80% of field moisture capacity, the growth is excellent and the fruit setting rate is high.

1.4 Soil

Pepper has strong adaptability to soil, and sandy loam soil with deep layer, loose, good fertility and water conditions, and pH value of 6.2−7.2 is suitable for pepper planting. Pepper root system is generally distributed in the soil layer of 20−40 cm deep, and loose, fertile, well-drained soil is conducive to the development of pepper root system. Plots with high groundwater level, poor drainage conditions, and sticky soil are not suitable for pepper planting.

1.5 Fertilizer

Pepper has a high demand for nitrogen, phosphorus, and potassium fertilizers. During the entire growth and development period, nitrogen is the most needed, accounting for 60%, followed by potassium at 25%, and phosphorus at 15%. At the same time, pepper also needs a variety of trace elements such as calcium, magnesium, iron, boron, molybdenum, and zinc. For pepper cultivated in Yunnan, adding boron and molybdenum fertilizers to the crop has an ideal effect on increasing yield and improving fruit quality. In addition, nitrate nitrogen fertilizer is more conducive to the absorption and utilization of pepper, and the appropriate ratio of nitrate nitrogen to ammonium nitrogen is 7 : 3.

2 Types and cultivars

2.1 Types

There are various types of peppers, which can be divided into three main categories according to consumption habits: fresh-eating type, processed type, and both fresh-eating and processed type. Fresh-eating peppers include wrinkled pepper, sweet peppers, cayenne peppers, and screw-shaping peppers; Processed peppers include dried pepper, beauty pepper, and industrial pepper; Line pepper, pod pepper, and bird eye's chili can be used both for fresh-eating and for processing.

2.2 Cultivars

The main production area of fresh-eating pepper is mainly Baoshan, with a wide range of cultivars, mostly guided by market demand; Pod pepper is distributed throughout Yunnan, with Wenshan Prefecture, Honghe Prefecture, Pu'er City, Xishuangbanna Prefecture, Lincang City and other places as the main production areas, mainly planting single-birth pod pepper varieties.

The dried pepper represented by Qiubei pepper and Leye pepper is one of the characteristic varieties in Yunnan Province. Qiubei pepper is well-known at home and abroad for its outstanding quality, which is distributed in Wenshan Prefecture with Qiubei County as the center. The main varieties include Yun dried pepper series, Qiu pepper series and Wen dried pepper series. Leye pepper has a long history, with a unique aroma and taste, mainly cultivated in Leye Town, Huize County; Yunnan bird eye's chili, as another characteristic variety, is well-known for its processed products such as chicken feet with pickled peppers and instant noodles with pickled peppers, which is cultivated in hot areas of Yunnan Province. Honghe Prefecture, Wenshan Prefecture and Pu'er City are the main production areas of bird eye's chili.

In recent years, the industrial pepper planting and the corresponding deep processing industry have sprung up, occupying a place in the pepper industry of Yunnan Province. At present, the variety of industrial pepper is relatively single, and the disease resistance and yield of varieties need to be further improved.

3 Cultivation and management techniques

3.1 Seed processing

3.1.1 Seed sterilization and disinfection techniques

Before sowing pepper seeds, sterilization and disinfection treatment are conducted to kill harmful microorganisms and pathogenic bacteria carried by the seeds, reduce seedling stage diseases, and help cultivate strong seedlings. The operation method is as follows. Step 1: the seeds are soaked in warm water at 60 °C for 15 min with constant stirring; step 2: the seeds are soaked in 1% copper sulfate or 50% carbendazim WP (1 : 500) for 1 h; step 3: the seeds are soaked in 10% trisodium phosphate solution for 1 h; step 4: the seeds are rinsed with clean water for 3–5 times, and then removed and dried for sowing and seedling raising.

3.1.2 Seed treatment techniques of moisture–absorbing and drying

The seeds treated by the above mentioned methods can be further treated by moisture absorption and drying techniques, which can enhance vitality, improve germination rate and germination potential to make the emergence of seedlings fast and uniform, and elevate drought resistance of young seedlings, transplantation survival rate and short-term drought resistance.

Treatment methods: ① Moisture absorption: the above sterilized seeds are soaked in water at 25–30 °C for about 6 h. ② Drying: the seeds should be dried until they reach their pre-disinfected weight, or grasping the seeds with hands and gently loosening, seeds will slip through fingers, which is the standard for drying. ③ After 3–5 d, the seeds should be repeated the moisture-absorbing and drying process, and then drying for sowing. It is necessary to note that the treatment should be carried out on the seeds 7–10 d before sowing.

3.2 Seedling raising techniques

3.2.1 Main methods of seedling raising

The current production of pepper mainly adopts plug-seedling, which has the advantages of using a small amount of substrate, developed seedling root system, not easy to damage the roots during transplanting, short or even almost no seedling recovery period.

3.2.2 Plug-seedling techniques

3.2.2.1 Seedling raising substrate and disinfection

Seedling raising substrate is very crucial for the cultivation of strong seedlings, which can be directly purchased commercial seedling substrate or prepare homemade seedling substrate by composting farmyard manure. The proportion of farmyard manure composting for homemade seedling substrate: 2 t of farmyard manure or humus soil+300 kg of chicken manure or sheep manure+20 kg of compound fertilizer (N : P : K=15 : 15 : 15) +10 kg of urea+50 kg of calcium superphosphate.

3.2.2.2 Sowing

The substrate should be loaded into the plug and then pressed moderately, a small hole with 1 cm deep is dug in the center of the substrate, and the pepper seeds are placed in this hole. After sowing, the seeds should be covered with 1 cm thick fine soil, not too thin, not too thick. If the covering soil is too thin, the seeds will be easily exposed after watering or capped seedlings will emerge, while if the covering soil is too thick, it will affect the emergence of seedlings, and then pouring enough water. According to the temperature of the sowing season, plastic thin film or shading net can be chosen to cover the plug, keeping it warm and moisture in favor of emergence of the seedlings. The plug must be placed 50 cm above the ground to avoid soil borne disease. Do not put the plug directly on the cement floor or plastic thin film, so as not to burn the root system.

3.2.2.3 Management before emergence

The temperature should be kept at 25−30 °C before emergence, and seedlings can emerge in 5−9 d. When about 70% of the seeds have emerged, the mulch should be removed in time, and watering should be conducted appropriately to keep the covered soil moist. After seedling emergence, the temperature should be kept at 20−23 °C during the day and no less than 15 °C at night by taking measures such as removing the film and ventilating.

3.2.2.4 Management after emergence

The temperature should be kept at 27−28 °C during the day and 18−20 °C at night, with soil relative moisture of 70%−80% and air relative humidity of 50%−70%. Water-soluble formula fertilizer should be applied every 3−5 d according to the growth of the seedlings. Watering should be conducted around 10:00 on sunny days, with sufficient watering at once to avoid watering too little and frequently. 0.2%−0.3% potassium dihydrogen phosphate can also be sprayed twice during the seedling stage.

One week before final planting, seedlings should be hardened off by gradually ventilating and cooling the temperature, controlling water and fertilizer, and the lowest night temperature can be reduced to about 10 ℃. The day before final planting, seedlings should be thoroughly watered.

3.3　Site preparation and cultivation pattern

3.3.1　Site selection and preparation

It is recommended to choose a plot with convenient drainage and irrigation, without solanaceous previous crops, rich organic matter, and a deep soil layer. After applying 3 000–4 000 kg of organic fertilizer per mu, the soil should be plowed about 30 cm, leveled and raked.

3.3.2　Cultivation pattern

Ridge cultivation is widely used in production. The ridge should be opened a 1.2 m ditch with 30 cm deep, 2 rows per ditch, 50 cm row spacing, 40 cm hole spacing, and cultivated in a wide and narrow row of 50 cm–70 cm–50 cm.

In arid and lack of rain areas, cultivation can be conducted by adopting the drought-mulching rain-harvesting cultivation technology. The specific operation method is: a hole pond should be dug and covered with mulch film. A sharp wooden rod with a diameter of 5–6 cm is used to perforate in the center of the hole pond to collect precipitation, and the porthole is covered with a small amount of loose soil. After the temperature rises in late April, the final planting can be conducted when there is a small amount of rain. The two dried pepper plants can be planted into each hole, which should be closed together and not separated, while other peppers should be planted individually.

3.4　Final planting and growing period management

3.4.1　Final planting

The best final planting time is when the seedlings are 10–15 cm high and have 7–8 true leaves. After final planting, it is advisable to sprinkle enough root water and cover the mulch with fine and dry soil around the seedlings, which can prevent the mulch from scalding the pepper seedlings and maintain and improve the earth temperature inside the much.

3.4.2　Growing period management

3.4.2.1　Pre-growth management

It includes the management from post-final planting to pre-harvest. After final planting, the soil temperature should be raised to promote the seedling recovery as soon as possible. For the seedlings are cultured by substrate, the seedling recovery period is generally only 2–3 d. After a week of seedling recovery after final planting, a seedling fertilizer should be applied, and 6–8 kg of 0.5% ammonium nitrate or ammonium nitrophosphate solution should be applied to per mu. Then, the seedlings should be hardened for 10–15 d to promote root system growth by controlling water and fertilizer to improve the soil temperature and make the soil ventilated.

After hardening of seedings, a large amount of water should be irrigated combined with fertilization, and 15 kg ammonium nitrate should be applied to each mu. At the same time, 0.3% potassium dihydrogen phosphate or amino acid fertilizer should be sprayed on the leaves for 1–2 times. Pepper is sensitive to boron and zinc, so 0.1% borax and 0.05% zinc sulfate trace element fertilizer can be sprayed on the leaves.

3.4.2.2 Mid-growth management

It includes the management from the beginning of harvest to full bearing period. The main goal is to promote the seedling and fruit growth and development, strive for the early arrival of full bearing period, and conduct closing of crop before the arrival of the high temperature season. Specific measures are as follows. ① Harvest should be conducted timely, once every 4–6 d during the full bearing period. ② Fertilizer and water management should be strengthened, topdressing and irrigation should be done after harvesting 1–2 times. Each mu should be applied with 10 kg ammonium nitrate+15 kg compound fertilizer, and leaf spraying should be performed every 10 d or so.

3.4.2.3 Post-growth management

The key point of this stage management is to promote the plant to grow new branches and reach the second fruiting peak. Specific measures are as follows. ① Pruning should be carried out to clean up diseased and damaged plants. Strong branches with branching ability and branches that have been extracted from the lower part should be retained, and old, weak branches that have lost branching ability should be cut off, and old and diseased leaves should be removed. ② Fertilizer and water management. Generally, readily available fertilizer is applied every 10–15 d, and 10–15 kg ammonium nitrate is applied to each mu to promote the growth of new branches. ③ Watering and drainage. In the hot and rainy season, it is advisable to keep the soil moist and water in time to cool down; The ponding water in the ditch should be drained in time after rain to prevent macerating roots.

4 Prevention and control of diseases and pests

Guided by the green policy of prevention first and combining prevention and control, the prevention and control principle is centered on agricultural prevention and control, physical prevention and control, and biological prevention and control, supplemented by chemical prevention and control.

4.1 Prevention and control of diseases

4.1.1 Damping-off

4.1.1.1 Main symptoms

Infected before emergence causes rotten seeds and buds; Infected after emergence causes yellowish-green water-stained lesions at the base of the stem, which soon turn yellowish-brown and develop around the stem. The diseased tissues rot and dry up, causing concavity and constriction. The water-stained lesions expand from bottom to top, and the seedlings fall to the ground. In the early stage of the disease, only a few young seedlings in the seedbed are affected. After a few days, the disease gradually expands and spreads outward from the diseased seedling as the center, which finally causes the seedlings to fall down and die in a large area.

4.1.1.2 Pathogenic conditions

The pathogenic bacteria are mainly transmitted through irrigation water or rainwater, and the disease is severe under high humidity conditions. During the seedling stage of peppers, if there is continuous rain, insufficient light, and weak growth potential of young seedlings, the disease is prone to occur.

4.1.1.3 Prevention and control methods

(1) Agricultural prevention and control. The stable manure or composing must be fermented at high temperatures to fully decompose; the ventilation is enhanced to prevent the seedbed soil from being too wet.

(2) Chemical prevention and control. Seedbed disinfection should be strictly carried out. In the early stage of disease, hymexazol, metalaxyl-M·hymexazol, and cymoxanil·mancozeb can be used for spraying. After spraying, the ash should be applied until the leaves are dry.

4.1.2 Wilt disease

4.1.2.1 Main symptoms

It usually affects peppers after the emergence of true leaves and before flowering and fruiting. The young seedlings wilt during the day and recover at night, and after several days, the plants wither and die. The base of the stem produces oval, dark brown lesions that slightly sag and expand around the base of the stem. The diseased parts shrink and dry up, the leaves turn yellow and wither, and the roots turn brown and rot, until the whole plant dies. When the humidity is high, the diseased parts produce sparse brown spider-web-shaped mold.

4.1.2.2 Pathogenic conditions

Too dense sowing, not thinning out seedlings in time, high humidity, and temperatures between 16 °C and 24 °C are prone to disease.

4.1.2.3 Prevention and control methods

(1) Agricultural prevention and control. The disease-free soil or substrate should be selected for seedling raising, and the application of phosphorus and potassium fertilizers should be increased to prevent the soil from being either dry or wet.

(2) Chemical prevention and control. Chlorothalonil or Fenaminosulf can be sprayed for prevention before the onset of disease; Ash or dried fine soil mixed with hymexazol and carbendazim·thiram agent can be sprinkled after the onset of disease; 38% metalaxyl·thiram WP (1 : 800), 30% hymexazol AS (1 : 1 500), 50% iprodione WP 2–4 g/m^2 or 30% metalaxyl·hymexazol AS (1 : 2 000) can be used for pouring and irrigation root.

4.1.3 Blight

4.1.3.1 Main symptoms

The diseased part of stem and branch begin with dark green water-stained lesions, then turn brown necrotic stripes, sag and shrink, and the upper part of the plant wilts and dies. The damaged leaf produces dark green water-stained circular or nearly circular lesions, with a diameter of 2–3 cm; The whole leaf rots when the humidity is high, while the lesions turn light brown and the diseased leaves fall off easily in dry conditions. The fruit damage begins at the stalk, producing dark green water-stained lesions, turning brown and soft rot when the humidity is high, growing white sparse mold layer on the surface, and forming stiff fruits remaining on the branches when they are dry. The root damage causes the leaves to turn brown and rot, and the whole plant wilts and dies, but the vascular bundle does not change color, which is different from the fusarium wilt caused by Fusarium.

4.1.3.2 Pathogenic conditions

Continuous cropping, low marsh, poor drainage, excessive use of nitrogen fertilizer, excessive density, and weak plants are all conducive to the occurrence and spread of the blight. The suitable temperature for the occurrence of the disease is 20–30 °C, and the soil relative humidity is above 95%. It takes 2–3 d to complete an infection cycle, so large irrigation or heavy rainstorms and high temperatures are easy to cause outbreaks.

4.1.3.3 Prevention and control methods

(1) Agricultural prevention and control. Paddy-upland rotation should be adopted;

Disease-resistant varieties should be selected; The application of nitrogen fertilizer should be appropriately controlled, phosphorus and potassium fertilizer should be increased, and composted farmyard manure should be applied; Deep furrows and high bed cultivation should be adopted to avoid pouring after a long drought, and strictly prohibit flood irrigation.

(2) Chemical prevention and control. Seed treatment: After the seeds have absorbed enough water, they can be disinfected by immersing them in 69% dimethomorph·mancozeb WP (1 : 1 000) for 5 min. In the initial stage of disease, 58% metalaxyl·mancozeb WP (1 : 600), 30% metalaxyl·hymexazol AS [1 : (600—800)], and 50% dimethomorph WP (1 : 1 500) and other agents can be used for root irrigation or rhizome spraying, with each plant root irrigated by 200—250 mL, continuously for 2—3 times, with an interval of 7—10 d.

4.1.4　Anthracnose

4.1.4.1　Main symptoms

Anthracnose mainly affects fruits, leaves, even stems. Infected leaves appear as water-soaked faded green lesions at first, and then gradually turn brown. The lesions are nearly round, with grayish-white centers and small black dots forming on the edges. After enlargement, the lesions become irregular with concentric rings, and the leaves are easy to fall off. Infected fruits appear as water-soaked yellowish-brown lesions at first, which then become oblong or irregular. The lesions are sunken with concentric rings. The edges are reddish-brown, and the centers are grayish-brown, with black dots. When in wet condition, red sticky substances are produced on the lesions, while lesions become filmy and easy to break when in dry condition.

4.1.4.2　Pathogenic conditions

During the peak season of fruit bearing, when the weather is highly humid, morning dew is heavy, or exposure to the sun after a heavy rain, which cause severe sunburn, the disease is prone to occur.

4.1.4.3　Prevention and control methods

(1) Agricultural prevention and control. Cultivation management should be strengthened, diseased residues should be removed, and diseased fruits should be picked and destroyed out of the field immediately. Planting density should be reasonable, so that the rows are not excessively overshadowed and the fruits are not exposed, reducing the risk of sunburn.

(2) Seed disinfection. Anthracnose is mainly caused by seed carrying bacteria, and seed disinfection treatment has an ideal preventive effect. Seeds can be soaked in warm water at 55 ℃ for 30 min, then transferred to cold water to cool, and dried before sowing; Seeds can be soaked in cold water for 10—12 h, then soaked in 1% cupric sulfate for 5 min. After washing,

germination-breaking can be done for sowing; 50% carbendazim WP (1 : 500) can be used for soaking seeds for 1 h. After washing, germination-breaking can be done for sowing.

(3) Chemical spraying prevention and control. In the initial stage of disease, 25% prochloraz EC (1 : 1 500), 10% difenoconazole WDG (1 : 800), 68% metalaxyl-M·mancozeb WDG (1 : 500) or 75% trifloxystrobin·tebuconazole WDG [1 : (4 000−6 000)] can be used for spraying, continuously for 2−3 times, with an interval of 7−10 d.

4.1.5 Powdery mildew

4.1.5.1 Main symptoms

It mainly infects leaves, and when the disease is severe, it can also infect the stem and stem. At the initial stage of the disease, white round molds with powdery spots are mainly formed on the leaves' surface or back, starting from the lower leaves and developing upward. In severe cases, there will be a layer of white mold on the leaves surface. In the later stage of the disease, the white mold on the leaves surface turns grayish-brown, and the leaves turn yellow and necrotic.

4.1.5.2 Pathogenic conditions

Warm environment, overcast weather, dense planting, and unventilated environment are prone to disease, and it is easy to spread under 25−28 °C and slightly dry conditions. Flood irrigation, high humidity and insufficient fertilizer can cause severe disease in the late growth stage of plants.

4.1.5.3 Prevention and control methods

(1) Agricultural prevention and control. Cultivation management should be strengthened to improve the disease resistance of pepper. When planting in greenhouses, pay attention to controlling the temperature and humidity to prevent the greenhouse from being too dry or too low in humidity.

(2) Chemical prevention and control. In the initial stage of the disease, 12% difenoconazole fluxapyroxad SC (1 : 1 000), 20% prochloraz EC [1 : (1 000−1 200)], 30% picoxystrobin·tebuconazole SC (1 : 2 000) or 10% difenoconazole WDG (1 : 1 000) can be used for spraying, continuously for 2−3 times, with an interval of 7−10 d.

4.1.6 Scab disease

4.1.6.1 Main symptoms

It mainly infects the leaves. At the early stage, water-soaked lesions are lumped on the back of the leaves, then expand to irregular shapes with slightly raised and dark brown periphery, light color and slightly sunken interior and rough scabby surface. The lesions can integrate and connect together to form larger lesions, causing the leaves to drop. When the fruit is infected, white dots can be seen on the surface of the fruit. The typical symptoms of diagnosing pepper scab disease are raised lesions with ring patterns.

4.1.6.2 Pathogenic conditions

Long-term dew and heavy rainfall will cause severe disease.

4.1.6.3 Prevention and control methods

(1) Agricultural prevention and control. The disease-resistant varieties should be selected, rotating with non-Solanaceae crops; drainage should be conducted timely on rainy days.

(2) Chemical prevention and control. The seeds can be soaked in 3% zhongshengmycin WP (1∶1 000) for 30 min to prevent the disease. In the initial stage of the disease, prevention and treatment can be conducted by spraying 30% copper (succinate+glutarate+adipate) WP (1∶800), 2% kasugamycin AS (1∶800) or 20% thiodiazole-copper SC (1∶700), continuously for 2−3 times, with an interval of 7−10 d.

4.1.7 Viral disease

4.1.7.1 Main symptoms

Common symptoms of pepper viral disease include the following 4 types.

(1) Mosaic. Diseased leaves appear obvious yellow-green mottled, wrinkled, or produce brown necrotic lesions.

(2) Leaf deformity or clustering. At the early stage, the veins of the tender little leaves that grow from the top of a plant retreat green, gradually forming uneven mottled shades, leaf surface wrinkling, and later thickened leaves, producing yellow-green mottled or large yellow-brown necrotic lesions, with leaf margins curling upward. Young leaves are narrow and linear in severe cases, and the upper internodes of the plant are short and clustered. Severely diseased fruits have uneven green mottled speckle and verrucous protrusions.

(3) Stripes. The main veins of the leaf are brown or black with necrosis, spreading along the petiole to the lateral branches and main stem, producing systematic necrotic stripes, often causing early leaf, flower, and fruit dropping, and severe cases of whole plant death.

(4) Top blight viral disease. The plant is dwarf, yellowing, unfruitfulness, or the fruit is small and stiff, and does not grow.

4.1.7.2 Pathogenic conditions

High temperature and dry weather, the occurrence of virus transmission insects (aphids and thrips) have seriously promoted the prevalence of diseases.

4.1.7.3 Prevention and control methods

(1) Agricultural prevention and control. Disease-resistant varieties should be selected,

isolation and pest control should be adopted for seedings, and wounds caused by farming operations should be reduced.

(2) Chemical prevention and control. The seeds can be soaked in clear water for 3-4 h, then conducted seed disinfection by putting them into 10% trisodium phosphate solution for 40-50 min; Virus transmission pests such as aphids, thrips should be prevented and controlled; 2% oligosaccharides AS (1∶300), 20% moroxydine hydrochloride WP (1∶500) or 20% ningnanmycin WP (1∶1 000) can be used for preventing and inhibiting the occurrence of disease, continuously for 2-3 times, with an interval of 5-7 d.

4.2 Prevention and control of major pests

4.2.1 Thrips

4.2.1.1 Damage characteristics

The thrips that harm pepper are mainly divided into two species: the flower thrips and the western flower thrips. The imagoes and nymphs usually gather in the flowers to feed and cause damage. The floral organ and petals become albino after being injured, and turn dark brown after being exposed to the sun. In severe cases, the flowers wilt. The leaves appear silvery white streaks after being infected, and become scorched and shriveled in severe cases. The imagoes overwinter in the forest floor and mantle of soil.

4.2.1.2 Prevention and control methods

(1) Agricultural and physical prevention and control. High temperature stuffy shed are adopted during summer fallow; insect-proof nets are used to block pests; blue sticky card traps are used to lure and kill pests.

(2) Chemical prevention and control. At the early stage, imidacloprid, spinosad, spinetoram, and cyantraniliprole can be used for spraying. Adding silicone as a penetrant will be more effective. The spraying should focus on the upper part of the plant, the back of the tender leaves, and the tender stems.

4.2.2 Spider mites

4.2.2.1 Damage characteristics

The spider mites that harm peppers include yellow tea mite, carmine spider mite (red spider), and two-spotted spider mite, etc. Imago and larva mites concentrate on the tender parts of the plant for piercing sucking. The back of the affected leaves is grayish-brown or yellowish-brown, oil-stained, and the edges of the leaves are curled downward; The tender stems and branches of the affected plants turn yellowish-brown and twist, and the tops of the plants dry up when in severe condition; The affected pericarp turns yellowish-brown.

4.2.2.2 Prevention and control methods

(1) Agricultural and biological prevention and control. The cultivation management should be strengthened, weeds around the plot or greenhouse should be cleaned up, predatory mites should be released to control pest mites.

(2) Chemical prevention and control. In the early stage, avermectin, spirodiclofen, chlorfenapyr, bifenazate can be used alternatively for spraying. The young and tender parts should be the focus of spraying, turning downwards while spraying upwards.

4.2.3 Aphid

4.2.3.1 Damage characteristics

The aphids that harm peppers mainly include green peach aphid and Cotton aphid. They live in groups on the leaf back, pedicel or tender stems, sucking plant sap and secreting honeydew. Affected leaves turn yellow and their surfaces become wrinkled and curled. The tender stems and pedicel become bent and deformed, affecting flowering and bearing, inhibiting plant growth, and even withering and dying. Aphids can also spread viral disease.

4.2.3.2 Prevention and control methods

(1) Agricultural and physical prevention and control. It is recommended to select varieties with resistance by interplanting and relay cropping, and strengthen field management; Yellow sticky card traps are used to lure and kill alatae, and silvery gray film is used to repel aphids.

(2) Biological prevention and control. Natural enemies should be protected and released, and Entomophthora aphidis should be sprayed.

(3) Chemical prevention and control. In the early stage of aphid occurrence, acetamiprid, pymetrozine, thiamethoxam, imidacloprid, and dinotefuran can be used for spraying, and smoke agent such as isoprocarb can be used for fumigating in greenhouse.

4.2.4 Tobacco whitefly

4.2.4.1 Damage characteristics

Tobacco whitefly is one of the main pests in tropical and subtropical areas, and is recognized as a super pest in the industry. It secretes honeydew on plants, causing sooty blotch, which seriously affects leaf photosynthesis.

4.2.4.2 Prevention and control methods

(1) Agricultural and physical prevention and control. Continuous cropping should be avoided; The prevention and control can be conducted by using 40-mesh insect-proof screen

in the greenhouse; Yellow sticky card traps can be used to lure and kill pests.

(2) Chemical prevention and control. In the early stage, the prevention and control can be conducted by spraying imidacloprid, thiamethoxam, acetamiprid, spirotetramat·thiacloprid, flupyradifurone and nitenpyram.

4.2.5 Cotton bollworm, oriental tobacco budworm

4.2.5.1 Damage characteristics

The larvae feed on buds, flowers, and fruits, often forming bore holes in the fruit pedicle, causing the fruit to rot and fall off.

4.2.5.2 Prevention and control methods

(1) Agricultural and physical prevention and control. Pruning and removing superfluous branches or shoots should be done timely to remove the egg mass on the tender leaves and tender branches; Watering should be done to kill the pupae during the peak of pupation; Weeding should be done in the field and at the edge of the field; Insecticidal lamps should be used to lure and kill pests.

(2) Chemical prevention and control. Before the young larvae have not yet eaten the fruit, the prevention and control can be conducted by spraying beta-cypermethrin, emamectin benzoate (D), *Bacillus thuringiensis* (Bt), HaNPV, chlorantraniliprole·lambda-cyhalothrin, etc. It is advisable to apply the agent in the morning, focusing on the upper part of the plant.

5 Timely harvesting

Pepper should be harvested when the weather is clear, avoiding rain. Generally, pepper should not be harvested on rainy days or when the dew is not dry, so as to avoid the breeding of bacteria. According to the commodity demand of different varieties, each batch should be harvested as soon as it is ripe. When harvesting, it is to be noted that the ripening degree of pepper. Most peppers can be harvested when the fruits are full-grown, the fruit surface is glossy, and the fruit size no longer changes. In order to ensure the freshness, it is advisable to harvest fruits in the morning and evening. After harvesting, the fruits should be stored in a cool and ventilated place timely, while those with storage conditions should be stored in the warehouse timely.

6 Post-harvest treatment

After harvesting pepper, the diseased and rotten fruits should be removed as soon as possible, and pre-storage treatment should be carried out after the cleaning process is done. Pre-cooling treatment is a conventional treatment method for fresh peppers before storage in the warehouse, which can reduce the respiration rate and prolong the storage period of fresh peppers. Heat-shock treatment at an appropriate temperature can reduce the rotting occurrence of fresh pepper fruits, so as to improve the quality. Short-wave ultraviolet treatment is a physical treatment method without chemical pollution, which can improve the disease resistance of fruits and vegetables, reduce the application of chemical preservatives and lower post-harvest rotting losses through irradiation induction, which is a green and environmentally friendly way of storage and preservation.

References

ZOU X X, 2002. Chinese Chili [M]. Beijing: China Agriculture Press.

ZOU X X, 2021. New Technology of Pepper Breeding and Cultivation [M]. Changsha: Hunan Science and Technology Press.

❖ Key Techniques for Eggplant Cultivation

1 General information

Eggplant (*Solanum melongena* L.) is an annual herbaceous plant in the genus *Solanum* spp. under the Solanaceae or nightshade family with berries as products, and is a perennial in the tropics. Native to Southeast Asia and India, eggplant was introduced to China as early as the 4th to 5th centuries It is generally believed that China is the eggplant's second place of origin. China's eggplant cultivation area of 804 381 hm^2, accounts for 41.0% of the world's total cultivated area in 2021. Eggplant is one of the most widely cultivated vegetables in China's south and south, with high yield, strong adaptability, and easier cultivation. Eggplants are mainly boiled and fried, but they can also be made into dried eggplant, eggplant sauce, or pickled.

2 Growth conditions

2.1 Temperature

Eggplant prefers warmth and has strong heat resistance. The suitable temperature during the fruiting period is 25–30 °C.

2.2 Light

Eggplant is a light-loving vegetable, insensitive to the response of photoperiod. The photosynthesis of eggplant is affected by the intensity of light exposure.

2.3 Moisture

Eggplant has lush branches and leaves, and requires a large amount of water during its growth period. Usually 70%–80% of the soil moisture capacity of the soil is appropriate. However, before the formation of the first branch's eggplant, it requires less water and should not be watered too much to prevent the seedlings from growing excessively, root stunting and blossom dropping rate increases. After the fruiting of the first branch's eggplant, the water demand gradually increases, and before and after the eggplant harvest, the water demand is the highest. Insufficient soil moisture can seriously affect yield and quality, but too wet soil can cause poor soil aeration and cause root rotting.

2.4 Soil

Eggplant has a strong adaptability to soil, and can be cultivated in both sandy and clay soils. Due to the poor drought resistance and fertile soil preference of eggplant, it is advisable to choose fertile or clayey loam soil with deep soil layers, strong water retention, and a soil pH value of 6.8–7.3 for planting, which is conducive to the growth of eggplant root systems and the formation of vigorous root groups. Land with high groundwater level, poor drainage, and shallow plough layer with heavy soil texture are not conducive to the development of eggplant roots and should not be selected.

2.5 Fertilizer

Eggplant grows simultaneously with its stems and leaves during the fruiting period, requiring a large amount of fertilizer. Their first requirement for mineral fertilizers is potassium, followed by nitrogen, and phosphorus is the least. Nitrogen plays an important role in plant growth, flower bud differentiation, and fruit enlargement. When nitrogen deficiency occurs, the growth of the plant is weak, branching decreases, flower buds develop poorly, there are many short columnar flowers and a high rate of blossom drop, fruit growth pauses, and the skin color is not ideal. Eggplant is a fertilizer-tolerant vegetable with a long growth and fruiting period, so it needs to be topdressed multiple times to ensure an increase in yield.

3 Types and cultivars

3.1 Types

According to botanical classification, eggplant cultivars can be divided into 3 types: round eggplant, long eggplant, and dwarf eggplant. According to maturity period, they can be divided into early, medium-early, medium-ripe, medium-late and late ripe varieties. According to the color of eggplant fruits, there are dark purple, purple, purplish red, green, and white.

3.2 Cultivars

Eggplant cultivation is greatly influenced by consumption habits and has strong regional characteristics. Therefore, variety selection should be based on market demand and production purposes. Before large-scale planting, seed introduction experiments should be conducted to fully understand the characteristics of the variety, and rational production plans should be arranged to achieve the goal of increasing production and income, not blindly.

4 Cultivation and management techniques

4.1 Sowing and seedling raising

Seedling raising is an important part of intensive vegetable production, and raising and applying age-appropriate seedlings is a vital guarantee for early maturity, high quality, and high yield of vegetables. At present, there are three commonly used seedling raising methods in production: plug-seedling, seedbed seedling, and nutritional bowl seedling. Among them, plug-seedling is conducive to factory seedling raising (to convenient management, transportation, etc.). Except for a few remote areas, large-scale production of vegetables' areas all adopt plug-seedling. The following is the introduction to the method of plug-seedling.

4.1.1 Preparation before sowing

Most regions of South Asia and Southeast Asia can be cultivated throughout the year, with the most suitable cultivation period being winter and spring. The sites for seedling raising should choose fields with excellent drainage and irrigation, and the field's previous crop should not have been planted with solanaceous vegetables (such as tomatoes, eggplants, and chili peppers). Plug-seedling can not only reduce the occurrence of soil-borne diseases, but also ensures neat and robust emergence, without the need for separate seedling operations. The ground for placing seedling trays should be leveled in advance to avoid uneven terrain that may affect the quality of seedling emergence. Where conditions permit, it is better to use seedling racks. The seedling substrate can be vegetable seedling substrate, tobacco seedling substrate, or self formulated. The seedling substrate can be mixed with decomposed pig manure and field soil in a 1 : 1 ratio, and 100 g of compound fertilizer and 10 g of carbendazim can be added per cubic meter and mixed evenly. The field soil can be selected from the surface soil within 15 cm of the field that has not been planted solanaceous vegetables within 1–2 years, or from red soil that has not been planted with vegetables. The day before sowing, fill the seedling tray with seedling substrate or soil, flatten it with a small wood block, and then use a sprinkler to pour water on the bottom of the tray for later use. Two seedling trays are arranged in groups, which is convenient for workers to operate during sowing and thinning out seedlings.

4.1.2 Seed soaking and germination

The sowing amount for eggplant seedling raising is generally 15–20 g per mu. In order to achieve the goal of early emergence, complete emergence, and raising strong seedlings, seed treatment is generally carried out before sowing, mainly soaking and disinfection (coated seeds do not need to be treated). The commonly used method is to soak the seeds in hot water. The specific method is to first put the seeds into a nylon mesh bag or gauze bag, immerse them in room temperature water for 15 min, and then soak them in hot water at 55–60 °C for 15 min, with a water volume of 10 times the seed volume. After 15 min, let it cool naturally and soak the seeds at room temperature for 4–6 h. The soaked seeds can also be soaked in 50% carbendazim WP (1 : 1 000) or 10% trisodium phosphate solution for 20 min. After fishing them out, rinse them with clean water and wrap them in gauze or a damp cloth to germinate.

Wrap the processed seeds in a wet towel and place them in a bowl. Cover the seed bag with a few layers of wet towels to maintain humidity. After that, place the seeds at a temperature of 25–30 °C for 3–5 d. As long as 50% of the seeds sprout, they can be sown. During the germination process, flip the seeds once a day to keep them loose. If the seeds are found to be sticky, immediately rinse the towel with clean water. Generally, they should be cleaned once a day then control excess moisture after cleaning, in order to avoid excessive water absorption by the seeds, which may cause oxygen deficiency and poor germination. If there is no condition for germination, the seeds can be sown after soaking directly.

4.1.3 Sowing

The day before sowing, water the seedling tray or bed thoroughly. Generally, one seed is sown per hole in the seedling tray. After sowing, it should be covered with 1 cm of fine soil which should not be too thin or thick. If it is too thin, the seeds are easily exposed after watering or capped seedlings may emerge. If it is too thick, it will affect the emergence of seedlings. If the seeds are exposed after watering, it is necessary to cover them with soil in a timely manner. It is best to sow on a sunny morning, which is beneficial for raising the temperature inside the seedbed after sowing. After sowing and covering with soil, use a sprinkler to irrigate thoroughly.

4.1.4 Seedbed management

From sowing to emergence, it is required that the bed soil has sufficient moisture and good ventilation. When the seedlings are unearthed and the cotyledons open, the temperature of the seedbed should be maintained at 20—25 °C to prevent excessive growth. In the process of seedling raising, it is necessary to topdress appropriately, usually combined with watering and 0.1%—0.2% urea. Water should be sprayed with a fine mesh watering can or sprayer, while avoiding pouring with a gourd ladle to cause excessive soil moisture and serious seedling stage disease. The main disease in the seedling stage is damping-off disease. Once a diseased plant is found, it should be immediately removed and carbendazim or original chlorothalonil powder should be sprinkled around the plant to prevent the spread of the disease. Start exercising 10 d before planting seedlings, strengthen ventilation, reduce watering, control growth, and enable seedlings to adapt to outdoor environmental conditions as soon as possible, shortening the time for seedling recovery.

The watering of the seedbed is generally carried out from 8:00 to 10:00 or from 17:00 to 19:00. If there are symptoms of poor growth caused by lack of fertilizer, such as yellowing of leaves, small leaves, and thin stems during the seedling stage, the method of top dressing outside the roots can be adopted by spraying 0.3% potassium dihydrogen phosphate+0.2% urea solution on the leaves.

4.2 Site preparation and open furrows

Loose, fertile, and well drained loam for final planting should be chosen and continuous cropping should be avoided, while timely plow and sun dry the soil clod after harvesting the previous crop. Apply 3 000—4 000 kg of decomposed organic fertilizer, 50 kg of compound fertilizer, and 30 kg of calcium superphosphate per mu, and level the ground to form a 1.3 m wide furrow (including a ditch) with a depth of about 30 cm.

4.3 Timely planting

Generally speaking, eggplants cultivated in the open air in spring start planting when the last frost period comes and the lowest temperature stabilizes at 12 °C. The planting density of eggplants should be determined based on factors such as variety characteristics, cultivation methods, and soil fertility levels. The early maturing varieties are often denser than late maturing varieties; Varieties with compact plant types are often denser than those with larger plant development; Soil with poor fertility is often denser than soil with high fertility. The spacing in the rows and spacing between rows of medium-early varieties is 30 cm × 60 cm, medium-ripe is 40 cm × 60 cm and late ripe varieties is 50 cm × 60 cm, planting two lines per row. Dig a hole with a small hoe, put in the seedlings, cover them with soil and compact the soil. The appropriate cultivation depth should reach the cotyledon node. After planting, water the roots thoroughly to facilitate the survival of the seedlings. Planting should be done on sunny days, and avoid planting on wet soil during rainy days, which is difficult for new root growth and slow seedling recovery. The standard cotyledons of excellent eggplant seedlings are intact and wide, with a height of 15–18 cm and 6–8 true leaves. The leaves are thick and wide, with dark green color, thick and sturdy stems, short internodes, and well differentiated flower buds. The first alabastrum has appeared, but it has not yet bloomed. The root system is developed with many roots, white and tender, and no diseases or pests.

The age of seedlings from sowing to planting varies with the sowing location, season, and management level. Generally in Kunming City, it takes 60–90 d to raise seedlings in winter and spring, and 30–40 d in summer and autumn.

4.4 Field management

4.4.1 Fertilization

The principle of top dressing eggplant is that lightly applying seedling fertilizer, steadily applying flower fertilizer, and heavily applying fruit fertilizer. After the seedling recovery process is done, further fertilizer should be applied, and generally 10–15 kg of urea per mu should be used. The sequential fruit setting and expansion period of quadrate-bearing shoots, which is a total of four eggplants grown after the third branching start to enlarge and fruit setting in succession have the highest water demand. At this point, the soil should be intertilled and hilled when the eggplant begins to enlarge, and the fertilizer should be topdressing, and the compound fertilizer of 25–30 kg or urea and potassium fertilizer of 15 kg each should be applied on per mu. The sequential fruit setting and expansion period of quadrate-bearing shoots should be heavily fertilized, once every 10 days. In the middle and late stages of growth, potassium fertilizer should be increased while the phosphate fertilizer should be decreased. Because of the lack of potassium, plants are prone to disease and lodging, and excessive phosphate fertilizer is easy to cause fruit stiffness. In the fruiting period, in addition to proper topdressing of the soil, it should be combined with foliar fertilization to make up for the lack of fertilizer absorption capacity of the root. Foliar fertilizer spraying should be carried out on sunny days, with a 0.2% urea and 0.3% potassium dihydrogen phosphate mixture, spraying once every 15 d, and the spray effect is best when it does not rain for 24 h. When spraying, it is best to carry out in the evening or in the morning when there is still dew, and should not be carried out at noon, so as to avoid speeding up the concentration rate of the liquid due to high temperature and causing drug damage.

4.4.2　Moisture management

The lack of soil moisture will cause slow growth of plants, even cause blossom dropping, and the pericarp is rough, dull, and poor quality. In the season of eggplant's final planting from March to April in Yunnan, the rainy season has not yet arrived, and the weather is dry. Therefore, it is necessary to water frequently in the early stage, once every 2–3 d until the eggplant seedlings are alive. When the first branch's eggplant flowering, appropriate hardening of seedling and water control should be done. The purpose of hardening of seedings is mainly to properly control the growth of above-ground plants, promote their continued development in depth, regulate the balance of nutritional and reproductive growth, so as to accumulate more nutrients to supply the development and enlargement of fruits, and obtain high yield. Attention should be paid to the drainage of water in June to August when the rainfall is abundant. In the period of "glaring stage" that the fruit has just begun to enlarge, and the hardening of seedling and water in time. The stems, leaves and fruits grow at the same time, then the growth rate is fast, and the moisture requirement is also significantly increased. The highest moisture requirement happens when the "couple eggplants" which are the two eggplants of the second branching and the sequential fruit setting and expansion period of quadrate-bearing shoots which are the four eggplants of the third branching are both enlarged in succession. After the "glaring stage" that the fruit has just begun to enlarge, depending on the weather and the growth of the plant, watering should be done once every 5–7 d to keep the soil moist and prevent sudden drying and wetting. Eggplant needs more fertilizer and water than tomato for its long growing season and lush foliage. However, stems and leaves growth are predominant before bearing fruit, requiring less fertilizer and water, while more fertilizer and water is needed after bearing fruit. Reasonable topdressing according to the characteristics of different growing stages is one of the main measures for high yield.

4.4.3　Intertillage and hilling

Combined with weeding, intertillage can be deeper in the early stage and shallower in the late stage. When the plant grows to about 30 cm, it should be combined with intertillage for hilling. Intertillage and hilling always combined with weeding can ensure that the soil is loose and breathable, so as to promote the new roots generating and enhance the absorption capacity of roots. Generally, intertillage should be done once every 10–15 d from final planting. Intertillage should be carried out when the soil is half dry and wet. Intertilling in the flowering and fruiting period should be combined with topdressing and hilling.

4.4.4 Pruning and set up supports

After flowering, remove the lateral branches that are lower than the first branch's eggplant, and set up supports in time. If the pruning is too early, it is not conducive to the roots, while too late, it will cause excessive growth and is not conducive to early fruit setting. Pruning should be done on sunny days to prevent the wound from getting infected with germs. There are two kinds of eggplant pruning methods, double-rod pruning method and four-rod pruning method. The double-rod pruning method is to leave only the main branch and the first lateral branch under the first branch's eggplant, and remove all the other lateral branches. The four-rod pruning method is to retain the main branch and the first lateral branch under the first branch's eggplant, but also to retain the lateral branches growing on them, and the rest of the lateral branches below the base of the the first branch's eggplant are all removed. At present, the double-rod pruning method is commonly used in production. When set up supports, the rod is inserted 10 cm away from the main stem of the plant, and other rods are erected every 2 m, then the cross rods are tied at a height of 30 cm from the ground. The main stem of the plant is tied by strings or cloth strips to the rack to prevent lodging. In the middle and late stages of eggplant growth, the yellow leaves, old leaves and diseased leaves below the first branch's eggplant can be appropriately removed to facilitate light transmission and reduce the harm of diseases and pests. The old, diseased and residual leaves removed should be cleaned out of the garden in time and buried deeply or burned.

4.4.5 Fruit and flower saving

The main factors that cause eggplant blossom dropping are insufficient light, malnutrition, and too high (above 38 °C) or too low temperature (below 15 °C). In production, in addition to strengthening water and fertilizer management and pruning, 25−30 mL of 2,4-D (2,4-Dichlorophenoxyacetic acid) or 4-Chlorophenoxyacetic acid 40−50 mg/L can also be used to soak flowers or apply on the flower stalks, and fruit and flower saving should be conducted every 3−4 d to promote rapid fruit enlargement. The concentration is higher when the temperature is low and the humidity is high, and the concentration is lower when the temperature is high and the humidity is low. 2,4-Dichlorophenoxyacetic acid is harmful to young leaves and growing points, so use it carefully.

4.4.6 Thinning of blossom and fruit

Some eggplant varieties' flowers are solitary, and thinning of blossom and fruit is not needed, while some varieties are inflorescences, which have 2−6 flowers on one branch, need to thin flowers and fruits. Only one flower is left, and excess flowers are manually removed to ensure nutritional supply and avoid uneven fruit size, deformed fruit and low commodity rate.

5 Prevention and control of diseases and pests

Guided by the policy of prevention first and combining integrated prevention and control, the principle of centering on agricultural prevention and control, physical prevention and control, and biological prevention and control, supplemented by chemical prevention and control should be adopted.

5.1 Agricultural prevention and control

First, selecting and using the disease-resistant varieties, cultivating strong seedlings, and crop rotation should be implemented. Second, deep furrow and high bedding cultivation and reasonable dense planting should be adopted. Third is to prune and set up supports. Fourth is to pull out the seriously sick plants and clean the field. Then is to deep plow the soil to reduce the source of diseases and pests and fertilize scientifically.

5.2 Physical prevention and control

Yellow and blue sticky card traps and frequency trembler grid lamps could be used to trap and kill imagoes. Laying silver gray film or hanging strips of silver gray film in the field could be used to repel aphids. Then, insect-proof screens can be used in greenhouses to prevent pests from entering, and manually removing pest eggs, catching and killing pests are also considered as useful methods. Using sex attractant for pest disinfestation is also useful.

5.3 Biological prevention and control

Selecting and using biological agents from botanical sources, microbial sources, and agricultural antibiotics and so on to prevent and control the pests and diseases.

5.4 Chemical prevention and control

5.4.1 Prevention and control of major diseases

The main diseases of eggplant are damping-off, verticillium, phytophthora rot, phomopsis rot, powdery mildew, etc.

5.4.1.1 Damping-off

(1) Symptoms. It is also called sprout tumble, small foot plague. After the seedlings were

unearthed, they were infected with the disease, and there was a yellowish to yellowish brown water-soaked lesion on the base of the young stem. The lesion then surrounds the stem and contracts into a linear shape. Before the cotyledon or young leaf withered, the seedlings fell to the ground and there was a phenomenon of damping-off. Seedlings can also be damaged before they are unearthed, causing rot. When the humidity is high, a layer of white flocculent hyphae can be seen near the diseased part.

(2) Prevention and control methods. When the seedlings are unearthed, spraying medicine, pulling out the sick plants in time, and sprinkling chlorothalonil powder in the area of sick plants can be used to prevent and control the disease. 58% metalaxyl mancozeb WP (1 : 600), 722 g/L propamocarb AS (1 : 600), 70% mancozeb WP (1 : 600), and 50% carbendazim WP [1 : (700−800)] and so on can be sprayed in the early stage of disease, once every 7−10 d, continuously for 2−3 times. After spraying, dry soil or plant ash can be scattered to reduce the soil moisture of the seedbed. Ventilation should be strengthened on sunny days.

5.4.1.2 Verticillium

(1) Symptoms. It mostly shows symptoms in the fruiting period. In the early stage of the disease, the symptom is the leaf veins turning green to yellow, and gradually develops to the whole leaves. Because most eggplant plants are half diseased and half normal, commonly known as "half insanity". In the early stage, the leaves wilt at noon on sunny days, and can be recovered in the morning and evening or on rainy days. After a few days, the whole plant wilts and the leaves fall off. Transversely cutting the stem, it can be seen that the vascular bundle turns brown. Extrusion discoloration does not exude turbid liquid, and can be distinguished from bacterial wilt.

(2) Prevention and control methods. First, grafting is the most effective method to prevent and control verticillium in current production. Second, it is better to apply 10 g thiophanate-methyl powder before the final planting, mixing well and then planting the seedlings, or pouring the second and third root fixing water after transplanting with 30% metalaxyl-hymexazol AS (1 : 800) +96% copper sulfate (1 : 1 000), which can effectively prevent the verticillium. At the early stage of the disease, the root can also be irrigated with 38% oxadixyl-Azoxystrobin [1 : (600−800)], or 50% carbendazim WP (1 : 500), and 50% Benomyl WP (1 : 1 000). Each plant can be irrigated with 300−500 mL of the prepared solution, once every 10 d, and continuously irrigated 2−3 times.

5.4.1.3 Phytophthora rot

(1) Symptoms. It generally affects the fruit, especially the part of the fruit's top that touches the ground. At first, it produces a water-soaked brown round lesion on the fruit with not obvious edges. The lesion is slightly concave, and may expand throughout the fruit. With

high humidity in the field, the flesh becomes black and rotten, and the white flocculent hyphae is produced in the moist diseased part, and the diseased fruit is easy to fall off or shrink into a stiff one. In the early stage of disease, the stem is water-soaked, then becomes brown and decayed, even narrows to break. The leaves have irregular or nearly circular water-soaked lesions with a wheel pattern that can expand rapidly. When it is wet, the edge of the lesion is unclear, there is sparse white mold, and when it is dry, the edge of the lesion is obvious and easy to crack.

(2) Prevention and control methods. It is recommended to choose disease resistant varieties and pay attention to crop rotation. Paddy-upland rotation is the best option when conditions permit. 58% metalaxyl mancozeb WP (1 : 600), 722 g/L propamocarb AS (1 : 600), 70% mancozeb WP (1 : 600), 72% cymoxanil mancozeb WP [1 : (500−700)], 50% dimethomorph WP [1 : (600−800)], and 75% chlorothalonil WP (1 : 500) and so on can be sprayed in the early stage of disease, once every 7 d, continuously for 2−3 times.

5.4.1.4 Phomopsis rot

(1) Symptoms. It mainly affects the leaves, stems and fruits. When they get infected, the symptoms usually appear on the lower leaves, first is small white spots, then the lesion become brown and near circular or irregular, and there are wheel patterns also small black dots. When the stem is infected, it initially appears as a prismatic lesions on the stem, with a dark purplish-brown margin and a grayish-white sunken dry rot ulcer in the middle, and black dots. When the fruit is infected, it initially appears as a light brown oval concave lesion then expands to dark brown, and the upper wheel pattern has small black dots.

(2) Prevention and control methods. 75% chlorothalonil WP (1 : 600) and 70% mancozeb WP (1 : 500) can be used for prevention and protection. The 50% iprodione WP (1 : 1 000), 25% pyraclostrobin SC [1 : (1 000−1 500)], 40% difenoconazole SC [1 : (1 000−1 500)] and so on can alternately be sprayed to prevent and control in the enlargement period of the first branch's eggplant, once every 7−10 d, continuously for 2−3 times.

5.4.1.5 Powdery mildew

(1) Symptoms. It mainly affects the leaves. On the leaf surface, irregular and various sized white powdery mildew spots are generated and spread throughout the entire leaf, causing the leaf tissue to turn yellow and then dry up.

(2) Prevention and control methods. It is recommended to plant closely and remove the old leaves in the lower part of the plant in time. At the early stage of the disease, it can be prevented and treated with 29% isopyrazam-azoxystrobin SC (1 : 1 500), 40% fuxing flusilazole EC [1 : (6 000−10 000)], 10% difenoconazole WG [1 : (1 500−2 000)], or 70% thiophanate-methyl WP (1 : 600), etc.

5.4.2 Prevention and control of pests

The main pests are thrip, whitefly, aphid, red spider, etc.

5.4.2.1 Thrip

(1) Damage characteristics. Imagoes and nymphs hide in the fuzzes on the back of tender leaves and in flowers, sucking sap. The fuzzes and flesh on the back of the infected leaves are taupe, becoming hardened and aging, and the buds are injured and shed. The infected plants grow slowly and their internodes shorten. Thrips reproduce extremely quickly, with over 10−20 generations per year, overlapping generations and reproducing year-round. The peak period is from May to September, with autumn being the most severe. Imagoes are active and fly well, afraid of light, and nymphs fall into the topsoil and pupate.

(2) Prevention and control methods. Using plastic mulching can greatly reduce insect-population density, timely remove old and diseased leaves, clean up weeds in the field, and clear branches and residual leaves for concentrated burning after the harvest period to reduce the source of pests. When the pest population of each plant reaches 3 heads, it should be sprayed for prevention and control, with 25% imidacloprid WP (1∶3 000), or 25% thiamethoxam WG [1∶(2 500−3 000)], 25% spinetoram WG [1∶(2 000−3 000)], 10% nitenpyram AS [1∶(1 000−1 500)], 10% beta-cypermethrin EC (1∶2 000), etc, once every 3−5 d, 3 times in a row. The focus of spraying is on the growth point of the plant, the back of the tender leaves, flowers, alabastrums, and other parts.

5.4.2.2 Whitefly

(1) Damage characteristics. The whitefly is an omnivorous pest with a fast reproduction rate, and its population reaches its peak in autumn. Imagoes and nymphs cluster on the back of the leaves, sucking sap with piercing-sucking mouthparts. The infected leaves turn green into yellow, causing slow growth and wilting of plants, as well as secreting a large amount of honeydew.

(2) Prevention and control methods. In the early stage, 10% imidacloprid WP (1∶1 000), 25% buprofezin EC (1∶2 500), and 20% fenpropathrin EC (1∶2 000) can be used by adding 10% pyriproxyfen EC (1∶1 000) or 10% lufenuron SC [1∶(2 500−3 000)], once every 3−5 d, 2−3 times in a row.

5.4.2.3 Aphid

(1) Damage characteristics. Aphids often gather on the back of leaves and tender stems to use piercing-sucking mouthparts to suck plant sap and secrete honeydew, causing severe dehydration and malnutrition in the plants. The young leaves are infected, curled and wrinkled, which may cause the leaves of fade green, speckled, and leaves turn yellow in mild

cases, while the leaves shrink and deform even wither in acute cases. Aphids can also spread viral diseases.

(2) Prevention and control methods. 50% pirimicarb WP (1∶2 000), or 10% imidacloprid WP [1∶(1 000–1 500)], 20% acetamiprid SL (1∶3 000), 50% pymetrozine WP [1∶(2 500–5 000)], 20% fenvalerate EC (1∶2 000), 20% fenpropathrin EC (1∶3 000) can be used for spraying, every 7–10 d, 2–3 times in a row, depending on the pest situation in the field.

5.4.2.4 Red spider

(1) Damage characteristics. Red spiders cluster on the back of the leaves to suck sap and fusule to make webs. The infected leaves appear small white spots at the beginning, then fade green into yellowish white, even become rust brown and burning-like in severe cases, resulting in early leaf fall, thicken pericarp, rigid fruit can not grow, and plants wilt and die.

(2) Prevention and control methods. The weeds around the edge of the field and withered branches and leaves after harvest can be removed to centrally dispose of to reduce the source of pests. The primary stage is the emphasis on prevention and control. 73% propargite EC (1∶1 500), 34% spirodiclofen SC (1∶4 000) + 43% bifenazate SC [1∶(1 800–2 500)], 20% avermectin-bifenazate SC [1∶(2 000–2 500)], and 5% hexythiazox EC (1∶3 000) can be selected to spray, once every 7 to 10 d, 2–3 times in a row.

6 Timely harvesting

After pollination, the fruits of eggplant enlarge rapidly, and it generally takes 20–25 d from blooming to fruit harvesting. Therefore, the harvest of eggplant must be timely, and harvesting too early results in low yields, while harvesting too late results in hard fruits and redundant seeds that are not edible and consume too much nutrients, which affects the blooming and fruiting and the normal growth of branches.

6.1 Harvest standards

The fruit should grow to the length and width of the variety, the color of the base is lighter, and the color of other parts becomes darker. Then the surface of the peel showed strong elasticity when pressed by hand. Also the pericarp becomes bright and shiny. The appropriate time for fruit harvesting can be judged according to the width of the joint of the sepal and fruits (called "eggplant's eye"). When the "eggplant's eye" is wide, indicating that the fruit is growing rapidly and is not suitable for harvesting, while when it is not obvious, indicating that the fruit growth becomes slow or has stopped growing, and can be harvested.

6.2 Harvest methods

The correct method of harvesting is to use a knife or branch cutting to the root of the fruit stalk, without the fruit stalk, in order to avoid pricking the pericarp during the packing and transportation and affecting the appearance and quality of the fruit. It is best to harvest in the morning, because the fruit is fresh and tender with high quality and excellent storage performance. After harvesting, it should be graded and packaged for sale in a timely manner.

References

Institute of Vegetables and Flowers, Chinese Academy of Agricultural Sciences, 2010. Olericulture in China [M]. 2nd Edition. Beijing: China Agriculture Press.

LIN J R, LIU S Y, QIU M Y, et al., 2004. Colored Illustrated Book of Pollution-Free Production of Tomato and Eggplant [M]. Guangzhou: Guangdong Science and Technology Press.

Key Techniques for Common Bean Cultivation

Common bean (*Phaseolus vulgaris* L.) is an annual plant in the pea family, also known as kidney bean. Originating in the Americas, after hundreds of years of spread and cultivation, it has become a widely planted and eaten vegetable crop worldwide, especially in tropical and subtropical areas. Common bean is a nutritious vegetable with a refreshing taste, rich in protein, carbohydrates, dietary fiber, vitamins and minerals, especially potassium, calcium, magnesium, iron and other minerals.

Yunnan's tridimensional climate environment has realized the annual production of common beans, making Yunnan a leading production area for winter and spring common beans, with products sold all over the country. In recent years, taking advantage of Yunnan's proximity to Southeast Asian countries, some enterprises and farmers have grown common beans in Laos, Myanmar and other countries for sale in China, driving the development of the vegetable industry in Southeast Asian countries.

1 Growth conditions

1.1 Temperature

The suitable temperature for growth and development is 10−30 °C, and it cannot tolerate frost injury. The suitable temperature for germination is 20−25 °C, and it is difficult for seeds to germinate above 35 °C or below 8 °C. The suitable temperature for young seedling growth is 18−20 °C, and for blooming and podding is 18−25 °C.

1.2 Light

The growth and development of the common bean is not strictly required for the length of sunshine, that is, it can bloom under longer or shorter sunshine. However, the growth, blooming and podding of common bean requires stronger light, and if the light is insufficient, excessive growth, blossom and pod dropping are prone to occur.

1.3 Moisture

Common beans have a well-developed root system with many lateral roots, and are more resistant to drought than flooding. Seeds need to absorb sufficient moisture to germinate, but excessive moisture can cause soil hypoxia and seed decay. The suitable soil moisture for the growth of plants is 60%–70% of the field moisture capacity. The flowering and podding stages are strictly dependent on moisture, and the suitable air relative humidity is 65%–80%. In the case of high temperature and drought during the podding stage, the growth of tender pods becomes slow, the pericarp of the pod wall is prone to harden, the division of endocarp cells is accelerated, the cavity between the subcells occurs in advance, and the endocarp becomes thin, resulting in reduced quality. During the flowering stage, heavy rain, excessive soil and air humidity will also affect the germination of pollen, and excessive moisture will reduce the concentration of mucus on the pistil stigma, making the pistil unable to pollinate normally even blossom dropping and pod dropping, and it is easy to cause the occurrence of diseases.

1.4 Soil

Common bean is most suitable for growth in sandy loam or loam soil with deep soil layer, soft, rich humus and good drainage, and the optimal pH value is 6–7. The absorption of nutrient elements is mainly nitrogen and potassium, with less phosphorus. The absorption of nitrogen and potassium gradually increases during flowering and podding stages, while the nitrogen and potassium in the stem and leaves are transferred to the fruit pod with the change of growth center, and a large amount of calcium is absorbed when the young pods rapidly extend.

2 Variety selection

Common bean varieties are divided into sprawl cultivar and dwarf cultivar. Currently, sprawl cultivar includes Red-flower-green-shell bean, Shuangqingyu bean, Chunfeng No. 4, Qiukang No. 6, Dayuqiubei, Thailand Jiadouwang, Shijijiadou and Lülong, while dwarf cultivar includes Shakesha, Meiguogongjizhe, 8916 dwarf bean and French kidney bean.

3 Cultivation and management techniques

3.1 Land preparation

Common bean is a crop that requires high soil fertility. The proper amount of base fertilizer can improve the growth rate and yield of common bean. Apply 1 500–2 000 kg of composted organic fertilizer, 30 kg of ammonium dihydrogen phosphate and 10 kg of potassium sulphate per mu. The soil should be mechanically ploughed to a depth of 25–30 cm, and the fertilizer should be fully mixed with the soil as a bed. Generally, high-ridge planting is adopted, and 40 cm deep drainage ditches are set around the ridge.

3.2 Cultivation season

Common bean production can be achieved annually in Yunnan. Sowing is carried out from March to July in Central Yunnan, Northeastern Yunnan, and Northwest Yunnan, while winter and spring shifting peak planting can be conducted from September to November in Western Yunnan, Southern Yunnan, and dry-hot valley areas. In tropical and subtropical areas of Southeast Asia, sowing is carried out from September to November. According to the local climate and precipitation, the specific sowing time can be adjusted and the appropriate variety and planting time can be selected.

3.3 Seed treatment

Seeds with smooth surfaces and plump grains should be selected and should be dried for 1–2 d before sowing to improve the activity of enzymes inside the seeds and promote uniform germination. In addition, seed treatment can improve the germination rate and uniformity of seeds and reduce disease.

3.3.1 Seed soaking by hot water

The seeds should be soaked in hot water at around 80 °C for 5–8 min, followed by rapid cooling, which can effectively kill pathogens on the seed surface and promote germination.

3.3.2 Seed soaking by agent

The seeds should be soaked in 0.3% potassium permanganate solution or 1% cupric sulfate solution for 20–30 min, and then washed with clean water to kill fungi and bacteria on the seed surface, so as to improve the germination rate and disease resistance at the seedling stage.

3.3.3 Seed dressing by agent

The seeds should be mixed with 50% carbendazim WP, which accounts for 0.2% of seeds weight, to prevent and treat various diseases such as wilt disease, damping-off, and anthracnose at the seedling stage.

3.4 Sowing and post-emergence management

3.4.1 Sowing

Direct seeding is generally used for common beans, with a seed amount of 1.0–1.5 kg per mu. Before sowing, bottom water is first poured, and after sowing, the bed surface is covered with mulch film with a width of 100 cm. Each bed is sown in 2 rows, with a bed width of 80 cm, and a furrow width of 40 cm and a depth of 20 cm. 3–4 seeds are sown in each hole, with a depth of 5–6 cm, and the soil is covered by 1–2 cm after sowing.

3.4.2 Seedling stage management

After sowing about 5 days later, film breaking and seedling releasing should be conducted in time, compacting the soil around the mulch film, which is conducive to heat preservation and moisture retention, and prevent the high temperature steam under the mulch film from damaging the seedlings during the day. After complete emergence, the seedlings should be checked and supplemented in time, and the seedlings should be replanted immediately if any shortage of seedlings is found. When there are 2 true leaves, the diseased and weak plants should be removed, and 2 strong plants are left in each hole. The amount of watering should be as little as possible during the seedling stage to promote root system growth and prevent the young seedlings from overgrowing and root diseases.

3.4.3 Setting up supports

When the plant has 5 leaves and the tip starts to jilt tendril, it is time to set up supports. The vertical or horizontal supports made of bamboo, wooden strips, iron pipes and so on can be used as the material of supports, so that the common beans can climb and grow. When setting up supports, enough space should be left for the picking and management of the common beans.

The climbing height of common bean and the form of the supports can be adjusted according to the variety and growth condition. Usually, 1.8–2.2 m high poles are selected and inserted into the ground about 15 cm deep with 15–20 cm away from the root of the plant, so as to avoid root injury. One pole is inserted for each plant, one A-frame can be set every 4 plants in the adjacent two rows on the bed, and every 4 poles are made into 1 bundle. After the supports are set up, the vines are guided to the supports in a counter clockwise direction. After the vines fall, the orientation of stem and vine should be adjusted in time to induce the stems and vines to the vacant position, which can make full use of the space, and avoid the stems winding, stacking or overlapping with each other. Some common bean varieties have too many branches, which will cause local crown closure or stem stacking in the field. Therefore, it is necessary to tidy up and remove the old and diseased leaves in the lower part in time for better ventilation and light penetration.

3.5 Water and fertilizer management

Water and fertilizer management follows the principle of less amount of water and fertilizer at the seedling stage, controlling amount at the vine-elongating stage, and more amount at the podding stage.

3.5.1 Seedling stage

After the emergence, the seedlings can be watered once according to the soil moisture. After that, the growth of seedlings should be controlled, while the growth of the root system should be promoted, aiming to improve the nutritional growth. Generally, watering is not started until near the time of flowering, which can be flexibly controlled according to the soil moisture. Before the beginning of flowering, appropriate phosphorus and potassium fertilizers should be applied to promote the formation of flower buds, also flowering and podding, ultimately increasing yield. Blossom and pod dropping is an important factor affecting the high yield of common bean, and boron deficiency is one of the main reasons for blossom and pod dropping. Supplementing boron can improve the flowering and pod setting rate of common bean, so boron fertilizer can be added in the seedling stage, and 2 kg of borax can be applied per mu.

3.5.2 Flowering stage

The flower bud formation and flowering stage require a large amount of phosphorus and potassium elements. Therefore, water and fertilizer application can be increased appropriately after the young pods of common bean are set stably, based on weather conditions and plant growth. If the soil is too dry during the flowering stage, pollen will prematurely age and fail to complete fertilization; If the soil and air humidity are too high, pollen adhesion will not be conducive to pollination, which will lead to poor pollination and fertilization, and finally result in flower and pod dropping. It is strictly prohibited to irrigate the plant by flood during the flowering period, lest the vegetative growth and reproductive growth compete for nutrients, causing severe flower and pod dropping. High temperatures above 30 °C and low temperatures below 10 °C directly affect the normal differentiation of flower buds, or make the floral organ develop incompletely then abort, indirectly causing flower and pod dropping, short or deformed pods, ultimately leading to reduced yield. Therefore, watering under the high temperature stage should be done in the evening as far as possible to reduce the ground temperature, so as to promote the coordinated growth of branches, leaves and pods, and pay attention to drainage after raining.

3.5.3 Podding stage

A large amount of nitrogen fertilizer is required during the podding stage. After the bean pods, an appropriate amount of nitrogen fertilizer should be applied according to the soil nutrient status. Generally, the method of combining organic fertilizer and chemical fertilizer can be adopted to maintain the balance of soil fertility, promote the development and growth of pods, and improve yield. Twenty kg of compound fertilizer should be applied per mu, and try not to fertilize or water before cold weather in winter and cloudy days to reduce the occurrence of diseases.

3.5.4 Harvesting stage

Combined with watering, fertilizer applying should be conducted once every 7–10 d. 10 kg of three-element compound fertilizer (N : P : K=15 : 15 : 15) and 5 kg of urea can be applied per mu, and 0.2% potassium dihydrogen phosphate and 500 times borax can be sprayed on the leaf surface 2–3 times, once every 10 d. After that, three-element compound fertilizer (N : P : K=15 : 15 : 15) should be applied once after 2–3 times harvesting, and 0.3% potassium dihydrogen phosphate can be sprayed on the leaf surface.

3.5.5 Post-harvest stage

Plant management should be adjusted according to the market price. If the market price is high, water and fertilizer management can be strengthened to promote blossom increasing and prolong the harvest period. Three-element compound fertilizer (N : P : K=15 : 15 : 15) can be applied 1–2 times to promote plant growth, with 10 kg each time. During the whole growth period, diseased plants and weeds should be pulled out in time, yellow leaves should be removed, light and ventilation should be strengthened, and pod vines should be straightened out to prevent the occurrence of malformed pods caused by mechanical barriers.

4 Prevention and control of diseases and pests

4.1 Prevention and control of diseases

4.1.1 Rust disease

4.1.1.1 Damage symptoms

It mainly affects the leaves, which are yellowish-green at first, and then gradually turn into the brown round lesions with yellow-green rings. The lesions have brown to black-brown small spots, and finally the brown part falls off, forming a perforation. On the back of the leaves, there are a large number of rust spores clustered together, resembling a coarse woolly mold of yellowish-white to light yellowish-brown. After the veins, petioles and stems are infected, the lesions are spindle-shaped or near spindle-shaped strips, slightly raised, faded green and water stained, and sometimes vertical cracks appear on the stems, with brown to black-brown small spots in the center. When the disease is serious, it can also infect stems and pods.

4.1.1.2 Occurrence characteristics

Rust disease is prone to occur and spread quickly in spring and autumn with large temperature differences and abundant fog and dew. It generally begins to occur in the middle stage of common bean growth, and is prone to occur in plants with weak growth or in those with partial application of nitrogen fertilizer.

4.1.1.3 Prevention and control measures

(1) Agricultural prevention and control. Crop rotation should be implemented; Disease-resistant varieties should be adopted; Ditches clearing and draining in vegetable fields should be done in the rainy season to prevent water pooling in low-lying areas and reduce field humidity; Planting density should be reasonable to ensure good ventilation, the vegetable garden should be well cleared, and diseased leaves should be removed and burnt after harvesting; Excessive application of nitrogen fertilizer should be avoided in the early stage.

(2) Chemical prevention and control. When sporadic infection occurs on the bottom leaf, 70% sulfur·mancozeb WP, 20% difenoconazole microemulsion ME, and 60% pyraclostrobin·metiram WP can be sprayed for prevention and treatment. When the middle and lower leaves are commonly affected, the agent combinations such as tebuconazole+difenoconazole·kresoxim-methyl, or difenoconazole·azoxystrobin+tebuconazole or hexaconazole, and isopyrazam+ethirimol+potassium dihydrogen phosphate can be sprayed for prevention and treatment. The interval is 5−7 d, and more than 3 continuously times are required.

4.1.2 Anthracnose

4.1.2.1 Damage symptoms

When the disease occurs, spindle-shaped or long stripes lesions appear on the stems and pods. At the beginning, the lesions are purplish-red, then the color becomes light, slightly sunken and even cracked, and a large number of black spots are densely covered on the lesions, with the pod being the most seriously affected. The disease occurs more in the rainy season, and the diseased parts tend to become black due to the growth of saprophytic bacteria, accelerating the disintegration of stem tissue. In mild cases, the growth is stagnant, while in severe cases, the plant dies.

4.1.2.2 Prevention and control measures

The latency of pathogen infection is long, so the focus should be on protection and prevention. Entering the fruit pod stage and the sporadic infection stage of leaf, stem and vine, 70% thiophanate-methyl WP, 70% mancozeb WP, 25% bromothalonil ME, and 50% iprodione WP can be sprayed for prevention and control; The occurrence of common infection in the field can be controlled by spraying with 25% prochloraz EC, or 20% flusilazole·prochloraz EW, prochloraz+triazole, and prochloraz+pyraclostrobin, with an interval of 7–10 d, continuously for 2–3 times.

4.1.3 Powdery mildew

4.1.3.1 Damage symptoms

In the early stages, small round white spots appear on the back of the leaves, which then expand and connect with each other, spreading throughout the entire leaf and expanding into a powdery band along the veins. The color changes from white to grayish-white to purplish-brown. In severe cases, diseased lesions form on the leaf surface, causing the leaves to yellow and fall off.

4.1.3.2 Prevention and control measures

In the initial stage of disease, 15% triadimefon WP, 10% difenoconazole WP, 12.5% diniconazole WP, 43% tebuconazole SC can be sprayed for prevention and control, with an interval of 5–7 d, and continuously for more than 3 times.

4.1.4 Root rot

4.1.4.1 Damage symptoms

The lower leaves of the plant turn yellow, and the diseased part produces punctiform lesions, which spread from the branching roots to the main root, causing the whole root system to rot or necrosis. The diseased plant is easy to pull up. The longitudinal section of the diseased root shows that the vascular bundle is reddish-brown, and the disease extends to the stem after expansion. After the main root is completely infected, the aboveground stems and leaves wilt and die. When the humidity is high, pink moldy substances are produced in the diseased part, which are conidia of the pathogen.

4.1.4.2 Prevention and control measures

(1) Agricultural prevention and control. Disease-resistant varieties should be selected; Crop rotation should be implemented, or rotation with non-pea family crops should be

implemented for more than two years; Deep furrows and high beds should be established to prevent water stagnation and timely drainage should be done after rainfall; Field management should be strengthened, and phosphorus and potassium fertilizers should be applied to improve plant disease resistance.

(2) Chemical prevention and control. In the initial stage of disease, thiophanate-methyl or fenaminosulf solution can be used to spray the root and stem base; hymexazol, metalaxyl-hymexazol, phenamacril can also be used to irrigate the roots, once every 7 d, continuously for 2-3 times.

4.1.5 Viral disease

4.1.5.1 Damage symptoms

The young leaves appear symptoms such as mosaic, vein-clearing, chlorosis or malformation, and the dark green parts of the new leaves are slightly protruding and verruciform; Some diseased plants produce brown and sunken stripe spots, necrosis of mesophyll or veins. The diseased plants grow poorly, are dwarf, have deformed floral organs, produce few pods, and form yellowish-green flecks on the beans. Some diseased plants wilt and die at the growing point, or necrosis begins from the tender shoots.

4.1.5.2 Prevention and control measures

(1) Agricultural prevention and control. Disease-resistant varieties should be selected; Continuous cropping with leguminous, solanaceous and gourd vegetables should be avoided; Cultivation management should be strengthened to improve plant disease resistance.

(2) Chemical prevention and control. ①Insect control and disease prevention should be implemented to cut off the transmission vectors. When there are aphids, thrips, whiteflies and red spiders in the field, imidacloprid, buprofezin, avermectin, acetamiprid, spinetoram, pyrethroids and bifenazate are selected for prevention and control to cut off the transmission and spread of the virus. ②Prevention and control by pesticides. Before or at the early stage of the disease, ningnanmycin, mushrooms proteoglycan, moroxydine hydrochloride·cupric acetate+ oligosaccharins+brassinolide can be sprayed for prevention and control, with an interval of 7-10 d, continuously for 2-3 times.

4.1.6 Common bean bacterial blight

4.1.6.1 Damage symptoms

Water-stained necrosis occurs at the leaf edge first, then expands, and the lesions are distributed along the veins, with yellow haloes around the lesions; The onset of the disease mostly starts from the lower bottom leaves and expands upward along the leaves on the same side; Low-lying waterlogged plots are prone to the disease, rainstorms, sudden sunshine, and hailstorms are also prone to the disease.

4.1.6.2 Prevention and control measures

(1) Agricultural prevention and control. Disease-resistant varieties should be selected; Pay attention to field drainage and waterlogging prevention to avoid flood irrigation; Cultivation management should be strengthened to improve plant disease resistance.

(2) Chemical prevention and control. When bottom leaves of sporadic plant in the field appear initial symptoms of disease, agents such as thiodiazole-copper, oxine-copper, copper hydroxide, zinc thiozole, kasugamycin, copper amine solution, zhongshengmycin, or chloroisobromine-cyanuric-acid can be sprayed for prevention and control, with an interval of 7–10 d, continuously for 2–3 times.

4.2　Prevention and control of pests

The main pests of common bean are bean pod borer, vegetable leaf miner, red spider, aphid, whitefly, thrips, beet armyworm, common cutworm, and tobacco budworm, etc. Guided by the policy of prevention first and combining comprehensive prevention and control, agricultural prevention and control, physical prevention and control, and biological prevention and control should be insisted as the foundation, supplemented by chemical prevention and control, preventing as much as possible, so as to achieve the purpose of economically, safely, and effectively controlling pests.

4.2.1　Physical prevention and control

4.2.1.1　Trap-killing by colored sticky cards

Imagoes of aphid, whitefly, thrips, and vegetable leaf miner have strong taxis to yellow or blue. Therefore, yellow or blue sticky cards can be hung in the field to lure and monitor them, with 30 cards placed per mu of vegetable field.

4.2.1.2　Trap-killing by light

Imagoes such as cotton bollworm, cutworm, common cutworm, beet armyworm, and tobacco budworm can be lured and eliminated by taking advantage of their phototaxis, one solar insecticidal lamp can be installed in 15–30 mu of vegetable field.

4.2.1.3　Trap-killing by sex attractant

Male moths seeking mating in the field can be attracted by female pheromones and trapped in traps, so that the females lose the opportunity to mate and cannot effectively reproduce offspring. According to different pests, different numbers of sex attractant traps are generally set per mu, such as one dedicated sex attractant trap for common cutworm per mu, and one dedicated sex attractant trap for beet armyworm per mu.

4.2.1.4 Trap-killing by poison bait

According to the characteristics of nutritional supplement and nocturnal activity of pests, poison baits made of sugar-acetic-acid liquid, bran and fresh grass are used to lure and kill moths, crickets, underground pests, etc.

4.2.2 Biological prevention and control

(1) Biological natural enemy. It is possible to protect and utilize predatory natural enemies such as ladybird beetle, lacewing fly, syrphus fly, and spider, as well as parasitic natural enemies such as trichogramma wasp and encarsia wasp. The pests can be prevented and controlled by taking advantage of biological natural enemies, for example, the encarsia wasp can be used to control the greenhouse whitefly, the lacewing fly can be used to control aphids and red spider, the ladybird beetle can be used to control aphids, and trichogramma wasp can be used to control soybean moth.

(2) Microbial pesticide. *Bacillus thuringiensis* (Bt) is used for the control of common cutworm and cabbage moth, etc.; Beauveria bassiana and metarhizium anisopliae are used for the control of whitefly, aphid, and scarab, etc.; nuclear polyhedrosis viruses (NPV) can be used for the control of common cutworm and other pests.

(3) Botanical pesticide. Matrine is used for the control of red spider, aphid and whitefly, etc.; pyrethrins can be used for the control of cabbage worm, red spider and vegetable leaf miner, etc.; Sabadilla is used for the control of whitefly, etc.

4.2.3 Chemical prevention and control

(1) Bean pod borer. 4.5% beta-cypermethrin EC, 10% chlorfenapyr SC, 35% chlorantraniliprole WG, 30% indoxacarb WG can be sprayed for the prevention and control.

(2) Aphid. 3% acetamiprid ME, 25% pymetrozine SC, 10% imidacloprid EC, 10% cyantraniliprole SC can be sprayed for the prevention and control.

(3) Spider mites. 15% pyridaben EC, 43% bifenazate SC, 1.8% chongmanke (avermectin) and etoxazole EC, 24% spirodiclofen SC can be sprayed for the prevention and control.

(4) Vegetable leaf miner. 1.8% avermectin EC, 70% cyromazine WP, 10% chlorfenapyr SC, and 2.5% deltamethrin EC can be sprayed for the prevention and control.

(5) Whitefly. 70% imidacloprid WG, 25% thiamethoxam WG, 3% acetamiprid ME, or 10% nitenpyram AS can be sprayed for the prevention and control. If multiple insect states coexist on the back of the leaves, ovicides such as pyriproxyfen, chlorobenzuron, or hexaflumuron can be added to the above-mentioned pesticides.

(6) Beet armyworm and common cutworm. 5% emamectin benzoate ME, 5% chlorantraniliprole SC, 2.5% deltamethrin EC, and 4.5% beta-cypermethrin EC can be sprayed for the prevention and control.

5 Timely harvesting

During the flowering and podding peak stage of common beans, the lower inflorescence has pods, and the middle and upper inflorescences bloom one after another, which requires a large amount of nutrients. If the plant is overburdened, it is easy to cause blossom and pod dropping due to nutrient imbalance. Therefore, in addition to paying attention to fertilization at the early stage of flowering, tender pods should be picked in time to reduce the nutritional burden of the plant. The harvest of dwarf cultivar begins 50−60 d after sowing, and the harvest of sprawl cultivar begins 65−80 d after sowing, which can be continuously harvested for 30−60 d or longer. The optimal harvest time is 10−15 d after blossom dropping. Harvesting too early affects yield, and too late degrades quality. Harvesting can be conducted once every 3 d during the peak podding stage. Common beans after harvesting should not be stored for a long time, but should be kept ventilated and dry, otherwise they are easy to mold and deteriorate. After harvesting, sorting, packaging, and transportation should be done as soon as possible, in order to avoid affecting quality.

References

DAI C, HE Y H, BAO S Y, et al., 2017. Morphology genetic diversity analysis on common bean (*Phaseolus vulgaris* L.) germplasm resources in Yunnan [J]. Southwest China Journal of Agricultural Sciences, 30(2): 256−261.

Institute of Vegetables and Flowers, Chinese Academy of Agricultural Sciences, 2010. Olericulture in China [M]. 2nd Edition. Beijing: China Agriculture Press.

QIN W, CHEN K, LIU Y Y, et al., 2017. High-yield and high-efficiency cultivation technology of common bean in solar greenhouse [J]. Anhui Agricultural Science Bulletin, 23(13): 61−62.

QU Y M, ZHENG S H, MA R F, et al., 2021. Technical regulations for reducing fertilizer and pesticide application in common bean cultivation [J]. China Cucurbits and Vegetables, 34(2): 92−94.

TIAN R X, 2020. Improvement of cultivation techniques for spring sowing of open—field common beans [J]. China Vegetables (6): 111−112.

WANG B G, DONG J Y, WANG Y, et al., 2022. Evaluation and genetic diversity analysis of common bean germplasm in Zhejiang province, China [J]. Acta Agriculturae Zhejiangensis, 34(11): 2416-2427.

WANG Q, TAO J, YUAN Y, et al., 2018. Paddy-upland rotation between paddy rice and kidney bean in the mountainous areas of North Laos [J]. Chinese Journal of Tropical Agriculture, 38(4): 27-30.

❖ Key Techniques for Fast-growing Leafy Vegetables Cultivation

Fast-growing leafy vegetables generally refer to leafy vegetables with a relatively short growth cycle after seed germination. They typically mature and can be harvested within 4–6 weeks after planting. The rapid growth of fast-growing leafy vegetables allows for a quick harvest, making them suitable for obtaining a bountiful yield in a short period. This also helps increase crop yield and improve land resource utilization and multiple crop index within a brief growing season. Common fast-growing leafy vegetables include cabbage mustard, choy sum, green pakchoi, loose-leaf lettuce (Italian lettuce), amaranth, Indian Lettuce, etc. With the expanding market demand, fast-growing leafy vegetables have become an essential part of human daily diet.

Yunnan takes full advantage of its favorable natural environment and unique three-dimensional climatic conditions, establishing three major vegetable production regions for year-round vegetables, summer-autumn vegetables, and winter-spring vegetables. This has realized a development model where different types of vegetables are produced and supplied throughout the year. Among them, the year-round vegetable production region primarily focuses on fast-growing leafy vegetables, forming advantageous production areas for fast-growing leafy vegetables represented by places like Luliang, Songming, Jinning, and Zhanyi. These regions have a North Asian tropical highland monsoon climate, with an average annual temperature of 14–18 °C and an elevation of 1 800–2 000 m, providing suitable conditions for year-round cultivation of fast-growing leafy vegetables in plastic greenhouses.

1 Common fast-growing leafy vegetables

1.1 Cabbage mustard

1.1.1 Nutritional value

Cabbage mustard, belonging to the Brassica of family Cruciferae, cabbage species, cabbage mustard subspecies, are also known as brassica albograbra bailey, collard greens, Chinese kale and other names. They are annual herbaceous plants. In a 100 g sample of fresh flower stalks of cabbage mustard, the water content is approximately 90%. Other components include 0.74−1.00 g of reducing sugar, 1.60−2.08 g of protein, 1.20 g of fiber, 0.96−2.0 mg of β-carotene, and 81−101 mg of vitamin C. Additionally, they contain various trace elements. The flower stalks of cabbage mustard are characterized by their tenderness, crispness, sweetness, and delightful taste. They can be used in stir-fries, soups, or as a side dish.

1.1.2 Varieties and distribution

Cabbage mustard originated in China, primarily in the southern part of the country. Cultivation areas include Guangdong, Guangxi, Fujian, Jiangsu, Zhejiang, Yunnan, with a major concentration in the southern regions. Cabbage mustard is also found along the Mediterranean coast in foreign countries. Main cultivated varieties include young leaf early cabbage mustard, willow leaf early cabbage mustard, Fujian yellow-flowered cabbage mustard, heat-resistant cabbage mustard, etc.

1.2 Choy sum

1.2.1 Nutritional value

Choy sum is a variety of the Chinese cabbage subspecies of *Brassica rapa* in the Brassicaceae family, an annual or biennial herbaceous vegetable plant, also known as Chinese flowering cabbage, green choy sum, and choy tip among others. Choy sum contains per kilogram of edible part: 13–16 g of protein, 1–3 g of fat, 22–42 g of carbohydrates, 410–1 350 mg of calcium, 270 mg of phosphorus, 13 mg of iron, and 1.0–13.6 mg of carotene. Choy sum has a unique fragrance characteristic of *Brassicaceae* vegetables, tender texture, and its rich fiber content helps promote intestinal movement, aiding digestion, preventing constipation, diluting intestinal toxins. It is an indispensable premium vegetable for residents in Guangdong, Hong Kong, Macao.

1.2.2 Varieties and distribution

Choy sum is mainly distributed in Guangdong, Guangxi, Taiwan, Hong Kong, macao, and other regions. It was successfully introduced in Japan in the late 20th century. Currently, choy sum is the vegetable with the largest cultivation area in Guangdong Province, with an annual planting area of over 270 000 mu, accounting for 30% of the annual market supply of vegetables. In recent years, choy sum has also been cultivated in Ningxia, Sichuan, Gansu, Yunnan, and other places, and is listed as a high-quality specialty vegetable. Generally, based on the growth period and adaptability to the cultivation season, it can be divided into three types: early maturing, mid-maturing, and late maturing. The main cultivated varieties include Si Jiu Choy Sum, Si Jiu Choy Sum No. 19, Guilin Willow Leaf Early Choy Sum, Baoqing 40 days, Green stemmed Willow Leaf Mid Choy Sum, March Green Choy Sum, Teqingchixin No. 4, and others.

1.3 Green pakchoi

1.3.1 Nutritional value

Green pakchoi, commonly known as non-heading Chinese cabbage, is a plant of *Brassica rapa* in the Brassicaceae family. Also called bok choy or pak choi, it possesses advantages such as heat and cold resistance, strong adaptability, and high economic benefits. Green pakchoi is rich in vitamins, proteins, carotenoids, as well as minerals such as calcium, phosphorus, iron, potassium, sodium, magnesium, chlorine, etc. According to measurements, its vitamin C content is more than 3 times that of Chinese cabbage.

1.3.2 Varieties and distribution

Green pakchoi, originating from China, has a cultivation history of over 1 000 years in our country. Fujian province is a major production area for green pakchoi. Conventional varieties are predominant, with the main cultivated varieties including Xia Di, Hua Guan, Jin Pin 1 Xia, Jin Pin 552, Su Zhou Qing, Si Yue Man, Wu Yue Man, etc.

1.4 Loose-leaf lettuce

1.4.1 Nutritional value

Loose-leaf lettuce, also known as Italian lettuce, is an annual or biennial herbaceous plant belonging to the Asteraceae family and Lactuca genus. It is a leafy lettuce with leaves as the edible organ. Loose-leaf lettuce includes various non-heading lettuces such as butter lettuce, red wave lettuce, sauteed lettuce, loose-leaf lettuce, upright lettuce, wild lettuce, and flower leaf lettuce. It is rich in nutrients, containing a significant amount of beta-carotene, antioxidants, vitamin B_1, vitamin B_6, vitamin E, vitamin C, dietary fiber, and trace elements such as magnesium, calcium, as well as small amounts of iron, copper, and zinc. It has a fresh and tender taste, boasting high nutritional and economic value, as well as unique health benefits.

1.4.2 Varieties and distribution

Loose-leaf lettuce is a semi-cold-resistant vegetable that prefers cool temperatures and is sensitive to high temperatures. The most suitable growth temperature is 16−22 °C during the day and 10–12 °C at night. It can withstand high temperatures up to 30 °C and low temperatures down to −1 °C. There are significant temperature requirements differences among varieties, with sauteed lettuce, butter lettuce, red wave lettuce, and wild lettuce being less tolerant to high temperatures and suitable for growth at lower temperatures. American California Dasu Lettuce and Italian Lettuce are heat-resistant and cold-resistant, with broad adaptability. Loose-leaf lettuce can be cultivated throughout the year, preferring sunlight and avoiding shade. It thrives in a humid environment and is not tolerant of drought. Adequate nitrogen fertilizer is essential, along with phosphorus, potassium, and trace element fertilizers. It is best suited for planting in deep, loose, and organically rich soil. Key cultivated varieties include Yinong Four Seasons Heat-resistant Lettuce, American California Dasu Lettuce, Rosa Lettuce, Roman Upright Lettuce, and Loose-leaf Lettuce.

1.5 Indian Lettuce

1.5.1 Nutritional value

Indian lettuce, belonging to the Asteraceae family, is an annual or biennial herbaceous plant primarily consumed for its leaves, which contain various nutritional components. Rich in vitamin A, Indian lettuce stands out among vegetables. It also boasts significant levels of ascorbic acid and folic acid. Ascorbic acid stimulates the body's hematopoietic function, facilitates the conversion of cholesterol in the blood, leading to a reduction in blood lipids, while folic acid plays a role in protecting the cardiovascular system. Indian lettuce is particularly suitable for stir-frying or serving as a cold dish, offering a delightful combination of tenderness, crispiness, and a pleasant fragrance.

1.5.2 Varieties and distribution

China stands as one of the primary producers of leaf lettuce. Leaf lettuce is extensively cultivated in the southern regions of China, with a focus on local varieties and heat-tolerant loose-leaf varieties. Some examples of these varieties include Bei San Sheng, Zi Yu, Salinas 88, Yun Cui, Hong Fan Purple Leaf Lettuce, and Lüqun Lettuce.

2 Main cropping pattern

Fast-growing leafy vegetables are predominantly cultivated in open fields, low tunnels, steel-frame greenhouses, and unit-building canopies, with an emphasis on facility-based cultivation. Due to the characteristics of fast-growing leafy vegetables, such as short growth cycles and quick maturation, different regions have conducted exploratory research on the annual high-quality and efficient production cycles of these crops. They have identified effective planting models that are suitable for various areas, with the goal of achieving year-round production and consistent supply of fast-growing leafy vegetables.

2.1 Year-round green pakchoi cultivation model

Taking the central region of Jiangsu Province as an example, a year-round planting model for green pakchoi has been explored, with Shanghai Green as the main variety. This model is implemented in cities such as Yancheng and Nantong, where six rotations are planted throughout the year, with harvests every 40 d in summer and autumn, and every 60 d in winter and spring. The production process incorporates a combination of core technologies, including mechanical tillage, seeding, harvesting, and integrated water and fertilizer management. Additionally, eco-friendly pest control techniques such as insect-attracting color boards and vibratory insect-killing lamps are used to enhance both the yield and quality of the produce.

2.2 Year-round production model for fast-growing leafy vegetables in Shanghai

Taking the Shanghai region as an example, the cultivation sequence of crown daisy cabbage mustard-choy sum-Indian lettuce-cabbage mustard has been proven effective through many years of production practice. In this sequence, crown daisy is sown in mid-January and harvested in early March; The first batch of cabbage mustard is sown in late March and harvested in late May; choysum is sown in early June and harvested by mid July; Indian lettuce is sown at the end of July and harvested in early October; And the second batch of cabbage mustard is sown in mid October and harvested in early January of the following year. This model significantly enhances land use efficiency, facilitates proper crop rotation, and ensures the quality of the produce.

2.3 Year-round production model for fast-growing leafy vegetables in Yunnan

Taking Yunnan Province as an example, a cultivation model of "Indian lettuce–yellow cabbage-green pakchoi-Italian lettuce-yellow cabbage-Indian lettuce" has been formed by reducing the occurrence of pests and diseases through crop rotation. Based on the growth characteristics of four types of fast-growing leafy vegetables and market demand, a cultivation model of six crops per year has been established. This not only improves land utilization but also generates good economic benefits. The use of crop rotation, soil solarization, and high-temperature soil disinfection techniques reduces the occurrence of pests and diseases, decreases the use of pesticides, and improves vegetable quality.

3 Cultivation and management techniques

3.1 Soil disinfection

Soil disinfection is carried out during the idle period of the greenhouse. Before disinfection, it is necessary to thoroughly remove any residues inside the greenhouse, as well as other vegetables and weeds, and apply an appropriate amount of farmyard manure in combination with land preparation work. By irrigating, the soil relative moisture is maintained at around 60%. Following this, the greenhouse is covered with mulch film, and maintains a high temperature inside for approximately 13 d to ensure soil temperature reaching 60 °C. This step is crucial for eradicating soil-borne pathogenic bacteria and parasites, thereby accomplishing the objective of soil improvement.

3.2 Seedling cultivation

In order to cultivate robust seedlings and enhance land utilization, fast-growing leafy vegetables adopt seedling cultivation and transplanting. This approach can shorten the growth period of young seedlings in the field by 20−25 d. Intensive seedling cultivation facilitates centralized pest and disease control and concentrated nutrient supply, ensuring the uniformity of field plants after transplanting. In general production, commercial substrate trays are used for seedling cultivation, with one seed per cavity and a seeding depth of 0.5−1 cm. Seedlings typically emerge 4−5 d after sowing. During hot midday periods in summer and autumn, shade nets are used, and ventilation is increased to reduce temperature. Seven days after emergence, a 0.1% solution of compound fertilizer (N : P : K=15 : 15 : 15, the same below) is applied, followed by the 0.2% solution after 15 d. Fertilization is carried out 2−3 times during the winter seedling period. In summer and autumn, the seedlings age is 20−25 d, while in winter and spring, the seedlings age is around 30 d. Exercising seedings before transplantation involves controlling watering frequency, increasing ventilation and light exposure, allowing seedlings to adapt to the external environment, and enhancing the survival rate of transplantation.

3.3 Land preparation and ridge management

Leafy vegetables with quick growth and small plant size are generally suitable for wide-spacing and dense planting. Before preparing the land, apply 2 000 kg of thoroughly decomposed farmyard manure or commercial organic fertilizer per mu, along with 30 kg of compound fertilizer and 200 kg of microbial fertilizer. After loosening and turning over the soil, level the planting surface with a rake. Treat the entire greenhouse as one planting surface without separately reserving operational pathways. Yellow cabbage should have a row spacing of 30 cm×30 cm, green pakchoi 15 cm×15 cm, Italian lettuce 25 cm×25 cm, Indian lettuce 15 cm×20 cm, choy sum 10 cm×10 cm, and cabbage mustard 10 cm×10 cm. Winter cultivation is done under plastic mulch.

3.4 Water and fertilizer management

Select a cloudy or sunny afternoon for planting seedlings to prevent leaf scorching during the midday heat. After transplanting, ensure timely watering with sufficient root-setting water to enhance the seedlings' survival rate. During the seedling stage, adjust the frequency of watering appropriately and use sprinkler irrigation for fast-growing leafy vegetables. Adapt the watering regimen according to weather conditions and plant growth. The short growth cycle of fast-growing leafy vegetables allows for a primary focus on base fertilization in terms of fertilization. Applying fertilizer once or twice during the growth period is adequate to fulfill their growth requirements. After transplanting the seedlings and allowing them to recover, it is advisable to apply a 0.3% urea solution for seedling enhancement. For yellow cabbage transplanting, the second round of fertilizer should be applied 20 d after transplantation, using 20 kg of compound fertilizer per mu. For green pakchoi transplanting, the second round of fertilizer should be applied 15 d after transplantation, using 15 kg of compound fertilizer per mu, or irrigating with a concentration of 0.5% water-soluble fertilizer. After transplanting Italian lettuce for 20 d, the second round of fertilization begins, with 15 kg of compound fertilizer per mu. Indian lettuce undergoes two rounds of fertilization 15 d after transplanting, with 15 kg of compound fertilizer per mu in each round, or spraying 0.5% water-soluble fertilizer. For choy sum and cabbage mustard, 15 kg of compound fertilizer per mu is applied 10 and 20 d after planting. For fast-growing leafy vegetables, the frequency of fertilization can be increased in winter based on the plant's growth conditions. Simultaneously, combine pesticide spraying and the foliar application of 1–2 rounds of foliar fertilizer for optimal results.

4 Prevention and control of diseases and pests

Diseases affecting fast-growing leafy vegetables include downy mildew, soft rot, black rot, black spot, sclerotinia rot, gray mold, anthracnose, and virus diseases. Common pests include yellow striped flea beetle, aphid, daikon leaf beetle, diamondback moth, *Pieris rapae*, *Spodoptera litura*, and beet armyworm.

4.1 Agricultural comprehensive prevention and treatment

Before planting, the land should be thoroughly cleaned, including the removal of weeds, plant residues, and the centralized collection of discarded materials. During the growth period of vegetables, promptly remove leaves affected by diseases or pests and uproot plants damaged by diseases or pests. After harvesting vegetables, the field should be thoroughly cleaned by taking diseased plant residues out of the field for centralized burning, deep burial, or placement in a dung cellar, aiming to reduce the occurrence and spread of diseases and pests. During the fallow period, it is advisable to plow deeply in winter, taking advantage of the freezing and thawing cycles. The upturned soil and land should be exposed to sunlight in summer, with lime spread before plowing to disinfect and eradicate pests and diseases in the soil. Mid-tillage weeding should be moderately reduced to prevent wounds from human activities and to protect against pathogen invasion. Refrain from cultivating the same type of vegetables in the same plot for consecutive years, thereby minimizing the transmission of soil-borne diseases.

4.2 Control of microbial pesticide

In the context of vegetable disease management, preference should be given to the utilization of biological pesticides. For the control of diseases such as soft rot and black rot, biopesticides like 3% zhongshengmycin WP, 2% kasugamycin AS can be selected. To prevent virus diseases, options include 8% ningnanmycin AS, 2% amino-oligosaccharin AS. In the case of anthracnose and wilt diseases, 1 billion spores/g *Bacillus subtilis*, and 0.15 billion spores/g *Trichoderma* can be employed. To control pests such as the *Pieris rapae* and diamondback moth, 2% abamectin·*Bacillus thuringiensis* WP can be applied. The WP of *Spodoptera litura* nucleopolyhedrovirus can be used to control the *Spodoptera litura*, and the WP of *Metarhizium anisopliae* with 8 billion spores/g can be employed to prevent pests such as wireworms, grubs, and cutworms.

4.3 Control of plant-derived pesticide

Use 0.5% physcione AS [1 : (500−600)] to control powdery mildew, downy mildew, gray mold, anthracnose, etc. Use 4% berberine AS [1 : (300−500)] to control powdery mildew, downy mildew, etc. Use 1% osthole EW [1 : (800−1 000)] to control powdery mildew, etc. Use 0.3% matrine AS [1 : (500−700)] to control aphids, whiteflies, etc. Use 0.5% veratrine AS [1 : (400−600)] to control *Pieris rapae*, aphids, etc. Use 3% pyrethrins EC [1 : (50−80)] to control *Pieris rapae*, diamondback moths, aphids, etc.

4.4 Chemical agents for disease control

4.4.1 Downy mildew

When the lesions first appear on the bottom leaves, it is recommended to use 80% zineb WP, 722 g/L propamocarb AS, 70% propineb WP, 70% fluorobacterium·frosmovir SC, 10% cyazofamid SC, and other agents for spray treatment.

4.4.2 Soft rot and black rot

In the early stages of the disease, agents such as 2% kasugamycin AS, 1% zhongshengmycin WP, 40% zinc thiazole SC, 20% benziothiazolinone SC, 47% kasumin·bordeaux WP, and 50% chloroisobromine SG can be used for spraying.

4.4.3 Black spot

In the early stages of the disease, agents such as 50% iprodione WP, 70% antracol WP, 10% difenoconazole WG, 25% azoxystrobin SC, 45% prochloraz EW, and 25% pyraclostrobin SC can be used for spraying.

4.4.4 Anthracnose

In the early stages of the disease, 10% difenoconazole WG (1 : 1 500), or 50% iprodione WP (1 : 1 000), or 45% prochloraz EW (1 : 2 000) can be used for spraying.

4.4.5 Virus diseases

To cut off the transmission and spread of viral diseases, it is crucial first to control the transmission vectors like aphids, whiteflies, thrips, red spiders, etc. Next, chemical control should be adopted by timely spraying 1.5% sodium dodecylsulfate WP or 20% moroxydine hydrochloride WP, 8% ningnanmycin AS, and other preventive sprays before the onset or during the early growth stage. Spraying should occur every 7−10 d and this treatment should be repeated 2-3 times continuously for best results. Pay attention to alternating pesticide use.

4.4.6 Sclerotinia rot

In cases where the pathogenic point and central disease pond appears sporadically in fields, agents such as 50% procymidone WP, 50% iprodione WP, 40% dimethachlone WP, 70% thiophanate-methyl WP can be applied. Spraying should occur every 5-10 d, continuously for 2-3 treatments, with careful rotation of pesticides.

4.5 Prevention and control of pests

4.5.1 Physical control

4.5.1.1 Control with insect-proof netting

When used independently, silver-gray or black insect-proof netting should be chosen. When used in conjunction with shade nets, it is advisable to choose white netting with a mesh size of 40 to 60. For summer and autumn cultivation, insect-proof netting is generally used to cover the entire greenhouse. Fast-growing leafy vegetables cultivated in summer and autumn, due to their short growth period and relatively concentrated harvest, can adopt mulching cultivation of little arch sheds. Before covering the insect-proof netting, it is essential to promptly eliminate any existing pests within the greenhouse. After covering the insect-proof netting, some pests and diseases may still be transmitted through soil, seeds, and seedlings. Therefore, it is recommended to treat the seedlings with insecticide and fungicide before planting, selecting robust plants free from pests and diseases.

4.5.1.2 Control with insect-trapping card

Yellow insect-trapping boards are effective in attracting and eliminating aphids, whiteflies, *Bemisia tabaci*, planthoppers, leafhoppers, liriomyza, while blue ones are useful against corn seed maggots and thrips. The insect-trapping board should be made of high-quality material, capable of trapping insects on both sides, non-toxic, resistant to sun exposure, and durable against rain wash. In each mu of a greenhouse, it is recommended to hang 30-40 insect-trapping boards (measuring 30 cm long and 25 cm wide), suspended 15-20 cm above the plant's highest growth point to ensure optimal pest control effectiveness. The height of the yellow boards should be adjusted as the plants grow. When the insect-trapping boards are fully covered with pests, they need to be replaced in a timely manner.

4.5.1.3 Control with light trap

In vegetable cultivation, frequency-vibrancy pest-killing lamps are commonly utilized. They effectively target a range of pests including cabbage armyworms, black cutworms,

Spodoptera litura, beet armyworms, helicoverpa assult, cabbage webworms, diamondback moths, leaf beetles, and more. Typically suspended at a height of 1.5 m, the lamps are strategically placed in a checkerboard pattern across the field, spaced 100–150 m apart. Each lamp's effective control area spans approximately 20–40 mu. Prompt and regular removal of pest corpses from the trapping bags is crucial.

4.5.1.4 Control with sex attractant

Sex attractants can be used to lure and kill *Spodoptera litura*, beet armyworms, diamondback moths, black cutworms and other pests. Sex attractants for diamondback moths are used in May—June and July—September, with one trap per mu, containing 3 lure cores. For *Spodoptera litura* and beet armyworms, sex attractants are used in July—October, with one trap per mu, containing 1 lure core. The trappers are spaced 30 m apart, hung on canopy frames or wooden sticks, elevated 30 cm above vegetables. In spring and fall, lure cores are replaced every 30 d, while in summer, they are replaced every 20 d.

4.5.1.5 Biological control with natural enemies

In the case of whitefly pests, Encarsia formosa is suggested for control. *Encarsia formosa* should be released 7–10 d after transplantation, if whitefly pests are detected, at a rate of 2 000 individuals per mu. This release should be repeated every 7–10 d, for a total of 3–5 times. For thrips pests, biological control of natural enemies is recommended to use *Orius minutus*, releasing them at a rate of 500 individuals per mu every 7–10 d, for a total of 2–4 times. To manage mite pests, the *Phytoseiulus persimilis* can be employed. These should be applied at 5–10 individuals per square meter on leaves, increasing to 30 individuals per square meter at the center of affected areas when infestation spots are observed. This release should be repeated every 2 weeks, for a total of 3 times. Aphid pests can be controlled by using ladybugs. Ladybugs (eggs) and *Aphidius rhopalosiphi* should be released at a rate of 2 000 per mu upon detection of pests, with the process repeated every 7–10 d for 2–3 times. For lepidopteran pests such as diamondback moths, beet armyworms, cotton bollworms, and *Spodoptera litura*, it is advisable to release trichogramma and apanteles sp. This should be done at a rate of 20 000 per mu every 5–7 d, for a total of 3 times, to attain effective pest control.

4.5.1.6 Chemical control

For controlling aphids, whiteflies and thrips, a spray of 3% acetamiprid microemulsion ME, 10% imidacloprid EC, 25% thiamethoxam WG, or 60 g/L spinetoram SC can be used. For controlling diamondback moths and *Pieris rapae*, a spray of 5% emamectin benzoate ME, 5% chlorantraniliprole SC, 3% abamectin EC, or 2.5% decamethrin EC is effective. To control *Spodoptera litura* and beet armyworms, use a spray of 2% emamectin benzoate ME,

10% chlorfenapyr microemulsion SC, or 30% indoxacarb WG. For leafminer flies, a spray of 5% abamectin ME, 70% cyromazine WP, or 10% chlorfenapyr microemulsion SC can be used. These treatments should be applied every 7–10 d, with 2–3 consecutive applications, and pesticides should be alternated to prevent resistance.

5 Post-harvest treatment

5.1 Harvesting criteria

Harvest promptly based on the characteristics of fast-growing leafy vegetables and market demands. Harvesting is generally done in the morning and afternoon, avoiding midday harvests. For leafy vegetables grown in open fields, harvest after the dew has dried. The vegetables are stored in plastic turnover boxes and promptly transported them to the cold storage after harvesting.

5.2 Storage and fresh-keeping

In locations where conditions permit, leafy vegetables undergo vacuum pre-cooling for 20–40 min to remove field heat and preserve their freshness. Following vacuum precooling, the leafy vegetables are stored in a cold room at 3–5 °C. Initial vegetable processing and grading can be conducted within the cold room, with grading standards established based on the requirements of the target market.

5.3 Packaging and transportation

After initial processing and grading, leafy vegetables are typically packaged in foam boxes. The packaging plastic wrap is placed inside the foam box. Two ice bottles are placed diagonally within the foam box, and the exterior of the ice bottles is wrapped with a layer of newspaper to prevent freezing. In locations with suitable conditions, cold trucks can be employed for transportation. When using a trailer for transportation, the vehicle is insulated with thermal blankets, and the entire vehicle is covered with a thin film to minimize vegetable losses during transit.

References

Institute of Vegetables and Flowers, Chinese Academy of Agricultural Sciences, 2010. Vegetable Cultivation in China [M]. 2nd Edition. Beijing: China Agricultural Press.

WANG Q, KONG L M, TAO J, et al., 2017. Cultivation technology of fast-growing leafy vegetables during the rainy season in Northern Laos [J]. China Vegetables (3): 96–98.

WANG Q, NIAN H Y, YUAN Y, et al., 2019. Application of multi-cropping cultivation technology for fast-growing leafy vegetables in Yunnan facilities [J]. Agricultural Engineering Technology, 39(4): 72–74.

❖ Key Techniques for Pumpkin Cultivation

Originating in South America, pumpkin, in the Cucurbitaceae family and *Cucurbita* genus, is an annual vine or trailing plant. To date, China has been the largest producer and consumer of pumpkin since it first came to China. There are 5 significant cultivars: Chinese pumpkin (*Cucurbita moschata* Duch.), Zucchini (*Cucurbita pepo* L.), squash (*Cucurbita maxima* Duch.), *Cucurbita ficifolia* B, and Silver Seed Gourd (*Cucurbita argyrosperma* Huber.). This chapter mainly introduces cultivating techniques in China.

There are many familiar names for *Cucurbita moschata* Duch., including Maigua (Wheat Melon), Fangua (zucchinis), Wogua (golden nugget), Fangua (melon with rice), Beigua (north squash). All stems, leaves, flowers, seeds, and greens are edible apart from the fruit. The edible ratio exceeds 80% or more. The fruit has a thin rind, thick meat, delicate tissue, a sweet flavour, and high nutritional value. It is rich in starch, fat, reducing sugar (RS), amino acids, vitamins, minerals, and cellulose. According to research, there are 0.7–2.0 g of protein, 0.1–0.5 g of fat, 3.3–11 g of sugar, 0.34–0.78 mg of Vitamin A, and 6–48 mg of in every 100 g of pumpkins. In addition, reveal research that hat polysaccharides, carotenoids and other substances in pumpkin can also reduce blood lipid, regulate blood sugar level and improve immunity.

1 Growth conditions

1.1 Temperature

Pumpkin seeds germinate when the temperature is 28–30 °C. It takes 7–10 d for cotyledons to extend. Seedlings grow well at 25–30 °C during the day. Flowering and fruits are setting well at 25–28 °C during the day and 13–18 °C at night. Higher than 35 °C or lower than 13 °C leads to unhealthy growth of plants. If the temperature exceeds 35 °C, it will lead to withered flower organs and pollen abortion.

1.2 Moisture

Pumpkins are resistant to drought due to well-developed roots. Relative moisture requirements remain 55%–70%. In general, watering demands little in the seedling period.

Just guarantee sufficient water for setting the root. From the extension to the flowering period, water once when the land is too dry to crack. If water is too much, overcrowded plants hinder pumpkinset in the later stage. In the fruit-developing period, enough watering is necessary. Otherwise, fruits will deform, causing productivity to diminish. Underwatering leads to plant wilt, causing the fall-out of flowers and fruits. Overwatering causes fruit and flower loss. Alternatively, too much or too little watering causes an outbreak of mildew, hindering flowering and pollination. Hence, soil moisture status is an essential indicator for watering in cultivation.

1.3 Sunlight

Pumpkin is one kind of short-day plant that needs full sun to grow. Under 8 h of short-day sunlight, female flowers differentiate. Nevertheless, the differentiation rate will be low if the sunlight is over 12 h. The more direct sun, the better the pumpkin crop. Weak lights lead to tender and slim plants, which are inclined to grow excessively. In that regard, melon setting ratings will be lower, and melons will even melt.

Control the light hours within 8 h by covering shade and screen, which can lower the node of female flowers, increase melon ratings and increase the productivity of melons.

1.4 Soil nutrition

Pumpkins prefer warmth and are drought-tolerant, suitable for planting in neutral or slightly acidic loam or sandy loam plots with deep soil layers with good air permeability and high fertility. Pumpkin needs little fertilizer before the vine-pumping period. Fertilizers are greatly needed in the fruit setting period. Phosphorus and potassium absorption increase, but nitrogen intake decreases.

2 Varieties

Select good quality, bountiful harvest, strong resistance and commercially viable varieties refering to local weather, cultivation purposes and market demand, such as Miben and Yunnan local Jiang Bing (pattypan squash).

3 Plough and fertilization

Plough the soil and expose it to the sun for one month before ploughing and transplanting. Apply fully decomposed organic fertilizer 3 000 kg/mu and 30 kg/mu nitrogen, phosphorus, potassium compound fertilizer before sowing or 2–3 d before transplanting. Be mindful to spray fertilizers evenly into plots three to four meters wide (including ditches). In terms of fertilizer reservation and efficiency, apply 1 kg of organic fertilizers and 50 g of three-nutrient compound fertilizer to each plot. After returning and recovering two to three meters of soil, mix it and save it for future use.

4 Sowing and seedlings nursery

4.1 Processing seeds

Selecting complete and big new grains guarantees the emergency rate, eliminating shrivelled and mouldy grains. The germination test should be carried out before sowing. Soak seeds in 50–55 °C lukewarm water for 15 min, during which keep string. After that, cast seeds to germinate in an incubator at 28 °C after 3–4 h of emergence.

4.2 Direct sowing

The direct sowing period is generally a frost-free period. For instance, if sorrowing in Yunnan, where the altitude is 1 800–2 000 m, it usually begins in the middle of April or at the end of April. Sow in plots. Before sowing, water the ponds until they seep out. Sow 2–3 seeds in each plot. Seeds should be separated and covered with 2–3 cm of fine soil. Sowing too deep hinders the emergence of seedlings, too shallow to come out with seed coats. Generally, the row spacing and plant spacing of climbing or creeping cultivation are 3–4 m and 0.6–0.8 m, respectively. The row spacing and plant spacing of frame cultivation are 1.0–1.5 m and 0.6–1.0 m, respectively. When there are 2–3 true leaves on the seedlings, the seedlings are fixed twice. Sort out the seedlings with mechanical damage, diseases, and insect pests for the first time. Select and reserve the seedlings with the strongest growth for the second time.

4.3 Transplanting seedlings

Put the seed on the soil's surface or insert seedlings with tips upward in the 32-hole trap, which has been covered by substrate-one seed in one hole. One seed is sown in each hole, covering about 2 cm of soil. Seedlings can emerge in 5–7 d by keeping the temperature of the seedbed at 25–28 °C. Keep the temperature at 20–25 °C during the day and 13–15 °C at night after seedlings. It should begin ventilating and cooling the temperature 7 d after planting. Keep the temperature down to 20–22 °C during the day and 10–12 °C at night. Keep the relative humidity of the seedbed at 60%–70%, and spray water in time when water is scarce.

In substrate seedling raising, water-soluble fertilizer should be supplemented appropriately after the emergence of true leaves, and 0.2% urea or potassium dihydrogen phosphate solution should be sprayed after the emergence of the first true leaf.

Standard of Robust Seedlings: aged in 20–25 d with three to 4 true leaves. Seedlings are 10–15 cm tall with thick leaves, dark green leaves, and short and thick internodes at the base of stems, free from diseases and insect pests.

Select seedlings one week prior to transplant. Keep the temperature from 15–25 °C. Enhance ventilation and proper control of water. Transplanted, strong and robust seedlings with soil on sunny days should be selected for fixation. It is better to cover the soil lump in case of direct exposure to the fertilizer. Enough root fixation water is needed after transplanting.

4.4 Seedling management

When raising seedlings in early spring, excessive watering will lead to a drop in ground temperature, which is not conducive to the growth of seedlings. When the temperature is high and the light is insufficient, excessive watering will lead to the overgrowth of seedlings, and it is easy to cause diseases at the seedling stage, such as downy mildew. Therefore, water little but water frequently is necessary for both getting wet and getting dry of the soil.

5 Fields management

5.1 Fertilization and water management

During the tendril extension period, water and fertilizer should be controlled to promote rooting, and less or no watering should be given to prevent excessive water and fertilizer from causing excessive plant growth. During the flowering and fruiting period, after the first melon starts to expand, water is poured once to promote the expansion of the melon, and fertilizer is applied once with the water. Generally, ternary compound fertilizer (N : P : K=15 : 15 : 15) or high potassium water-soluble fertilizer 10–15 kg/mu is used. Pay attention to the flowering and fruiting period, and avoid topdressing a large amount of nitrogen fertilizer, which will quickly lead to overgrowth of plants, falling flowers and fruits, etc. If tender melons are harvested, after each batch of melons is harvested, it is necessary to topdress once with water when new shoots grow.

5.2 Branches arrangement and melon reserving

Generally, only one main vine is left for each plant, and all other lateral vines are removed. If eating old melons, only one melon will be left for each plant. After the melons are firmly established, top them at a distance of 40–50 cm from the melons, leaving more than six functional leaves. If eating tender melons, 4–6 tender melons can be left according to the habit of melon bearing, and the growing point can be removed after the last melon is firmly seated.

If frame cultivation is adopted, the old leaves and diseased leaves at the root should be knocked out during the period, and the tendrils should be appropriately lowered according to the length of the main tendrils. For climbing cultivation, the soil should be cultivated at the leaf nodes before the melon seedlings stretch their vines. When the melon vines are 1 m long, they should be pressed for the first time, and then every 1 m long or so, and the vines should be pressed at the second node after the melon. Method of pressing tendrils: Open a shallow ditch at the edge of the tendrils, gently move tendrils into the ditch, and press them with earth. The tendril pressing time is generally carried out at noon or afternoon when the temperature is high.

5.3 Mid-tillage and weed removal

Weeds are easy to spread due to the large row spacing. Therefore, mid-tillage weeding should be carried out. Pay attention to hurting roots when weeding. The first mid-tillage weeding should be done after slow seedling, and the tillage depth should be 3–5 cm. The second mid-tillage should be carried out in combination with tendrils when tendrils are extended, and the tillage should be carried out according to the growth of weeds in the later stage.

5.4 Manual-assisted pollination

In order to improve the fruit setting rate, auxiliary pollination should be carried out. First is insect-assisted pollination: By planting flowers with long flowering periods and bright colours in melon fields, honey-collecting insects or artificially fostered bees are attracted to pollination. Second is manual-assisted pollination: If the flowering period encounters low temperatures and rainy weather or no insect pollination in the facility, artificial-assisted pollination is required. Generally, bagging is carried out one day before flowering, and pollination is finished before 10:00 the next day. Pollination method: Remove the petals of male flowers, align the stamens with the female flowers, and gently smear them on the stigma of the female flowers. Pay attention not to touch small melons during pollination, which can easily cause bacterial infection and melons. After pollination, gently bind the petals with a rope to prevent the rain from washing away the pollen.

6 Nutrient defficiency symptoms and the prevention

6.1 Nitrogen deficiency

Symptoms: Stunted leaves and light green new leaves. They turn yellow from the downside to the upper part. Veins become yellow, and fewer fruits are set after flowering, which is slow in size expansion.

How to prevent: Apply decomposed organic fertilizers referring to specific demand for nitrogen, phosphorus and potassium to prevent nitrogen deficiency. Nitrate nitrogen is used when the temperature is low; Nitrogen fertilizer is immediately sprayed on the roots or foliage when the symptoms are found in field pumpkins.

6.2 Potassium deficiency

Symptoms: Abnormal colour or yellow spots on older leaves on the lower parts of plants. Brown or burnt-like spots, golden gilt, are found on the leaf edges. Later, the edges are damaged and dead.

How to prevent: Apply potash fertilisers or spray potassium sulfate and 0.2% potassium dihydrogen phosphate on foliage.

6.3 Magnesium deficiency

Symptoms: Begin with interveinal chlorosis between the leaf veins on older leaves, from the middle to the edges. The discolourration or yellowing later turns up.

How to prevent: Spray 0.2% magnesium sulphate solution.

6.4 Zinc deficiency

Symptoms: Begin in leaves and fruits. Affected plants have stunted growth on nodes. At the same time, the discolouration or whitening turns on veins.

How to prevent: Do not use too much phosphate fertilizer. Selectively apply acid fertilizer to lower the soil pH. Spray 0.2% zinc sulfite solution in the field.

6.5 Calcium deficiency

Symptoms: Seen in robust and young leaves, such as newer leaves, tipping boots, and fruits. Leaves are distorted, yellowing, or whitening at the edges, where they fold and twist upward or downward (in parachute shape). Calcium deficiency may also result in cracking and blossom-end rot. Those symptoms are found in the blossom end or the top of plants.

How to prevent: Spray calcium fertilizer on foliage. For instance, spray 0.3% calcium chloride solution plus 0.2% potassium dihydrogen phosphate or apply 10 kg of fused calcium magnesium phosphate and potassium fertilizer on 1 mu of field.

6.6 Iron deficiency

Symptoms: First seen on young leaves, they are akin to calcium and magnesium deficiency. In comparison, the difference is yellowing stars between the veins (interveinal chlorosis), which are still green. The leaves later take up yellow or whitening.

How to prevent: Keep the soil pH value at 6–6.5 in drought or wet conditions. Furthermore, spray 0.1%–0.2% ferrous sulfate solution.

6.7 Boron deficiency

Symptoms: Include flowering but no fruit set. The internodes near the growing point shortened significantly, the growing point stopped growing, and the root system did not occur. The upper leaf edge curls upward, the leaf edge part turns brown, the leaf unfolds slowly, and the upper leaf veins shrink at the same time. There are stains and signification on the surface of the fruit.

How to prevent: Apply organic fertilizer. Based on reasonable application of organic fertilizer, nitrogen fertilizer is controlled, and 0.1% borax solution is sprayed on the leaves.

7 Prevention and control of diseases and pests

7.1 Diseases control

The most common pumpkin diseases are powdery mildew, virus disease, downy mildew, phytophthora blight and so forth.

7.1.1 Powdery mildew

It is the most common and severe disease of pumpkin, which is emergent in hot and windtight summer. The disease is caused by fungi, fairly host specific, mainly including *Erysiphe cichoracearum* and *Podosphaera xanthii*, which are rarely to be isolated.

7.1.1.1 Symptoms

The pathogen weakens leaves first, then stems and petioles. When it is severe, fruit surfaces can be affected. The white and powdery mould first appears on foliage, stems, and petioles. They are expanding into white spots as the disease continues to develop. White powder covering the entire leaves and white spots on stems and petioles signal a severe outbreak, leading to wilt leaves.

7.1.1.2 Managements

(1) Agricultural Control. Select and grow disease-resistant varieties. Normal rotations with noncucurbit crops are recommended. Appropriate fertilization is needed but it should be avoided using nitrogen fertilizer only. Apply organic fertilizer and potassium phosphate fertilizer. Enhance plant management by timely removal of infected leaves on the base and older leaves. Trim branches in case of overcrowded plants for good air circulation.

(2) Chemical Control. Immediate and timely treatment is effective against powdery mildew. It is prioritized to spray the undersides of leaves. At the initial stage of occurrence, spray 15% triadimefon WP (1 : 1 500), 25% azoxystrobin SC [1 : (1 000−2 000)], 25% bupirimate EC (1 : 1 500), and 250g/L pyraclostrobin EC (1 : 1 500). Once every 7−10 d, continuously for 2−3 times.

7.1.2 Virus disease

It is complicated in symptoms, barely being distinguished between deficiency or pest insects. Pathogen spreads quickly with a wide range of hosts in culture through insect vectors and mechanical operations that disturb plants and bruise leaves and vines.

7.1.2.1 Symptoms

Infected leaves and fruits will be distorted or tend to be mottled, with areas of tonal differences in green colour. Light yellow, barely observed stripes will appear on foliage when plants are inflicted at the beginning. Later, these patterns consist of irregular patches and are dark green and yellow, giving the leaves a mottled appearance. At the same time, leaves are stunned, with edges curling, puckering and crisp. Some tender leaves on shoot tips will be brown and burn-like, developing into dieback. Some nodes shorten, with leaves on branches clustering. Green stripes or blotches are on mature pumpkins. Severe affected foliage and pumpkins will be misshapen. Mosaic viruses will lead to delayed growth and smaller pumpkins in size.

7.1.2.2 Managements

(1) Agricultural Control. Select and plant disease-resistant varieties. Rotations with non-cucurbit crops and any crops but other cucurbit family are recommended. Timely clean and control weeds in the fields. Remove infected plants once they are found. Bury and burn those plants somewhere out of the fields.

(2) Chemical Control. Spray 5% oligosaccharides AS (1 : 2 000), 1% lentinan AS (1 : 600), 80% moroxydine hydrochloride WP (1 : 1 000), 5.9% cinnamaldehyde·moroxydine hydrochloride AS (1 : 300), every 5–7 d, continuously for 2–3 times.

7.1.3 Downy mildew

The causal agent of pumpkin downy mildew, a kind of infectious disease, is *Pseudoperonospora cubensis* (Berk. et Curt) Rostov. The pathogen sporangia attaches to the leaf surface, reproducing and spreading via air, rainwater, machines and cultural activities.

7.1.3.1 Symptoms

Both seedlings and adult-stage plants may be victims of the pathogen, mainly leaves. At the initial infection, water-immersed spots, whose edges are vague, appear on the undersides of leaves. Yellowing lesions and spots have been formed on the surface, expanding in angular yellow and borrowing spores down to the plants' upper part. The infected plants will be burnt yellow and killed to the end.

7.1.3.2 Managements

(1) Agricultural Control. Select and grow disease-resistant varieties. Rotations with non-cucurbit crops and any crops but other cucurbit family are recommended. Timely clean up diseased leaves. Trellis plants and on-time trim to improve air circulation. Increase the concentration of carbon dioxide in utilities.

(2) Chemical Control. Spray 0.5% chitosan WP (1 : 600) to prevent. Spray 58% metalaxyl WP (1 : 1 000), 75% chlorothalonil WP (1 : 600), 72% manganese zinc and creamy urea (1 : 800), 80% dimethomorph WG (1 : 2 000) to treat diseased plants once in 5–7 d, continuously for 2–3 times at the initial stage of occurrence.

7.1.4 Phytophthora blight

The causal agent of phytophthora blight, a vegetative disease, is *phytophthora* spp. The disease emerges and spreads when the temperature is between 25 °C and 30 °C, and the relative air moisture exceeds 85%, or within and among airtight fields. It is prone to break out in the rainy season or a sudden sun-lit after heavy rain when the temperature climbs to the highest.

7.1.4.1 Symptoms

The pathogen may affect plants throughout the growing period, affecting stems, vines, leaves and fruits. Lesions readily form on the stems and fruits of pumpkins, causing rotten fruits and stems. Chlorosis may be found on foliage until the whole plants wilt and die. The diseased part of the stem seems water-soaked and light brown at the beginning and then gradually turns brown and wet rot, and the diseased part has cotton mould. At the beginning of the disease, the leaves appeared round, dark, water-immersed spots, soft rot, drooping, greyish brown when dry, and easy to crack. The fruit is mainly the pumpkin cultivated on the ground, which is vulnerable. The disease spot is dark green and waterlogged at the beginning and then gradually wet and rotten. There is cotton mould on the surface of the diseased part.

7.1.4.2 Managements

(1) Agricultural Control. To select and grow early ripening and resistant cultivars. Four to five years of rotation with non-host crops, such as graminaceous crops. Do not plant too much for good air and well-drained soil conditions, which lower the chances of an outbreak. Timely clean off weeds. Trim side branches, cover vine with soil and pull over pumpkins. Remove infected plants once they are found. Bury or burn those plants somewhere out of the fields.

(2) Chemical Control. In the initial stage of the infection, spray 722 g/L propamocarb hydrochloride (D) AS (1 : 600), 58% metala WP (1 : 600), 30% metalaxyl·hymexazol AS [1 : (600−800)], 50% dimethomorph WP (1 : 1 500), 2−3 times and once every 7−10 d.

7.2 Prevention and control of pests

The common pests that are most likely to show up in pumpkin patches are whiteflies, aphids, cutworms, thrips.

7.2.1 Whiteflies

7.2.1.1 Symptoms

They suck up pumpkin plants' juices, in turn, produce sticky honeydew, causing fungal disease. The honeydew hinders plants from carrying out photosynthesis. The excrement of adult whiteflies will be detrimental to plants' respiration, later inflicting diseases with shrivelling fruits and leaves.

7.2.1.2 Managements

(1) Agricultural Control. Clean patches and timely remove weeds. Frequently trim and prune—To select the main vine, cut off tertiary vines and trim overgrown secondary vines. Remove older and diseased leaves.

(2) Physical Control. Use yellow sticky traps.

(3) Chemical Control. In the initial stage of whitefly occurrence, spray 10% imidacloprid SC (1 : 1 000), 25% buprofezin WP (1 : 2 000) to prevent. As whiteflies are active, spray is not easy to detect. It is better to spray water to dislodge them first and spray pesticides. Apply once a week, continuously for 3 times.

7.2.2 Aphids

7.2.2.1 Symptoms

They are detrimental to all stages of pumpkin growth. Usually, aphids or their nymphs gather in vast numbers and feed on the underside or tender leaves with their sucking mouth parts. That victim leaves and tenders' shoots will be distorted, wilted and dead, which weakens fruit development. Moreover, the honeydew, a by-product of aphids' feeding, vectors viruses to pumpkin leaves and fruits.

7.2.2.2 Managements

Spray 3% acetamiprid EC (1 : 1 000), 10% imidacloprid WP (1 : 1 500), and 25% thiamethoxam WG (1 : 4 000) to prevent and manage aphids.

7.2.3 Cutworms

7.2.3.1 Symptoms

Cutworms, moth larvae, are known as underground worms and soil warms. The 1–2-year-old larvae often destroy spear and young plants, causing white spots and small holes. Since they are three years old, those larvae cut down pumpkin stems or plants nearest the soil. Because of the worms, seedling loss and cultivar damage appear.

7.2.3.2 Managements

(1) Agricultural Control. Timely remove and take away any weeds, which will dislodge nymphs and eggs.

(2) Physical Control. Make an insecticide with sugar, vinegar, liquor and water to the ration of 3 : 4 : 1 : 2 and add small amount of trichlorfon. To place the solution at a height of 50–100 cm with 10–15 sets per mu.

(3) Chemical Control. Spray 2.5% deltamethrin EC (1 : 1 500), 40% cypermethrin EC (1 : 2 000), and 5% lambda-cyhalothrin ME (1 : 2 000) to prevent and manage.

7.2.4 Thrips

7.2.4.1 Symptoms

Both thrips adults and nymphs pierce and suck contents of spear leaves, tender shoots, flowers, and young pumpkins. After their feeding, the tender leaves are damaged and shrink, and their pubescence becomes grey-brown or blackish-brown. The plant growth is stunned as nodes shorten. Suppose they feed on young pumpkins, fruits harden, whose fluff turns black and later falls off. If they feed on tender leaves, they will be marked with silvering or grey-brown stippling alongside the vein. The leaves look like black spots on the surface, distorted, curled and stunned.

7.2.4.2 Managements

(1) Agricultural Control. Timely clean off weeds, which will dislodge eggs. Clean and sanitize soils before ploughing.

(2) Physical Control. Thrips are attracted to the colour blue. Place blue sticky traps in the fields.

(3) Chemical Control. Spray 25% thiamethoxam WG [1 : (5 000–6 000)], 20% acetamiprid WP [1 : (5 000–8 000)], 28% abamectin·spiromesifen SC (1 : 4 000) to prevent and control.

8 Timely harvesting

Chinese pumpkin is generally eaten as a mature fruit, which can be harvested when the rind hardens and presents the inherent colour of this variety. The fruit powder increases and the fruit stalk is lignified about 60 d after pollination of female flowers. Tender melons, which can be harvested when they reach commercial maturity about 20 d after pollination, are welcomed as well. When harvesting, leave the melon handle 2–3 cm, and be careful not to damage the melon rind. After harvesting, store in a cool, ventilated and dry place, which can be stored for 2–3 months in general and 6 months at most.

❖ Key Techniques for Cucumber Cultivation

The cucumber (*Cucumis sativus* L.) is an annual climbing or trailing herbaceous plant in the cucurbitaceae family, and is an important globally popular vegetable. Its cultivation area ranks fourth, only followed by tomatoes, cabbage, and onions. Cucumbers are predominantly cultivated in Asia, accounting for approximately 50% of the global cultivation area, followed by Europe, North America, and Central America. China's cucumber cultivation area accounts for approximately 28% of the world's total, ranking first among all countries. Different types of cucumbers are cultivated in various regions of Yunnan, and plastic greenhouses are commonly used for production.

Cucumbers are rich in nutrients, with every 100 g of fresh fruit containing 1.6–2.4 g of carbohydrates, 0.4–0.8 g of protein, 10–19 mg of calcium, 16–58 mg of phosphorus, 0.2–0.3 mg of iron, 4–16 mg of vitamin C. Cucumbers are also rich in dietary fiber, which accelerates the elimination of waste from the body and lowers cholesterol. Additionally, they have the functions of clearing heat, detoxifying, promoting diuresis, and quenching thirst. Modern medical clinical practice has proven that cucumber vines have a good effect on lowering blood pressure and reducing cholesterol.

1 Growth conditions

1.1 Temperature

Cucumbers prefer warmth and are not cold-tolerant. They will wither in the event of frost. The optimal temperature for growth is 20–25 °C during the day and 12–16 °C at night. Physiological disorders occur when temperatures drop below 10 °C or exceed 35 °C, causing growth to halt.

1.2 Light

Cucumbers are short-day plants and thrive in well-lit conditions. They can also tolerate low light. Flowering and fruiting are promoted with 8–11 h of sunlight. Insufficient light can negatively impact cucumber yield and quality. The light saturation point for cucumbers is 55 000–60 000 lx, the light compensation point is 15 000 lx, and the optimal light intensity is

20 000−60 000 lx. Plant growth slows when light intensity falls below 20 000 lx.

1.3 Moisture

High cucumber yields require significant water. The optimal soil relative moisture ranges from 60% to 90%, with lower moisture levels during the seedling stage at 60%−70%, and adequate moisture during fruiting at 80%−90%.

1.4 Soil and nutrition

Cucumber roots have weak regenerative abilities and low nutrient uptake, making them susceptible to waterlogging. Therefore, loose, fertile, and well-ventilated sandy loam soil with a pH value of 5.5−7.2 is preferred. Cucumbers have strict requirements for nutrients and water, especially in cases of nitrogen or potassium deficiencies, which can lead to flower drop and bitter-tasting fruits (due to glucoside). For every 1 000 kg of cucumbers produced, 1.7 kg of nitrogen, 0.99 kg of phosphorus, and 3.49 kg of potassium are required, with more than 80% of the required nutrients needed during the fruiting stage.

2 Variety selection

It is recommended to select cucumber varieties that are tolerant to low temperatures and weak light, have strong disease resistance, can set fruit on both main and lateral vines, have predominantly female flowers, or have strong parthenocarpy. Currently, most of the cucumber varieties promoted in Yunnan Province are introduced from other provinces or overseas, including Jinza No. 1, Jinza No. 2, Jinza No. 3, Jinchun No. 1, Zaochun No. 2, Jinchun No. 3, Jinyan No. 7, Zhongnong No. 5, Zhongnong No. 7, Liangtiao Wang, Jindun 10−12, Yameite 2188, Chunwang F1-A Cucumber King, Jinpi cucumber (golden skin cucumber), Naire Wang (heat-resistant king), improved super (454) spineless cucumber, Laifu 13−18 spineless cucumber, and Lüyougua king (green excellence cucumber king).

3 Cultivation and management techniques

3.1 Land preparation and ridges

After the previous crop is harvested, it is necessary to plow the land to a depth of 20−25 cm, apply 2 000−3 000 kg of well-rotted farmyard manure and 25−30 kg of calcium superphosphate or 10−15 kg of diammonium phosphate per mu near the planting period.

Thoroughly mixing the soil with the fertilizer should be done. It usually needs to ridge the soil before planting, with ridge width of 80 cm, height above 20 cm, and furrow width of 30 cm, planting in double rows. After ridging, covering the soil with plastic film and tightly sealing it with soil should be conducted.

3.2 Sowing and seedling

3.2.1 Conventional seedling

3.2.1.1 Substrate selection

It is advisable to use a mixture of peat, vermiculite, and perlite as an alternative to nursery soil. The ratio of peat, vermiculite, and perlite is 3 : 1 : 1, and 1 m^3 of substrate should be mixed with 1.2 kg of compound fertilizer (N : P : K=15 : 15 : 15) and 25 g of 50% carbendazim WP by thoroughly being mixed and then resting for 2−3 d before use.

3.2.1.2 Seed disinfection

Cucumber seeds often carry various pathogenic bacteria such as wilt disease, black spot disease, blight, damping-off, brown spot disease, and angular leaf spot. Necessary disinfection of seeds before seedling cultivation can effectively prevent diseases caused by seed-borne pathogens. Cucumber seed disinfection is usually done through the methods of soaking seeds in a solution or coating them with a disinfectant. Typically, 0.1% hydrochloric acid solution of carbendazim is used for soaking seeds for 1 h, followed by rinsing with clean water. Subsequently, the seeds are soaked in clean water for 4 h before germination. Depending on the seed contamination, nethanal (300-fold dilution) can be used to prevented wilt disease and black spot disease. For virus diseases, 10% solution of trisodium phosphate can be used for soaking for 30 s, resulting in effective disease prevention. Coating seeds with fungicides like captan, fenaminosulf, and carbendazim, with a dosage of 0.3%−0.5% of the seed weight, is another common method. The process involves placing seeds in a container, adding the fungicide, covering, and shaking until the powder adheres evenly to the seed surface.

3.2.1.3 Soaking and germination

After disinfection, the seeds proceed to the soaking and germination stage. The procedure involves rinsing and soaking the disinfected seeds for 5−6 h, followed by drying and packaging in cloth. They are placed in a well-ventilated and shaded area with a controlled temperature of around 26 °C. Regular turning is recommended to avoid uneven temperature distribution, and germination usually begins approximately 24 h later.

3.2.1.4 Sowing

(1) Sowing time: The optimal sowing and seedling time is typically 30 d before

transplantation, considering the characteristics of the variety, local climate, cultivation mode, and market demand.

(2) Sowing method: It is suggested to plant one seed per hole, lay the seeds flat or with the sprout tip facing downwards, at a depth of 1−1.5 cm. After sowing, it should be covered with substrate, flattened by using a wooden ruler or stick, and then thoroughly watered with a spray bottle.

3.2.1.5 Temperature control

Temperature control is crucial in cucumber seedling cultivation. From sowing to emergence, the bed temperature should be maintained at 25−30 °C during the day and 22−23 °C at night to promote germination. After emergence, the temperature can be slightly reduced to 20−25 °C during the day and 14−16 °C at night. After slowing down the seedlings, the temperature should be appropriately regulated to prevent futile growth and promote flower bud differentiation, especially by lowering the night temperature. Conditioning the seedlings 5−7 d before transplantation involves daytime temperature of 20−23 °C and nighttime temperature of 10−12 °C. This enhances disease resistance and adaptability to the environment after transplantation.

3.2.1.6 Strong seedling standards

Seedlings are considered robust when they reach a height of around 15 cm, with 3−4 true leaves at one node, intact cotyledons, short and stout internodes, dark green and glossy leaves, developed and healthy roots without diseases, and an age of approximately 30 d.

3.2.2 Grafting and seedlings cultivation

Cucumbers should be sown 3−5 d earlier than the rootstock. The optimal grafting period is when the rootstock has fully expanded its two cotyledons and the first true leaf is about to unfold, while the scion's cotyledons have turned from yellow to green and are fully expanded, just before the emergence of the first true leaf.

Generally, the inarching method is employed. Using a blade, the growing point of the pumpkin is removed from the cotyledon to prevent the development of lateral shoots. A downward oblique cut is made about 1 cm below the pumpkin cotyledon at an angle of 35°−40°, with a depth of half the stem's thickness. Subsequently, an upward oblique cut is made about 1.5−2 cm below the cucumber cotyledon at an angle of approximately 30°, with a depth of 3/5 of the stem's thickness. The two cuts are interlocked, pressing the two cucumber cotyledons onto the pumpkin cotyledons, and a grafting clip is used to secure them. During the grafting process, precautions should be taken to avoid infection and damage to the seedlings, as this could lead to tissue necrosis. Grafting should be carried out in conditions of controlled temperature and humidity, avoiding direct sunlight to prevent desiccation of the scion, which could affect survival rates.

Post-grafting management: The temperature and relative humidity are crucial for the survival of grafted seedlings. Adequate water should be supplied to well-grafted seedlings, and a small arch should be placed over them. Within 1–4 d, the relative humidity inside the arch should be maintained at around 95%, with daytime temperatures of 25–28 °C and nighttime temperatures of 15–18 °C. These conditions promote wound healing under high temperature and humidity. After 4 d, gradual ventilation and exposure to light should be introduced. One week after grafting, once the union is complete, the cucumber roots are pruned with a blade, the grafting clip is removed, and a disinfectant like chlorothalonil is applied. Generally, transplantation can be done 25–35 d after grafting.

3.3 Field management

3.3.1 Planting

3.3.1.1 Planting time

For spring cucumber seedlings with 4–5 true leaves and a height of 5–10 cm, transplantation should be done as early as possible, ensuring that there is no risk of frost after transplanting. The minimum nighttime temperature should be above 5 °C, and the soil temperature at 0–10 cm depth should be higher than 12 °C. For autumn cucumber seedlings with 2–3 true leaves and an age of around 20 d, transplantation should be done.

3.3.1.2 Planting density

Plant density should be 4 000–4 500 plants per mu, with both small and large rows. The small row spacing is 40 cm, the large row spacing is 80 cm, and the plant spacing is 25–30 cm.

3.3.2 Trellising

Trellising should be done soon after transplanting to prevent wind damage and facilitate vine growth. Trellising can use either flower racks or an A-frame, with a height of 1.8–2 m and a distance of 10 cm from the base.

3.3.3 Tying vines

When cucumber seedlings reach 30 cm, the "8"-shaped method should be used for tying vines to prevent stem abrasion and downward drooping of the vines. It is recommended to tie the vines every 3–4 leaf nodes. The operation should be performed in the afternoon as the vines are prone to break in the morning. The tightness of the vine tying should be adjusted to support the weak and restrain the strong. For vigorously growing plants, it is advisable to tie them tightly to maintain a uniform height for cucumber seedlings.

3.3.4　Pruning and pinching

For varieties that mainly produce fruit on the main stem, removing lateral shoots promptly should be conducted. For varieties that can bear fruit on both main and lateral stems, it is recommended to remove the lateral shoots below the first fruit (root fruit) promptly, leave one fruit on the upper lateral shoot and two leaves before the fruit, and then pinch the tip. When the main stem reaches the top of the trellis, it should pinch the tip, and timely remove lower, yellow, diseased leaves, and deformed fruits to save nutrients, enhance ventilation and light penetration, and reduce the occurrence of diseases and pests.

3.3.5　Water and fertilizer management

3.3.5.1　Irrigation

Timely irrigation and inter-row cultivation should be based on weather conditions and growth stages. From transplanting until the first fruit is stable, the focus should be on "control," with appropriate hardening of seedings, more inter-row cultivation and soil aeration, and less watering to improve the growth environment, promote root development and make flower buds differentiate in large numbers. After entering the fruit-setting stage, the focus should be on "promotion," and follow the rule of "light, heavy, light." The amount of water applied during the root fruiting stage should not be excessive, maintaining the soil surface at a consistently moist level. During the middle fruiting stage, when temperatures are high and sunlight is abundant, plants exhibit vigorous growth and require more water.

3.3.5.2　Fertilization

Fertilization should adhere to the principles of using fertilizers and follow the guidelines of "light in the beginning, heavy in the end, small amounts and multiple times, and additional phosphorus and potassium fertilization". Generally, it should be applied with water, with clear water once and fertilizer water once, and water can be used with fertilizer when fruit-sitting reaches to the peak stage. Organic fertilizer, slow-release fertilizer, and quick-acting fertilizer should be combined during fertilization. The total amount of nitrogen fertilizer should be regulated, with not exceeding 40 kg of pure nitrogen per mu. It's recommended to apply urea at 10−15 kg and 10 kg of potassium sulfate per mu each time, along with 0.3% potassium dihydrogen phosphate for spraying leaves.

4 Prevention and control of diseases and pests

4.1 Major diseases

4.1.1 Downy mildew

4.1.1.1 Symptoms

In the early stages, water-soaked small spots form on the underside of the leaves. Later, the lesions gradually expand, taking on a polygonal, water-soaked appearance due to the restriction by leaf veins. In humid conditions, a purple-black mold layer develops on the lesions' surface. On the front side of the leaf, the lesions are initially yellow with indistinct edges, turning yellow-brown later. Severe cases result in extensive lesions, which causes the entire leaf to curl and wither, with only the heart leaf remaining.

4.1.1.2 Prevention and control measures

(1) Agricultural control: It is advisable to choose disease-resistant varieties and practice crop rotation with non-cucurbitaceae crops or alternate between wet and dry cultivation.

(2) Physical control: High-temperature closed greenhouses are effective against cucumber downy mildew. On a sunny morning, it is suggested to water the plants and close the greenhouse, raising the temperature to 43–45 °C for 1.5–2 h. And then slowly opening the vents should be performed to gradually lower the temperature. After the high-temperature treatment, enhancing fertilizer and water management are needed.

(3) Chemical control: In the early stages, it is recommended to alternatively use 80% dimethomorph WG (1 : 2 000), 68.75% triflumizole·propamocarb SC (1 : 1 000), 75% chlorothalonil WP (1 : 500), and 58% metalaxyl mancozeb WP (1 : 400) for spraying every 7 d, continuously for 2–3 times.

4.1.2 Anthracnose (Anthrax)

4.1.2.1 Symptoms

In young seedlings, brown semi-circular or circular lesions appear at the edge of cotyledons. The base of the stem is affected, leading to shriveling and discoloration, which causes the seedlings to collapse. The leaf lesions are reddish-brown with a yellow halo. Immature fruits are less susceptible, but when mature, they exhibit light green and water-soaked spots that quickly turn black-brown, enlarge and become concave with a darker center and small black dots on the affected parts. Infected fruits become distorted, a common occurrence in seed cucumbers.

4.1.2.2 Prevention and control measures

In the early stages, it is recommended to use 70% thiophanate-methyl WP [1 : (1 000–1 500)], 50% prochloraz-manganese chloride complex WP [1 : (1 000–2 000)], 70% thiophanate-methyl·thiram WP (1 : 600), 35% triflumizole-tebuconazole SC (1 : 2 000) for spraying every 7 d, continuously for 2–3 times.

4.1.3 Powdery mildew

4.1.3.1 Symptoms

Powdery mildew primarily affects leaves, stems, and petioles but generally spares the fruit. In the early stages, white and nearly-circular powdery spots develop on the front or back of the leaves, gradually expanding into large, contiguous powdery patches with indistinct borders. With these diseased spots covering the whole leaf surface, these powdery patches turn gray-white or reddish-brown, and the leaves become yellow and brittle, though generally not shedding.

4.1.3.2 Prevention and control measures

In the early stages, it is recommended to use 250 g/L azoxystrobin SC (1 : 600), 250 g/L pyraclostrobin EC (1 : 1 500), 15% triadimefon WP (1 : 1 500), 50% thiophanate-methyl WP (1 : 1 000) for spraying every 7 d, continuously for 2–3 times.

4.1.4 Bacterial angular spot

4.1.4.1 Symptoms

Bacterial angular spot affects leaves, presenting light brown lesions on the front side and angular lesions on the back, limited by leaf veins. In the early stages, leaf lesions appear water-soaked and later dry up and fall off. Fruit and stem lesions start as water-soaked spots with visible milky bacterial pus. Young seedlings may develop water-soaked circular spots on cotyledons and cotyledons are slightly sunken, which later turn brown and dry up, potentially causing seedling softening and death if the infection spreads to young stems.

4.1.4.2 Prevention and control measures

Treatment options include 77% copper hydroxide WP (1 : 500), 30% copper (succinate glutarate adipate) WP (1 : 200), 47% kasugamycin·dicopper chloride trihydroxide WP (1 : 600), 3% zhongshengmycin WP (1 : 600), and 14% cupric-amminium complexion AS (1 : 300) for spraying every 5–7 d, continuously for 2–3 times, alternately using different drugs.

4.1.5 Blight

4.1.5.1 Symptoms

The typical symptom of blight is wilting. In the early stages of the disease, plants exhibit a gradual wilting of leaves from bottom to top, resembling a water-deficient condition. This becomes more pronounced around noon, with partial recovery in the morning and evening. After a few days, the entire plant's leaves wilt, droop, and no longer return to their normal state. The base of the stem may show slight constriction, and some affected plants may exude amber-colored colloidal material. The roots of diseased plants become brown and rotten. The base of the stem often longitudinally splits, and under humid conditions, a white or pink mold layer may appear on the surface of the affected area. Seedlings affected by the disease may exhibit wilting of cotyledons or complete plant wilting, with the base of the stem turning brown and constricted, often appearing as if suddenly collapsed.

4.1.5.2 Prevention and control measures

(1) Agricultural control: It is advisable to practice crop rotation with non-cucurbit crops, select cucumber varieties with stronger resistance, cultivate robust cucumber seedlings, use grafted seedlings, and promptly remove and burn infected plants.

(2) Chemical control: In the early stages of the disease, a mixture of 80% metalaxyl mancozeb WP (1 : 800) and 80% zineb WP (1 : 1 000) can be applied as a root drench and foliar spray. Additionally, a 2% kasugamycin WP at a rate of 673−900 g per mu can be applied as a root drench, and a 3% metalaxyl-hymexazol AS (1 : 500) can be sprayed every 7 d, continuously for 2−3 times.

4.1.6 Scab of cucurbi (*Cladosporium cucumerinum*)

4.1.6.1 Symptoms

Scab of cucurbi can infect various parts of the plant throughout its growth period, including leaves, stems, tendrils, fruit stems, and growing points. The disease is most severe on tender parts such as young leaves, stems, and young fruits, while older leaves and fruits are less sensitive to the pathogen. Infected seedlings show yellow-white circular spots on cotyledons, with the cotyledons rotting, and severe cases may lead to the complete decay of the seedling. When young leaves are infected, initially small round greenish spots appear on the leaf surface, which later expand into 2−5 mm light yellow lesions with star-shaped margins. The lesions dry up and turn yellow-white, forming star-shaped holes with yellow halos on the edges in the later stages. Infected young stems initially exhibit water-soaked dark green rhomboid lesions, which later darken, become concave and cracked, and develop a gray-black mold layer in humid conditions. When the growing point is infected, the central leaves wither, resulting in the formation of a bare stub.

Infected fruit stems initially show circular or elliptical greenish fading spots, and a translucent yellow-brown jelly-like substance overflows from the lesions, congealing into masses. The lesions gradually enlarge and become concave, with an increase in the jelly-like substance, which accumulates near the lesions and eventually falls off. In high humidity, the affected areas are densely covered with a black mold layer.

4.1.6.2 Prevention and control measures

(1) Agricultural control: It is recommended to choose disease-resistant varieties, treat seeds with disinfectants, practice water and dry rotations or crop rotations with non-cucurbit crops. As scab of cucurbi is a disease favored by low temperatures and high humidity, it often occurs in early spring greenhouses and winter greenhouses. What's more, it is necessary to strengthen field management, practice appropriate planting density, increase greenhouse temperature, use techniques such as plastic film covering and drip irrigation to save water, ventilate timely to reduce humidity inside the greenhouse, and shorten the dew time on leaf surfaces.

(2) Chemical control: In the early stages of the disease, 400 g/L flusilazole EC (1∶4 000), 50% carbendazim WP (1∶500), 75% thiophanate-methyl WP (1∶600), and 20% myclobutanil-thiram WP (1∶500) can be used. It is advisable to spray every 7 d, continuously for 2–3 times.

4.2 Major pests

4.2.1 Aphids

4.2.1.1 Symptoms

Aphids use piercing-sucking mouth parts to extract a large amount of sap from plants, causing stunted growth, curled leaves, buds that cannot open, and premature aging and senescence of plants. Excess sap extracted by aphids is excreted, attracting ants, promoting fungal infection, and triggering sooty mold disease.

4.2.1.2 Prevention and control measures

In the early stages within facilities, spraying with 70% acetamiprid WG [1∶(7 000–10 000)], 10% imidacloprid WP (1∶1 000–1 500), or 25% thiamethoxam WG [1∶(3 000–5 000)] is recommended.

4.2.2 Yellow Tea Mites [*Polyphagotarsonemus latus* (Banks)]

4.2.2.1 Symptoms

Infested leaf undersides exhibit a gray-brown or yellow-brown color with an oily sheen or oil-soaked appearance. Leaves become stiff and thickened, with edges curling downward. Affected tender stems and branches turn yellow-brown, become distorted, and severe infestations can lead to the disappearance of cucumber growth points, resulting in top wilting.

4.2.2.2 Prevention and control measures

In the initial stages, it is suggested to control yellow tea mites by spraying 25% thiamethoxam WG (1 : 1 500) or a 10% imidacloprid WP [1 : (1 000−1 500)].

4.2.3 Greenhouse Whiteflies (*Trialeurodes vaporariorum*)

4.2.3.1 Symptoms

Adult greenhouse whiteflies congregate on the undersides of leaves and feed on leaf sap, causing leaves to fade, wither, and lose their green color. This leads to the deformation and stiffening of fruits, causing premature plant senescence. Additionally, greenhouse whiteflies secrete large amounts of honeydew, resulting in contamination of leaves and fruits and often causing extensive sooty mold.

4.2.3.2 Prevention and control measures

It is recommended to control greenhouse whiteflies by spraying 20% dinotefuran SG (1 : 1 500), 25% thiamethoxam WG (1 : 7 000), 25% buprofezin WP (1 : 2 500), or 2.5% lambda-cyhalothrin EC (1 : 3 000).

4.2.4 Thrips

4.2.4.1 Symptoms

After thrips damage leaves, they turn yellow−white after losing their green color. Affected fruits exhibit slow growth and can become deformed. Severe cases lead to dropped fruits, significantly impacting cucumber quality and yield.

4.2.4.2 Prevention and control measures

It is recommended to control thrips by spraying with 20% dinotefuran SG (1 : 1 500), 5% imidacloprid EC [1 : (1 500−2 000)], 1.8% avermectin EC [1 : (4 000−5 000)], 10% fenpropathrin EC [(1 : 1 000−1 500)], or 40% 1-Methyl-3-phenyl-5-(3-trifluoromethyl) phenyl)-4(1H)-pyridinone and pymetrozine WG (1 : 3 000).

5 Timely harvesting

Harvesting timely should be performed when the cucumbers meet the commercial standards based on their variety characteristics. During the initial stage of fruit setting, it is recommended to harvest once every 3–4 d, while during the peak fruiting period, it should be harvested daily or every other day. Single cucumber weight ranges from 100–150 g in the early stage and 150–250 g in the middle to later stages. When picking, it is advisable to leave a 1 cm-long fruit stalk at the cucumber handle, use scissors to cut it off gently and handle carefully to avoid mechanical damage, aiming to maintain the thorn at the flower tip. After harvesting, avoiding exposure to sunlight or rain, placing them in a shaded or cold storage area for pre-cooling, and promptly packaging should be done.

Timely harvesting is crucial. Harvesting too early results in lower yields and substandard products with compromised flavor, quality, and color. On the other hand, late harvesting not only increases the burden on seedlings and affects yields but also results in products that are less storable and transportable. It is advisable to retain the fruit stem during harvest. After harvesting, the cucumbers should be placed in a shaded area or pre-cooled in a cooling facility to dissipate heat, avoiding exposure to direct sunlight, which helps maintain product quality and prevents quality degradation.

6 Post-harvest treatment

6.1 Grading and packaging

After harvesting, cucumbers need to be sorted by removing their stems and cleaning any dirt off the skin. They should then be graded and packaged, arranging them neatly in the boxes or baskets according to the same variety, grade, and size specifications. The height of cucumber packaging boxes should not be excessive to prevent bruising of cucumbers at the bottom layer during transportation. The packaging boxes should indicate basic product information such as variety, grade, specifications, net weight, place of origin, etc.

6.2 Transportation

During transportation loading and unloading, precautions should be taken to prevent mechanical damage. For transportation between producing areas, it is suggested to choose regular vehicles. Attention should be given to sun protection, moisture preservation, and ventilation. Cool measures in summer and frost prevention in winter should be conducted. Cucumbers that have undergone pre-cooling can be transported for up to 10 hours by using insulated vehicles. Beyond 10 h, refrigerated vehicles should be used, maintaining a temperature of 12 °C.

6.3 Storage and preservation

Cucumbers are prone to spoilage, necessitating strict storage conditions. To extend the supply period and enhance economic benefits, unsold cucumbers should be stored and preserved at the producing sites or sales location.

References

CAO R R, 2023. Key points in facility cucumber cultivation techniques [J]. Xiandai Nongcun Keji (5): 29−30.

GUO H X, 2022. Characteristics and cultivation techniques of cucumbers [J]. Henan Agriculture (22): 40.

KANG X N, WU Y H, ZHOU Y W, et al., 2023. Key points in greenhouse cucumber compound substrate cultivation techniques [J]. Xiandai Nongcun Keji (6): 31−32.

LIU Y F, 2022. Common disease prevention and control techniques for cucumbers [J]. Xiandai Nongcun keji (4): 31.

XIAO Y C, 2004. Post-harvest standardized treatment techniques for pollution-free cucumbers [J]. Jiangsu Agricultural Science (4): 97−98.

YIN D X, ZHAO J J, LI Y, et al., 2023. Simplified cultivation techniques for cucumber vine regrowth and summer survival [J]. Journal of Changjiang Vegetables (1): 29−30.

❖ Key Techniques for Bitter Gourd Cultivation

Bitter gourd (*Momordica charantia* L.) is an annual vine-like herbaceous plant of Momordica in the Cucurbitaceae family. Bitter gourd is also known as bitter melon, bitter cucumber, balsam pear, balsam apple, bitter apple, red lady, etc. Bitter gourd is usually consumed when it is young. In many Southeast Asian countries, people also consume the tender shoots, leaves and flowers, and its juice is used for medicinal purposes.

Bitter gourd has high nutritional value, with abundant minerals, amino acids, and various vitamins in its young fruits. Bitter gourd is a treasure in its entirety, with its roots, stems, leaves, flowers, fruits, and seeds all having medicinal uses. It is recorded in the *Compendium of Materia Medica* that bitter gourd has the effects of "dispelling evil heat, relieving fatigue, and clearing the mind and improving vision". It has been reported that bitter gourd also has the functions of lowering blood sugar, antibacterial, anti-inflammatory, antiviral, and improving the body's immune system. Bitter gourd is not only rich in various active ingredients such as plant insulin, bitter gourd extract, peptides, sugars, and amino acids, but also contains a large amount of vitamin C, crude protein, and soluble sugars. Bitter gourd flesh is bitter yet slightly sweet, fresh and fragrant, making it a highly nutritious and health-promoting vegetable. The contents of trace elements such as calcium, magnesium, and iron which are beneficial to human health far exceed those of vegetables such as tomatoes, eggplants and peppers. The content of vitamin C in bitter gourd is 14 times that of cucumber and 5 times that of winter melon, making it the highest among melon vegetables.

With the re-recognition of the nutritional value and various therapeutic effects of bitter gourd in recent years, bitter gourd production in China has developed rapidly, with the cultivation area expanding year by year. It has become a popular vegetable in the market.

1 Growth conditions

1.1 Temperature

Bitter gourd requires relatively high temperatures for growth, being heat-tolerant but not cold-tolerant, with strong adaptability to temperatures ranging from 10 °C to 35 °C. The optimal temperature for seed germination is 30–35 °C. Although the seed coat of bitter gourd is thick, soaking the seeds in 40−50 °C warm water for 4−6 h can promote germination, and germination can start after 48 h at the optimal temperature, with over 70% of the seeds germinating after 60 h. Germination is slow below 20 °C and difficult below 13 °C. Germination is fast and plant growth is vigorous at around 25 °C, and true leaves can reach 4−5 leaves after 15−20 d. The optimal temperature for flowering and fruiting is around 25 °C. Within the range of 20−25 °C, higher temperatures are more favorable for the growth and development of bitter gourd plants, resulting in earlier fruiting and higher yields.

1.2 Light

Bitter gourd is a short-day crop that prefers sunlight and is not tolerant of shade. Spring sowing of bitter gourd often encounters low temperatures, rainy weather and insufficient light, which leads to leggy seedlings, yellow leaves and weak plants. During the flowering and fruiting period, bitter gourd requires strong sunlight, as sufficient light promotes photosynthesis, accumulation of more organic nutrients, increased fruit set rate, higher yield, and improved quality.

1.3 Water

Bitter gourd prefers moisture but is susceptible to waterlogging. During the growth period, it requires relative air humidity and soil relative moisture to reach 70%−80%. However, if there is prolonged overcast and rainy weather or poor drainage, the plants will have poor growth, and are prone to root rot and outbreaks of diseases.

1.4 Soil

Bitter gourd has less stringent soil requirements and is adaptable to a wide range of soils. It can be cultivated in various regions of Yunnan Province. However, the root system is sensitive to waterlogging and lack of oxygen, so it is recommended to cultivate in well-drained, fertile loamy or sandy soil with good aeration. This facilitates strong plant growth and higher yields.

1.5 Nutrition

Bitter gourd is tolerant to fertilizers but not to infertile and nutrient-deficient soils. It primarily absorbs potassium, followed by nitrogen, and the least amount of phosphorus. It is important to apply sufficient basal fertilizer before planting, follow up with appropriate supplementary fertilization during the seedling stage to promote stem and leaf growth, and continue fertilization during the flowering and fruiting stages. With an adequate supply of organic fertilizers, plants will grow vigorously with abundant branches and leaves, along with prolific flowering and good quality fruit. Particularly in the later growth stages, insufficient nutrient supply can lead to weakened plants, yellow-green leaves, limited flowering and fruit setting, smaller fruit size, intensified bitterness, and reduced quality. Hence, timely supplementary fertilization is crucial, especially during the peak fruiting period, requiring adequate nitrogen and phosphorus fertilization.

2 Variety selection

Depending on the cultivation purpose, market demand, and local microclimate, it is important to select bitter gourd varieties with strong growth potential, strong branching ability, heat and disease resistance, and high yield. Representative local varieties in Yunnan include Yuxi white bitter gourd, green skin bitter gourd, and small bitter gourd, while introduced varieties include Xiafeng bitter gourd, Chuanlü No. 1 bitter gourd, Nongyou No. 6 bitter gourd, and Green Gem bitter gourd. There is also a tradition of wild bitter gourd cultivation in the hot regions of Yunnan.

3 Cultivation and management techniques

3.1 Land preparation

3.1.1 Site selection and plowing

For bitter gourd cultivation, it is advisable to choose higher ground with convenient irrigation and drainage, and fertile soil with high organic matter content. Bitter gourd should not be planted after other gourd crops. After plowing and sun-drying the ridges, proper spacing is crucial for achieving high yields. Typically, double-row planting with 1.5–1.8 m width of ridge (including furrows), plant spacing of 0.3 m, and row spacing of 1–1.3 m is adopted. After harvesting the previous crop, it is suggested to remove any remaining plant debris and weeds from the field and plow the soil deeply for sun-drying.

3.1.2 Base fertilization

Before transplanting, it is advisable to apply 50 kg of calcium phosphate tribasic per mu, mix thoroughly with 500–1 000 kg of well-rotted organic fertilizer, and add 40 kg of compound fertilizer and 1.5 kg of borax evenly into the planting hole. Following this, it is suggested to transplant the seedlings into the holes, ensuring that the roots do not directly contact the fertilizer, and provide sufficient water for root establishment.

3.2 Seedling cultivation

3.2.1 Sowing period

Bitter gourd requires a temperature range of 15–30 °C for normal growth and development. Due to the unique microclimate in different regions of Yunnan Province specific sowing time varies according to the local climate conditions. Generally, bitter gourd is sown in March–April for spring planting and in July—August for autumn planting.

3.2.2 Seed treatment

Bitter gourd seeds have a hard seed coat and can be manually scarified before sowing, by using pliers or other tools to make a small incision about one third of the length of the seed at the hilum.

Additionally, the seeds should be disinfected before sowing. They can be soaked in clean water for 3–5 h, followed by treatment with 10% solution of trisodium phosphate for 20 min or 50% carbendazim WP (1 : 500) for 60 min. After rinsing with clean water, the seeds can be germinated in a constant temperature of 25–30 °C or directly sown. Alternatively, the seeds can be heated in water at 55 °C until the temperature drops to around 30 °C, then stop stirring and soak for 4–5 h, finally put on clean and moist gauze in a constant temperature of 25–30 °C to germinate. They are ready for sowing when 75% of the seeds show white tips.

3.2.3 Seedling preparation

To prevent damage to the roots and stems during transplanting and reduce the incidence of diseases, seedling cultivation is commonly carried out using nutrition bags or seedling trays. The substrate can be composed of burned soil or vegetable garden soil where gourds have not been previously planted, mixed with about 30% well-rotted organic fertilizer and an appropriate amount of compound fertilizer. This mixture should be thoroughly mixed and sprayed with 50% carbendazim WP (1 : 500) and 70% thiophanate-methyl WP (1 : 1 000).

3.2.4 Sowing

Sowing is generally carried out on sunny days. Before sowing, it is advisable to thoroughly water the substrate and then plant the seeds at a depth of 1.5–2 cm in nutrition bowls or seedling beds, covering with a thin layer of soil. After sowing, covering with plastic film and setting up small arches are needed. Before seedling emergence, it is necessary to ensure that the temperature of the seedbed is maintained at 30–35 °C to provide favorable conditions for germination.

3.2.5 Seedling management

After seedling emergence, it's proposed to remove the plastic film timely and lower the temperature of the seedbed to maintain 25–30 °C during the day and 15–20 °C at night. After 2 true leaves emerge, spraying the surface of the seedlings leaves 2–3 times with 0.3% solution of potassium dihydrogen phosphate should be conducted, which promotes the vigorous growth and enhances the cold tolerance of the young seedlings. It is necessary to harden off the seedlings 7–10 d before transplanting, maintaining daytime temperature around 25 °C and nighttime temperature at 12–15 °C, while strengthening ventilation and gradually removing the covering plastic film to acclimate the young seedlings to external environmental conditions. Generally, no watering is done before seedling emergence, and after emergence, it needs to keep the substrate moist but not waterlogged, watering on sunny afternoons.

3.2.6 Grafting

To prevent bacterial wilt and blight, it is recommended to use grafting technique for seedling cultivation by using loofah (*Luffa aegyptiaca* Miller) or black-seeded pumpkin (*Cucurbita ficifolia*) as the rootstock. When the rootstock seedlings have grown to the stage of having 2 true leaves at one node, and the scion has grown to the stage of having one leaf at one node, it is time to remove the central leaf and cut one leaf from the tip of the rootstock. Then, cut the axis of the scion diagonally to match the cut surface of the rootstock, and secure them together using a grafting clip. After grafting, it is suggested to spray with disinfectant and place the grafted plants in a small greenhouse covered with shading net to avoid direct sunlight, ensuring a relative humidity of over 95% and a temperature of 25–28 °C.

3.3 Field management

3.3.1 Mid-tillage and weeding

After transplanting bitter gourd, timely mid-tillage, weeding, and soil loosening should be carried out to prevent soil compaction. Generally, the first mid-tillage and weeding should be done around 10 d after transplanting, when weeds begin to appear and the soil surface starts to compact. The second cultivation and weeding should be carried out 10–15 d after the first one, taking care to protect the new roots and keeping the cultivation shallow and not too deep.

3.3.2　Training the vines on the trellis

When the seedlings have reached 3–4 leaves, trellising or using strings to support vine growth should be implemented. When the plant has grown to 10–12 true leaves, timely pruning of the vines should be conducted to promote adventitious root formation and expand the range of root absorption, thereby promoting stem and vine growth. When the vine reaches about 30 cm, it should be tied up every 4–5 nodes, lifting the vines onto the trellis. Generally, the vines should be tied up at 9:00 after the morning dew has dried to prevent breaking the vines.

3.3.3　Plant adjustment

Bitter gourd has strong branching ability on the main vine. Without intervention, it will produce excessive lateral branches, affecting flowering and normal growth of the main vine. Therefore, it is necessary to remove the excess lateral branches timely. Usually, 1–2 lateral branches that grow from the ground should be left, and the nutrient side branches above the ground surface up to the 15th node should be removed. After leaving 1–2 leaves above the female flowers, the lateral branches should be removed, and it is important to regularly remove dense, senescent stems and weak lateral branches to ensure good ventilation and light transmission.

No fruits should be allowed to develop below 50 cm on the main vine. It should remove the female flowers to promote robust growth, and after the main vine has set 6–7 fruits, the vine should be pinched at 5–6 leaves.

3.3.4　Fertilization and water management

Bitter gourd is tolerant to fertilizers, but cannot thrive in infertile soil. It prefers moist but not waterlogged conditions and requires a relatively high amount of water. It is necessary to maintain the air relative humidity and soil relative humidity of 70%–80% throughout the growth period. Organic fertilizers should be the main source of base fertilizer, with 1 500–2 000 kg of well-rotted organic fertilizer applied per mu. Nitrogen-based topdressing should be carried out 6–8 times during the growth period for spring-planted bitter gourd, with the frequency reduced in hot and humid regions during the summer and autumn seasons due to the shorter growth period. The topdressing schedule is as follows. Ten d after transplanting, it is recommended to apply 8 kg of urea and 5 kg of potassium chloride per mu, then repeat every 10 d. When the female flowers appear, it is advised to apply 25 kg of 45% compound fertilizer per mu. Additional topdressing during peak fruiting can include 5 kg of potassium sulfate, 15 kg of calcium superphosphate, and 15 kg of urea per mu. It is applied at 10-day intervals to promote vine and leaf growth and extend the harvest period.

3.3.5 Artificial pollination

Bitter gourd has a high rate of parthenocarpy, but artificial pollination can increase the fruit-set rate and ensure the development of the fruits. Pollination should be carried out on the day when both male and female flowers are in bloom. When the temperature is stable at 15–30 °C, the peak flowering period is 6:00–10:30 each day, with minimal flowering occurring after midday. Typically, one male flower can be used to pollinate four female flowers. After the bitter gourd flowers, the male flowers are picked, and their pollen is transferred to the stigma of the female flower. After pollination, the ovaries gradually enlarge, and fruits can be harvested after around 20 d.

4 Prevention and control of diseases and pests

The main diseases of bitter gourd include sudden wilt, anthracnose (anthrax), brown spot, powdery mildew, and virus disease, while pests mainly include aphids (*Aphis Gossypii* Glover), thrips, greenhouse whiteflies (*Trialeurodes vaporariorum*) and melon flies [*Bactrocera cucuribitae* (Coquillett)].

4.1 Major diseases

4.1.1 Damping-off

4.1.1.1 Symptoms

Sudden wilt is prone to occur in the early seedling stage. Initially, the base of the young seedling stem becomes water-soaked, and then the affected area turns light brown. The young seedlings near the ground surface exhibit obvious wilting, with the cotyledons still green but falling over. The growth of the pathogen is favored by low soil temperatures and high humidity, and severe outbreaks occur when nights are cool, daylight is insufficient, and the seedbed is humid.

4.1.1.2 Prevention and control measures

Options for treatment include root drenching with 72% propamocarb hydrochloride AS (1 : 750), 75% mancozeb WP (1 : 800), 58% metalaxyl-mancozeb WP (1 : 700), and 32% metalaxyl-hymexazol WP (1 : 300).

4.1.2 Anthracnose (Anthrax)

4.1.2.1 Symptoms

The disease can affect the melon strips, leaves, stems and vines. The lesions on the melon strips are round or irregular, initially light yellow-brown, later turning reddish-brown to pale brown and slightly sunken. Lesions on the leaves are round or irregular, ranging from gray-brown to brown, with slight wet rotting. Lesions on the stems and vines are long-oval, brown, and sunken, with severe cases showing cracking.

4.1.2.2 Prevention and control measures

(1) Agricultural control: It usually adopts resistant varieties and disinfects seeds. It is advisable to practice crop rotation with non-cucurbit crops. It should avoid excessive and imbalanced nitrogen fertilization, increase the application of phosphorus, potassium, and medium-micro nutrient fertilizers, meanwhile it should control water usage appropriately, and drain water after rainfall. Promptly removing diseased leaves, branches, and fruits, and maintaining good field ventilation and light transmission are needed.

(2) Chemical control: In the early stages of the disease, spraying with 25% prochloraz EC (1 : 1 500), 10% difenoconazole WG (1 : 800), or 75% trifloxystrobin-tebuconazole WG at [1 : (4 000−6 000)] should be conducted, continuously for 2−3 times at a 7−10 d of interval.

4.1.3 Brown spot

4.1.3.1 Symptoms

The leaves are mainly affected. Small brown circular spots initially appear on the leaves, which gradually expand into round or irregular yellow-brown lesions, often with a fading greenish halo. Under suitable environmental conditions, the lesions rapidly expand, coalescing into patches and eventually causing the entire leaf to wither.

4.1.3.2 Prevention and control measures

(1) Agricultural control: It usually adopts resistant varieties. It is advisable to plant in higher areas with good drainage. It should enhance field management improving field ventilation and light transmission. It is recommended to apply organic fertilizers for base fertilization. It is necessary to drain accumulated water after rain, control field humidity and practice crop rotation in heavily affected areas.

(2) Chemical control: In the early stages of the disease, spraying with 25% prochloraz EC (1 : 1 500), 10% difenoconazole WG (1 : 800), 75% trifloxystrobin-tebuconazole WG [1 : (4 000−6 000)], 32.5% benzyl methyl-azoxystrobin SC (1 : 1 500) should be conducted, continuously for 2-3 times at a 7−10 d of interval.

4.1.4 Powdery mildew

4.1.4.1 Symptoms

The disease can affect leaves, petioles, and stems. When leaves are affected, small white powdery spots appear on the upper surface, which gradually expand into round or irregular sparse white powdery patches. With the development of the disease, the powdery spots are contiguous and the leaves are covered with white powder. In severe cases, the leaves gradually turn yellow and eventually dry up, hindering plant growth and fruiting and shortening the reproductive period. The petioles and vines also develop sparse white powder when affected.

4.1.4.2 Prevention and control measures

(1) Agricultural control: It usually adopts resistant varieties. It is recommended to practice proper planting density, timely trellising, pruning, and leaf removal to increase plant ventilation and light penetration. Employing balanced fertilization with additional medium-micro nutrient fertilizers should be conducted. It is advisable to maintain appropriate soil moisture.

(2) Chemical control: Before disease onset, it is recommended to use 325 g/L azoxystrobin-difenoconazole SC (1 : 2 000) as preventive treatment. In the early stages, it can use 10% difenoconazole WG (1 : 1 500), 80% sulfur powder SC (1 : 600), 25% triadimefon WP (1 : 1 000), 250 g/L azoxystrobin SC (1 : 600) for spray treatment, continuously for 2–3 times at a 7–10 d of interval.

4.1.5 Virus diseases

4.1.5.1 Symptoms

Infected plant leaves exhibit a mosaic of yellow and green, and the plants appear stunted with distinct symptoms on their tender stem and leaves. New leaves fail to unfold properly, showing wrinkling and yellow-green mottling, later developing necrotic yellow spots. In the early stages, affected melon seedlings exhibit poor growth, shortened internodes, and yellowing progressing from lower to upper leaves leading to withering.

4.1.5.2 Prevention and control measures

(1) Agricultural control: It usually employs disease-resistant varieties and performs seed disinfection. It is advised to practice crop rotation, foster robust seedlings, and promptly control aphids, thrips, and greenhouse whiteflies to break the chain of infection. Promptly removing and destroying diseased plants upon identification should be done.

(2) Chemical control: In the early stages of infection, it is recommended to conduct spray treatments with 2% ningnanmycin AS (1 : 300), 20% moroxydine-cupric acetate monohydrate

WP [1 : (1 500–2 000)], and 1% lentinan AS (1 : 600), continuously for 2–3 times at 5–7 d intervals.

4.2 Major Pests

4.2.1 Aphids (*Aphis Gossypii* Glover)

4.2.1.1 Symptoms

Aphid clusters suck sap from the leaf undersides, tender stems, and shoots, which causes curling of young leaves, death of growing points, wilting of seedlings, and severe cases leading to complete plant death. The excretion of "honeydew" by aphids contaminates the leaf surface, leading to sooty mold, which affects photosynthesis. Most importantly, aphids transmit viruses, which causes symptoms such as mosaic, deformation, and stunting, leading to premature aging and significant losses in affected plants.

4.2.1.2 Prevention and control measures

(1) Physical control: Protected areas promote the use of 24–30 mesh, 0.18 mm silver-grey insect-proof nets or yellow boards for trapping and killing.

(2) Biological control: During the initial stages of an aphid infestation, it is advisable to release 1 500 ladybird beetles per mu to control the population density.

(3) Chemical control: Options include spraying with 3% acetamiprid EC (1 : 1 500), 10% imidacloprid WP (1 : 2 000), and 25% thiamethoxam WG (1 : 4 000).

4.2.2 Thrips

4.2.2.1 Symptoms

Infested bitter gourd plants exhibit withered and blackened growth points, which results in clustering, folding the heart leaves and preventing the normal setting of fruits. Affected young fruits show blackened trichomes, rust-brown skin, slow growth, and deformities. Severe infestations cause fruit drop, significantly impacting yield and quality.

4.2.2.2 Prevention and control measures

(1) Agricultural control: Timely removal and burning or deep burial of weeds in fields are needed, which can eliminate overwintering adult and nymph thrips and reduce the source. It also should enhance water and fertilizer management to promote robust plant growth.

(2) Physical control: Thrips are attracted to the blue color. Therefore, it is recommended to hang blue boards between crop rows for detection and trapping and promptly clear pests from these boards.

(3) Chemical control: Options include spraying with 40% imidacloprid WP

[1 : (2 000−3 000)], 20% dinotefuran WDG (1 : 1 500), 40% 1-Methyl-3-phenyl-5-(3-trifluoromethyl) phenyl-4(1H)-pyridinone and pymetrozine WG (1 : 3 000), and 18% bisultap AS [1 : (250−400)].

4.2.3 Greenhouse Whiteflies (*Trialeurodes vaporariorum*)

4.2.3.1 Symptoms

Adults and nymphs of greenhouse whiteflies feed on sap from tender stems, shoots, or leaf undersides, causing discoloration, yellowing, and curling of leaves, stunting plant growth, and transmitting viruses.

4.2.3.2 Prevention and control measures

(1) Physical control: It is suggested to hang 30−40 yellow boards per mu, use silver-grey films to repel aphids, and employ devices like frequency vibration insect-killing lamps, black light lamps, and high-pressure mercury lamps to attract and kill pests.

(2) Biological control: It is available to release *Encarsia formosa* gahan (a parasitic wasp) for aphid control, releasing them once every 10 d, repeating this process 3−4 times.

(3) Chemical control: Options include spraying with 20% dinotefuran WDG (1 : 1 500), 3% acetamiprid EC (1 : 1 500), 10% imidacloprid WP (1 : 2 000), and 25% thiamethoxam WG (1 : 4 000).

4.2.4 Melon Flies [*Bactrocera cucuribitae* (Coquillett)]

4.2.4.1 Symptoms

Affected melons initially exhibit localized yellowing, followed by complete rotting and a foul smell, resulting in a large number of fallen fruits. Injury sites show congealed exudate, deformation, depression, hardened fruit skin, bitter taste, and diminished quality.

4.2.4.2 Prevention and control measures

(1) Agricultural control: It is advisable to remove and collect affected and fallen melons from the vegetable garden promptly, and dispose of them by burying deeply, soaking in water, or incineration. It should avoid consecutive cropping with cucurbit vegetables.

(2) Physical control: It is recommended to utilize frequency vibration insecticidal lamps, black light lamps, and high-pressure mercury lamps for attraction and extermination.

(3) Bagging for bitter gourd protection: In severely affected areas, when the melon fruit has fully withered after flowering, bagging should be conducted. Before bagging, it is necessary to spray the melon once with pesticides to prevent other diseases and pests, ensuring the quality of the melon after bagging.

(4) Chemical control: Options include spraying with 1.8% avermectin EC (1 : 2 000),

4.5% cypermethrin EC (1 : 1 000), and 5.2% avi-perchlorate EC (1 : 1 000).

5 Timely harvesting

Bitter gourds grow rapidly. Generally, within 10−12 d after flowering, the melon fruit is fully developed. Its flesh is tender, slightly bitter with sweetness, exhibiting an excellent taste. Delayed harvesting increases the crude fibrous content in the flesh, which is detrimental to subsequent crop growth. During harvesting, melon fruits should have plump, elongated or nodular protrusions, glossy skin, and a beginning fade in color at the fruit top.

6 Post-harvest treatment

6.1 Storage

After harvesting, the bitter gourds should be placed in a cool and shaded area promptly. If available, pre-cooling them in a cold storage facility is needed, which brings the melon temperature close to the appropriate storage temperature for transportation. The suitable storage temperature for bitter gourds is 10−13 °C. Temperatures below 10 °C may cause chilling injury. The relative humidity in the storage environment should be maintained at 85%−90%.

6.2 Transportation and packaging

Specific requirements for bitter gourd transportation and packaging should be determined based on factors such as distance between harvesting, distribution, or sales locations, and prevailing climatic conditions. For immediate or short-distance transportation, normal temperature transportation is generally suitable. In hot weather or during consecutive rainy days, sunshade or rain protection measures should be taken. During severe winters, it should cover the melons with blankets or straw to prevent freezing damage. Moreover, for long-distance transportation, a low-temperature method should be employed, using internally lined cardboard boxes or bamboo baskets as packaging materials.

References

HU Q H, 2012. Health Vegetables: Pumpkin Cultivation Techniques [M]. Beijing: Scentific and Technical Documentation Press.

LIANG J L, 1996. Practical Color Atlas of Vegetable Diseases and Pests [M]. Zhengzhou:

Henan Science and Technology Press.

LONG R H, ZHANG S Z, 2014. Cultivation of Health Vegetables in Yunnan [M]. Kunming: Yunnan Science and Technology Press.

LU G Y, 2003. Year-Round Production Techniques of Melon Vegetables [M]. Beijing: Jindun Publishing House.

SONG S H, 2002. Cultivation Techniques of 14 Kinds of Specialty Melon Vegetables [M]. Beijing: China Agricultural Press.

WANG G Y, ZHANG J W, 2000. Study on seed germination characteristics of bitter gourd [J]. Chinese Agricultural Science Bulletin, 16(1): 45−46.

WANG J X, 2009. Color Atlas of Diagnosis and Control of Diseases and Pests in Melon Vegetables [M]. Beijing: Jindun Publishing Press.

XIE H Y, HUANG S X, DENG H N, et al., 1998. Study on chemical components of *Momordica charantia* [J]. Journal of Chinese Medicinal Materials, 21(9): 458−461.

❖ Key Techniques for Onion Cultivation

Onion (*Allium cepa* L.) is a biennial herbaceous plant in the monocotyledonous lily family, Alliaceae, known for its fleshy bulbs. It is also called Scallions, round onion, or bulb onion. Onions are native to Central Asia and the Mediterranean coast. The cultivation of onion has a long history, spanning nearly a century in various regions across China. Onions are rich in nutrients, containing a significant amount of protein, vitamins, as well as various minerals such as sulfur, phosphorus, and iron. Onions are resistant to storage and transportation, and besides being consumed domestically, they are also a major vegetable exported to countries such as Japan, Southeast Asia, and Russia. Onions have strong disease resistance and are relatively free from pests. They can be intercropped with grains, cotton, and other crops. For example, intercropping with barley or peas can help suppress barley smut and pea black spot diseases. It is an ideal crop for crop rotation and alternation with vegetables from the Solanaceae, Cucurbitaceae, and Brassicaceae families. In addition to being consumed fresh, onions can also be processed into onion powder, onion sauce, onion oil, dehydrated onion, and onion juice. Onions contain phytoncide, as well as prostaglandin-like substances and components that activate fibrinolytic activity, which have the effects of dilating blood vessels, lowering blood pressure, and reducing blood lipids.

Yunnan onion is cultivated year-round on 100 000 mu of land, with the main cultivation areas including Honghe prefecture, Chuxiong prefecture, Kunming City, Yuxi City, Baoshan City, Dali prefecture, and other regions. Among them, Honghe Prefecture and Chuxiong Prefecture have the largest cultivation areas, accounting for 90% of the entire province. Taking advantage of Yunnan's diverse climate, onions are available in the market from January in Yuanyang County to June in Yao'an County, making it the earliest marketable onion region in China with the longest supply period. Honghe Prefecture mainly cultivates red onion varieties and enjoys location advantage in the markets of Guangdong and Guangxi. The climate in Yuanmou County is dry with significant day-night temperature differences, abundant sunlight, low winter rainfall, which is conducive to the growth of yellow and white onion varieties. Onions grown there have sufficient moisture, a sweet taste, and high quality, providing a significant advantage for export. In general, Yunnan onions are favored by consumers in Japan and Korea for their pleasant flavor and fresh, crunchy texture. They are exported to these markets after undergoing processes such as peeling, processing, and vacuum packaging.

1 Growth conditions

1.1 Temperature

Onions are semi-cold-resistant vegetables. The minimum temperature for seed germination is 4 °C, the maximum is 33 °C, and the optimal temperature is 12−25 °C. The optimal temperature for seedling growth is between 12 °C and 20 °C, but the seedlings have strong cold resistance and can tolerate low temperatures of from −7 °C to −6 °C. The plant's vigorous growth period is best around 20 °C; growth is poor if the temperature exceeds 25 °C. The root system essentially stops growing below 5 °C, and its optimal temperature range is slightly lower than that of the above-ground parts. Soil temperature above 26 °C can promote aging of the root system. Before the bulbous enlargement growth of some varieties, a temperature condition of 15−25 °C is required for the subsequent bulbous enlargement growth. The temperature requirements for the bulbous enlargement growth period vary significantly; for short-day early-maturing varieties, the suitable temperature for bulbous enlargement growth is between 15 °C and 20 °C; for long-day mid-to-late maturing varieties, the bulbous enlargement growth period requires temperatures between 20 °C and 26 °C.

1.2 Light

Onion seeds do not need light during the germination process. The requirements for daylight duration during the bulbous enlargement phase vary: long-day varieties need 13.5−15 h, while short−day varieties require 11.5−13 h. Additionally, some varieties are not very strict in their light requirements during bulb formation. Generally, northern varieties are mostly long-day late-maturing types, while southern varieties are mostly short-day early-maturing types, so careful attention is needed during introduction. For instance, Da Shuitao and Bi Qibian in Tianjin are long-day varieties, and when introduced to places like Chongqing and Shanghai, reduced sunlight duration often leads to decreased yields. Onions thrive in moderate light intensity during their growth period, with an optimal light intensity of 20 000−40 000 lx.

1.3 Moisture

Onions require ample moisture during the germination, vigorous growth, and bulbous enlargement phases. After autumn planting, water should be controlled appropriately, but sufficient watering is essential before winter to facilitate winter survival. Watering should be stopped 1-2 weeks before harvesting to induce physiological dormancy and enhance storage resistance.

1.4 Soil and nutrition

Onions are highly adaptable to environmental conditions and are not demanding in terms of soil texture. If the soil is relatively clayey, it can be detrimental to root growth, but it results in tightly formed bulbs. Loose, sandy soil is beneficial for root extension but has weak water and nutrient retention. For early-maturing cultivation, sandy soil is preferred; For onions stored before sale, loamy soil or clay loam is suitable. The optimal pH value for onions is between 6 and 8, but the seedlings are sensitive to saline-alkali conditions.

For onion seedling cultivation, it is advisable to increase the application of phosphorus and potassium fertilizers while controlling the amount of nitrogen. To produce 1 000 kg of marketable onions, 2.37 kg of nitrogen, 0.7 kg of phosphorus, and 4.1 kg of potassium are needed. Onion cultivation requires high nutrient concentration in the soil to support its rapid growth over a short period. When soil fertility is insufficient, appropriate topdressing should be applied.

2 Main cultivars

Onion cultivars are classified into common onions, tillering onions, and top bulb onions, based on their bulb morphology and growth traits.

2.1 Common onion

Typically, each plant forms one large bulb with a large size, good quality, and high yield. It has relatively strong cold resistance and is widely cultivated through seed propagation. The bulb colors include purple-red, coppery-yellow, pale yellow, and white; The shapes of the bulbs include flat-round, spherical, high-spherical, and spindle-shaped. Based on the daylight requirements for bulb enlargement, they are categorized into long-day, short-day, and mid-day ecotypes; And according to the different growth and maturity periods, they are classified into early-maturing, mid-maturing, and late-maturing varieties.

2.2 Tillering onion

Tillering onions develop multiple irregular-sized bulbs per plant. The bulbs are typically copper-yellow with inferior quality and lower yields, yet they are storage-tolerant. These plants exhibit extreme cold resistance, rarely bloom or produce seeds, and are predominantly propagated through the tillering of small bulbs.

2.3 Top bulb onion

Top bulb onion (*Allium* L. var. *viviparum* Metz.), also referred to as top-set onion (*Allium cepa* var. *viviparum* Merg.). This variety of onion bulb can normally undergo bolting, but does not commonly flower or bear seeds. Instead, it forms 7–8 or more aerial bulbs on the flower stalk. These aerial bulbs are used for propagation without the need for seedling cultivation. Topset onions have strong and cold resistance, making them suitable for planting in very cold regions. They can be used for processing and pickling.

3 Cultivation and management techniques

3.1 Sowing and seedling raising

Onions are suitable for autumn sowing in subtropical regions such as western and southern Yunnan, as well as in arid-hot valleys. Generally, they are sown in September to October, while in other regions, they are sown around the Spring Festival. Timely sowing is a crucial aspect of onion production. If sown too early, the seedlings may grow too large before winter, leading to early bolting the following spring and hindering the normal development of large bulbs, thereby affecting onion yield. On the other hand, if sown too late, the nutritional growing period is shortened, resulting in small bulbs and a sharp reduction in yield. Seedling cultivation can be categorized into tray seedling and seedbed seedling methods. In the case of seedbed seedling, it is crucial to choose soil that is loose, fertile, and exhibits strong water retention properties. It is advised to opt for a land parcel that has not hosted the cultivation of onions or garlic in the past 2–3 years, and avoid areas susceptible to waterlogging. The designated area for seedbed cultivation should be around 1/15 of the total cultivation area. For every 100 m^2 of the seedbed, apply 300 kg of thoroughly decomposed, finely crushed farmyard manure. The recommended seeding quantity is 0.6–0.7 kg.

3.2 Seed treatment

The seeds should undergo a 3–5 h soaking in water at 50 °C. Following this, germination is to take place under conditions of 20–25 °C. Throughout the germination process, it is essential to rinse the seeds once daily with clean water. Planting should be initiated promptly when the seeds exhibit white shoots.

3.3 Seedling management

After the first true leaves have grown, it is advisable to regulate the water supply appropriately. When two true leaves have emerged, the watering routine should integrate the application of nitrogen-containing fertilizers. Typically, for every 100 m^2, an application of 3.4–5.1 kg of ammonium sulfate or 1.7–2.5 kg of urea is recommended. Additionally, foliar spraying of 0.2%–0.4% potassium dihydrogen phosphate should be administered 1–2 times. Prior to applying the fertilizer, thinning out seedlings should be conducted in conjunction with weed removal.

3.4 Soil Preparation for ridge planting

During land preparation, the plowing depth should not be less than 20 cm. The first plowing must reach the required depth, followed by the application of base fertilizer. Afterward, shallow plowing 1–2 times is carried out to make the furrow ground fine and ensure the even distribution of organic fertilizer and soil. For each mu, apply fully decomposed compost, barnyard manure, or other farmyard manure at a rate of 1 500–2 000 kg, and 40 kg of calcium superphosphate. Onions are generally cultivated on raised ridges with a ridge width of 1.3–1.5 m, a furrow depth of 0.3 m, and the length determined to accord to the plot.

3.5 Planting

3.5.1 Planting period

The planting period varies depending on climate and variety. It should be done when the monthly average temperature is 4–5 °C, and the seedlings are 20–28 cm tall. In spring, planting should be done as early as possible for increased yield. The size of the seedlings during planting is closely related to overwintering bolting rate and yield. If the seedlings are too large, they are more likely to undergo the vernalization phase during winter. Consequently, they tend to bolt early in the subsequent spring, resulting in smaller onion bulbs. Conversely, if the seedlings are too small, their growth is weakened. Although they are less prone to bolting, the overall yield is diminished. The recommended planting specifications include row spacings of 10 cm × 20 cm or 15 cm × 20 cm. Moderate planting density does not affect the size of individual onion bulbs, but excessive density can reduce the weight of the bulbs.

3.5.2 Planting method

Mulching film is to pierce holes in the film at predetermined plant intervals using bamboo sticks or similar objects. After creating the perforations, seedlings are inserted into the holes, and the film is securely sealed with soil around the seedlings. In the absence of mulching, dig furrows at the specified row intervals and plant the seedlings according to the recommended plant spacing. During planting, for larger seedlings, it is advisable to trim 1/3 to 1/2 of the leaves and approximately 1/3 of the root system. Plant the seedlings in an upright position, avoiding any tilting. The appropriate planting depth is crucial. Planting too deep can hinder growth, resulting in elongated bulbs instead of a desirable spheroid shape. On the other hand, planting too shallow makes them vulnerable to drought, hampers root development, and as the bulbs enlarge, excessive exposure above the soil surface may lead to cracking. A planting depth of 3–5 cm is recommended. After planting, timely watering plays a crucial role in establishing the root system, facilitating optimal contact between the roots and the soil to foster swift growth recovery.

3.5.3 Irrigation

After transplanting, the seedlings require frequent watering for about 20 d during the slow growth period. Once the plants have grown sufficiently and shift to bulbous growth, restrict water for about 10 d to restrain the growth of seedlings. When the outer leaves exhibit a deep green color, with an increased wax on the leaf surface and a corresponding deepening of the inner leaves' color, it indicates the time to conclude the seedling restraint phase and ensure ample watering. Generally, from planting to harvest, water the crop 12–15 times. Cease watering when individual plants in the field start lodging, and the bulbs reach maturity. Additionally, attention should also be given to drainage to promote the full maturation of onion bulbs, enhancing their storage resilience.

3.6 Fertilization

Onions have a shallow root system and require a lengthy growth period. Beyond the application of ample base fertilizer, frequent topdressing is also essential. For onions planted in spring, the first topdressing should occur once the seedlings have fully recovered. For those planted in late autumn, the first topdressing should be administered after reviving. Incorporate irrigation with top-dressing of 10–15 kg of diammonium phosphate and 8–10 kg of potassium sulfate per mu. Subsequently, an additional application of seedling-promoting fertilizer, involving 10–15 kg of ammonium sulfate per mu, is carried out to support the growth requirements of the functional foliage above ground. During the bulb expansion phase, conduct 2–3 sessions of topdressing (bulb-boosting fertilizer), with a focus on the mid-period of bulb enlargement. 10–15 kg of ammonium sulfate and 5–10 kg of potassium sulfate in topdressing should be applied per mu.

3.7 Mid-tillage and ridging

In the absence of film-mulching, inter-row cultivation is necessary, particularly before transplanting seedlings, with a recommended cultivation depth of 3–4 cm. Integrating intertillage with ridging contributes to the improvement of crop yield.

3.8 Prevention of premature bolting

Poor cultivation conditions and improper management can lead to premature bolting in onions, which directly affects the yield, quality, and storability of the onions. To prevent premature bolting in onions, it is advisable to select varieties that are less prone to early bolting and have strong winter hardiness. Planting at the right time, fostering sturdy seedlings, and applying a 250 mg/L ethephon solution during the seedling stage contribute to a reduced bolting rate. For plants that exhibit early bolting, timely removal or breaking off of the flower stalks can reduce nutrient loss and mitigate significant yield reduction to some extent.

4 prevention and control of diseases and pests

4.1 Prevention and control of diseases

4.1.1 Purple blotch

4.1.1.1 Symptoms

Purple blotch can occur throughout the entire growth period, primarily affecting leaves and flower stalks. Initially, it presents as water-stained small white spots, which then expand into circular or spindle-shaped depressed areas. The lesions progress from small to large, displaying a dark brown to dark purple color and forming a concentric wheel patterned black mold layer. In severe cases, multiple lesions can merge, leading to the withering of entire leaves, flower stems, and pedicels. The disease is prevalent in temperatures of 20–30 °C during the summer and autumn seasons.

4.1.1.2 Preventive measures

(1) Agricultural control. It is recommended to implement crop rotation. By removing diseased and disabled plants, and simultaneously applying organic fertilizers, especially phosphorus and potassium fertilizers, the goal is to enhance the plants' resistance to diseases.

(2) Chemical pest control. Seeds can be disinfected with a 40% formaldehyde solution (1 : 300). After soaking for 3 h, they should be taken out and rinsed with water for later use. In the early stages of the disease, sprays such as 75% chlorothalonil WP [1 : (500–600)],

50% iprodione WP (1 ∶ 1 500), 64% hymexazol·mancozeb WP (1 ∶ 500), and 58% metalaxyl·mancozeb WP (1 ∶ 500) can be used. Spraying should be adopted every 5−7 d and this treatment should be repeated 2−3 times consecutively for best results.

4.1.2　Downy mildew

4.1.2.1　Symptoms

Infection affects onion leaves, stems, and flower stalks, producing oval-shaped light yellow spots with unclear borders. In high humidity, a white mold layer appears on the surface, later turning pale yellow or dark purple. Lower leaf portions become diseased, gradually drying and drooping above the affected area, ultimately causing severe withering and yellowing. Pathogenic oospores overwinter on seeds, soil, and diseased plant remnants.

4.1.2.2　Preventive measures

For prevention, early-stage spraying with 75% chlorothalonil WP (1 ∶ 600), 64% hymexazol·mancozeb WP [1 ∶ (600−800)], and 72% cymoxanil·mancozeb WP (1 ∶ 500) is recommended. Spraying should be adopted every 7−10 d and this treatment should be repeated 2−3 times consecutively for best results.

4.1.3　Rust disease

4.1.3.1　Symptoms

Rust disease tends to occur during the humid conditions of spring and autumn, primarily affecting leaves, flower stalks, and green stems. In its initial stages, the leaves display small white spots with raised lesions. Subsequently, the epidermis ruptures, releasing brown fungal spores into a powder, especially under high humidity conditions, exacerbating the situation.

4.1.3.2　Preventive measures

In the early stages of the disease, effective prevention and control can be achieved through the application of certain compounds. Spraying with 15% triadimefon WP [1 ∶ (2 000−2 500)], 25% hexaconazole SC (1 ∶ 4 000), 50% carboxin EC [1 ∶ (700−800)], and 70% mancozeb WP (1 ∶ 500). Spraying should be adopted every 7−10 d and this treatment should be repeated 2−3 times consecutively for best results.

4.1.4　Virus disease

4.1.4.1　Symptoms

The infected plants show no obvious symptoms in the early stage, but later, the green leaves exhibit faint, short streaks that gradually develop into long streaks alternating between yellow and green. In severe cases, the leaves become wrinkled and change from round to flat.

The elongation of new leaf sheaths is hindered, resulting in short and wrinkled leaves. Dense leaf clusters form, leading to a significant decrease in yield, quality, and storage capacity. Currently, there are no effective medications to treat viral disease. The primary preventive measures involve large-scale regional crop rotations, meticulous selection of scallion seedlings, prompt removal of central diseased plants, and the extermination of vectoring pests such as aphids and thrips to prevent viral infection.

4.1.4.2 Preventive measures

The key focus in preventing and treating this disease is on controlling pests like aphids and thrips. To achieve this, insecticides such as imidacloprid, acetamiprid, and thiamethoxam are employed. It is advised to use 5% bacteriophage AS (1 : 500), 1.5% sodium dodecyl SE (1 : 1 000), 2% lentinan AS (1 : 1 000), 80% moroxydine hydrochloride·copper acetate WP (1 : 1 000), 5% chitosan SL (1 : 600), among other pharmaceutical sprays. Apply the spray once every 5–7 d, continuously for 2–3 times.

4.2 Prevention and control of pests

4.2.1 Thrips

4.2.1.1 Symptoms

Thrips, both in their adult and nymph stages, feed by rasping on onion leaves and extracting sap from flower stems, causing gray-white stripes or spots at the puncture site. In severe cases, numerous lesions can coalesce, leading to the withering of leaf blades.

4.2.1.2 Preventive measures

(1) Agricultural control. Early spring field cleaning, regular watering, and weed removal can help alleviate the damage.

(2) Chemical control. During the initial period of thrips occurrence, it is recommended to spray with a 60 g/L spinosad SC (1 : 1 000), 10% imidacloprid WP (1 : 1 000), 10% cypermethrin EC (1 : 1 000), and 25% thiamethoxam WG [1 : (1 000–1 500)] for prevention.

4.2.2 Liriomyza

4.2.2.1 Symptoms

The adult of the liriomyza lays eggs beneath the epidermis of the leaves. The hatched larvae then feed on the leaf tissue, creating winding tunnels and narrow, band-shaped necrotic spots on the leaves. When the insect density is high, necrotic spots can coalesce, leading to leaf withering, hampering growth, and rendering the onion leaves inedible.

4.2.2.2 Preventive measures

Spraying with 1% abamectin·cypermethrin EC (1 : 1 500), 10% Chlorfenapyr AS (1 : 1 500), and 20% abamectin-monosultap ME (1 : 1 500) is recommended for prevention and control.

4.2.3 Onion maggot

4.2.3.1 Symptoms

Onion maggot larvae burrow into the soil, posing a threat to the base of the young onion plants' pseudo stems. This can damage the growth point of onion plants, causing bulb rot, yellowing and wilting of leaves, and even widespread death of the plants.

4.2.3.2 Preventive measures

(1) Agricultural control. The organic fertilizer needs to become thoroughly decomposed and uniformly incorporated into the soil, promptly tilled deeply, and should not remain exposed on the soil surface to prevent attracting insects for egg-laying purposes.

(2) Chemical control. Root irrigation with pesticide solutions to eliminate larvae. Options include 90% crystal dipterex [1 : (800−1 000)], 50% phoxim EC (1 : 800), etc. To prevent adult insects, spraying with 90% crystal dipterex (1 : 800) or 2.5% decamethrin EC (1 : 1 000) can be effective.

4.2.4 Beet armyworm

4.2.4.1 Symptoms

The beet armyworm is a polyphagous pest that poses a threat by its larvae either consuming or stripping leaves. In the early stages, they tend to cluster, forming webs on the heart leaves, leading to damage, and subsequently dispersing to harm other leaves.

4.2.4.2 Preventive measures

For prevention and control, it is advised to employ 10% cypermethrin EC [1 : (2 000−3 000)], spray with 2% emamectin benzoate EC (1 : 1 000), 20% tebufenozide SC (1 : 800), or 10% chlorfenapyr ME (1 : 1 500).

5 Post-harvest treatment

5.1 Harvesting criteria

The thinning and shriveling of the onion leaf sheath neck serve as indicators of onion bulb maturity. This signals that the bulb has reached full growth, and harvesting can commence when the leaves have wilted. Varieties with a brief dormancy period and low storage resilience should be harvested promptly when 30%−50% of the plants have lodged. Meanwhile, mid to late−maturing varieties with strong storage resilience can be harvested when the lodging rate reaches 70%. Subsequent to harvesting, it is essential to place the onions in a well−ventilated and dry area for approximately 3−5 d to facilitate drying. This process aids in removing surface moisture and minimizes the likelihood of pest and disease infestations. The specific duration of drying should be determined based on prevailing weather conditions.

5.2 Grading

After harvesting, onions need to undergo grading based on size and quality, and then be placed in different packaging bags accordingly.

5.3 Storage

Storage is a crucial step for onion bulbs before long-term preservation. To ensure that the tubular leaves and outer skin of the bulbs are in a dry state, they should be allowed to sun-cure in the field. During drying, the rear leaves should cover the front bulbs to prevent direct exposure to sunlight. Subsequently, the onions should be braided or tied together, with the bulbs facing downward for further drying. Afterward, they can be stored by hanging indoors or outdoors or by creating elevated stacking for storage It is essential to regularly turn them over during the initial storage period. When the nighttime temperature drops below −5 °C, it is necessary to store onions indoors. The storage location should be a cool and dry place. Traditional storage methods include stacking, piling, heaping, and hanging.

References

CHENG Y Q, XU J, 2003. Pollution-free and Effective Cultivation of *Allium* vegetables [M]. Beijing: Jindun Publishing House: 71−101.

Institute of Vegetables and Flowers, Chinese Academy of Agricultural Sciences, et al., 2010. Vegetable Cultivation in China [M]. 2nd Edition. Beijing: China Agriculture Press, 394−402.

WANG J X, SUN C Y, 2005. Field Guide to Diagnosis and Prevention of Vegetable Pests and Diseases (*Allium* Vegetables) [M]. Beijing: Scientific and Technical Documentation Press: 15−35.